PENGUIN BOOKS

NO IMMEDIATE DANGER
VOLUME ONE OF CARBON IDEOLOGIES

William T. Vollmann is the author of ten novels, including *Europe Central*, which won the National Book Award. He has also written four collections of stories, including *The Atlas*, which won the PEN Center USA West Award for Fiction; a memoir; and eight works of nonfiction, including *Rising Up and Rising Down* and *Imperial*, both of which were finalists for the National Book Critics Circle Award, and *No Good Alternative* (Volume Two of *Carbon Ideologies*). He lives in California.

* * *

Praise for *No Immediate Danger*
(Volume One of *Carbon Ideologies*)

"*Carbon Ideologies* is an almanac of global energy use . . . a travelogue to natural landscapes riven by energy production . . . a compassionate work of anthropology that tries to make sense of man's inability to weigh future cataclysm against short-term comfort. . . . One of the most honest books yet written on climate change."
—Nathaniel Rich, *The Atlantic*

"*No Immediate Danger* tussles with the comprehension-defying nature of climate change. . . . Terrifying insights are to be found . . . It embodies the confusion of our current moment, the insidiousness of disbelief, and the mania-inducing reality that our greatest threat is the hardest to act upon. It is a feverish, sprawling archive of who we are, and what we've wrought."
—*The Washington Post*

"In the face of complex, contested data, Vollmann is a diligent and perceptive guide. He's also deeply mindful of those who've been sacrificed in the name of profits and political expediency. Amid the Trump administration's rollbacks of environmental protections, these are incontestably important books."
—*San Francisco Chronicle*

"Vollmann's many fans . . . will not be disappointed. . . . He packs research and voice into his impassioned works. . . . Reading these two books did have an effect on me; I became even more conscious of the resources I waste in my own life."
—John Schwartz, *The New York Times Book Review*

"One of the enjoyable things about this massive work is the way Vollmann employs irony, and that bluntest of irony called sarcasm, throughout the volumes. He can be quite humorous. You might even call this the *Infinite Jest* of climate books.... There's something admirable, even noble, about the sheer time and effort—and sheer *humanity*—that went into these volumes." —*The Baffler*

"Equal parts gonzo journalism, hand-wringing confessional, and one hot mess ... the books document Vollmann's quest to understand how capitalism, consumerism, and fossil fuels are ruining the planet." —*Sierra*

"The best part of the books [are] the conversations Vollmann had during his travels, the sensitive histories he gives of the places he visited, and the moral impressions those conversations and places have made on him. It's these parts that made *Carbon Ideologies* a unique, lasting, definitive contribution to the global warming literature." —*The Humanist*

"An elegy to our damned epoch that's also a work of enlightenment and education ... The book is a performance of the vexations involved in trying to understand our energy reality. ... [Vollmann's] project—not unlike that of his historical fiction—is to show with utmost fidelity what it was like to be a human involved in terrible things." —*Los Angeles Review of Books*

"[Provides] profound insights into both Japanese society and universal themes regarding the human response to and preparation for major disasters and tragedies." —*International Examiner*

"Vigilant in his precision, open-mindedness, and candor, Vollmann takes on global warming.... [His] careful descriptions, touching humility, molten irony, and rueful wit, combined with his addressing readers in 'the hot dark future,' makes this compendium of statistics, oral history, and reportage elucidating, compelling, and profoundly disquieting." —*Booklist* (starred review)

In *No Good Alternative*, the second volume of *Carbon Ideologies*, Vollmann turns to an examination of coal mining and oil and natural gas production, in investigative journeys that take him from West Virginia, Kentucky, and Colorado to Bangladesh, Mexico, and the United Arab Emirates.

CARBON IDEOLOGIES

VOLUME I

NO
IMMEDIATE
DANGER

WILLIAM T. VOLLMANN

PENGUIN BOOKS

PENGUIN BOOKS
An imprint of Penguin Random House LLC
penguinrandomhouse.com

First published in the United States of America by Viking Penguin,
an imprint of Penguin Random House LLC, 2018
Published in Penguin Books 2019

"When the Wind Blows from the South" was first published in different form by *Byliner* in 2011.
"Harmful Rumors" first appeared in different form in *Harper's* in 2015.

Maps and illustrations by the author

I will always remain grateful to *Harper's* for helping to underwrite my documentary projects,
on which I rarely break even.—WTV

ISBN 9780399563515 (paperback)

THE LIBRARY OF CONGRESS HAS CATALOGED THE HARDCOVER EDITION AS FOLLOWS:
Names: Vollmann, William T., author.
Title: Carbon ideologies / William T. Vollmann.
Description: New York, NY : Viking, 2018- |
Identifiers: LCCN 2018013155 (print) | LCCN 2018019016 (ebook) |
ISBN 9780399563508 (ebook) | ISBN 9780399563492 (hardcover)
Subjects: LCSH: Power resources—Social aspects. | Energy policy—Social
aspects. | Climatic changes—Social aspects.
Classification: LCC HD9502.A2 (ebook) | LCC HD9502.A2 V65 2018 (print) |
DDC 333.79—dc23
LC record available at https://lccn.loc.gov/2018013155

Printed in the United States of America
1 3 5 7 9 10 8 6 4 2

Set in Garamond Premier Pro
Designed by Nancy Resnick

Note to the Reader

Carbon Ideologies was envisaged as a single work. When the manuscript arrived, several times longer than its contractually stipulated maximum, my publisher asked me to cut it. But for some reason, I just didn't want to. After anxious negotiations, Viking finally agreed to indulge me once more, and it was decided to break the book into two. (My gratitude is expressed in the acknowledgments section.)

This first volume, *No Immediate Danger,* contains a primer on global warming, and reportage about the wondrous effects of nuclear power in Japan. A subsequent volume, *No Good Alternative,* will deal equally cheerfully with coal, natural gas and oil.

As originally laid out, *Carbon Ideologies* ended with a section of definitions and conversions. What is the difference between a gamma ray and an alpha particle? How many gallons of Mexican crude oil would it take to do the work of a ton of bituminous Appalachian coal? Some of this information, particularly the items on heat, energy, efficiency and power, gets referenced throughout the whole work. Where then should it go? Viking and I decided to place it all at the end of the first volume, where it could find immediate use. To reduce expenses, it will not be reprinted at the end of the second volume.

The primer and the conversions section are both references. Anyone wishing to skip those can read between pages 219 and 514.

Carbon Ideologies also contains about 129,000 words of source notes, citations and calculations. I am sorry to say that Viking could not justify the cost of printing these. Therefore, *Carbon Ideologies* will be the first of my books to contain a component which exists only in the electronic ether (see https://www.penguinrandomhouse.com/carbonideologies). I will deposit a copy of that section in my archive at the Ohio State University.

All this renders the first volume particularly uninviting. It also makes that volume more practically useful, in exact proportion to its dreariness. Perhaps some unborn desperate generation will be helped by those pages in finally crafting a social code of thermodynamic work.

The title of Volume I was proposed by my editor, Mr. Paul Slovak. It refers to an official phrase often deployed by the Japanese authorities after the accident at

Fukushima. I consider this an inspired title. The title for the second volume, *No Good Alternative,* was one of several that I proposed, and while the final result was not my first choice, it now rings aptly in my mind's ear. Meanwhile, please let me reiterate that *Carbon Ideologies* is all one work.

WTV
Sacramento, April 2017

Near Dhaka, Bangladesh

A crime is something someone else commits.

John Steinbeck, 1961

Barcelona port facility

I dedicate this book to my daughter Lisa,
who cannot live anywhere else but in our future.

Carbon Ideologies is a companion to *Poor People* and *Rising Up and Rising Down,* which I wrote some years ago. All three volumes use induction to generalize from subjective case studies into analytical categories of the phenomenon under investigation. Although I do take sides (against the ill-regulated abuses of energy, and not practically *for* anything), I strive not to judge individuals,* not even myself: Coal, oil, natural gas and nuclear power, haven't I consumed them all?—Better an honest muddler than a carbon-powered hypocrite.

---

* Certain politicians do get singled out, to satisfy their need for attention.

Contents: VOLUME I

———

* The "About" sections were inspired by their namesakes in my old *Joy of Cooking.* After all, we should have known the various recipes through which our goose was being cooked.

For source notes to both volumes of *Carbon Ideologies,* please see:
https://www.penguinrandomhouse.com/carbonideologies

List of Maps and Illustrations

Nuclear

1. Lower Than for Real Estate Agents

When the Wind Blows from the South

Harmful Rumors

The Red Zones

Definitions, Units and Conversions

CARBON IDEOLOGIES

VOLUME I

NO IMMEDIATE DANGER

0

When We Kept the Lights On

We all have in us the ghosts of long-vanished things, of fallen cities and marvelous machines.

Gene Wolfe, 1982

1

Someday, perhaps not long from now, the inhabitants of a hotter, more dangerous and biologically diminished planet than the one on which I lived may wonder what you and I were thinking, or whether we thought at all. This book is for them.

When I read another embrittled document predicting the disappearance of bison from the American Plains, my melancholy is untainted by urgency. Captive bison do survive, but the great herds have been gone since 1884. And as I write this book about coal, oil, natural gas and atomic power, I do my best to look as will the future upon the world in which I lived—namely, as surely, safely *vanished*. Nothing can be done to save it; therefore, nothing need be done. Hence this little book scrapes by without offering solutions. There were none; we had none.* All the same, it may not be uninteresting to learn what went on in the minds of buffalo hunters, Indian killers, coal miners, freeway drivers, homeowners and nuclear engineers.

In the time when I lived, it was still possible to meet Americans who disbelieved in global warming, although the ones I knew became shyer and rarer in about 2013. In 2016, they helped elect Donald Trump President, upon which their various carbon ideologies naturally came roaring back.

"We sure need a good Sierra snowpack this year," said a contractor friend of mine. "Skiing was lousy last year and the year before. If we can only get some snow, that will make those global warming people shut up."—That was just before Christmas. Come spring, the snowpack was 6% of what we had been calling normal. But why not call California a special case? Up in Washington the snowpack was a full 16% of what it should have been; and by May, "seeing things happen at this time of year we just have never seen before," Governor Inslee declared a "statewide drought emergency."—Fortunately, my contractor friend was vindicated, and those global warming people utterly foiled, for after a long dry year, the subsequent winter blew flurry-rich, and by January the Sierra snow level had reached 115%!

February turned unseasonably warm. The leftwing hoaxers got impudent

---

* If, like me, you wish to pretend otherwise, see what we should have done, on II:627. (Henceforth, cross-references will be as follows: A page preceded by a roman numeral II, e.g., II:123, directs you to Volume II of *Carbon Ideologies*. A page number by itself cites the volume that you hold.)

again. As for the skeptics, they took strength in the fact that carbon forecasters of other stripes had been wrong before, in token of which I quote from my grandfather's *Mechanical Engineers' Handbook,* copyright 1958: Petroleum would soon run out! *The peak of production in the United States should come about 1965 . . . World shortage of petroleum may be expected to begin about 1960.*—If only!—As for coal, in predicting that American production would reach its height in about 1975 the *Handbook* was not far wrong, but it anticipated that a world shortage of *total world fossil fuels . . . would be noticeable* around that same year, which is precisely when a build-your-own-alternative-style-house primer warned us all: *A Federal Power Commission staff study, released in January of 1975, concluded that natural gas production from the forty-eight contiguous states has reached its peak and will decline for the indefinite future.*—We were all on the verge of getting cold!—But in 1993 the National Coal Association announced that "at present rates of use" our coal reserves *can be expected to last nearly 250 years. There are about 1,000 tons of recoverable coal for every man, woman and child in the United States.* Then came fracking, which afforded gas enough to toast us in our planetary oven.

In each of these dull and distant comedies, we got condemned to future deprivation, and then the diagnosis brightened! (I myself got cynical; I didn't care; I chalked it up to financial manipulation.) In 1999 my atlas advised me that oil might last another 40 years if we were lucky. And in 2015, as I sat beside the

automatic gas fire, writing *Carbon Ideologies,* Iran announced intentions to increase her gas output by 40% in the next five years, while coal prices fell farther; oil prices had just decreased again: There remained enough fossil fuels to choke us all! So why not deal sharply with pessimists, or refrain from dealing with them at all? Mr. Jonathan Lee, whose company rented out supertankers, felt *as excited as a*

On display at the Kentucky Coal Mining Museum*

---

* I had hoped to interview someone in this organization. My intermediary inserted here: "I think you should mention that the Kentucky Coal Association NEVER got back to me even though they promised."

rookie because *you are seeing history change before your eyes*! China and India meanwhile began seizing *a bargain opportunity to top off their petroleum reserves.* The prior errors of prophecy proved that no one knew anything about anything; therefore, climate change was the merest hot air.

Not far from the disbelievers dwelled those who couldn't be bothered about "an ecosystem somewhere."* In 2016 a kindly barber told me: "I don't really think none about it, although I have to say that when I see people pick up cigarette butts from the sidewalk I appreciate their caring, and people that care about the earth, I mean, that's nice, and when I think about the polar bears losing their land, I do feel touched about that, because I care a lot about animals." For him, an ecosystem was something to watch on television while he ate takeout pizza. He was a decent fellow who had never been consulted by the carbon vendors—whose systems of extraction and delivery had long since become invisibly ubiquitous.

In 2014 my friend Philip, a cheerful, hardworking realtor in his early 40s, allowed that global warming might exist, but that it was natural and "evolutionary"; the human race had little to do with it. For years we had drunk together and listened to each other, so I asked him to tell me more. "Why should I concentrate on anything that stresses me out?" he demanded, and when I saw that the subject might dent his cheerfulness, I changed it.

Kindred sorts reassured me that our new weather was "natural" and cyclical, and therefore required no action. Indeed, precious little action was taken. *For more than 40 years, Homer City* Generating Station in Pennsylvania *has spewed sulfur dioxide from two of its three units completely unchecked, . . . because it is largely exempt from federal air pollution laws . . . Last year, the facility released 114,245 tons of sulfur dioxide, more than all of the power plants in neighboring New York combined.* This pollutant was both a killer of many organisms and a dangerous "precursor" gas with unpredictable effects on the climate.† In 2011, the Environmental Protection Agency finally demanded that Homer City clean up. After threatening *immediate and devastating consequences* and losing a lawsuit, the utility found a way to comply—without even raising its electricity rates. When I read this tale in the newspaper, my first emotion was happy astonishment that mitigation had proved so practicable—after which I felt all the more

---

* Linda Tirado, 2014: "It's not that I don't care about global warming or the environment; it's that there's only so far out of my way I'm willing to go . . . Overconsumption is a concern for people who've made it to regular consumption . . . I do not care about the whales . . . Once we've hit the part where my own species is mostly taken care of, I'll start to worry about African rhinos. Until then, I'll just keep restraining myself from punching people when they look me in the face and argue that an ecosystem somewhere is more important than homelessness."

† See "About Sulfur," II:26.

amazed that Big Coal kept digging in its heels against reducing harmful emissions elsewhere—and was allowed to do so—while Big Oil and Big Frack behaved much the same. (As for Big Nuke, its mantra, as we shall see, was: *No immediate danger.*)—In 2017, a fellow who had repeatedly sued the EPA was appointed to run the agency. Well, after all, who gave a damn about some old ecosystem somewhere? (A newspaper item: **Pounded Again, Coastal Town May Consider a Retreat**. *Just repairing and repairing the sea walls—it isn't a permanent solution with the ocean coming ever closer to us.* Those words used to be exotic, back when I was alive.) Who could say whether *somewhere* might be *here*? That ecosystem's peculiarities had always lain beyond our ken. Although odorless methane might be accompanied by the scent of crude oil, and while carbon dioxide, that *colorless gas with a faintly pungent odor and acid taste,* sometimes heralded itself in jet trails and smokestack-clouds, both of these quickly vanished into our all-accepting sky. Then what? As a West Virginian pastor told me (you will meet him in the coal section): "Here you do see the smokestacks and you know that they do put off the smoke and everything, but it seems to me that the earth is so large and there are so many trees and everything that how could manmade equipment put up enough smoke to make a difference?" His question was absolutely reasonable. Answering it would have required the help of scientists, instruments and historical records. Even then, *causation* could never be proved. All one might hope to establish was plausible correlation with predictive value. Most of us were non-correlators; to us the clashing claims felt wearisome, complicated, inscrutable. It took me all my life merely to understand aspects of myself—and why shouldn't the latest scaremongers be as wrong as the Cassandras of "peak oil"? In Bangladesh I met coal mine workers and even a labor union leader who had never heard of global warming; of course they asserted that there was "no alternative" to coal extraction. Indeed, they proved their own point.

I remember another courteous old West Virginian who had just been speaking cogently about his childhood, his coal miner father and the decline of coal extraction in Appalachia; he grew vaguer when I requested his views on climate change: "I'm sure it's got something to do with the situation on TV"—meaning that he had seen television footage of weather-related disasters, and supposed that global warming might be part of the cause. "That's part of the pollution problem," he allowed. Then he added, and I failed to follow his logic, "That's why the EPA is doing what they are, stepping beyond their authority, I think, in a lot of ways."

"Do you think coal contributes to global warming?"

He reassured me: "They've got technology now that can cut all the pollution out."

Had it only been so!—In fact it was so *a little bit,* as Homer City unwillingly proved—but our captains of fuel and electricity resisted even that little for all they were worth, which was plenty.

Anyhow, that good old man, who cannot be faulted for the incompleteness of his knowledge on greenhouse gases; and the disdainers of somewhere-nowhere ecosystems, the gloom-and-doom handwringers like me, the climate change deniers and the consoling weather-cycle asserters, we were all outnumbered by ordinary practical folks for whom cheap energy and a paycheck incarnated all relevance. One fellow wrote in to the newspaper: *My son works in the coalfields of southern West Virginia[;] he supports my two grandchildren by mining the coal that keeps the lights on in America*—this last phrase being often used by my nation's carbon ideologues. *So I know firsthand the importance of winning the war on coal that Obama declared five years ago.* Whether it was truly Obama who started it, who our enemy was, and whether America's lights might beneficially be dimmed here and there, failed to encumber him. The maintenance of his two grandchildren trumped other arguments; their needs caused him to *know firsthand* the small selfish thing that he knew. I will not celebrate him, but I decline to blame him, either. Why should his kin go hungry? (You in our future can go hungry instead; after all, we don't even know you.)

In *The Wall Street Journal*'s "Notable & Quotable" section, a so-called "environmental writer" enlarged his argument into a carbon eulogy:

> In 1971 China derived 40 percent of its energy from renewables. Since then, it has powered its incredible growth almost exclusively on heavily polluting coal, lifting a historic 680 million people out of poverty ... A recent analysis from the Centre for Global Development shows that $10 billion invested in such renewables [as solar energy] would help lift 20 million people in Africa out of poverty. It sounds impressive, until you learn that if this sum was spent on gas electrification it would lift 90 million people out of poverty. So in choosing to spend that $10 billion on renewables, we deliberately end up choosing to leave more than 70 million people in darkness and poverty.

In other words, back when I lived, some of us believed that *heavily polluting coal* could somehow *lift people out of poverty* without impoverishing us in any more fundamental way. We believed that because it was convenient to believe it. So we kept the lights on.

2

A Russian woman opined that the winters of her city, Saint Petersburg, were often just as severe as formerly, but their duration had diminished from five to four months. Well, but what could such anecdotes prove? (*Encyclopaedia Britannica*, 1911: *As concerns the popular impression regarding change of climate, it is clear at the start that no definite answer can be given on the basis of tradition or of general impression.*) I went to Mexicali for half a week, and found the nights hotter than I remembered, but a taxi driver who had lived there all his life insisted that temperatures were unchanged, although the humidity might have increased. (Across the border, another cab driver told me that his relatives in Algiers were now experiencing snow, which they had almost never seen; they loved it.) A fellow Californian who had backpacked the Sierras for decades returned from a hike in June, when he would have expected to see wet green meadows, and then snow at the high elevations; this time he told me: "A lot of trees dead and dying. Kind of alarming, actually."—Again, these observations described only *weather*. They proved nothing about *climate*.—A businessman in Bangladesh asserted that 20% of his country would lie underwater within 35 years. (You from the future might know that more of it sank faster.) And as we waited to board our Amtrak train in Sacramento, I asked the old lady who stood beside me on the platform whether she believed in global warming, to which she said: "Oh, absolutely. Even the children are getting sunburned more easily. When you wear black pants you feel the sun sooner. We've done too many things not knowing what the effect would be."

"What should we do about it?"

"Pray. That's all we can do."

That was how so many of us felt! We couldn't make any practical difference. As the Unabomber calculated in his manifesto: *When a decision affects . . . a million people, then each of the affected individuals has, on the average, only a one-millionth share in making the decision.* So we couldn't do anything. Or else we just didn't want to. In 2015, when the price of crude oil decreased by half, the U.S. Congress might have increased the fuel tax, however modestly, and employed the proceeds on funding renewable energy (or, if nothing else, on building seawalls). In its customary elevated style, *Time* magazine announced how everything would actually play out:

CASHING IN: THE DROP IN OIL PRICES IS EQUIVALENT TO A $125 BILLION TAX CUT. HERE'S WHAT THAT MEANS: . . . BIG CARS ARE BACK.

Accordingly, solar energy stocks decreased in value; sales of hybrid vehicles declined. How could we help it that we liked big cars? There was nothing to do but watch the buffalo get slaughtered.

3

For awhile our powerlessness got represented and recapitulated up through various levels of government. In 2014 the Attorney General of West Virginia announced: *The Office's end goal is to stop the EPA from hurting our state, and if we can't do that, we plan to at least gum up the works enough so that it limits the damage that the Obama administration can inflict on our citizens.* The defiant frustration in his utterance, which played well to certain interests, must have been felt in greater measure by the White House's feeble occupant. (Near the end of his second term he sadly remarked: "There's this notion that there's something I might have done that would prevent Republicans to [*sic*] deny climate change . . .") You see, we were *all* feeble! A West Virginian delegate (who happened to be a coal miner) complained that Obama "speaks as though the Earth's climate ends and begins at the shores of the United States, while willfully ignoring the fact that China is revving up its economy by using coal"—some of which came from the United States, a fact *he* willfully ignored. In other words, we Americans could hardly make a difference; therefore, nobody should make us try.*

Accordingly, carbon's ideologues empowered themselves when and where they could. In 2015, a Wisconsin *agency that manages thousands of acres of state land . . . banned its employees from working on climate change issues while on the job.* The only person who voted against that ban, Wisconsin's Secretary of State, was a scientist. In other words, he actually possessed the competence to determine whether or not our actions could "put up enough smoke to make a difference." Remarking that climate change was in fact altering forests, he decried "the trend of public officials who, either out of ignorance or out of political expediency, deny climate change."—And so denial led happily to silence.†

Meanwhile the Pope, who had also unfortunately been encumbered with scientific training, prepared to release *a key policy document which is expected to blame mankind for climate change, a view which has enraged US sceptics who say religious leaders have no right to take part in a scientific debate.*

---

* And how wonderful *not* to try! The West Virginian politician added: "The life of coal miners might not be an important factor to Obama's most ardent supporters, but it is our way of life and one of which we are proud."

† Wisconsin's Attorney General, who of course voted for the ban, refused to discuss the matter with *The New York Times*. His descendants died choking like the rest of us.

In that year the Norwegian Parliament, the Church of England, the Rocke-feller Brothers Fund and the French insurance entity AXA all decided to divest in varying degrees from coal-associated businesses; fortunately for order and tra-dition, these measures would have little or no impact on the vast market capital-ization of most companies.*

So the War on Coal rolled on, until we all lost.

4

The future for which I write will most likely also be a more radioactive time. Just as we continued mining the coal, fracking the shale and drilling the oil that kept the lights on in America, not to mention Bangladesh, even as carbon dioxide levels crept up and up in our atmosphere, so it was that even after Hiroshima, Chernobyl and Fukushima, plenty of Japanese kept assuring me that nuclear power was "necessary." (You will hear from them in this book.) It might have been "better," at least. According to the first Chairman of the Atomic Energy Commission, *low level radiation is probably less "dangerous" than the emissions from burning coal.* Anyhow, I hope that you in the future have learned how to make your spent fuel rods safe.

5

As I shuttled here and there to write this book, I came to find certain effects of oil, coal, natural gas and nuclear power comparable if not exactly equivalent. All of them undermined you in one way or another—a matter invariably minimized by the corporate entities that produced them—and all were tolerated or even celebrated by people whose children they endangered, because how else could the lights stay on?

I who send this letter to the future hereby plead that we were no more evil or even selfish than anyone else. As a friend wrote to me: "Yes, I do want to have

---

* *Little or no impact* was also the argument made by those who resisted other carbon reduction mea-sures. Consider the President of the Missouri Farm Bureau, who disliked wind farms. His terms of complaint resembled those of the few fracking dissidents I interviewed in Weld County, Colorado: Pads and pipes rose up in their back yards against their will! I could not blame one farmer's resentment when he learned that a 140-foot wind energy transmission tower "was to be located not in one of his open fields, but in the corner of his yard, about 250 feet from his front door." But the Farm Bureau's President did not stop with that case. He informed us that "Missouri landowners" placed no stock in the notion of "permanently changing the countryside to theoretically shave a fraction of a degree off global temperatures sometime in the next century." In other words, they folded their hands along with me.

children—and I hope to leave them some kind of world to inherit, one where there are still birds and rivers and salmon." Wasn't her heart as good as yours?

Some of us lived in a fairly robust democracy of opinion, but lacked any democracy of ideas, let alone of policy. Our various educational systems failed to impart the minimum knowledge which a citizen would have needed to *judge* coal, nuclear power and other methods of keeping on the lights. This knowledge would have entailed some competence in the following procedures: carrying out simple mathematical conversions, marshalling facts, comparatively quantifying energies, emissions and efficiencies; performing risk-benefit analyses, deducing the specific material interests of each carbon ideologue, recognizing omissions, inaccuracies and outright lies, positing and testing relationships between facts, verifying and disproving all claims, including our own—and, most crucially, deciding what we needed to know, and how to seek that information.* Some apparent phenomena would still have resisted measurement and much would have remained arguable. But the less we measured, the more conveniently we could argue—while the threat continued to become a calamity. Some lonely soul in Florida whose basement kept flooding might blame the landlord instead of the climate—and who was I to dispute his theory? In President Trump's time it got even better; his administration coined the phrase *alternative facts*. Meanwhile, each cool day disproved global warming anew—because in the time when I lived (and maybe every time), official communicators hindered us from learning the semantics of "experts," while encouraging us to pretend that information could in and of itself be "entertainment," so that any nuanced statement became inferior to an easy answer. The "experts" liked it that way, of course; so did the spin doctors and cost-cutters, the anti-secularists, politicians, industrialists, tired "consumers," schoolchildren, wildcatters, go-getters, marketing consultants, advertisers and bought technicians.

To be sure, there were some experts, whose knowledge and integrity I will honor in these pages. We should have listened to them.

In Japan the people relied on their leaders great and small. An understanding of, say, radioactivity was not required. The officials in contaminated areas discouraged what were called *harmful rumors*. Citizens tried to believe in the goodness of corporations and the sincerity of cabinet ministers, or else shut out of mind what could not be helped. They lacked comprehension of the various waves and particles that threatened them, not to mention the units of measurement used in media pronouncements. We all learned to live with what we could not see. In the "red" zones, the word of the day was "invisible." And in America,

---

* *Carbon Ideologies* attempts to do some of this work.

morning television kept the climate crisis invisible by emoting on and on about a man whose house collapsed in a sinkhole, or devoting 20 minutes of live aerial footage to a high-speed police chase. Should they express themselves on those urgent subjects, my fellow citizens might be rewarded with the momentary half-attention of strangers. They were rarely troubled for their opinions on anything that mattered.

Of course we did it to ourselves; we had always been intellectually lazy, and the less asked of us, the less we had to say. Again, we were no worse than others. (If anything, we might have been less rigid in our ideas than the Assyrians.) In our time the sky never stopped raining claims and counterclaims. We came to think that we had heard them all. Had we in the teeth of our complacent mis-education arrived at any common conclusion, we lacked the power to enact it. So we lived private lives, not worrying about the unpreventable, while the "experts" kept cashing in.

So did I. When I lived, the profit motive was unanswerable. An unemployed West Virginian told me: "Well, I can't blame them coal companies for going away. I mean, business is business."

For those who couldn't aspire to profit, mere survival was even harder to argue with. An art teacher in that same county (the high school where he taught sat right on top of a mountaintop removal mine) remarked: "It's cultural devastation to lose families in the coal mines and it's cultural devastation to have families break up when men can't feed their loved ones. When you make a product, and you base it on the labor of men's backs, and then you take it away, you turn us into a Third World country." Naturally he did not want the coal companies to take it away. Coal was poison, sure. It poisoned the rivers when it got cleaned. His solution: "Why do you clean coal? It's got dust on it. Why not make the Chinese clean their own coal?"—He was another good man, who couldn't see that what hurt the Chinese would eventually hurt us. Well, maybe he saw it and didn't care. Were I in his shoes, or in his pupils' shoes, I might have behaved the same. I might have said: "It's going to be a hungry winter, and the baby's sick, and I can't pay the electric bill. Let me worry about my own."—Isn't that how it must be for you in our starving future?

We all lived for money, and that is what we died for.

Now that we are all gone, someone from the future is turning this book's brittle yellow pages. Unimpressed with what I have written so far, he wishes to know why I didn't do more, because when I was alive there were elephants and honeybees; in the Persian Gulf people survived the summers without protective suits; the Arctic permafrost had only begun to sizzle out methane; San Francisco towered above water, and there were still even Marshall Islands; Japan was barely radioactive, Africa not entirely desertified.

As I said, this little book makes not even the beginning of a solution. It is simply an attempt to give the main questions their due. When I began *Carbon Ideologies* I walked away from my nonfiction publisher, because my research had already cost more than the editor proposed to pay. So I spent my own money, and occasionally other people's, to hike up strip-mined mountains, sniff crude oil, and occasionally tan my face with gamma rays. Had I been richer, I might have accomplished more—but what, exactly? To tell the truth, there were times when I longed to forget about some ecosystem somewhere. I knew I'd find no adequate *personal* answer to the question "What should we do?" But I felt ashamed of doing nothing.

Well, in the end I did nothing just the same, and the same went for most everyone I knew. This book may help you in the hot dark future to understand why.

WTV
2011–2017

*Here I pause for one moment to exhort the reader never
to pay any attention to his understanding
when it stands in opposition
to any other faculty
of his mind.*

Thomas De Quincey, 1823

Transmission towers near Ruwais, Abu Dhabi, United Arab Emirates

About the "Primer" Section

Actually, this book was not of the slightest value to human progress. Climatic change was not a practical problem. In any case, the book had been superseded.

George R. Stewart, *Earth Abides,* 1949

promised to do my share to facilitate risk-benefit analyses. Now I will try to keep that promise.

Units of power consumption, measures and categories of greenhouse emissions and even compositions of fuels may change. However, the basic relationships between such quantities will not.

Being one of those pathetic creatures called "literary" writers, I never before got called upon to quantify peak load capacity or ponder the carbon content of dirty diapers. For much of *Carbon Ideologies* I will follow my custom, and tell stories of people and places. In the next 200 pages I have stepped back a little, until the trees became a forest. To endeavor to see *what is* is to abstract; this kind of seeing comes unnaturally to me, an in-the-moment-sighted fellow who would rather admire leaves than wonder how they grew. That I was not the only such being may be inferred from the fact that we called economics *the dismal science;* while those who did not get their livings from them frequently considered physics, arithmetic, ecology, engineering, chemistry, politics and waste disposal in the same light. All I could do was my blinking, nearsighted duty—which, at least as I understood it, meant taking hold of dismalness, until familiarity made it not only tolerable, but remarkable; the fog became a blur, then a pattern whose complexities rewarded me with a different kind of vision.

As Buddha once said, enlightenment confers no special benefit; nor is unenlightenment bad. Accordingly, I chose unenlightenment, which is to say attachment to the inhabitants of this world; and so I worried once it came to me that their position was deteriorating "faster than the worst-case scenario."* There is nothing like imminent doom for sharpening a person's interest in practicalities. Indeed, I may not be going too far to call the topics in this section necessities, at

---

* See II:650.

least for my contemporaries; in your sad future day they must partake more of the might-have-been and should-have-done. Thoreau called for books *such as an idle man cannot read, and a timid one would not be entertained by, which even make us dangerous to existing institutions,—such I call good books.* I would hope beyond my means, to imagine that *Carbon Ideologies* could do anybody the favor of making him dangerous, but maybe I can take credit for rendering this section as soporific to the idle as would be any other crowd of mathematical quantities. Following Buddha, I assure you that there will be no harm in skipping to page 219.

About Tables

Human eyes can perceive things only in the forms that they know.

Montaigne, "Apology for Raymond Sebond," 1575–80

Again, not every reader likes them. Because I kept wondering what we should be doing, back in the days when I was alive, I began to ask such subsidiary questions as: "How dangerous is this? How wasteful is that?" It seemed best to answer quantitatively when I could. I tried to keep my numbers simple and clear. Wherever possible, I expressed comparisons *in multiples of something near at hand,* such as the radiation level in my studio, or the number of bricks that a Bangladeshi whom I had just interviewed could bake from a given amount of coal. Thus in these tables the multiples matter more than the units—which may be unfamiliar but are all explained, both as we go and in the end matter.[*] To see that pure methane contains by weight 1.7 times more heat energy than pure carbon may suffice you. If not, please turn to page 534 for a definition of high heating value.

Please be advised that some of this book's assertions and calculations depend on approximate numbers. For instance, the energy content of a ton of crude oil may vary by as much as 8%.[†] Estimates of power consumption in different nations, particularly over time, reflect differing assumptions. Climatologists alter their descriptive quantifications. A few very large figures might be "fuzzy" up to plus or minus 12 or even 15%.

In my own battle against spurious precision, I simplify numbers where I can. Thus, 105,775,826,160 tons of West Virginia coal becomes 105.78 billion tons—a procedure which will loosely be called "rounding to two significant digits right of the decimal point." In the unlikely event that you crave more digits, check the source notes.

And again, since *Carbon Ideologies* is primarily a record of people's experiences, if you skip my tables and their numbers, my point will remain clear enough; better yet, any mathematical errors might then escape your censure.

---

[*] My scales create their own problems. Measurements of, for instance, very low radiation levels may be less accurate than those of higher ones. Thus the value I call "1" may be off, and everything else with it. I address this point on p. 244.

[†] See p. 577.

About Photographs

Survey... the most remarkable events that have shaken the earth and decided the fate of men. Alas! What remains of these... great exploits? The most real signs... are the traces they have left on our canvases in forming these pictures.

Tiphaigne de la Roche, *Giphantie,* 1760

always thought it advantageous to our species that seeing was believing. The unseeable might require numbers to gain credence, as was the case with radiation, which all too many of the Japanese I met at Fukushima consoled themselves was *invisible.* Holding my pancake frisker* to some weed or drainpipe, and learning from the digital readout what stood so nastily at hand, thrilled me with healthy horror, and so I have included photographs of the frisker as it samples some object, in order that you, too (with or without the horror), may see and believe.

Much of this book is an attempt to witness. I believe with all my heart that the recollections of Mrs. Glenna Wiley, aged 89, who survived the Buffalo Creek flood of 1972—a disaster caused by a coal company's negligence—will mean more to you when you see her portrait as she sits at her kitchen table with the door open and light shining in from the humid greenery; her dark eyes, her lean, wrinkled face, and her half-smile, which I call both polite and melancholy, may tell you something more than I have expressed; her slender, freckled arms and the hard-working hand in the foreground show that this woman has been much outdoors; the table's bare wood, occluded in part by a coarsewoven placemat, is somehow of a piece with the "folk art" rooster on the paper towel dispenser behind her. Here is a gentle, trusting old lady who let me into her house although she had never met me nor had any notice of my coming; she offered me a soda and answered all my questions about an event whose associations must have been painful. I am hoping that this photograph† will help you believe and remember her.

Then there are the photographs of ordinary things. On the next page you will see part of a glassfronted building in Sharjah, United Arab Emirates. Your first thought may be that this prospect (or dead end) exudes such familiarity that members of my generation would have lacked any apparent cause to look at it. As

---

* A kind of scintillation meter. See p. 397.
† See II:54.

you will presently read,* glass, while not one of the "big five" energy-ravenous construction materials, still needed enough power, and gave off enough carbon dioxide, to call into question its unbridled manufacture. Meanwhile, this is a pretty facade, designed by someone who knew and cared about aesthetic factors. I took pleasure in recording its existence. You from the future may enjoy ogling its bygone luxury, which in its own way resembles the gold-and-white ornateness of Marie Antoinette's boudoir, and which your world cannot afford.

Glass facade, Sharjah

And now, let us wander together through our predicament.

---

* See p. 132.

PRIMER

Overleaf: Retired derrick at Devon Oil and Gas Exploration Park, Oklahoma City

What Was the Work For?

We are the soul, of which railroads, copper mines, steel mills and oil wells are the body—and they are living entities that beat day and night, like our hearts, in the sacred function of supporting human life . . .

Ayn Rand, *Atlas Shrugged,* 1957

An internal combustion engine is *a machine for converting chemical energy to mechanical energy by burning a fuel with air in a confined space and expanding the products of combustion, extracting energy as work.* This definition, which combines straightforwardness and completeness into a kind of elegance, comes courtesy of a World War II primer for U.S. Navy midshipmen. The world in which I lived had long since been remade by just such machines, and by their cousins, which converted nuclear and calorific energy to electrical energy, likewise doing work. But what was the work for?

When my generation was alive, I used to wake up, reach out and turn a switch to light up my bedside. To you who can no longer live this way, this must sound as incredible as if I were to claim that I called on angels or that I could wander anywhere without a radiation meter. And I do admit that every now and then the bulb would burn out, while on rarer occasions the power would even fail, at which my essentially windowless dwelling grew rapidly unpleasant—too dark to read in; humid in summer, clammy in winter—but such perturbations remained so unlikely that each morning I flicked that switch with confidence: Somewhere, unknown mechanisms would do work on my behalf. My dressing table now being illuminated, I picked up the plastic pocket clock whose two slender silver batteries converted chemical energy into electricity for a year or more at a time, after which I removed them, threw them "away" and inserted new ones. That was how I knew the time. If it was already rising hour, I strolled to the bathroom and showered to my heart's content. In those days I saw nothing wrong in letting a stream of hot water tingle the back of my neck even when I was clean; it did wonders for grogginess. One of my womanfriends loved to shower just to think. If you from the future are the losers for this, I am sorry. I think we felt a kind of grandness to have so many energies at our call, even if we rarely thought about our situation. Why shouldn't they serve us faithfully? The sweet open friendliness of a young child, which can charm a smile out of an impatiently busy adult, derives from no sublimer cause than inexperience: The child has not yet learned that the world will refuse and reject him, then plow him under. And when my generation flourished, why should we have imagined that electric power would ever decline to fulfill our whims? That is why we were happier than you. I certainly enjoyed showering to warm up, showering to cool down, showering before and maybe again after making love, washing my hair whenever it felt greasy. My water heater, by the way, was electric-powered, because the contractor who installed it had explained that electricity would cost me less than gas. He was another one whose magic powers brought him joy. I used to like his smile when he fired off that hissing battery-powered gun of his: Great nails flew out, slamming themselves deep into boards, panels and beams, instantly in place as if they had been there forever, except that the nailheads were warm. There went his arc-welder, strengthening my window-bars and improving the evening with fireworks of sparks. He was a man of almost incredible endurance and physical strength, who helped me over and over. Sometimes he burned or gashed himself on my behalf. I loved to watch him fix things.* Had someone inquired as to whether a gas or electric heater was "better for the planet," neither he nor I would have known the answer. (Here it is: *Using electricity to generate . . . heat is extremely inefficient,* said a solar energy handbook.) In any case, the shower proved nearly as reliable as a bedside lamp. I let the water run and run until its temperature suited me; then I stepped beneath that comforting flow. It is true that when someone else took a shower right before me, the hot water ran short, but that was on account of my economically small heater (58 gallons), not because of any lack of electricity in the world.

My stove employed gas because I had heard that electric stoves take forever to heat. Anyhow, gas was cheap; I used to prepay my bill with a check that (like my clock's batteries) kept me at peace for two or three years. So here too I had something to anoint me with confidence. When did the gas ever not come on?—Had I compared my gas bill with my electric bill, perhaps I might have learned

---

* Like my realtor friend, he described global warming with the word "evolutionary." He said that the poles of the Earth had rotated several times, and this might now be occurring. He said that even Al Gore had stopped referring to global warming and now spoke of climate change. That the climate was changing he was inclined to acknowledge, but he suspected that humans were only a small part of the cause. How this would ameliorate our situation if it were true I could not tell at first, but then he went on to mention epidemics, and the great meteor which might have exterminated the dinosaurs, and remarked that some such disaster might befall us at any time, so that I began to comprehend his reaction as a kind of fatalism near to mine. I had begun to suspect that global warming was inevitable, so what should I be but fatalistic?

something about relative efficiencies, but probably not, with the pricing so arcane—therms *versus* kilowatt-hours—and the various appliances hardly comparable. All I can tell you is that showering and cooking were both easy. Whenever I fried an egg, warmed a can of soup or boiled water for my tea, I would open the valve on the wall, turn the burner dial all the way counterclockwise, strike a flame with my longhandled lighter (powered by a disposable natural gas cartridge, of course), and watch the ring of blue fire magically kindle itself! Then I reached into my refrigerator, which was a convenient machine that you in your day must lack; it saved me from going to the supermarket more than once a week; all one needed to do to prevent food spoilage was pay the electric bill. I took out eggs, chilis, tomatoes, spinach, herbs and cheese—I could buy them all year long!* A potato and an onion already sat diced up on the cutting board. The burner hissed; in my frying pan olive oil began sizzling. Another switch on the wall activated a fan, to evacuate the smell of cooking into the outside air. Making breakfast was delightfully labor-free, at least for me; my carbon-powered slaves did the work!—Afterward, unfortunately, I had to scrub out the frying pan myself.

Whenever my sheets and clothes got dirty, I threw them into a machine that spun them around in soapy heated water. Then I tossed them into a gas dryer. About 4% of the energy consumed in American homes went for clothes drying. How much energy it took to wash my clothes I never learned.†

On my roof sat a pair of great square machines; one was a two-ton and the other was a ton-and-a-half. They could cool or warm my place in almost no time. Although I was supposed to get them serviced every month, I let them go for years on end, and they went on faithfully just the same. I also enjoyed all the ceiling-light I wished for; I could even plug a plastic machine into the wall and call up music by artists living and dead. Unlike you from the future, I lived independent of nightfall and winter. Having camped outside from time to time, I knew enough to prize this home of mine whose electric light and warmth allowed me to anytime do anything one could at high noon in a mild summer. I often loved to sit at my kitchen counter reading or writing, while dusk drew in

---

* I will not soon forget the first time (it must have been in the 1990s) that I went to the only store in a certain impoverished Canadian Inuit town and saw for sale not only frozen hunks of local marine mammals, not only canned and powdered convenience foods from Down South—but fresh kiwi fruits from Australia. And I wondered how it possibly made sense to fly them there. But at home I rarely thought twice about buying some exotic food that caught my fancy.

† From a work on Victorian-era Britain: On washday (every two weeks), "servants might rise as early as 2:00 a.m. to begin heating large cauldrons of water . . . Major physical effort was required to wring out sheets and other heavy items; and drying not only depended on the weather but also, in urban areas, might leave the newly washed clothes grey with soot from the outside air."

around me, until suddenly I would realize that no more sunset light showed itself through my blinds. The outer darkness made my carbon-powered fortress all the more delightful. For several more hours, internal combustion engines continued to sigh past me on the highway, then grew infrequent, while I sat on as long as I liked, nourished by heat and light.—I wanted another cup of tea. Why not? A turn of the valve and dial, a click of my igniter, and once more the burner hissed; in two minutes water was bubbling in my saucepan.

Disdaining to write drafts of this book on paper and then recopy them, which might have made my hands tired, I employed my slender, bright-screened and ever so convenient portable computer. Too bad that I had to buy a new one every three or four years! Whenever the power ran low, I plugged it into the wall.

Now I was lonely; I felt like chitterchatting with a woman I loved. Picking up my cordless phone, I dialled her up, then lay in bed at my ease, enjoying her voice. That was pretty pleasant, I can tell you! She detailed her cat's latest tricks, and the book she was reading; we had time for those trifles when we were alive. Then before sleeping I returned the phone to its charging cradle, which drank endlessly from the wall. Do you from the future think me such a bad person, that I lived as many others did, in that easy way?

I slept at a temperature that suited me. (I liked to keep cool.) Another awakening, and I walked outside to get my morning coffee. I enjoyed that walk, because unlike you from the future, I lacked any necessity to move my limbs; I walked only for pleasure and because it was good for my heart.

Confess it, reader: Don't you wish you could have what I did? Wouldn't you have enjoyed a life of equally luxuriant selfishness?

What was the work for? Once upon a time, not long before or after the beginning of the 20th century, a little man named Lucien Lucius Nunn proved that sending high-tension alternating current over long distances could be safe, reliable and cheap. Shall I tell you his purpose? In an undated letter, he wrote:

> Lessening man's toil by the use of a wire in lieu of transporting coal in sacks over steep trails on pack trains is but one of many results already accomplished. To raise man's efficiency—to reduce man's toil—to give him time and means to love his family, his country, and his soul is the work to be accomplished . . .

These words touch me. He meant well, and I think he did well.

In India *the work to be accomplished* was more urgent than Nunn's. A paper from 1876 on "Oudh Affairs" laid out the situation:

It has been calculated that about 60 per cent. of the entire native population . . . are sunk in such abject poverty that unless the small earnings of child labor are added to the small general stock by which the family is kept alive, some members of the family would starve. With the bulk of them education would be synonymous with starvation.

If electric power could somehow render their food more plentiful and hence cheaper, then educating their children would help rather than endanger them. A hundred and ten years later, a treatise on *The Power Sector in India* firmly insisted: *Energy is a vital component of development. Together it ranges with the classical factors of production—land, labour and capital.* And in 2016, when I was finishing *Carbon Ideologies,* India's "development" continued, which meant that Indians drank in more and more power. A twilight of smog pressed down on Delhi; sometimes children struggled to breathe—but some of that derived from the illegal burning of crop waste; one couldn't blame it on "development."

In Africa, energy production between 1980 and 2011 increased nearly tenfold. Surely that benefitted agriculture, transportation, health services, education . . .

In 1965, the average time spent cooking and cleaning in American homes was 65 minutes. In 1995, it was 31 minutes. *To raise man's efficiency—to reduce man's toil,* that wish came true, thanks to electricity.

As for the *hired* toiler, the one who dug the coal our civilization burned, or the one who burned coal into coke, or shoveled coal into blast furnaces so that we could all build our pleasures on foundations of steel, I thought that the socialists got a great deal right when they said that our useful work defines the best part of us. That was why I felt fond of Kievsky Station's murals of grains and of kerchiefed workers, and Ukrainsky Station's mosaicked coal miners who overgazed the glass-smooth floor of circular and triangular insets as the rounded gilded portraited ceiling rose up and up.*—Like capitalists, Communists frequently turned work into misery and murder—one more reason why several of the friends who loomed largest in my respect were blue collar workers whose strength, craft, judgment and confident power got magnified by electric power and by the tools that it spawned. (Wherever possible,[†] capitalists replaced them with fossil fuels, which burned without sleeping or going on strike—but isn't it happier not to consider that?) In his great poem "Smoke and Steel," Carl Sandburg writes that at the heart of a steel bar lies smoke, *the slang of coal and steel,*

---

* See II:16.

† As will be especially apparent in the West Virginia chapter.

This Bangladeshi woman was harvesting corn and carrying it un-
assisted. Who would have begrudged her some electric help?

of sheet mills speaking to rolling mills—and along with that smoke, the worker's
blood. When West Virginian ideologues insisted that coal miners suffered and
died to "keep the lights on," they, too, got something right. The pipefitters, oil
drillers, truck drivers, engineers, frackers, inspectors, welders and other soldiers
in my world's myriad armies of resource extractors, their strenuous, risky lives
deserved to be celebrated by poets.—Whenever I thought of the ancient Greeks,
images of bronze spears and sacrificed cattle were prominent among my associations. And what might we remind you of? When I was alive, the smoke of carbon
and the sweat and blood of those who burned it in order to bring our machines

alive were too often forgotten by the so-called "consumers," but you from the future who cannot forget our deeds, please blame our workers less than the rest of us. They sold their sweat for smoke, and we bought their smoke for the sake of our marvelous toys.

For instance: *The modern American suburbanite of the late 1930s and early 1940s was leading an increasingly efficient lifestyle made more convenient by new household appliances and especially by the automobile.*

Efficient—convenient—well, even if that might in some cases have meant "trivial," where was the harm?

Another for instance (from the phone company in Abu Dhabi, 2016): *Stay connected and entertained with the new eLife packages and enjoy special rates on TV, Internet and Telephone.* Well, wasn't staying connected and entertained worth cooking the planet for?

Even as we taught each other to burn more carbon fuels, we were reshaping the places where we lived. In one primer on the petroleum industry I saw a scale drawing of oil tankers, from the childishly tiny bauble of 34,000 deadweight tonnage, *circa* 1945, to the man-sized 540,000 DWT affair of 1975, "on order." In case you ever wondered what that work was for, let me inform you that *a 276,000 DWT tanker can carry enough oil in one trip to power a Volkswagen-sized car for 10 round trips to the sun.* Reader, can you imagine anything more worthwhile? To satisfy demand, we had to set our tankers to ferrying oil back and forth. But how could such hulking ships avoid scratching their steel bellies against the bottoms of indigenous waterways? Of course we solved that difficulty. *Many of our largest rivers have actually been turned into canals by the channelization work of the Corps of Engineers.* These projects not only burned much carbon, at some profit to the vendors; they also (and this will be a leitmotif in *Carbon Ideologies*) gainfully employed armies of deserving, hardworking people. Furthermore, they consolidated the power of alert politicians. On this subject the petroleum primer deadpanned: *Few members of Congress are likely to vote against multimillion dollar appropriations for construction in their home districts.* And so we found more ways to cash in on carbon. The more of it we carried, the more we sold and used. We laid pipelines under every ocean on the planet. In electronic-eyed flying machines we spied out deep-buried coal, oil and natural gas. We refined them into new fuels and feedstocks. We turned them into electricity, for whose consumption we kept inventing new machines.

Although some of these devices sipped power with laudable frugality—for instance, the laptop computer on which I wrote this book required half the energy per given time that my body would have spent to walk upstairs*—their motors, silicon chips and generators frequently came to us with a robust thirst built in, because the sturdiest slaves tend to be voracious . . . and it often happened that in our hurry and lazy wastefulness, we let the fossil fuels they fed on go up in useless smoke:

POWER WASTAGE BY GROUP-DRIVEN MACHINE TOOLS, *ca.* 1945

(DEDUCTING IDLE MACHINES)

"In the group drive, the motor drives a length of lineshafting from which, in turn, the machine tools are driven by belts. The best arrangement for group drives is to divide the machine shop into small units, having a motor for each department or each type of machine."

"Power available" is the "average power per machine"; "power used" is the "average power used in doing actual work." Both figures are expressed in watts, the appropriate local unit for this rate of electrical consumption. [For BTUs per minute, the default unit of *Carbon Ideologies,* multiply × 0.056884.]

I calculate wastage by dividing the first number into the second, subtracting the quotient from 1.0 and multiplying the result by 100%.

24-inch Cincinnati drill presses, for small drilling on forgings.
[Power available: 360 watts. Power used: 70 watts.] Wastage: 80.56%.

No. 2 Horizontal Rockford boring mills, for boring bearings in aluminum cases.
[Power available: 1,620 watts. Power used: 300 watts.] Wastage: 81.48%.

1⅜-inch Gridley Automatics, for machining cast iron pistons.
[Power available: 1,520 watts. Power used: 270 watts.] Wastage: 82.24%.

Source: Machinery's Handbook, 1946, with calculations by WTV.

I have read that in 2012, 61% of the energy generated in the United States accomplished no useful work whatsoever.[†]

---

* 3.43 *versus* 6.28 British Thermal Units per minute. See the table beginning on p. 67. I will use BTUs in this book for historical, sentimental and locally expedient reasons. One BTU (contained in one match tip) is the amount of energy needed to warm a pound of water by 1° Fahrenheit.

† The worst sector was transportation, at an efficiency barely above 20%, with significant emissions of greenhouse gases (see p. 143).

About Waste

To condemn a thing thus, dogmatically, as false and impossible, is to assume the advantage of knowing . . . the power of our mother Nature; and . . . there is no more notable folly in the world than to reduce those things to the measure of our capacity and competence.

Montaigne, "It is folly to measure the true and false . . . ," 1572–74

I n such terms you from the future will surely tell our tale. But when we were alive we had our excuses.

Driving down the long dirt road between the airport and the Arctic town of Resolute Bay, Northwest Territories, my late friend David gentled the engine, and when we stopped for a few minutes he left it running, because the temperature was 20° below zero (Celsius), and if he shut it off it would not easily start again. (By the way, emissions of carbon monoxide, a greenhouse gas,* *occur especially in idle, low speed, and cold start conditions.*) He was a wise survivor and a generous helper whose advice minimized my frostbite. Was he wasteful, or simply prudent?

On a corner in my home town, a patrol car idled watchfully. The policeman saw a speeder, and rushed after him. How could he have managed that if the engine were off?

And what about the times when inefficiency could not be helped, and any production was heroic? In one foundry at the Kirov Works in Nazi-besieged Leningrad a war correspondent observed the following:

> The girls, with patched cotton stockings over their thin legs, were stooping under the weight of enormous clusters of red-hot steel they were clutching between a pair of tongs, and then you would see them— and as you saw it, you felt the desperate muscular concentration and willpower it involved—you would see them raise their slender, almost

---

* See the table of Comparative One-Century Global Warming Potentials on p. 176.

child-like arms and hurl those red-hot clusters under a giant steel hammer.*

Under such conditions, what the work was for was quite simply resistance, defense, survival; therefore, its absolute wastefulness—of the workers themselves, who sometimes died for lack of calories; and of the equipment and procedures, which surely could have accomplished greater output had it not been for German bombs and undernourished, incompletely trained personnel—was no strike against it.—Do we cry "waste" when something can be done more efficiently somewhere, but not here?—In Leningrad the work struggled along, at the upper limit, not of its thermodynamic but of its *known possible* efficiency.

Commemoration of mobilized war workers, Hiroshima

As for those group-driven American machine tools of 1945 (immortalized for you in the work-stained pages of my grandfather's *Machinery's Handbook*), they certainly squandered power, all right—whose generation gave off extra global warming agents. Of course back then they could laugh off climate change

---

* To fill out this picture of manufacturing bottlenecks I now quote Comrade Puzyrev, factory director: "In a workshop that's had a direct hit, production slumps heavily . . . especially if many people have been killed or injured. It's a horrible sight, all the blood, and makes even our hardened workers quite ill for a day or two . . ."

in good faith—and as *Carbon Ideologies* will sadly show, even in the 21st century financial gains and losses usually defeated more fundamental considerations. So whenever that group drive came on line, probably in the strenuously ad hoc conditions of World War II, "waste" defined itself only monetarily. What was the work for, but profit? With electricity cheap (in 1947 it cost less than a cent's worth of wall current to produce each dollar of retail value in manufactured goods[*]) and factory capital expensive, switching from group drive to individual power stations would have been not only wasteful, but anticompetitive, ruinous.

A 21st-century case may or may not fall into the same category: In 2002 the Germans (not that they were the only ones) began producing optical glass fiber with the aid of sulfur hexafluoride—whose molecules exerted the most active greenhouse effect we knew.[†] *According to experts, 70% of the input SF_6 quantities escape*—why I cannot tell you. Would a more efficient trapping procedure have been cripplingly expensive, or for that matter mechanically impossible? If so, would "waste" be the appropriate description for our actual procedure? (You from the future might have preferred to call it negligent ecocide.)

What about when the primary aim of the work was socialist emulation? In 1958, as part of his Great Leap Forward, Chairman Mao directed the nation *to overtake Great Britain in steel production within fifteen years.* Whether or not the objective would be possible, it was certainly plausible—unlike the means: backyard steel furnaces! When an inspection tour arrived in Anhui Province, Mao's private doctor grew "astounded": *The furnace was taking basic household implements and transforming them into nuggets called steel,[‡] melting down knives into ingots that could be used to make other knives.* And the "mobilized" peasants kept cramming more pots, pans and doorknobs in, rushing around with armloads of fuel, so that at night, all along the railroad tracks when the Chairman's special train came by, the furnaces shone *as far as the eye could see.* Once all the coal had been consumed, in went peasants' beds and chairs! There was no time either to harvest the produce, which rotted, or to refashion the new ingots into pots and pans; hence, people could not cook what little food they had. Starvation came. Even after Mao realized that the backyard steel was no good, for some time he kept quiet, because he *still did not want to do anything to dampen the enthusiasm of the masses.* Finally he told the residents of his old home town: "If you can't produce good steel, you might as well quit."

How much thermodynamic work had by then been accomplished by coal and

---

[*] See p. 58, "Comparative Energy Requirements," header **2,706.**

[†] See p. 183.

[‡] They were probably pig iron.

wood in melting down useful objects into useless ingots? How much more work would have been required to recast them into something useful? How many tons of carbon dioxide and carbon monoxide did their own invisible work? The latter was mostly a "precursor" which enhanced the warming capabilities of other gases, then faded away. But the former's contribution to warming would continue for at least two millennia.—Mao's backyard furnaces worsened our planet's situation for nothing. (Had *the enthusiasm of the masses* led to something other than famine, I might feel kinder.) Here was waste of the militantly ignorant sort.

When I was writing *Carbon Ideologies,* it remained the shocking case that half the liquid metal which we processed into sheet metal ended up as scrap—which, to be sure, generally got recycled—but that squandered additional BTUs in order to melt it down again. Better designs and improved manufacturing technologies could have reduced some of these "yield losses."

Sometimes we called upon electricity to carry out aesthetic work. I consumed considerable amounts of paper (whose manufacture required a surprising amount of electric power[*]) in making my so-called "art." Then there was the "art" of interior design. Even when we were alive we knew perfectly well that clear glass globes absorbed 5 to 12% of the light within them, while cobalt-blue globes absorbed 90 to 95%. Would you from the future have had us outlaw blue globes? They were wasteful—but they certainly accomplished their predetermined effects in high-class restaurants and brothels.

When I made tea, I dumped the extra boiling water down my sink, believing that to be "better for the planet" than deploying harsh alkaline drain cleaners. Was that wasteful or not?

In 2016 two economists found that *one-quarter of the food that is purchased in the state of Hawaii is wasted.* Only 11% of the state's edibles originated locally; the rest got flown or shipped in, thanks to internal combustion engines which produced greenhouse gases en route. The discarded food mostly ended up in landfills: 237,000 tons per year. So far as I can calculate, if the landfills were anaerobic the food in them might give off something like 68,730,000 annual pounds of methane (CH_4),[†] which over each 20 years would heat our planet as well as 5.9 billion pounds of carbon dioxide. (Over 100 years the relative mischief would lessen—because CH_4 broke down after 12 years, while CO_2's warming effect kept growing.) Why was this permitted? The economists explained: *The*

---

[*]See p. 135.

[†] The carbon emissions of anything are proportional to the amount of carbon in it. In 2014, Japanese food waste was 43.4% carbon—slightly higher than the percentage for human waste. I assume that Hawaiian waste was comparable.

root of individual food waste is personal preference. In short, the thermodynamic work of producing, packing, shipping, storing and disposing of uneaten matter was accomplished not for nothing, but for the sake of "freedom."

You from the future, who must scrape by in a state of scarcity unimaginable to me, how would you judge the following recollection of ignorance and abundance, published in 1911? *Forty years ago kerosene was refined for use in lamps, while gasoline was a by-product. Gasoline of 76° to 85° B[aumé] was disposed of in enormous quantities by burning it in the open air.** Was this more "wasteful" than dumping it into a river?

(Speaking of which, one history of the boom in Drumright, Oklahoma, reminisces about the *glory days* when surplus oil *flowed down Tiger Creek* and the Cimarron River *became a lake of oil.*)

Of the energy generated in an engine cylinder, from 2 to 24 per cent. may be lost as compared with the work which will be done at the fly-wheel. An engineer could tell us how much of that loss was thermodynamically inevitable. As for the residuum—the waste, that is—if your era discovered how to capture that energy, would you fault us for having "lost" it in ignorance?

And if when I was alive certain desperately undercapitalized people had burned fossil fuels cheaply and dangerously, further blackening our atmosphere in order to, as that *Wall Street Journal* editorial put it, *lift themselves out of poverty,* how would you judge them?—Perhaps it depended on what their work had been for.—If you declined to condemn them simply for profiting, would you do so if they lied about what they burned?—I have in mind the poverty-ridden territory of Puerto Rico, whose Power Authority kept burning (for decades!) the lowest-grade, filthiest fuel oil, appropriately called *sludge,* all the while falsifying air quality tests, and charging utility users for high-grade oil. The pollutant in question, sulfur dioxide,[†] is *a colorless gas with a suffocating, choking odor,* and eventually the Authority had poisoned the neighbors with 100 million pounds of it—more than any comparable set of power plants in New York or New Jersey. Adsorbed by dust,[‡] some of the stuff got breathed in by people and animals, where on contact with moist lung tissue it became sulfuric acid. The gas itself impaired the photosynthesis of plants. Nor was that necessarily sulfur dioxide's worst failing; one scientist went so far as to say:

---

* Much had improved by 1977: "Natural gas, the major raw material for nitrogen fertilizers, is being burned off in large quantities in the oil wells of the Middle East because there are only a few fertilizer conversion plants located near the oil fields; yet the world has a shortage of nitrogen fertilizers."

[†] This was the pollutant of Homer City Generating Station in Pennsylvania (above, p. 5). See also "About Sulfur," II:26.

[‡] That is, concentrated on the surface of the dust particles.

Global warming during the 20th century was primarily initiated by a rapid increase in the rate of . . . emission of sulfur by man . . . The SO$_2$ reduced the oxidizing capacity of the atmosphere, leading to an increase in methane, water, and other greenhouse gases.

And so I would call the work of electric power generation undertaken by the Authority criminally as well as *wastefully* done. But since those profit-seekers in the Fuel Procurement Office did *lift themselves out of poverty,* would you call them inefficient in relation to their purpose?

You from the future, who understandably despise us, and might well conflate these examples into one string of abuses—could you have intervened in our time, I'd bet you'd have subjected us to harshly sweeping simplifications—please let me ask you to fairly consider the perplexities in these matters of power and emissions. You now perceive, I trust, how limited must be the judgments of even good-willed, thoughtful minds in the absence of expert knowledge. To best evaluate each efficiency, we would have required briefing on precisely how each process was being organized, which improvements were possible and which dangers were known at that time. (Sulfur dioxide's were known quite well.) Sometimes our acts demand accusation, which this book promises to provide, but when I am unsure what else we could or should have done, I will leave the accusations to you.

To better consider this matter of what was possible, let us consider one last case of considerable apparent similarity to that of the group-driven machine tools in 1945:

POWER WASTAGE DURING MACHINING OPERATIONS AT AN UNSPECIFIED TOYOTA FACTORY, *ca.* 2000

(BEFORE PLANT REDESIGN)

"Note that most of the energy is consumed even when the machine is 'idling.'"

CONSTANT ENERGY USE:	**82.2%**
Coolant	31.8%
Oil pressure pump	24.4%
Cooler, mist collector, etcetera	15.2%
Centrifuge	10.8%
ENERGY USE BY MACHINING:	**17.8%**

Source: *Journal of Cleaner Production,* 2005.

I suppose that the constant energy usage of 82.2% would not have counted as "wasteful" had the machining apparatus never idled. (Of course this procedure would have spent more power in total, and worn down parts more rapidly.) In fact, was it wasteful at all? Could uninterruptedly productive "constant energy use" ever enter our realm of deficient actualities?

The professor of mechanical engineering who supplied this data was named Timothy Gutowski, of MIT. You will frequently find his articles cited in *Carbon Ideologies*. When I sent him this table, he replied:

> I would not characterize these energy usages as waste[;] each has a function: the centrifuge separates coolant from the chips, ... etc. If you want to talk about waste, I suggest you consider the idea that there is a thermodynamic minimum to achieve a certain task. Then you could compare the amount of energy used to that thermodynamic minimum.*

He was surely right; I had been hasty to cry waste. I should have known whether the oil pressure pump and other constantly operating parts were themselves efficient, whether or not they could be idled or at least slowed in the absence of active machining work—in short, whether or not the whole thing could be done better.†

In addition, wrote Gutowski, *you could consider the utility of a given task.*

In other words, *what was the work for,* and *how efficiently and intelligently was it done?*

Not being an expert on anything, I came to realize that asking this question over and over would be my main contribution to *Carbon Ideologies*. Do you remember that German optical glass fiber manufactory? Why shouldn't a 70% loss of sulfur hexafluoride be comparable to an 85.2% "constant energy use"?—But I rejected that; I had to, because I was a carbon ideologue, and sulfur hexafluoride was "dangerous." No doubt the people who made Toyota automobiles and optical glass fibers would say that they accomplished useful work. So would the

---

* For instance, as he explained in another paper: "A number of manufacturing processes shape the work piece [*sic*] material primarily by plastic deformation. These would include machining and grinding processes as well as forming processes such as sheet metal stamping [and] ... forging ... In all [such] cases ... the plastic work is converted in large part to thermal energy and not recovered. Hence these processes are irreversible." Thus one can hardly call them wasteful—always provided that they are worth doing at all.

† From what I could tell, Toyota seriously tried to reduce energy expenditure, for which I must praise the company. For the process mentioned in the preceding table, "a minimization of coolants could ... save twice." In general, the factory's "equipment was redesigned to reduce energy, particularly when there was no production." Whether this equipment included the machining devices, and how much energy was actually saved, remained unstated.

buyers and users of those products. Perhaps I could have been persuaded into feeling the same. Meanwhile, hordes of other products came into being through thermodynamic work—too many of them or too much of it, if one believed in climate change. Therefore, I had to conclude that even if some particular process was thermodynamically wasteful, waste might lurk within the purpose of the work.

From 1975 to 1985, the American chemical industry managed to accomplish *an annual reduction of 4.2% in energy input per unit of output.* I found this commendable; for a moment I even hoped again that someday everyone on earth could spend energy without limit or repercussions. In elementary school they had taught me that atomic power plants would facilitate everything, probably by about A.D. 2000, at which point there would be a moon base and we would flitter about in personal air cars. If power generators and industrial manufacturers kept getting more efficient, why couldn't our demand continue to grow? Experts had identified for inspection and probable improvement many specific points in various production pathways: Add more insulation, precision test one's compressors, convert pumps from single to variable speed drives, preheat air before it entered one's boilers and furnaces, increase the accuracy of one's energy use measurements.—All the same, absolute energy inputs kept going up.

I would argue that wherever thermodynamic efficiency was low, energy input was high, and the question "What was the work for?" lacked any better answer than "for short-term profit or pleasure," then a given manufacturing process should have been branded *wasteful.*—You from the future would say: "It should have been prohibited."

Example: In 2014 a science whizz breezily concluded: *If I am making zero kilograms of ice per second, my ice maker [sic] would still draw 88 watts of power . . . Really, most machines still use electricity when they are not doing anything useful.* Running that icemaker for a year would have called on a power plant in some ecosystem somewhere to burn 425 pounds of medium-grade commercial fuel oil—or nearly a barrel and a half of it, at 42 gallons per barrel—to enable that sole purpose. That meant that every active icemaker consumed nearly a barrel and a half of oil each year.

I knew a woman whose luxuriously high-capacity refrigerator had been embellished with a built-in icemaker. I never saw her use it. When I told her about the 88 watts, she refused to believe it. How could her icemaker consume electricity to do nothing? I asked whether she could turn it off while leaving the refrigerator running. She allowed that this would be impossible, and indeed I could see no separate switch for the icemaker, which accordingly went on doing its idiotic work.

POWER WASTAGE BY DEVICES IN STANDBY MODE, 2000–2010

This phenomenon is sometimes known as "vampire draw." In this table I am overstating the actual wastage to an unknown extent, because some devices, such as telephone answering machines and motion detectors, must be ready to go into action at any time, and therefore are not wasting power. However, many machines consume power needlessly.

"Unplugging a device constantly consuming standby power saves a yearly 9 kWh for each watt of continuous consumption."

All figures expressed in [% of total power consumption for appliances].

Japan, 2008.   6%. *This equates to 285 kilowatt-hours [972,705 BTUs] per year.*
France, 2000.   7%.
Britain, 2004.   8%.
California, U.S.A., 2002.   "Over 15%" ["of statewide residential power consumption"]. This equates to "nearly" 1,000 kwh [3,413,000 BTUs] per year. "Our future estimate increases low power mode consumption [= standby mode] by 13%."

Sources: Energy Conservation Center, Japan; ACEE, France; Department of Trade and Industry, U.K.; Lawrence Berkeley National Laboratory, Calif.; with calculations by WTV.

In this book I pay less attention to production than to consumption, at which I excelled when I was alive. And on that topic, the figure on American primary energy consumption in 2012—61% of it spent without doing useful work! (or, to quote the book in which I found this statistic, 61% *wasted*)—seemed, if not indefensible, at least demanding of investigation, given the future we were leaving you. What if the true figure had "only" been 50% or 20%? The neighbors should still have been worrying about it. Mine, at least, did not.

About Demand

Between 2000 and 2005, UK consumers on average increased the number of garments they purchase[d] annually by 33% . . . It is cheaper to buy a new pair of trousers than repair a hole in their pockets . . .

Julian M. Allwood et al., "Material Efficiency: A White Paper," 2011

WHEN IN VEGAS SHOP AS THE ROMANS DO.
The opulence. The selection. The experience.

Where Las Vegas, 2016

I n our time we heard a certain story so often that its peculiarity had worn down into a kind of active platitude. There was no questioning it because there was hardly any *seeing* it. Let me now pull into view one of its banal incarnations.

Sales promotion in Dubai

In 2016, when the Italian energy firm ENI was trolling off the Egyptian shore, a huge field of natural gas revealed itself. Very possibly, more than half a year's entire global production could enrich someone.

This galvanized potential vendors, as you may imagine. A *visiting researcher* from Georgetown University informed us: "All around the region you are seeing activity. Israel doesn't want to miss the boat."

Fair enough. Didn't everybody "all around the region" deserve to make an income? And why indeed should Israel forgo her share?

But to cash in, Israel would need to create a market. As the newspaper

42

explained: *Operators may need to find new destinations for the gas . . . Demand in Israel, while growing, is still relatively modest.*

In other words, it would be necessary to stimulate demand. Israel would have to sell natural gas to people who had formerly gotten by without it. The necessary result: more global warming.

What Was the Work For? (continued)

A Battery of eleven Guns discharged by an electrified Phial of Water . . .
Spirits kindled by Fire darting from a Lady's Eyes (without a Metaphor) . . .
Animals killed by [the Electric Fire] instantaneously (if any of the Com-
pany desire it, and will be pleased to send some for that Purpose).

Benjamin Franklin, 1751

The first So Truly Real *baby doll so lifelike, she "breathes!" . . . A mecha-*
nism tucked inside the doll raises and lowers her tiny chest . . . Ashley looks
and feels so real, from her baby-soft RealTouch *vinyl skin to her delicate*
wisps of hand-applied baby hair.

Advertisement in National Enquirer, 2016

Because electricity was as inexhaustible as the buffalo herds, needless consumption of power provoked fewer complaints than you from the future might suppose. For exactly the same reason that we rarely studied where our water came from and whence our wastes departed, we "consumers"—who like other organisms went about our business of self-aggrandizement, reproduction and survival by taking whichever energies lay ready at hand—mostly used machines without comprehending how they worked, so that their efficiencies escaped our concern. A Japanese lady turned on her electric space heater for me because I had the flu; how extravagantly or frugally that glowing device spent current in order to accomplish its assigned thermodynamic work meant less to her than being a caring hostess. Meanwhile a tired Pakistani guest worker in Abu Dhabi, dwelling in a "bedspace" with six or ten other men* and driving a taxi for 12 hours at a stretch, in order to support his distant family and maybe see them once a year, operated his assigned internal combustion engine without any thought, much less ability, to improve its mileage, never mind reducing its carbon emissions; when I asked him what to do about global warming, he replied, as had several Indian colleagues, that we should stop burning petroleum

---

* See II:519.

and start using "electric" cars, as if "electricity" had nothing to do with the fuel-hungry power plants that generated it. (*Magic,* explained a science-fiction author, *is a practical science, or, more properly, a craft, since emphasis is placed primarily upon utility rather than basic understanding.*)—As for you, you may not care why we ruined you (which is the subject of this book).* Everybody's indifferent to something. On the production side, for instance, who cared about burning extra BTUs of carbon when they claimed so trivial a fraction of the total financial expense? In 2011 two technologists (one of them Professor Gutowski) wrote that *the seemingly extravagant use of materials and energy resources by many newer manufacturing processes is alarming and needs to be addressed alongside claims of improved sustainability from products manufactured by these means.* For instance, *carbon nanotubes are one of the most energy intensive materials humankind has produced. But the electricity cost in this case is on the order of 0.2% of the price.*

None of my neighbors craved to own carbon nanotubes. Almost nobody could have told me what they did or in which products they could be found. (I'll tell you one: scanning probe microscopes.)† Well, out of sight, out of mind!

When we were alive, we often heated residential buildings by means of single-setting steam boilers which could only bring the top floors up to a comfortable temperature if they made the lower apartments so hot that even in midwinter their inmates needed to open windows!—Why didn't we install locally controllable radiators? Because waste was "cheaper." (This happened in New York; likewise in western Siberia.)—The managers of workplaces did perhaps draw up energy budgets, for the sake of profit; as for us "consumers," so long as the heating bill got included in the rent, we didn't care. And why notice whether the new refrigerator drew more or less energy than the old? For that matter, how could we figure that out from a utility bill which typically broke down charges only into peak and off-peak kilowatt-hours, fees and taxes? If this month's bill was higher than last's, the reason might have been a change in electric rates, two extra dryer-loads of clothes because our nephew was home from college, or, yes, the new refrigerator—which possibly indeed drew more, because

---

* J. G. Ballard, 1996: "Without motives our investigation would be so much easier." Indeed, wouldn't you rather just damn my generation to hell?—Now, what about the generations before and after mine? Again Ballard had the answer: "Guilt is so flexible, it's a currency that changes hands . . . each time losing a little value."

† Nanotubes "are among the stiffest and strongest fibers known, and have remarkable electronic properties . . ." We created them by inserting hexagonal rings of carbon atoms into sliced-open molecules of buckminsterfullerene (see "About Carbon," p. 106).

our manufacturers could sell more toys more often by building them stronger, stranger, louder, brighter, cooler, faster-cycling, with capabilities we'd never thought to long for. And so a certain radio in 2012 drank in 11 times more current than its 1975 ancestor.*—We were willing enough to be seduced—since, as Dr. Johnson expressed it back in 1750, when our greenhouse gases first began their rise, *something more is needed to relieve the long intervals of inactivity, and to give those faculties, which cannot lie wholly quiescent, some particular direction. For this reason, new desires, and artificial passions are by degrees produced.*

Perhaps our new desires could have been satisfied without voltage. But any such notion was futile, hence absurd. *Energy is vital to our modern way of life,* explained the National Coal Association. *Very often, the energy we use daily is in the form of electricity.*

And there were always new ways to spend electricity!—Before sockets, plugs and generators, we'd spent energy just the same. In 1849, a single Cape Cod lighthouse burned 800 gallons of oil each year—what kind I cannot say, whale most likely, but whatever it was, you can trust that it gave off carbon dioxide when it combusted. *That* purpose was worthwhile enough (for humans if not for whales): the prevention of shipwrecks. Meanwhile, other work demanded to be accomplished as soon as we'd conceptualized it. For much the same reason that we used to keep slaves, we now ran machines, whose thermodynamic work got cheaply and conveniently disposed of.—Looking about him in 1904, the historian Henry Adams *could see that the new American,—the child of incalculable coal-power, chemical power, electric power and radiating energy, as well as of new forces yet undetermined—, must be a sort of God compared with any former creation of matter,* including Dr. Johnson. And existence kept getting better: Between 1870 and 1990, per capita energy use in America more than tripled![†] Maybe it should have, or at least harmlessly might have; perhaps (although I doubt it) everyone who inhabited the planet in 1870 could easily and pleasantly, without burdening our atmosphere, have increased energy consumption by a factor of five, eight, or at least two. But these factor-calculations were never made; the vendors let us go on trusting—not merely because it enriched *them,* but because neither we nor they perceived the far-off limit—that we could spend as much current as we could buy.—And thanks to population increase, between 1870 and 1990 our *absolute* energy consumption rose almost 21 times.

---

* See pp. 68, 70, headers **2** and **22**.
† See p. 72, headers **105** and **270**.

Mural in Cushing, Oklahoma

Therefore, how efficient or irresponsible we might have individually been made no difference: Greenhouse gas concentrations increased absolutely.—And you know why: *Electricity is essential to economic and human development,* explained an Australian government report.* You see, *it supports industrialisation, improved access to clean water, sanitation and basic health as well as better education services.* Some of that was debatable. Industrialization had a way of polluting water. By concentrating people in cities, it created slums. Nuclear and fossil-fueled power plants were actually hazardous to their workers and neighbors. But most of us did prefer electricity whenever we could get it. Andrey

---

* From what I could tell, its purpose was to bless Australian coal sales to India.

Platonov, 1939: *We saw a light in the gloomy dark of a destitute and barren space—the light of man . . . we saw wires hung on old wattle fencing; and our hope for the future world of communism, . . . a hope which alone made us human—this hope of ours turned into electrical power . . .* And so it was that in each two days of 2009, the world burned the entire oil output of 1990.

What was all that work for, and what *should* it have been for? In 1991, had we devoted all of our electric power to this task,* we Earthlings could have electroprocessed nearly 9.7 million pounds of titanium every minute—or else left 131.8 trillion icemakers on standby mode as they busily vampirized our current. Meanwhile, as I sat in my climate-controlled room writing *Carbon Ideologies,* one out of four people in India lacked access to electricity. *This is clearly a major barrier to economic and social progress, and . . . providing universal access has been a major priority for policy-makers.* Accordingly, the policymakers provided access. Between 1980 and 2015, India's power consumption quadrupled, making that nation the world's third largest energy user. But blackouts there still lasted for hours or days. What could the policymakers do but generate more power?

You from the future may consider my "what could they do?" a sarcastic rhetorical trick—as it sometimes is, but not always. Professor Gutowski and his students once computed the annual energy budgets of various American lifestyles as of 1997. A very earnest Buddhist monk (estimated income $8,500, estimated expenditures $13,000) could get by using 113.8 million BTUs. A homeless person lived nearly as frugally at 118.5 million. Next came a five-year-old child at 123.3 million BTUs (she loved her toys). They were all doing pretty well compared to the coma patient and the senator (2.37 and 3.51 *billion* BTUs!). Gutowski now pointed out the sad fact that even though the monk's energy use constituted a mere one-third of the 1997 national per capita average, it was still *almost double the global average energy use.*[†]

> Furthermore, such a level, we believe, is not obtainable for the average American on a voluntary basis. Which brings us to our second point; due to the combined effects of [energy] subsidies and rebound, the magnitude of possible reductions in energy use for people in the United States by voluntary changes in spending patterns appears limited.

---

* And (fat chance!) generated that power at 100% rather than 30% efficiency.
[†] That is, 113.8 *versus* 60.7 million BTUs.

The coma patient certainly lacked any say in the matter. Most likely, so did the nursing staff. As for the senator's energy use—nearly 58 times the global average (never mind his jet-setting carbon emissions)—perhaps that was also necessary to do whatever politicking he did. What about the homeless American? Fast food and plastic were not necessarily what he desired, although he became habituated to them. He took what he could get, growing fat and losing teeth. Police tore down his camps and destroyed his possessions—no use owning anything of value. He lived in trash and left trash behind him. Shall we blame that on demand?

Gutowski's findings disheartened but also comforted me, because I could reiterate to you from the future: See how they *made* us do it! It was never our fault!

In 1967 the President's Science Advisory Committee found that American farmers could grow three times more food per unit of area than their Asian and African counterparts—but at a cost of 10 times more mechanical energy, which came (of course) from fossil fuels. Wasn't feeding multitudes well and cheaply a worthwhile use of thermodynamic work?

If so, then why not, for instance, super-size the growing, harvesting, processing, packing and cooking of our edibles? Between 1997 and 2002, this form of American energy consumption ballooned more than six times faster than total domestic energy use! We powered, for instance, *frozen, canned, and snack food technologies*—because when I was alive, it would have been un-American not to keep things ready and cold.—In 1993, 14.9% of my fellow citizens owned two or more refrigerators; in 2005, 22.1% achieved that distinction, which we commemorated with greenhouse gases.

Shall we blame capitalism, then? Well, the Communists sought something comparable—to be established from the top down:

> The rate at which the population is supplied with refrigerators will increase from 32 per 100 families in 1970 to 64 in 1975 . . . By the end of the five-year-plan, the sale of automobiles to the population will have increased more than sixfold . . . In the next few years, the production and sale of ready-to-cook products, pre-cooked items, concentrates and other items that make the home preparation of food easier should be developed on a broad scale.

In fact, what we all aspired to was simple: *an unlimited supply, which could therefore accommodate an unlimited increase in demand.*

In 1920 an American correspondent had asked Lenin: "When do you think Communism will be complete in Russia?" The latter replied that 10 years of national electrification would be required. "All our industries will receive our motive power from a common source, *capable of supplying them all adequately* [italics mine]. This will eliminate wasteful competition in the quest of fuel, . . . without which we cannot hope to achieve a full measure of interchange of essential products in accordance with Communist principles."

Capable of supplying them all adequately! That's what they all wanted, back when they were alive.

So let's not invoke national peculiarity alone as we wonder why it was that in 2002, Americans spent 287,604 more BTUs per capita than in 1997 on the manufacture, distribution, preparation and disposal of sugar and sweets. That was the equivalent of an extra 69 pounds of coal apiece. Diabetes had long since become commonplace among us, but sweets tasted so delicious!—I used to like to take my daughter out to the ice cream parlor. And at Halloween it was just plain neighborly to buy giant bags of individually wrapped candies to make the trick-or-treaters happy. It never crossed my mind that I was doing wrong.

By 2007, American grain and bakery products were getting trucked a significantly greater average distance than a decade earlier: from 122 to 262 miles. As for "meat, fish and preparations," they used to travel 137 miles; now they had to go 243.* What was *that* work for? It must have benefitted drivers, mechanics, highway maintenance crews, refrigerant manufacturers, tire salesmen and the managers of large supermarket chains.

We accomplished other wonders during those years. For instance, in our homes and even in restaurants we *increasingly outsourced manual food preparation and cleanup activities to the manufacturing sector, which relied on energy-using technologies.* As a result, 16,000 food service preparation positions disappeared, although I grant that "food manufacturing industries" offered 4,800 new positions. In other words, the net loss was 11,200 jobs. Well, well, but hadn't Lucien Lucius Nunn called for a reduction of toil? (Who was toiling for us?—Mr. Carbon.)

And even as our toys put more people out of work, whoever could bought *more* toys, because back when I was alive, economies needed to expand without limit! The people who sold could hardly stay in business unless they kept selling. Products ought to wear out; new demands had to arise—Gross Domestic Product must go up!

---

* Of all food categories, only "fresh produce, oilseeds and other horticulture" traveled less, their average harvest-to-sale trajectories decreasing from 438 to 374 miles.

In general, said the National Coal Association, *each percentage increase in Gross Domestic Product . . . has resulted in just over a 1 percent rise in the demand for electricity, a trend that is expected to continue for the foreseeable future.*

Were that fundamental to the human condition, ratios between electric demand and GDP* would everywhere approach some constant value. In fact, when I considered the countries mentioned in this book (adding Afghanistan and Saudi Arabia for benchmarks), it turned out that their ratios of per capita power consumption to per capita GDP varied considerably. Saudi Arabia's was more than 40 times Afghanistan's:

RATIOS OF PER CAPITA POWER CONSUMPTION TO PER CAPITA GROSS DOMESTIC PRODUCT,

from Central Intelligence Agency Data,
Inset with United Nations Human Development Indicators,

2008–09 and 2012,

arranged in increasing order of per capita GDP

All population figures are July 2012 estimates. All GDP per capita figures are 2011 estimates. All gross power consumption figures (which the CIA calls "electricity consumption") are from 2008 or 2009, converted by WTV from kilowatt-hours into BTUs. The figures are rounded here, for the reader's convenience. Unrounded figures, which appear in the source notes, were used in actual calculations.

Per capita power consumption was computed by dividing total power consumption by population.

The ratio "Power : GDP" means "BTUs consumed per capita in proportion to GDP per capita" and was computed simply by dividing "Per capita power consumption" by "GDP per capita."

The two multiplier headers in large type both set Afghanistan at 1, the first larger numeral for GDP per capita, the second for ratio of "Power : GDP." Thus for Bangladesh <1.70>, <1.89> means that that country's per capita GDP was 1.70 times Afghanistan's, and its Power to GDP ratio 1.89 times Afghanistan's.

Had the National Coal Association's aphorism been correct, all the "Power : GDP" headers should have approached the same value.

---

* You will note that here and in the table above I have reversed the two terms of the NCA's ratio. Since this was supposedly close to unity, the reversal should not matter; and I thought it more appropriate to *Carbon Ideologies* to privilege total power demand in the numerator.

The United Nations indicators of human development come from a 1997 report, energy use being only one of several indicators. "Low development" calculation excludes India. "Medium development" excludes China. All express per capita power consumption [again, converted by WTV to BTUs per minute].

<1>, <1>

Afghanistan

Population:	30,419,928
Total power consumption:	788.74 billion [BTUs]
Per capita power consumption:	25.93 thousand [BTUs]
GDP per capita:	$1,000 = <1>
Power : GDP:	26 [BTUs per $1 of national income] = <1>

<1.70>, <1.89>

Bangladesh

Population:	161,083,804
Total power consumption:	13.45 trillion
Per capita power consumption:	83.48 thousand
GDP per capita:	$1,700
Power : GDP:	49

---

*U.N. indicator of "**low human development**": 501,711 BTUs per capita.* Producing this amount of electricity in a power plant would require 120 pounds of coal, 84 pounds of oil or 63 pounds of natural gas per person.

---

*U.N. indicator of "**medium human development**": 3,532,455 BTUs per capita, or 7.041 × "low development."* 84 pounds of coal, 588 pounds of oil or 441 pounds of natural gas.

<11.80>, <14.44>

World

Population:	7,021,836,029
Total power consumption:	31.02 quadrillion [or 31,024,170,000,000,000]
Per capita power consumption:	4.42 million
GDP per capita:	$11,800
Power : GDP:	374

<15.1>, <13.76>

Mexico

Population:	114,975,406
Total power consumption:	619.46 trillion
Per capita power consumption:	5.39 million
GDP per capita:	$15,100
Power : GDP:	357

U.N. indicator of **"high human development": 7,884,030** *BTUs per capita, or 15.71 × "low development."*
1,878 pounds of coal, 1,314 pounds of oil or 984 pounds of natural gas.

<24.0>, <40.18>

Saudi Arabia

Population:	26,534,504
Total power consumption:	663.49 trillion
Per capita power consumption:	25 million
GDP per capita:	$24,000
Power : GDP:	1,042

<34.3>, <25.90>

Japan

Population:	127,368,088
Total power consumption:	2.934 quadrillion
Per capita power consumption:	23.04 million
GDP per capita:	$34,300
Power : GDP:	672

<48.0>, <32.62>

United States of America

Population:	313,847,465
Total power consumption:	12.768 quadrillion
Per capita power consumption:	40.68 million
GDP per capita:	$48,100
Power : GDP:	846

<49>, <35.67>

United Arab Emirates

Population:	5,314,317
Total power consumption:	240.89 trillion
Per capita power consumption:	45.33 million
GDP per capita:	$49,000
Power : GDP:	925

Sources: United Nations Development Programme, 1997; Central Intelligence Agency, 2013; with calculations by WTV (verified by Ben Coleman, Marshall University)

This table does bear out the National Coal Association's fundamental premise: The higher the per capita GDP, the greater the per capita power consumption.

In his glowingly self-defensive treatise on fracking, the CEO of Breitling Energy Corporation asserted that between 1950 and 2011, American energy consumption per real dollar of Gross Domestic Product declined by 58%. How this sort of GNP compares with the per capita kind, I, a non-economist, am unfit to tell you.* Presumably the rich would *consume* the greatest proportion of GDP; and since American inequality increased in those years, this trend would grow ever less relevant to any dependence of average prosperity on electric power usage.

Well, even so, when we were alive it was certainly the case that we expected our economies to "grow"—and we expected new toys. In the words of *Motor Trend* magazine: *We all have at least one dream car, and probably several.*†

In this connection the International Energy Agency offered an axiom not quite parallel to and more alarming than the National Coal Association's: *CO_2 emissions have been clearly linked to economic growth.*

---

* According to the U.S. Environmental Protection Agency: "Due to a general shift from a manufacturing-based economy to a service-based economy, as well as overall increases in efficiency, energy consumption and energy-related CO_2 emissions per dollar of Gross Domestic Product (GDP) have both declined since 1990." Meanwhile the *BP Statistical Review* concluded: "The good news is that carbon emissions were essentially flat in 2016[,] . . . the third consecutive year . . . [,] in sharp contrast to the 10 years before . . . [,] when emissions grew by almost 2.5% per year. Some of this slowdown reflects weaker GDP growth, but the majority reflects faster declines in the carbon intensity of GDP . . . driven by accelerating improvements in both energy efficiency and the fuel mix . . . China's carbon emissions are estimated to have actually fallen over the past two years, after growing by more than 75% in the previous 10 years." This was one of the very few encouraging data I discovered while researching *Carbon Ideologies.*

† The ratio *circa* 2013 was one car for every eight Earthlings. Rate of car ownership did not closely correlate to GDP. In 1996, for instance, Lebanon ranked first at 732 cars per 1,000 people.

From this historical fact it was certainly convenient to deduce, as so many carbon ideologues did, that *economic growth cannot continue without continuing the growth of CO_2 emissions.*

A textbook case was *China, where strongly increasing electricity production is the main driver for record high CO_2 emissions.**—But the Mexican government put it most grandly of all: *As shown in Figures II.24 to II.26*—and if three figures show a thing contiguously, how can it not be true?—*a higher level of human development is associated with a higher level of emissions per capita.*[†]

How dare you deny us our high level of development! We approached perfection, and that is why you now retreat from rising acid seas. We accomplished great works.

What was the work for? Well, where did the work go? Here are some answers:

COMPARATIVE ENERGY REQUIREMENTS,

in multiples of 1 British Thermal Unit

All levels expressed in [BTUs per pound.] I have used the high heating value* where it is available.

• = One of the "big five" materials, which used the lion's share of energy in manufacturing. See p. 134.

Electroprocessing energies were originally expressed in kilowatt-hours. To consume this amount of current [1 kWh = 3,413 BTUs] a power plant at typical ⅓ efficiency would have needed to burn 3× as many BTUs as indicated here (see p. 150). Thus, to electroprocess 1 lb of chlorine with 1952 technology would have required [3 × 5,120] BTUs of fuel—or, for instance, 0.8 lbs of medium grade fuel oil, whose energy (see p. 213) is 18,570 BTUs per lb. Figures over 5,000 rounded to nearest 10. Over 1 million rounded to [decimal +] 2 significant figures.

Multiples above 10 are rounded to the nearest whole number.

* For a definition, see below, p. 534.

---

* I am happy to say that in one 24-year period, American industrial output rose 64% while industrial carbon emissions fell 3.5% The apparent causes were changes in technologies and manufacturing processes. The "clear link" between CO_2 emissions and economic growth allowed for some exceptions. But rather than shout out hopeful news I timidly reiterate that during those 24 years *total* American emissions rose 7.4%.

[†] And Mexico practiced what she preached: "The average per capita emissions for the country are 5.4 [metric] tonnes of CO_2 equivalent per inhabitant, with an average annual growth of 0.4% between 1990 and 2002." It may be worth pointing out in response that (as of 2013) "four of the top ten manufacturing carbon emitters are not in the top ten manufacturers based on monetary output. And six of the top ten carbon emitters are not in the high-income group."

< 1

Energy to keep warm 1 lb of water for a minute in a 40-gallon electric heater, purchased *ca.* 2011.* [0.065 BTU. *But assuming a typical ⅓ efficiency in the power plant that produced this electricity, 0.195 BTUs of combusted fossil fuels would be required.*]

< 1

Energy to move each poundweight of an unspecified "semi" moving truck 1 mile in 2016, loaded weight 70,000 lbs, average speed unspecified, average fuel consumption 6 miles per gallon. [0.40 BTUs per pound.]

< 1

Energy to move each poundweight of an unspecified "new" mid-1970s heavy rail train 1 mile. "Propulsion energy" per "vehicle-mile" stated at 75,000 BTUs. Weight of an Amfleet I coach car, introduced into service in 1975, was 106,000 lbs. [0.7075 BTUs per pound.]

1

Definition of 1 BTU. See header **3**.

1.5

Energy to move each poundweight of a winning-class American stock car 1 mile in the 1952 Mobilgas Economy Run from Los Angeles to Sun Valley, loaded weight 4,000 lbs, average speed 40.7 miles per hour, average fuel consumption 21 miles per gallon. [1.1516 BTUs per pound.]

1.6

Energy to move each poundweight of a 2017 XT5 AWD 3.6 "Platinum" Cadillac, average fuel consumption 21 miles per gallon, curb weight 4,025 lbs. [1.586 BTUs per pound.]

1.6

Energy to move each poundweight of a 2017 "F-Pace" Jaguar, average fuel consumption 20 miles per gallon, average of its 2 possible curb weights 4,025 lb. [1.586 BTUs per pound.]

---

* See "Comparative Power Requirements" (p. 67) for more information on this same appliance.

2.6

Energy to move each poundweight of a typical mid-1970s automobile 1 mile, average fuel consumption 12 miles per gallon. "Propulsion energy" per "vehicle-mile" stated at 11,000 BTUs. (Average weight in 1975 was 4,060 pounds.) [2.62 BTUs per pound.]

3

Energy to raise the temperature of 1 lb of water by 1° Fahrenheit. [Definition of 1 BTU.] *[Assuming that same ⅓ efficiency in the power plant, 3 BTUs of combusted fossil fuels would be required.]*

11

Energy to perform 1 "delivered horsepower per hour" of farmwork using a one-cylinder gasoline-powered tractor, *ca.* 1911, per pound of tractorweight: 1.273 pounds of 70 specific, 64 Baumé gravity gasoline, at period HHV of 20,000 BTUs per pound. Estimated tractorweight: 7 tons. [Energy needed for entire tractor: 25,460 BTUs or about 13 BTUs per pound.]

60–155

Specific cutting energy* for machining 1 lb of various aluminum alloys, *ca.* 2011. *Note: From here to end of table, I rely on stated manufacturing energies. To estimate actual power plant inputs, multiply × 3.*

245–344

Specific cutting energy for machining 1 lb of various nickel alloys, *ca.* 2011.

415

Energy required in 1975 to produce a pound of sulfuric acid (concentration unspecified).

426

• Energy to make 1 lb of cement (type unspecified), 2013. An astonishing improvement over the process of 1950. See **4,131.**

---

* Gutowski again [with Dusan Sekulic]: "The actual machining process involves significant friction at the tool–work piece interface such that the actual work requirement is considerably more than the calculated plastic work [of deformation such as stamping]. This work is provided by the spindle motor of the machine tool and is tabulated for various work piece materials as the so called specific cutting energy . . ."

The values within these lines [430 to 13,000 BTUs per pound] reflect typical manufacturing energies for plastics, metals and composites, ca. 2010.

500

- Average energy required *ca.* 1958 to heat from 2,000–2,400° F 1 lb of steel in an ingot heating furnace. Parroted from my grandfather's *Mechanical Engineers' Handbook*. Gutowski kindly ran this calculation for 2016 technology and got 276 BTUs/lb.

1,750

- Average heat required *ca.* 1958 from fuel per lb of steel in a batch in-and-out furnace, 2,000–2,400° F.

1,750

Energy to make a pound of "common brick," 1975.

2,464

- Energy to make 1 lb of Portland cement, U.S.A., 1990.

2,706 [= 0.793 kWh]

Average energy needed to manufacture $1 worth of American products, 1947. *Again, to make this much electricity, a power plant would have had to burn 3× as many BTUs, or more than ½ lb of Appalachian coal. The fact that the customer paid only 0.7452 cents for this thermodynamic work, with its associated economic, social and environmental costs, goes far to explain what we did to our planet.*

3,600

- Energy to heat 1 lb of iron [from air temperature?] to 5,000° F and vaporize it, *ca.* 1958.

4,131

- Energy to make 1 lb of Portland cement, U.S.A., 1950.

5,120

Energy to make 1 lb of chlorine by electroprocessing in 1952.

8,600–10,750

- Energy to make 1 lb of "ordinary" steel from pig iron in 2010.

8,900
Energy to make 1 lb of rayon by electroprocessing in 1952.

9,900
• Energy to make 1 lb of paper, 2013.

10,350
Energy to make 1 lb of liquid chlorine in 1975.

13,760
• Average energy to make 1 lb of plastic, 2013.

30,720
• Energy to make 1 lb of aluminum by electroprocessing in 1952. Probably a U.S. figure.

39,980
• Energy to make 1 lb of aluminum, 2013. Probably a global average figure.

68,260 [= 20 kilowatt-hours]
Energy to make 1 lb of titanium by electroprocessing in 1952. Each American "consumer" used about this much energy per day in 1975.

122,000
• Energy to make 1 lb of aluminum ingot in 1975.

429,920
Energy to complete chemical vapor deposition process, "an important step in the production of electronic grade silicon," upon 1 lb of material, *ca.* 2011.

15.48 million
Energy to make 1 lb of carbon nanotubes, *ca.* 2011. *This is equivalent to the high heating value of 121 gallons of gasoline.*

About Power

We stand naked and exposed in the face of our ever-increasing power, lacking the wherewithal to control it.

Pope Francis, Encyclical Letter of 2015

[Subsequent to the Chernobyl fiasco of 1986,] [c]ountries of the former USSR have been encouraged by the International Atomic Energy Agency . . . and the United States to shut down . . . [Chernobyl-style] reactors, but as of early 1995 demands for electrical power have prevented such action.

Encyclopedia of Chemical Technology, 1994

Those 68,260 BTUs required to electroprocess that pound of titanium back in 1952 could have been applied slowly or quickly;* either way they would have performed their work.—But had the factory staff set out to produce so many pounds per hour, their machines would have needed to do work *at a given continuous rate.* The same would apply had they wished to listen to the radio in the front office:

> Luxury is right. Her mother brings her ironing over every Tuesday, because we have the electricity. The last baby, her mother stayed with us a week and she never went to bed till two o'clock in the morning, listening to the radio . . . The old lady won't believe it that the music comes from Chicago, through the air.

What the old lady "consumed" was not a commodity but an experience: electric power creating fleeting sounds, moment after moment, at her will.

Power is amount of work done per unit of time. When I was alive, my homeland's unit of power was the watt.

---

* Within limits. There was some level of energy loss inherent in that system. A temperature gradient between the electroplating apparatus and the surrounding air would have allowed heat to escape at a certain rate. Energy needed to enter the system at a rate greater than the rate of energy loss.

1 watt = 1 joule* per second = 0.056884 BTUs per minute

In heat-equivalents, a watt expresses the energy in one match tip entirely burnt every 17 and a half minutes.

If none of its inherent energy were lost to electricity generation, and we could combust it with preposterous slowness, a pound of "average" Appalachian coal [at 12,500 BTUs] could give off 1 watt continuously for nearly 153 days.

By definition, a 100-watt lightbulb consumed 100 joules per second, which worked out to 5.6884 BTUs per minute. Twelve thousand simultaneously shining 100-watt bulbs would burn in one minute the same amount of energy as was needed to make that pound of titanium.†

Since the watt is an open-ended measure of *continuous energy consumption,* we require a different measure for *absolute energy consumed.* This latter is the watt-hour.‡ Two bulbs rated at the same wattage will draw the same amount of power. But the one that shines for a shorter interval will use less energy—fewer watt-hours.

An American radio from the 1950s might be rated at 60 watts. If someone played it for 15 hours, the power it spent would equal (60 × 15) or 900 *watt-hours*—a figure which in our day would be expressed in an electric bill as 0.9 *kilowatt-hours.*

A kilowatt is 1,000 watts. A kilowatt-hour inventories the result of 1,000 watts being consumed for the space of an hour, or 500 watts spent for two hours, or any other equivalent of 1,000 joules:

1 kilowatt-hour *is equivalent to* 1 kilowatt × 1 hour, *or*
1,000 × [0.056884 BTUs/minute] × 60 minutes

Hence this simple conversion:

1 kilowatt-hour = 3,600,000 joules = 3,413 BTUs

---

* A unit equally expressive of electrical energy and mechanical force. As Professor Gutowski wrote me, "Pick up an orange off the table and raise it over your head. The work you do is about one joule." For a scientific definition, see p. 524.

† Since 68,260 divided by 5.6884 = 11,999.859.

‡ So counterintuitively named because it means "watts multiplied by hours," or (substituting the definition of a watt) [joules divided by seconds] × [60 seconds]. Hence the two seconds units cancel each other out, and 1 watt-hour = 60 joules pure and simple—no time qualification.

But what does a kilowatt-hour really "mean"? How can one conceptualize it?

Since one BTU indicates the amount of energy needed to warm a pound of water by 1° Fahrenheit, then 1 kilowatt-hour could correspondingly warm 3,413 pounds of water—more than 409 gallons.

Should you happen to have 2,650 hundred-pound sandbags on hand, please open your trap door and let them all tumble 10 feet. The work thereby accomplished on them by gravity adds up to 1 kilowatt-hour. This quantity is rather staggering. Imagine what would be left of you after a 265,000-pound weight fell 10 feet onto your head!—But our electric-powered toys could slurp up a kilowatt-hour almost effortlessly—which cost us next to nothing (never mind the future). A children's illustrated book enthused: *Today, electricity is such a familiar and convenient form of energy that it is simply called "power"* . . .

In 1952, one kilowatt-hour could power a sewing machine for about eight hours or a vacuum cleaner for three. It could percolate 40 cups of coffee, or keep

Pastry-making machine, Kyoto

a 40-watt bulb shining for 25 hours.* At our typical ⅓ efficiencies, a pound of Appalachian coal burned in a power plant would produce 1 kilowatt-hour, with a few BTUs left over.

We liked to keep the lights on in 1952.—Back then, we had carbon to burn. A pound of kerosene contained within it enough energy to generate 5.8 kilowatt-hours—nearly 20,000 BTUs.†

PER CAPITA POWER CONSUMPTION, *ca.* 1925 and *ca.* 2014,

in multiples of the 1925 Japanese average

All levels expressed in [kilowatt-hours and BTUs]. *1 kWh = 3,413.0 BTUs.* I provide both units since kWh are the default in power reports.

All figures over 10 rounded up to nearest whole digit.

ca. 1925

1

Japan [88 and 300,344] (kWh and BTUs).

1.60

Germany [141 and 481,233].

5.36

United States [472 and 1,610,936].

5.60

Norway [493 and 1,682,609].

6.95

Canada [612 and 2,088,756].

7.96

Switzerland [700 and 2,389,100].

---

* All American figures.

† See p. 215, header **215**. This excludes efficiency losses.

ca. 2014*

77

Japan [6,764 and 23,084,798]. An increase of 77×.

82

Germany [7,192 and 24,545,108]. An increase of 51×.

83

Switzerland [7,315 and 24,966,095]. An increase of 11×.

139

United States [12,186 and 41,590,610]. An increase of 26×.

163

Canada [14,351 and 48,979,963]. An increase of 23×.

267

Norway [23,486 and 80,157,718]. An increase of 48×.

Satisfying each Canadian's electric appetite throughout 2014 would have required:†

Michigan brown peat: 4,885 pounds; or,
West Virginia bituminous coal, Pocahontas No. 3 bed: 3,666 lbs; or,
diesel fuel: 2,343 lbs = 362 gal; or,
gasoline: 2,174 lbs = 384 gal; or,
methane (primary constituent of natural gas): 1,890 lbs = 49,309 cu ft; or,
nuclear fuel: 0.00012374 lb = 1/505 of an ounce.

Sources: U.S. Department of Commerce, 1925; Central Intelligence Agency; 2014, with calculations by WTV.

---

* All population estimates were from July 2014. Energy consumption figures, however, ranged from 2010 through 2012. I divided the latter by the former with my customary nonchalance.
† These calculations were made using the table of Calorific Efficiencies (p. 208) headers **109, 158, 201–217, 226,** and **3.96 billion.** For power plant efficiency losses, multiply × 3.

Although the tools, appliances and toys that we owned all did measurable amounts of work (as when we ran a vacuum cleaner over the hall rug), we never had to worry (unless they were battery-powered) how much electricity they required.* We just plugged them into the wall! They drank in current, each according to its rated need, and as long as we wished them to work, they would, hour after hour, without limit, until we got tired of them or they wore out. Then we bought new ones, which were built even "better," because they could do more work.

Not all of us were careless and thoughtless. The government regulators who imposed efficiency standards and the engineers who met them deserve praise. Some appliances truly did become better, accomplishing more work while using less energy. Here is one such story: Between 1947 and 2008, American refrigerators grew in size by 159%—and by 1974 they were bolting down more than 400% more electric power than in 1947. Four ecologically conscious engineers from Stanford accordingly advised us: *In general, old small refrigerators use less energy than the big new ones.*—Then (in part because of what we called "the Arab oil crisis") manufacturing standards kicked in. Of course most "consumers" hardly knew or cared—but the 2008 models drew less power than the dinosaurs of 1947—so much less that keeping older models running often became the wrong choice.

Building a refrigerator took energy—this much to smelt and shape the steel, that much for the glass, so much for the refrigerant, etcetera. Call the sum of all such the *manufacturing energy.*

Once some "consumer" bought it and plugged it in, the machine began drinking electric power, as it would do almost without letup throughout its entire working life. Let that consumption be named the *use energy.*

The sum of these is the *embodied energy.*

A refrigerator made in a certain year might stay in service for a decade, consuming ever more embodied energy. The manufacturing energy remained fixed forever, while the use energy grew second by second. If the householder who owned it cared about "some ecosystem somewhere," she might strive to do good by nursing it as long as possible, delaying the time when the climate-changing CFCs or HCFCs from its dismantled guts rose up into the atmosphere,† and likewise putting off the necessity to smelt more steel and produce more glass for the sake of a new toy. All the while, her present refrigerator's ratio of use energy to manufacturing energy increased. Oh, she meant well—but she might be doing

---

* In my country at the turn of the millennium, "residential energy consumption" drank up a third of all the energy we used, and "major home appliances" consumed almost a third of that.

† See p. 175.

harm. And if the intuitively laudable approach of retaining one's workhorses, living frugally, declining to escalate demand, were *wrong,* who could blame her for throwing up her hands? When I was writing *Carbon Ideologies,* my friends said, "Bill, what's your solution?" and "Bill, what's the point of all this arithmetic?" and "Bill, if there's no hope why even think about it?" (Of all the excuses I heard for doing nothing about climate change, the last one actually touched my heart.)

Four technologists (one of them Professor Gutowski) who studied *the relative contribution of each lifecycle stage of the product from cradle to grave* concluded, as one might have predicted, that the *use phase of [a] refrigerator is the largest contributing phase in regards to energy consumption.* Better, then, to junk a 1974 model refrigerator and buy one from 1983, writing off the double expense of manufacturing energy for the two appliances, than to replace failing components, keeping the old one loyally humming and throbbing—and spending significantly more carbon-emitting electricity over its nine-year average useful life than would the model from 1983.

But keeping an old fridge running was the right thing to do in 1956 and the wrong thing in 1992.

If a householder's room air conditioner, washing machine, refrigerator and dishwasher were all 1981 models, and if she replaced them with 2008 models, she might save herself—and our atmosphere—from the effects of 213 million wasted BTUs: many dollars, 4,900 gallons of heavy grade fuel oil burned in a power plant, and more than 20 tons of carbon dioxide.

Woman walking past the Flamingo substation, Las Vegas

Wearing our clothes out, right into rags, was always best; while replacing an electric motor usually saved more energy than repairing it. As for retreading our automobile tires, that *strongly depends on the boundary conditions of the analysis.*

Meanwhile we lived our own lives, buying toys and plugging them in. This was our glory—in the developed world, at least, where we still had all the power we could pay for. Where that power came from was no business of ours. (Where *did* it come from? I quote from *Power and the Plow,* 1911: *Possibly the most stupendous discovery in the history of the world was that heat from burning materials could be made to do the work of plants and animals.*)

What an apt word, power!—Here are some ways we spent it:

COMPARATIVE POWER REQUIREMENTS AND ENERGY USAGES,

in multiples of what was needed per minute ca. *1975 to operate a plug-in vibrator*

[From American sources, unless otherwise stated.]

All levels expressed in [BTUs per minute]. Early-21st-century U.S. electricity assumed to be 115 volts. Note that some devices such as refrigerators cycle on and off, while others receive only intermittent use; hence while these comparisons are accurate per minute of operation, they prove nothing in regard to longer-term power consumption. Unless otherwise stated, these are *rated* figures. "Typically, the input power for a microwave oven is 50% higher than its rated power."

These estimates come from many different sources and are not utterly consistent. For instance, compare **1.87–2.49** to **4.0**, and **289.48 billion** to **299.39 billion**.

The symbol **<H>** indicates that a longer block of energy consumption has been broken up into assumedly equal increments of BTUs/min. This may be an oversimplification. For instance, the refrigerator in **2.4** must do more thermodynamic work than usual after the door has been opened in order to put away groceries.

The symbol **<C>** is a reminder that in a given case some or all thermodynamic work was supplied or could be supplied non-electrically—by combustion. For instance, much of the total U.S. energy consumption for 1990 went not to electric power but motor power: burning gasoline in vehicle engines.

All comparative headers over 10 rounded up to nearest whole digit. Absolute figures over 200 rounded likewise. Absolute figures and comparative headers over 1 billion rounded to 2 significant digits.

<1
<C> 1 watt [0.056884 BTUs/min]. *Please recall that a watt is a unit not of energy, but of energy consumption. 1 watt = 1 joule per second.*

<1
Operating a certain medium-quality rechargeable drugstore electric toothbrush, purchased 2015 [0.051].

<1
<C> The amount of heat to be withdrawn from 1 pound of water in order to form ice in 24 hours. [Based on definition of a refrigeration ton.] [0.100 BTU per minute].

<1
<H> Operating a passive infrared sensor LED motion detector light, purchased 2015 [1.74].

1
Enjoying a plug-in vibrator, *ca.* 1975 [2.28 BTUs per minute].

1.6
<C> Sleeping: metabolic requirement of a human body [3.67].

1.5
Operating a popular brand of laptop computer, purchased 2014 [3.43]. Compare with **359**.

1.87
Operating a "consumer" grade sewing machine, *ca.* 1975 [4.27]. Compare with **629**.

1.87–2.49
<C> "A typical adult male at sustained labor is estimated to produce 75 to 100 watts of power." [4.26 to 5.69]. Compare with **2.95** and **4.0**.

2
Operating a radio, *ca.* 1975 [4.56]. Compare with **22**.

2.4

<H> Operating refrigerator, *ca.* 1961 [5.53]. Compare with **3.8, 5.9, 419.**

2.5

Powering a 100-watt lightbulb, by definition [5.69].

2.95

<H> <C> "A moderately active human adult requires about 3 million cal[ories] [= 3,000 "food calories" or kcal] every 24 hr." [8.27].

3.7–4.7

<C> Removing 1 cubic inch [per minute] of brass, using a round-nosed lathe tool, *ca.* 1945 [8.48–10.61 BTUs per minute].

3.8

<H> Operating refrigerator, 1969 [8.69]. Compare with **2.4, 5.9, 419.**

4.0

<H> Sweeping the floor: metabolic requirement of a human body [9.2]. Compare with **1.6, 1.87–2.49.**

5.9

<H> Operating refrigerator without freezer, *ca.* 1975 [13.37]. *With frostless freezer: [25.31].* Compare with **2.4, 3.8, 419.**

7.5

<H> Running home freezer, hair dryer or color television,* *ca.* 1975 [17.05].

9.3–14

Activating 15-pound power hammer, *ca.* 1945 [21.09–31.82].

9.5

<H> Operating a 40-gallon electric water heater, purchased *ca.* 2011 [21.7].†

---

* Over time the freezer would obviously use the most power of the three, since it could not be switched off without spoiling its contents. Respective contemporary monthly power estimates for these appliances were: 76, 0.5 and 37.5 kilowatts.

† Note: Calculating as usual by simply converting this device's rating [4,500 watts] to BTUs would yield 255.978 BTUs/ min. But here I defer to Professor Gutowski, who has supplied figures on actual use.

13

Operating a certain brand of electric corn popper, 1956. "It's 25% bigger and much faster than many poppers of this type. Full 500-watt heating element (not just 400 watts) gives faster popping and fluffier, more tender popcorn. No stirring or shaking!" [28.44].

13

<H> Average per-minute power consumption by William T. Vollmann during December 2014 [28.92]. Compare with 16.

13

Operating a vacuum cleaner, *ca.* 1975 [30.72]. Compare with 34, 1,048.

15

Operating a cell phone charger, *ca.* 2012 [34.13].

16

<H> Average per-minute power consumption by William T. Vollmann during December 2013 [36.26]. Compare with 13.

19–33

<C> Removing 1 cubic inch of cast steel, using a round-nosed lathe tool, *ca.* 1945 [42.42–76.35].

19

<H> Operating a heavy-duty blender, purchased 2012 [42.51].

21

<H> "American average" household power consumption (about 20 kilowatt-hours per day), *ca.* 1975 [47.4]. Compare with 31.

22

Listening to a radio, *ca.* 2012 [51.2]. Compare with 2.

25

• <C> 1 kilowatt [56.884].

26

<H> Average power consumption "for a single family residence" in Sacramento, California, 2016 [59.3]. "S[acramento] M[unicipal] U[tility] D[istrict], through its energy efficiency programs for customers, set an aggressive goal in 2007 to accomplish 15 percent energy reduction in 10 years. SMUD is on track to meet or exceed that goal next year."

29

Operating .75 horsepower air compressor, purchased *ca.* 1999* [66.33].

30

Running a dishwasher, *ca.* 1975 [67.69].

31

Operating a 11 × 14" photographic dry mount press, mfg. *ca.* 1970s [71.105].

31

Average American household power consumption, 2010 [71.7]. Compare with **21**.

32

<H> Operating a window air conditioner, *ca.* 1975 [73.95].

33

<H> Operating a "Fan-forced Non-Automatic Electric Heater," 1956 [75.08].

34

<H> Using a home electric broiler, *ca.* 1975 [78.22].

34

Operating a hypoallergenic vacuum cleaner, purchased 2010 [78.50]. Compare with **13, 1048**.

35

<H> <C> **Per capita American "food-related energy flow,"**[†] **2002 [78.92].** Compare with **284**.

---

* Power consumption figure calculated from multiplying rated voltage × amperage. However, converting advertised horsepower (0.75) directly to BTUs halves the result to a mere 31.81 BTUs per minute. I suspect the first calculation is more accurate.

[†] An "energy flow" is an attempt to total up all various energy expenditures in the manufacture, packaging, shipping, preparation, consumption and disposal of a certain item. Thus the energy

81

<H> <C> Per capita American energy use, including burning wood for fuel, 1870 [184.87].

93

<H> <C> Accomplishing average work with a bulldozer of 29-inch width and 14-inch head movement, *ca.* 1945 [212].

93

<C> Operating a horizontal boring, drilling and milling machine with a 2.5-inch-diameter spindle, *ca.* 1945 [212].

100

<C> Rated capacity of Parsons steam turbine dynamo, 1884: 4,000 watts at 100 volts d.v. [228].

105

<H> <C> Per capita American energy use, 1900 [240].

132

Operating a LG front-loading dryer, electric model, purchased *ca.* 2015 [301]. *"Tumble driers [sic] are hugely energy inefficient and should be avoided if at all possible."*

161

Operating a Samsung front-loading dryer, DV42H5 series, gas model, purchased *ca.* 2015 [367].

180

Operating a 24-inch LCD television, *ca.* 2012 [410].

194

<H> <C> Per capita American energy use, 1950 [443].

270

<H> <C> Per capita American energy use, 1990 [616]. This is 3.33 times the comparable figure for 1870; see **81**. For 1990 per capita figure, see **67.84 billion.**

---

flow for a food item would include the energy units (such as BTUs) in its planting, fertilization, harvesting, cooking, waste collection, etcetera. See p. 522.

284

<H> <C> **Per capita American overall "energy flow," 2002 [647].** Compare with **35**.

292

Cooking on a home electric range, *ca.* 1975 [667].

359

Operating a desktop personal computer, *ca.* 2012 [819]. Compare with **1.5**.

372

<C> Accomplishing average work with a bulldozer of 63-inch width and 20-inch head movement, *ca.* 1945 [848].

372–465

<C> Operating a horizontal boring, drilling and milling machine with a 9.5-inch-diameter spindle, *ca.* 1945 [848–1,061].

419

<H> Operating a 12-cubic-foot refrigerator, *ca.* 2012 [956]. Compare with **3.8, 4.8, 5.9.**[*]

465–651

Activating a 1,000-pound power hammer, *ca.* 1945 [1,060–1,485].

629

Operating a light commercial sewing machine, purchased *ca.* 2010 [1,434]. Compare with **1.85**. This figure, calculated per usual procedure, strikes me as far too high.

688

Flying the first Piper Cub two-seat airplane at full engine capacity ("a whopping 37 horsepower"), 1938 [1,569].

897

Operating a hedge trimmer, *ca.* 2012 [2,044].

898

Powering the 5 screens of a multiplex movie theater, *ca.* 2009 [2,047].

---

[*] The sizes of those other three refrigerators were unspecified.

1,048
Operating a vacuum cleaner, *ca.* 2012 [2,389]. Compare with **13, 34**.

1,347
Operating a small microwave, *ca.* 2012 [3,071].

1,497
Operating a hair dryer, *ca.* 2012 [3,413].

1,796
Operating an espresso machine or a dishwasher, *ca.* 2012 [4,096].

2,096
Operating a large microwave, *ca.* 2012 [4,779].

7,442
<C> Capacity of an Illinois Central 1930 model 100,000-lb electric loco-motive, 400 continuous horsepower [16,967].

11,102
<H> <C> Average U.S. electric production, week of June 17, 2015 [25,313].

126,510
<C> Capacity of a Virginian 1948 model 1,033,800-lb electric locomotive, 6,800 continuous horsepower [288,442].

2.29 billion
<H> <C> U.S. energy consumed in paper manufacture, 1988 [5.23 billion].

5.44 billion
<H> <C> U.S. energy consumed in refining petroleum, 1988 [12.39 billion].

67.84 billion
<H> <C> Total U.S. energy consumption, 1990 [154.68 billion]. For corresponding per capita figure, see **270**.

74.85 billion
<H> <C> World energy consumption rate, *ca.* 1950: 3 terawatts [170.65 billion].

The following two entries vary by under 4%. Presumably we actually used more energy in 1991 than in 1990. Hence this small lesson in approximation errors.

289.48 billion

\<H> \<C> World energy consumption **amount,** 1991, converted as usual to the necessary per-minute rate, and computed from a different source than the next entry [660.01 billion].

299.39 billion

\<H> \<C> World energy consumption **rate,** *ca.* 1990: 12 terawatts [682.61 billion].

498.98 billion

\<H> \<C> World energy consumption rate, *ca.* 2015: "About" 20 terawatts [1.14 quadrillion].

748.47 billion

\<H> \<C> World energy consumption rate, *ca.* 2050 (*ca.* 2015 projection): 30 terawatts [1.71 quadrillion].

Circa 2010, per capita manufacturing power consumption was increasing by a steady 1.2% per year all around the globe.

And so a little picture-book on electricity, written for the benefit of American first-graders, did well (or at least appropriately) to end as follows: *People everywhere need more and more electricity each day. Scientists keep looking for new ways to make electricity.*

What Was the Work For? (continued)

Machines kept our thoroughfares clean when I was alive. Before dawn, streetlights protected me; day and night I could push a button whose signal would halt traffic so that I could cross even the busiest street. At all hours I would spy my fellow humans riding to and fro, enabled by internal combustion engines that perfumed the neighborhood with gasoline and diesel. It was marvelous how many of us there were. We were experienced in our way, much like cockroaches, and our experience assured us that we could keep going nearly forever in just this fashion; so too our children. When my daughter turned 16, I took her to the Department of Motor Vehicles so that she could hazard her driver's test on a touchscreen; once she had passed and I had paid, she took a different test and I paid again—after which she was allowed to burn all the gasoline she could afford.

On this subject, please do give credit to the automotive engineers of my time: In little more than half a century they made gasoline perform almost 35% more apparent work* than before:

ENERGY REQUIRED TO MOVE AN AMERICAN CAR ONE MILE, 1949 and 2010

Assuming gasoline's calorific efficiency = 112,500 BTUs/gal.

---

* I say "apparent work," because most cars were made lighter in 2010 than in 1949, so that there was less thermodynamic work to perform. Absolute efficiency increases also played a part.

1949: **8,654** BTUs
        [**13 miles per gallon**]
2010: **6,429** BTUs
        [**17.5 miles per gallon**]

Sources: Prentiss, 2015; *Mechanical Engineers' Handbook,* 1958, with calculations by WTV.

And what was that work for?—My daughter drove to school and back; she picked up her mother at the airport; how could I label this unworthy use?—I myself quite often let carbon power take me to a place for no better reason than that I felt like travelling there. (Perhaps you think that I should avoid this subject, but when I lived we were none of us ashamed.) Whenever I wished (and had enough money), I flew to Japan in a great metal airship that spewed carbon dioxide as it went. How much of that greenhouse gas should I have taken responsibility for?* There were many of us all together in this vessel, and mostly we sat bored and annoyed during those nine or 16 hours, while the power that carried us through the air labored unobserved. If the flight grew turbulent, then someone's baby might cry; and when the turbulence worsened, *then* perhaps a few frail or phobic passengers turned pale, remembering that we all continued aloft only through some mystery, and that should the mystery fail, we would plummet screaming into the ocean. The air grew calm again, and we resumed our indifference to jet propulsion. I suppose that sort of ride has been discontinued.

In middle age I happened to grow curious about nuclear accidents (in my day they were novelties), so when I got to Japan I would hire a $600 taxi ride into this or that red zone. A battery-powered dosimeter kept me conveniently abreast of the harm I was inflicting on myself. Sometimes when I wandered past ruined houses, photographing whatever I liked with a battery-powered camera that even rewound the film for me, or measuring radiation with my scintillation counter (after six months that excellent device still had plenty of chemical energy in its first pair of batteries), the white-gloved driver became anxious, especially at night; he would sit in the car with the engine running; at half a block's distance I could smell the exhaust. Well, that was his lookout. I didn't pay for the gas.

There were bullet trains in Japan, and glowing vending machines with ice

---

* On a recent business trip from San Francisco via Zürich to Barcelona and then back to San Francisco, my personal share of emitted CO_2 was nearly 1 and ¾ tons. For this I excuse myself, because I saw my mother and also slightly advanced my career.

coffee or piping hot soup. I have been told that up to 10% of that nation's electricity went to ready lights on various machines and appliances. In the convenience stores the freezers were doorless, so that we could feel the coolness beside the hotness and be tempted. Toilet seats warmed themselves at the touch of a button. (Thoreau: *We will sit on a mound and muse, and try not to make those skeletons stand on their legs again. Does nature remember, think you, that they were men, or not rather that they are bones?*) There were escalators, too. I loved all that.

And sometimes I went to Mexico to be entertained by oil extraction. At night the taco stands and tiny *sope* restaurants shone blue with gas-jet flames. Women sat ready to pound out tortillas with their hands and throw them on the griddle. Cauldrons of lard kept themselves on the boil. Often no customers were present, but one had to be ready, so the gas burned merrily on. In the center of town rose towers of peach-orange refinery flame.

There were movies wherever I went, thanks to electricity. And in West Virginia, where it pleased my fancy to look at coal mines, I often got to ride around in a car. (Walter J. Boyne, 1988: *A curious aspect of the automobile has been the manner in which men and women of all countries view it—not in terms of need, but of desire.*) I myself was never one to emote about *the way the 5.2 liter, twin-turbo V-12 storms to its 7,000-rpm redline or the barrel-chested boom from the exhaust on the way there* (that hymn celebrates the 2017 Aston Martin DB11), but without a vehicle dedicated to my personal goals I could not have written vivid descriptions of places for *Carbon Ideologies*.—It was wise to pay attention to the fuel gauge, but there would always be a gas station, and as long as I had money the wheels kept turning my way. Every motel room I checked into was warmable or coolable to taste, no matter how poorly insulated the building might be. Oil stayed cheap, and natural gas also; who wouldn't rather burn carbon than put in solid walls?*

So that was how I lived, and you who live after me may well agree that whether it was good or bad, it was certainly a pleasant life.

---

* In my day, the lords of commercial real estate frequently built flimsy; a Vegas casino magnate once told me that "remodeling" his establishment every 10 years generated higher profits than constructing for the ages—because places went in and out of fashion when we were alive. Big machines rolled in, wrecking and then pouring, pounding and piling. We got jaded, you see; we wanted something new; that was what the work was for.

Carbon Ideologies Approached

Life as we know it is not limitless in its capacities . . . Actually, its manifestations are confined to a small range on the thermometer.

Loren Eiseley, 1969

Was this Earth in any sense entrusted to us, or merely, coincidentally, made our available prey?

Once upon a time, the issue under contention between these opposing views was moral, spiritual, aesthetic. Hence in my youth I heard environmentalists called elitist, sentimental, selfish and racist, while loggers, miners and trawler-fishermen were accused of insensitivity; their ways of making a living deprived us of "wilderness" which hypothetically enriched our secret "natures." So the pleasures of a forest trekker got weighed against the projected needs of power generation; the practical advantage of situating an observatory atop a certain mountain might be litigated against by the Indian tribe to whom that mountain was sacred; jobs offered by the local paper mill did or did not measure up against tourist dollars driven off by the stink. And as usual, money generally weighed greater than anything, and competing moneys confused us all.

What a country it must have been before the lumber men came! Sherwood Anderson is writing about Appalachia. He goes on: *. . . The mountain men were always talking of the spruce forests of former days. Many of them worked in the lumber camps. They speak of soft moss into which a man sank almost to his knees, the silence of the forest, the great trees.* Meanwhile the coal men investigated the ground where *the great trees* had been. Wherever it paid them, they sliced off mountaintops, choking the hollows and streams with poisonous rubble, and if the result was uglier than before, well, at least somebody made a profit.

Presently arose two sharper queries. The first was: Were humans guilty of outright ecocide? If so, to what extent might this be a crime? What "value" was, for instance, the life of the spotted owl?—This got easily settled in economic terms; the owl's worth was zero.

However, the issue of financial value failed to go away. Hence the second query: What if unregulated taking were damaging somebody's investment? In

The pleasures of electricity in Hiroshima

1876 John Muir warned: *Strip off the woods with their underbrush from the mountain flanks, and the whole State, the lowlands as well as the highlands, would gradually turn into a desert.* This effect being so evident, at least to him, toward century's end he drew an optimistic conclusion: *The slow-going, unthrifty farmers, also, are beginning to realize that when the timber is stripped from the mountains the irrigating streams dry up in summer, and are destructive in winter; that soil, scenery, and everything slip off with the trees: so of course they are coming into the ranks of tree-friends.*

But were they? And if they were, was that good enough to meet a future-dweller's definition of stewardship?

More decades of happy material accumulation passed by, until the issue further sharpened itself: Might our actions actually be imperiling *us*? Who could believe such a ridiculous thing? And yet this worry asserted itself ever more pressingly—although we wriggled to avoid it.

In 1968 a scientist wrote: *Increasingly, there is but one way into the future: the technological way. The frightening aspect of this situation lies in the constriction of human choice.* In fact we had more choices than ever before.

What mainly fretted us about our fossil fuels was that they might run out. In 1981, a good decade after my biology textbook noted that *highways, gas stations and parking lots now cover more of the surface area of the U.S. than do homes, stores and schools,* a report on automobiles in America's urban zones advised:

> Transportation is heavily dependent on the use of oil derivatives* . . . In contrast, a variety of energy sources can generate electricity. A rational allocation of energy resources, therefore, might be to limit the use of liquid hydrocarbons in electric power generation and place major reliance instead on coal and uranium, augmented by hydroelectric and solar energy.

Yes, that was the "rational" way to get by, all right! By then several dozen of us had heard about contamination in "some ecosystem somewhere," but we had to get on with our lives. Don't say we didn't try to be good; several commuters even rode "rail rapid transit" powered by electrical energy! (More than 30 years later, as I flittered from Poza Rica to Oklahoma City to Dubai for *Carbon Ideologies,* I kept meeting people who thought we should abandon fossil fuels in favor of "electricity"—as if electricity emerged new and clean from nowhere all by itself!) To tell the truth, our lives were unthinkable without internal combustion engines—whose fuels polluted the atmosphere as follows: With respect to both carbon monoxide and nitrous oxide pollution, not to mention sulfurs and particulates, coal came out the worst in grams emitted per car-mile, followed by residual oil, then natural gas. As for carbon dioxide, how could *that* be a pollutant? In its "rail rapid transit" report (1979) the U.S. Department of Transportation did not mention it. For in those days there were so many other ways that we could have done for ourselves that climate change hardly stood out!—In 1972, in a science-fiction novel called *The End of the Dream,* the saturnine Philip Wylie asked what our work was for and expressly disliked the answer: *It was politically impossible in the late 1970s to compel even part of industry to suspend production for a mere twelve to eighteen months to make essential changes in its techniques,* even to save the environment. *The American citizenry . . . was addicted to consumerism.* So were other nationalities.—What happened next? As Wylie imagined it, children got cooked to death when an overheated nuclear reactor had to cool itself with massive quantities of river water (the "power interests" hushed it up because *so far it involves, mostly, a small number of slum kids, most*

---

* At this time, transportation consumed only a quarter of our national energy budget.

of them Negro or Puerto Rican); then came the nitrogen oxide inversion layer in Manhattan that killed 1,200,000 . . .—and *then,* because *the seas were like a chemical warehouse factory, with a million compounds in stock that were stirred together, outcome unknowable,* and because *certain manufacturing processes used world-wide in the making of cheap dyes* killed some obscure marine protozoa that had kept them in check, came *the vibes:*

> A great shadow, lightish, goes under my boat and past it for, maybe, a hundred feet . . . And the vibes start squirming out by the million. Looks like macaroni—. . . a vibe gets Amy while she's just ready to scram . . . Then the follow-upper types arrive and you can't see nothing, no dress, even, for the white wriggling mess is all over her.

They struck all over the world:

> Japan managed fairly well, its population already reduced to a tenth of its peak by the Rice Blast . . . Some cities, such as Cincinnati, where it was believed the invasion could not occur, were hit at night and virtually depopulated before dawn, amidst scenes of incredible panic and rout.

But never mind *the vibes,* who were presently undone by other pollution; for *by 2010 the seas had risen more than two hundred feet . . . Meanwhile the atmosphere, staggering under a cloak of coal-mine-fire pollution, began to deposit a rising volume of toxic compounds produced by volcanic activity.**

Near the end of my life the predictions in that final paragraph seemed likeliest. We had been using the oceans for sewers; we polluted land and air left and right . . . and for all I know, you from the future will conclude that climate change was not the worst of it. (From *The Japan Times,* 2013: *Dense clouds of yellow dust from China were forecast to begin reaching Japan on Friday.* From the same paper, 2014: **Yellow dust linked to more emergency room trips**. Maybe you've even met *the vibes.*)—In that era we used to entertain ourselves with all kinds of post-apocalyptic visions. George R. Stewart's *Earth Abides* imagined a plague that wiped out most of us; Neville Shute's *On the Beach* described our

---

* In exactly the same year that this prophecy was published, a paleontologist reiterated his longstanding anxiety about returning glaciers: "The swarming millions who now populate the planet may mostly perish in misery and darkness, inexorably pushed from their own lands to be rejected in desperation by their neighbors." He was *almost* on the money.

extinction after a thermonuclear war; John Brunner's *The Sheep Look Up* retailed the destruction of America from pollution, greed and technological incompetence. *A Stanford University biologist says the race between food production and population growth needs is already lost: by 1985 hundreds of millions of people will starve to death.*—One or more of these events could still happen in your time. As for other scenarios, they mostly derived from our failure to answer that question: What was the work for?

A hazy reply did come from the advertisements in *Inc.: The Magazine for Growing Companies* (you see, our companies had to grow). Trying to sell us a 1985 Toyota cargo van (2,500-pound payload), some genius urged: *Pack it full. Fuller. Pile it high. Higher.* Why not? We'd let carbon do the work! Meanwhile some other Pied Piper was peddling computers: *COMPAQ could have stopped here . . . but we didn't. Introducing the new COMPAQ DESKPRO 286. More features, more speed, more power*—courtesy of a fossil or nuclear plant in some ecosystem somewhere.—We liked that; we piled it all higher.

As climate change progressed, for a very long time it masked itself. The "coal-mine-fire pollution" imagined by Wylie would have been immediately unpleasant; but those wide white plumes I saw rising from the skirt-shaped smokestacks of the John E. Amos coal-fired power plant in Nitro, West Virginia, did not so much as tickle my throat; and a lady who lived with them nearly in her back yard said: "Well, I been here all my life and it never bothered me."* And for a long time it did not bother the rest of us, either.

What "it" was the 17th-century seeker Van Helmont used to call *gas sylvestre*, while Black a century later referred to it as "fixed air." Of course "it" was carbon dioxide.

In my 1976 *Encyclopaedia Britannica* lay two adjoining entries. The one on "climatic change," which presented our planetary situation much as would the literature at the Kentucky Coal Museum almost 40 years later, spoke of warming and cooling cycles. Then it told our fortunes:

> The cool-moist trend of the 1960s was of sufficient magnitude and universality to suggest that the next few decades will continue to be on the cool side in high latitudes and on the moist side in the tropics. Winters will more often than not be colder and snowier than average, and more summers will be either cool or wet.

---

* See II:46, 51.

But the entry on "climate" guessed in the other direction:

> It seems that the 10 percent increase in the amount of carbon dioxide in
> the atmosphere from the 1890s to the 1960s should tend to maintain a
> somewhat warmer climate (on the average, the effect should be about
> 0.3° C or 0.5° F) than that of the last century.

When I happened upon this passage in 2016, I was disagreeably amazed. In 1976 we knew that CO_2 concentrations had been increasing for *seven decades*—a long opportunity to have thought something and perhaps even done something.—Well, but as the first entry showed, even in 1976 we did not quite understand what those concentrations entailed.* The prophet of *the cool-moist trend* might not have been an ideologue at all; he resembled one of those period science-fiction writers who could penetrate Venus's clouds only with speculations. Why shouldn't Venus be warm and moist, rainy and jungly and crawling with, for instance, intelligent amphibians? *Being an Earth-like planet with a thick atmosphere, Venus was presumably well endowed with oceans . . . Until the 1960s, the commonest picture . . . was one which resembled the primitive Earth: steaming coal forests, swamps and—sometimes—even dinosaur-like reptiles of great size. In many ways it is unfortunate that the concept of Venus as a new Earth had to die,* as it did once we'd linked the microwave radiation emitted by Venus to hideous heat. Mariner 2 verified these readings; then came the Soviet landers, most relevantly Venera 7, which reported a surface temperature of nearly 900° Fahrenheit before falling silent.† Wondering and hesitating, the 1976 *Britannica* finally took a chance: *Perhaps the most promising explanation for the high surface temperature is . . . the so-called greenhouse effect*—for the Venusian atmosphere contained not only significant sulfur dioxide,‡ which can increase global warming, but also, coincidentally enough, carbon dioxide (to the tune of 96.4%). The clouds let the sunlight in, but not out, and so the planet became, whether or not it had ever otherwise been, unliveable. *Preliminary calculations suggest that such a runaway greenhouse effect could have occurred on Venus but not on the Earth because of the difference in distance from the Sun of the two planets and the consequent difference*

---

* In about 1976 I attended a lecture by the famous scientific popularizer Isaac Asimov. He assured us that the equilibrium between carbon dioxide and particulate pollution—whose cooling effects had been observed after volcanic eruptions and hypothesized in the case of a "nuclear winter"—would protect us from climate change. It made me happy to believe him.

† But let's not be greenhouse alarmists! "Searing temperature" was assisted in its murder of the probe by "crushing pressure." Besides, two more recent estimates chilled down Venus to "more than" 750° and to 840° F.

‡ See "About Sulfur," II:26.

*in the rate of temperature buildup.** Well, then, we could go back to sleep—especially since our ordinary thermometers and shortlived terrestrial weathermen could prove no global temperature increase.

In 1980, when it happened to be said that *the United States accounts for roughly one half of the world's total consumption of energy and produces substantially less than it consumes,* the United Nations and Petro-Canada held a symposium. They invited resource extractors from around the world, and even a few bleeding hearts. It must have been the end of November in Montréal, whose famous winter chill would have set reassuringly in, when the man who would soon publish all their papers under the happy title *Long-Term Energy Resources* addressed them:

"Very often the point is made that the increase of CO_2 in the atmosphere will lead to a 'greenhouse' effect, an increase in temperatures worldwide, and the melting of the polar ice caps. This theory is frequently mentioned, including in some of the papers submitted to the Conference, but it is a theory that is out of date."

And he invoked the climatologist Dr. George Kukla, who had opined that carbon dioxide *may have a cooling effect.* Unfortunately, Dr. Kukla could not be present (I would have loved to read his remarks), but his name must have been as good as gold—for thanks to him the greenhouse effect was now forever disposed of!—The moderator continued: "I mention the latest research . . . in order, if possible, to eliminate discussion of CO_2 from the Conference."

To eliminate discussion of CO_2 from the Conference! To you from the future that must ring high-handed, and worse! You might accuse him of censorial ignorance, or of covering up for his petrochemical pals; in hindsight it does ring to his discredit that *very often* the point about CO_2 was being made in his hearing, and yet he determined *to eliminate discussion* of it.

Well, please forgive him. In his defense, and ours, I would remind you that environmental change, which must be accelerating in your time, remained deniable to, and denied by, our multitudes who lived their own lives, being served by unseen sectors of a far-flung, hyper-specialized economy, whose unintended effects were correspondingly invisible; so that even in the second decade of the 21st century, when I was writing *Carbon Ideologies,* it seemed not at all preposterous to follow ancestral tradition, and make my own short life the measure. Had I done so, my daughter would have been happier—for how tedious I was, to go on and on about gloomy abstractions! Yes, cities were growing and forests were slimming out; we could see that, but how could we be expected to grasp these slower

---

* See II:643n.

alterations, especially when our most ironclad climatological records went back only a century or so? As you will see when we get to the sixth-grade science standards of the West Virginia Board of Education,* such changes could be stoutly denied, and the denials maybe even believed in by some of the deniers. In 1969 a paleontologist who certainly tried to take the long view concluded:

> If the cause of these glacial conditions . . . is directed by recurrent terrestrial or cosmic conditions, then man, unknowingly, is huddling memoryless in the pale sunshine of an interstadial spring . . . He is a survival from a vanished world, a denizen of the long cold of which he may yet be the returning harbinger.

Meanwhile our lives ran on, as yours probably will, with little concern for someone else's dark future.

That moderator of the *Long-Term Energy Resources* symposium was, as were we all, a devoted specimen of faith. And just before this he had recited our common creed:

"Modern industry cannot operate without electricity, and . . . electricity is today a necessity for our communications system and social life."†

What was the work for?—I refuse to reply: Merely for consumerism, profiteering and reptilian greed! For when L. L. Nunn wrote so long ago of *lessening man's toil by the use of a wire,*‡ his heart had been right. (Does that photograph of the old Bangladeshi woman sweatily and wearily clutching her shocks of corn mean nothing to you?)—And at the symposium in Montréal another good man now rose to speak. He was the Chief Scientist of the United Kingdom Department of Energy. And he said: "Energy planners must never forget that hoped-for increases in the standard of living of the world's poor will result in increased energy consumption to replace physical labor and prevent the exhaustion of limited wood resources."

You from the future, are you listening? Please don't say that we ruined you entirely out of selfishness! Now hear the pleadings of some altruists from Oak Ridge National Laboratory, published in 1982: *Not to use oil and gas for as long*

---

* See II:222.

† I should add that in introducing the symposium he had graciously remarked that "there is no consensus on this view . . . much work remains to be done on climate change." And then, to address the astronomically unlikely event that the greenhouse effect should turn out to be real, he offered a solution in the best tradition of Adam Smith: "It may also be possible to take advantage of the increasing commercial value of CO_2 by capturing it from the combustion gases of power plants."

‡ Above, p. 28.

as they are readily available would be an act of irresponsibility and inhumanity on the part of those governments struggling to provide for poor, expanding populations.

During the previous year, our planet's citizens had produced *285 quadrillion BTUs* of primary energy (we casually called them "quads"). Does that figure bear more weight in italics? I could truthfully equate it with *57 billion barrels of oil,* but would that assist any merely human understanding? A single barrel of oil (42 gallons, 159 liters, 296 pounds internal weight) is barely comprehensible at something like 6.5 million BTUs*—just enough (supposing the impossibility of perfect efficiency) to have manufactured 95 pounds of titanium in 1952. Why not say that we refined Himalayan mountains of carbon-based substances, which our high priests offered up in smoke?

The Oak Ridge altruists now admitted: *Burning fossil fuels will increase the level of CO_2 in the earth's atmosphere. This could become a problem a half century from now . . . Intensive research is needed to increase our knowledge about the factors that affect accumulation of CO_2.*

(From the National Academy of Sciences, 1979: *At present, the most serious man-made impact on climate in the next few centuries is believed to be produced by the increasing concentration of CO2 in the atmosphere, as a result of burning fossil fuels . . .* Fortunately, *it appears that the temperature changes due to increased CO2 will not be detectable as such before the year 2000.*)

So they knew! Reader, when you get to the coal, oil and fracking chapters of *Carbon Ideologies,* where you will hear, for instance, a certain lobbyist expressing his "reservations as to man's contribution" to climate change, which "has always occurred and will always occur; I'm intrigued and even find it a little humorous when I hear people say that the winters today are not what they were when I was a child," please remember that more than 30 years before, back in 1982, these brains at Oak Ridge wrote in black and white: *Burning fossil fuels will increase the level of CO_2 in the earth's atmosphere.*

In their defense, *intensive research is needed to increase our knowledge.*

(From a biology textbook published in 1967: *The rapid, very voluminous release of CO_2 by man-made combustions today may increasingly affect global climates . . . Carbon dioxide . . . has a "greenhouse effect," which in some measure probably contributes to the present warming up of the earth.* So even in 1967— 49 years before I went to Oklahoma and found people still denying it—a professor at Brown University was referring to "present warming"—but *in some measure probably* did not equal *certainly.*)

---

* My computation using the energy value of 10 API petroleum oil. But the U.S. Department of Energy estimated a barrel of American crude at 5.8 million BTUs. Each fossil fuel has a wide range of innate energies, based on source type, refined purity, etcetera.

To comfort us, the Oak Ridge boys now prepared two curves of projected carbon dioxide concentrations over time, the higher-sloping one (which would have thrilled the ideologues you will meet in West Virginia) labeled **COAL USE TO MEET ALL ENERGY DEMAND** and the lower one **1975 FUEL MIX**. Even the all-coal curve stayed obediently below 400 parts per million in the year 2000—as in fact happened. The caption read: *It is very improbable that atmospheric CO_2 will reach 500 ppm before 2030.*

(In 2030 I was by some definitions still alive, but 71 years old. So what did I matter?)

The Oak Ridge boys further calmed the waters: *And even in the high CO_2 case, a level of 700 ppm (which may prove to be acceptable) is not exceeded in 2050.*[*]

(Translation: They did not care about anyone alive after 2050. You were not in their department.)

It is fairly clear that no serious climactic changes will result from fossil fuel use in this century. After that they would be safely off the job. *For the present, fossil fuels will continue to be used while the potential consequences of the CO_2 problem are investigated.*

So we kept right on investigating. Then we died.

---

[*] In 2013, with a level exceeding 400 ppm, the Intergovernmental Panel on Climate Change modeled a "medium scenario" (what optimists!) of 700 ppm, and concluded that it would uplift our oceans by nearly 23 feet.

About Data

The literature review supports the oft-heard assumption that conversion of grassland to cropland leads to losses of soil carbon stocks, and that conversion of cropland to grassland enriches soil carbon stocks. Nonetheless, results can be adduced that support the opposite assumption.

Greenhouse Gas Inventory, Germany, 1990–2007

I was the first one back in 2002 to tell the truth about the global warming stuff and all of that. And my own granddaughter came home one day and said "Popi . . . , why is it you don't understand global warming?" I did some checking and . . . the stuff that they teach our kids nowadays, you have to un-brainwash them when they get out.

Sen. Jim Inhofe* (Republican–Oklahoma), 2016

As for that CO_2 . . . well, do you remember Wylie's ocean of *a million compounds in stock that were stirred together, outcome unknowable*? Our atmosphere was another such ocean. With respect to pollution a chemical technologist advised us: *Modeling only the gas-phase dynamics over an area like Los Angeles requires solving about 500,000 simultaneous, nonlinear equations.* So how could you complain if we didn't get it right?

Hindsight is always sharpest, they say—especially once the seer wakes up to the fact that past actions are cooking him alive! In this book I shall doubtless make plenty of errors, for which I ask your forgiveness. My data sources will never be in perfect concord; quantifications of the menace of methane altered but remained unsettled during the writing of *Carbon Ideologies;* coal's categorizations varied down the decades—and I, well-meaning but sometimes clumsy with numbers, may have dropped a zero or three in my conversions, even though I reviewed them twice. My handheld radiation meter evinced during calibration a tendency to underreport low-level energy fields by 4% (in the red zones of Japan it probably did better). Underfunded, ageing and absent-minded, crossing my fingers that my paid math-checkers and unpaid science Samaritans would save me, I did what I could; and if time proves my data wrong, I hope that my comparative method of cost-benefit analysis will remain useful to you.

---

* "In 2015 he brought a snowball onto the Senate floor to claim [*sic*] global warming was a hoax."

Why did I believe in global warming? Regarding its tally of slowly but re-morselessly increasing annual mean temperatures, the 2006 *CRC Handbook of Chemistry and Physics* swore that *the 95% confidence interval . . . since 1951 is +/− 0.12° C.* The compilers were, after all, scientists. Had they assured me with 95% confidence that the Earth was flat, well, I would have had to believe that, too.

And so, at least on my own behalf, I ask you to regard the follies of the 20th century with some tolerance. The two opposing entries in the 1976 *Britannica* are evidence that we "didn't know."

But then . . . *I mention the latest research . . . in order, if possible, to eliminate discussion of CO_2 from the Conference.* For your sake (and mine), the bias should have been the other way.

Moreover, *it is fairly clear that no serious climatic changes will result from fossil fuel use in this century. For the present, fossil fuels will continue to be used . . .*

What did the data actually reveal, back in that day? If you like, tell the story differently than I! Asking these ghosts from our past for certainty and accuracy about climate change is demanding too much. Their data disturb me, in a blurred and muted fashion. Their *comments* on the data disturb me more.

About Data Suppression

We can be silent witnesses to terrible injustices if we think that we can obtain significant benefits by making the rest of humanity, past, present and future, pay the extremely high costs of environmental deterioration.

Pope Francis, Encyclical Letter of 2015

Someone of goodwill might adduce that up until a certain date (if you like, 1967, 1982 or—stretching the case—2007), the dangers of greenhouse gases remained debatable even to future-thinking educated persons. Had *resolute certainty* come earlier rather than later, the human race might have preserved many of its privileges. Every now and then a question of a certain mournful interest used to hover around my face: How much was known *and deliberately concealed* by carbon ideologues?

In this book you will read about serious accidents involving all of our four major fuels.* In each case, the magnitude of the pollution was grossly understated in the newspapers. Had it been overstated half the time, I would have been less suspicious. Even as matters stood, that hypothetical individual of goodwill might have sighed: "Oh, well; they probably did it to prevent panic."

But selfishness is near about as reliable as gravity.

For 18 years, a Mr. Honma Ryu worked on behalf of Japan's second largest advertising agency, Hakuhodo. Here was the procedure allegedly followed by organisms of his pay grade:

> If an incident occurs in a factory or a plant and the press reports it, Dentsu† directly intervenes and visits the business department of the newspaper in question . . . We ask them politely to try to speak less about the case, not to put the article on the front page, or [else] to publish it in the evening paper which is less read.

---

* See "What Did Coal, Nuclear and Oil Have in Common?" (II:171). For natural gas, see II:319 and II:337.

† This was the nation's largest agency, about whose dealings Mr. Honma remarked: "I know very well how this happens . . . I did the same thing when I was at Hakuhodo."

Mr. Honma asserted that Dentsu *holds 80% of the market for nuclear advertising in Japan.*

Five years after the Fukushima disaster, which *Carbon Ideologies* will discuss in detail, pro-nuclear advertisements began to appear in the Japanese press. *The message of all these advertisements is identical, revealing the hand of Dentsu behind the scenes . . . According to Honma Ryu, . . . they are part of a campaign to closely monitor all information published on nuclear power, as well as to quasi-guarantee that local newspapers will limit the voice of opponents.*

What if carbon ideologues of a profit-minded stripe had made visits to opinion-shapers, decade after decade, and asked them politely to speak less about this book's theme?

If they did a good job, their efforts would have passed unrecorded.

About Disbelief

None of us really believe it's ever going to happen—not to us.

Nevil Shute, *On the Beach,* 1957

B ut let me exculpate those concealers a little, by further implicating myself: Even as the data aligned with ominous consistency—and, yes, even when the actual *effects* upon the climate began to obtrude themselves upon our coarse perceptions, my heart went out to the ones whom we cruelly called "global warming deniers"—not to all of them, of course, not to those who were bought and paid for, but unhesitatingly to people, who, like me, had not "done anything wrong." I wished them to be correct! *I mention the latest research . . . in order, if possible, to eliminate discussion of CO_2 from the Conference.* Having denounced that procedure because it propagated a dangerous error, let me say again: If only we could have burned carbon harmlessly forever!

A physicist once explained: *No theory can ever be* proved *to be correct; it can only be proved to be* incorrect.

And indeed, the deniers and disbelievers kept saying: "You can't prove that global warming is real!"

But that physicist also remarked: *A theory is of little value if it cannot make precise and unambiguous predictions.* And if it could, then it did have value. The theory of interest to us predicted that if atmospheric carbon levels rose, mean temperatures would go up.—But if they went up trivially, or imperceptibly, why not *eliminate discussion of CO_2 from the Conference*?

In August 2017, Hurricane Harvey struck Houston, whose flooding was shocking, Biblical, overwhelming, unprecedented, etcetera. Science had predicted that a warming atmosphere could hold more moisture, increasing the likelihood of heavy rains. That prediction was unambiguous but not necessarily precise. Had global warming *caused* Harvey, or merely enabled it—or did global warming remain a "theory" with plausible but not definite correlations to this particular storm?

If the theory ever grew aggressively convincing, then Houston's harried city planners might be compelled to limit new "development" across runoff-absorbing wetlands. You can't blame them for avoiding that! Although they would surely avoid the term *climate change,* they might start implying that there were changes in the climate. (Robert Louis Stevenson, 1889: *It is a strange art that can thus be practised; to talk for hours of a thing, and never name nor yet so much as hint at it.*) They might even have to dike and retrench.—You from the future must be laughing, to read that they hoped to pay so little!—But for a fact, there remained no immediate danger, not even then in 2017, and certainly not 37 years earlier, when members of that sympiosium in Montréal had been soothed by Dr. George Kukla to the effect that carbon dioxide *may have a cooling effect.*

The measurements at Mauna Loa Observatory, far off and irrelevant, were of a piece with the farther-off wearisome geometries deriving from a brain which once mathematized in a cruder observatory at Torun; I am thinking of Copernicus, that cautious solitary who built on the geometries of long-dead stargazers in a treatise (1543) whose iconoclasm was ambiguous and in fact has been called a last effort to "save the appearances," of planets and stars, including our own sun, orbiting around us in perfect circles (or, if need be, circles within circles), for the sake of God's glory, in which we, thus centered, had up until now been comfortingly included. Copernicus uncentered the Earth, setting it in explicit motion around the sun—and the inconspicuous unraveling of Aristotle and Scriptural literalism became shockingly manifest. Cautious and pious, he preserved what little he could: *And so the sun, as if resting on a kingly throne, governs the family of stars which wheel around.* In other words, the universe's center remained reassuringly near at hand. Then came Galileo, Kepler, Newton—and in 1851 Foucault's pendulum finally provided visible proof of our world's rotation. It is not for me to retell that long story here. But at the end of the 20th century, as the implications of the Mauna Loa measurements foamed in greater and lesser temperature-waves to attack our sandcastles, we became characters in a counterpart comedy: Another of our complacencies found itself condemned to die!—Well, it declined to. For the rest of my days and longer it struggled on, in the customary and even laudable fashion of any life-loving thing. It defied and rejected doom!

Before Copernicus, and for a good while after, we had believed in our sure and static place in the cosmic order. Determined to keep hold of that, some refused to look through Galileo's telescope. We accused Copernicus of heresy (fortunately for him, he had already died), and burned Giordano Bruno at the stake. Damaged and angry, we lived on, still grasping at a kindred notion: that "nature" was ours, for our use, without repercussions.—What was the work for? That was nobody else's business! And so in 2016 the Republican nominee for President, who like his predecessor *the decider* knew what needed to be true for the sake of inertia and short-term corporate welfare, explained to the electorate: "The concept of global warming was created by and for the Chinese in order to make U.S. manufacturing non-competitive."

If he were wrong, and our old way of working and being had to perish to save us from doing the same, how could we face the future? I lay in bed writing these words at dawn in my hotel in Dubai, watching the dusty sky glare ever more greyly in at me while my laptop, having sipped electricity on the other side of the world, made my words shine upon the screen as if I had asserted something important, and the air conditioner pulsed soothingly overhead. How could I write *Carbon Ideologies* if I didn't stay cool? Especially since I felt so jet-lagged just now—and don't tell me about the carbon dioxide I had spewed jetting over here—because I was officially on assignment for a magazine!

It would take awhile—far too long, I fear—for that convenient, selfish optimism of ours to die; longer still before the stink of its decay grew so intolerable that we had to bury it somewhere, after which we turned, resolutely I hope, to the ever more daunting requirements of our continued existence on Earth.

And I did not want to. I always hated funerals! Far easier to refute Copernicus, whom hardly anyone read, than to resist the evil temptation of peering through Galileo's telescope! Like the "global warming deniers," or any patient with a diagnosis of "terminal," I preferred to put off the bad days.

"Consider It Good Fortune"

We may be moving now towards a warmer phase. Half a million years hence this may be a much sunnier and pleasanter world to live in than it is to-day.

H. G. Wells, *The Outline of History*, 1920*

. . . and he felt reassured by the thought of New York in its ring of sacred fires, the ring of smokestacks, gas tanks, cranes and high tension lines.

Ayn Rand, *Atlas Shrugged*, 1957

So we believed what suited us.—If winters were going to be colder while the climate would get warmer, on the average life would remain average, which is to say accompanied by a reliable rise in Gross Domestic Product. And somewhat as GDP went up, which was usually gently and arithmetically, so upslanted that graph of monthly average carbon dioxide concentrations at Mauna Loa: a line, more or less, not a curve.

You from the future know all this, so I will trouble you merely with the first number and the last:

CARBON DIOXIDE CONCENTRATIONS IN OUR ATMOSPHERE,

as Recorded at Mauna Loa Observatory in Hawaii,

1959 and 2004,

in multiples of the 1959 value

All levels expressed in [parts per million].

---

* "If so, there will be less room for us, . . . If our present-day ice caps were to melt, . . . London, New York, and all ports and lowland cities . . . would be drowned."

> ## 1
> ### 1959: 315.98
>
> ## 1.068
> ### 2004: 337.38
>
> *Source: CRC Handbook,* 2006, with calculations by WTV.

Quite innocuous! A 7% increase over nearly half a century!—I admit that by this point the United Nations "Millennium Ecosystem Assessment" had announced:

> A consistent decline in average species abundance of about 40% between 1970 and 2000 . . . Between 12% and 52% of species within well-studied higher taxa are threatened with extinction.

But why worry about some ecosystem out there somewhere? And who could finger greenhouse gases? Doomsters had already fingered deforestation, pesticides, overgrazing, acid rain and other such improvements to our planet, which, yes, we were turning into *a chemical warehouse factory, with a million compounds in stock that were stirred together, outcome unknowable.* And even if carbon dioxide *could* someday, somehow injure us, well, *it is fairly clear that no serious climatic changes will result from fossil fuel use in this century.*

Unfortunately, that figure for 2004, courtesy of the latest volume of the *CRC Handbook of Chemistry and Physics* available in my hometown library, was nearly 147% higher than the planetary average over the last 800,000 years.—Projecting back to 1750,* scientists worked out *preindustrial chemical equilibrium values of 280 ppm.* From 280 to 337 parts per million gets only to 120%! Well, the two climatologists who cried out their 147% had set preindustrial levels at 230 parts per million. So was the 1750 baseline *actually,* and what were we going to call acceptable? This question will occupy us when we visit the contaminated zones of Fukushima: What was "normal background" radiation supposed to be, and how high a dose would prove "dangerous"? Some ideologues called making such determinations practical and necessary; counter-ideologues called them fiddling while Rome burned.—But we could breathe as well as ever before, and the climate seemed hardly altered beyond the bounds of everyday weather fluctuations . . .

---

* Thanks to core samples drilled out of polar ice. The heavy compression of millennia trapped bubbles of ancient air, whose concentrations of greenhouse gases were measurable. Moreover, the relative abundance of certain oxygen isotopes in those bubbles corresponded to specific atmospheric temperatures.

Back in 1965, when some people still believed that Venus might be cool enough to support the existence of swamp monsters, a certain Roger Revelle advised the American government that atmospheric carbon dioxide levels might increase by 25% at the century's end. What an alarmist! For the third time I insist that the rise didn't even reach 7%!—Unfortunately, after 2004 it steepened. In a decade it achieved 26.5%—400 parts per million. The last time our planet had seen such concentrations, the oceans were 82 feet higher.—And so Revelle was correct after all, although to be sure he had gotten 10 years ahead of himself.*

But if CO_2 concentrations had been rising since before I was born, why was there no perceptible warming during most of my lifetime? Two climatologists explained: *In earlier years, other factors were stronger than the relatively small forcing from carbon emissions.*

It could have been as early as 1951, although it might not have been until 1970, or even 1971, that the Earth began to absorb more heat than it sent back into space. But *it is* virtually certain, concluded climate scientists, *that the Earth has gained substantial energy from 1971 to 2010. The estimated increase in energy inventory between 1971 and 2010 is 274 . . . × 10²¹ J[oules] (high* confidence), *with a heating rate of 213 × 10¹² W[atts] . . . over that time period.*

By my calculation, 274×10^{21} joules equals something like 260,000 quads. But why worry? As you may remember,[†] in 1981 alone, global energy production had been 285 quads. So this *increase in energy inventory* totaled merely 919 times more than the entire global energy production of 1981! It was as if we had burned 10½ trillion tons of West Virginia coal in the course of those 39 years—174,000 tons each second—well, thank goodness global warming remained entirely debatable!

CARBON DIOXIDE EMISSIONS FROM FUEL COMBUSTION, WORLD AND SELECTED COUNTRIES, 1971 AND 2014,

in multiples of the U.S. percentage increase over that period

All figures are in [metric] megatonnes (MT). 1 MT = 2.2 billion lbs. To approximate total carbon dioxide emissions, add 30%.

---

* Another prediction: "Revelle estimates that the sea level would rise 400 feet if all the antarctic [*sic*] ice should melt, and this could occur within as short a time as 400 years." But how could such a pessimist be right twice? To avoid causing apprehensions, I hereby bury his gloomsaying in this footnote.

† Above, p. 87.

All listed nations will figure significantly in this book.

Each header is a multiple of the American increase of 21%. All figures over 10 rounded to nearest whole digit.

1
U.S.A., +21%

1971:	4,288.1
2014:	5,176.2

+2.76
Japan, +58%

1971:	750.7
2014:	1,188.6

+6.29
World, +132%

1971:	13,942.2
2014:	32,381.0

+17
Mexico, +360%

1971:	93.7
2014:	430.9

+98
Bangladesh, +2,048%

1971:	2.9
2014:	62.3

+329
United Arab Emirates, +6,916%

1971:	2.5
2014:	175.4

−1.23
Germany, −26%

1971:	978.2
2014:	723.3

Source: International Energy Agency, 2016, with calculations by WTV.

In 2016, when I was finishing *Carbon Ideologies,* the atmosphere's CO_2 concentration struck 407.42. That was nearly 29% over 1959—not to mention 43% over 1750. As you know far too well, it kept climbing.

No, the graph of annual global mean temperature from 1856 to 2004 was definitely not a line anymore, but a rising curve—although it had goodheartedly consoled us with an awful lot of jigjaggings up and down, varying from below 13.5 to as high as 14.6° Celsius, so that we could plausibly postpone certainties and decisions.—In 1856, the world average was 13.621° C. Not until 1938 did it first cross the line above 14°, after which it dipped lightly back down into the 13s no less than 10 times, sometimes for years on end, the last occasion being 1978, two years after that pair of *Britannica* entries were published.—How could I blame the *Britannica* anymore?

A.D. 2004 found us averaging 14.455° Celsius—less than a 6% temperature increase over 148 years! Like carbon dioxide concentrations, that was pretty negligible; why not keep believing in cool summers and snowy winters?

In Tokyo, annual mean temperatures between 1876 and 2011 inclusive rose tremblingly when plotted as "11-year running means" and steadily when idealized into a "long term trend"; either way, they had gone up from about 13.25 to 16.5° C. The considerable time over which these observations had been logged gave them worrisome credibility. (To be sure, one must give fair credit to the famous "heat island" phenomenon of Japanese urbanization: dark pavements, concentrations of body and engine heat, local leaks of waste heat from air conditioners, power plants, etcetera, all made central Tokyo warmer than the ricefields, villages and managed forests of the periphery.) In 2015 the Meteorological Observatory predicted that by century's end *about [a] 3° C increase will be seen in ... average temperature ... Tokyo (15.4° C) will become [like] Kagoshima (18.6° C).*—Well, but which figure should we name correct? Was Tokyo's present temperature 16.5° or 15.45°? Asserters of snowy winters loved such contradictions.

At the Massachusetts Institute of Technology they compiled *a world manufacturing CO_2 plot,* concluding that *the pattern shows a period of gradually increasing CO_2 emissions from 1970 to 2002, followed by a relatively sharp rise after 2002.* What if mean temperatures followed suit?

Very possibly we were ruining ourselves! Grieving on your own behalf for the best of reasons, you from the future might (if you are insufficiently embittered) make the mistake of grieving for us. But as a great novelist pointed out:

> We are inclined sometimes to wring our hands much more profusely over
> the situation of another than the mental attitude of that other, toward his
> own condition, would seem to warrant. People ... suffer, but they bear it

manfully . . . We see, as we grieve for them, the whole detail of their blighted career, a vast confused imagery of mishaps covering years, much as we read a double decade of tragedy in a ten-hour novel. The victim, meanwhile, for the single day or morrow, is not actually anguished.

That describes our period's optimistic American President, George W. Bush, the one who had styled himself *the decider*. When I was alive we mostly remembered him for launching the wicked and disastrous second Iraq War. (He bore that victory manfully, all right.) But another of his decisions proved yet more brilliant. In 1997 the U.S. had agreed to join 36 other developed nations in reducing greenhouse gas emissions per the famous Kyoto Protocol. Reneging on his campaign promise, and blindsiding his own nominee at the Environmental Protection Agency, *the decider,* having been warned by his political "base," decided against ratification. As he explained: *I did not believe . . . that the government should impose on power plants mandatory emission reductions for carbon dioxide, which is not a pollutant under the Clean Air Act.* No, he wasn't anguished at all! The *Times* of London concluded: *The country that emits 25 percent of the world's carbon dioxide with less than four percent of its population is not going to slow down.* Well, and why should we? Who could prove that scientists knew more about science than *deciders*?

For one thing, global warming continued to mask itself. One might suppose that if temperatures were creeping up all over the world, then new heat records should be trumping the old ones. But for a long time we managed not to go the way of Venus. Some heroic meteorologist had measured the world's highest thermometer reading—58° Celsius, or 136° Fahrenheit—in Aziza, Libya, on September 13, 1922. The highest reading in California*—134°—occurred in 1913. At that time there were 2 billion humans on Earth. Come the new millennium, with our population between 6 and 7 billion, that Libyan reading remained unsurpassed.[†] Some stretches of cool years scientists later *associated with a transfer of heat from the upper to the deeper ocean*—but that was merely some hypothesis about some ecosystem somewhere—and two climatologists at the conservative Cato Institute reassured us: *Consider it good fortune that we are living in a world of gradually increasing levels of atmospheric CO_2. The effects of this increase on food production are far more important than any putative change in climate.*[†]

---

* As of 2013.

[†] In fact, not long before *Carbon Ideologies* went to press this number actually *went down*—for due to "potentially problematical instrumentation," the World Meteorological Organization revoked this first prize, giving it instead to that reading of 134° at Furnace Creek, Death Valley, 1913.

[†] As Pope Francis remarked in his encyclical: "We can note the rise of a false or superficial ecology which bolsters complacency and a cheerful recklessness. As often appears in periods of deep crisis which require bold decision, we are tempted to think that what is happening is not entirely clear . . .

Indeed, it was good fortune for maize, beans, rice, wheat, soybeans. Those crops loved carbon dioxide. Meanwhile, in 2015 and then again in 2016, delightfully gentle winters in the state of Maine extended the season for maple syrup production. Now, wasn't *that* lucky?—Of course the day might come when Earth was too hot for any plants at all. But not yet, my friends—not (we hoped) while *we* were alive! And so one highly successful fracker and CEO (who for all his evident enjoyment of public pontifications declined to be interviewed for *Carbon Ideologies*) cited Richard Tol, *a prominent economist* with Sussex University, who had conveniently concluded that climate change would be beneficial up to a warming of 2.2° C, as measured from 2009.*

Just the same, our farmers began to watch for aphid-transmitted viral diseases. Five more generations of aphids every year (the result of shortening winters) would keep our agronomists on the defensive. Tomato necrosis and citrus tristeza were already worsening on the Iberian Peninsula, *where warming is double the global average.*—Well, who were we to complain, compared to you? I wonder what you find to eat—while in my day the weather was frequently droughtless and stormless, and we could still call on bee-pollinators . . .

Gradually, other signs loomed up. Agricultural statisticians noted: *November 1999 will go down in the record books as the warmest November ever in many Minnesota communities.* But what was Minnesota to the rest of us? So we laughed that off.—In Canada, six main glaciers *have been shrinking since standardized measurements of their mass began at various times during the 1960s and 1970s.* Now it was 2010, and they were still receding.[†] What that country's chief statistician delicately called "changes" in glacial mass had come to be *considered among the most robust indicators of climate change.* Why, those shivering Canadians ought to be grateful!—The Central Intelligence Agency's *Factbook* for 2012 dared to define desertification as *the spread of desert-like conditions in arid or semi-arid areas, due to overgrazing, loss of agriculturally productive soils, or climate change,* and presently *Time* magazine saw fit to run the following little item in the "Trending" column: *The contiguous U.S. experienced its warmest winter on record in 2015–16, with average temperatures 4.6° F above the 20th century average* . . . But that was not so bad, either! In January I went for coffee and the bakery girl remarked happily on the pleasant sunny weather. Her life was getting better, not worse.

This is the way human beings contrive to feed their self-destructive vices: trying not to see them, trying not to acknowledge them, delaying the important decisions and pretending that nothing will happen."

* Perhaps he moonlighted as a prominent climatologist.

† Coincidentally, from 1979 to 2007 the world's carbon dioxide emissions from fuel combustion alone rose by 105%. And out of neighborliness Canada had done her mite: 339 million metric tons in 1979, 573 million in 2007.

May 28, 2017, marked *potentially the hottest temperature ever recorded in Asia:* 129.2° Fahrenheit, datelined Turbat, Pakistan. But that still fell short of the Libyan reading from 1922! *Sixteen of the 17 hottest years ever recorded have occurred since 2000, scientists say, and 2016 was the hottest since modern record-keeping began in the 19th century* ... Well, but if they were the hottest, why didn't they leave Libya in the shade? Our arithmetic being democratic, we could declare that a lonely maximum proved less than an average, and so we got along in life.

And *consider it good fortune that we are living in a world of gradually increasing levels of atmospheric* CH_4. Well, not even the Cato Institute said that. The way to talk about methane was: *Consider it nature's fault.*—In 2001 a British chemist comforted us that while rising carbon dioxide (and monoxide) concentrations did come *from fossil fuel burning,* methane, which molecule for molecule was far more dangerous, derived *from paddy fields and cows.* Now, how could *those* be problematic? And the chemist summed up:

> Human contributions ... are still a minor component compared with natural sources: most carbon dioxide comes from plants, microbes and animals, while methane is given off by swamps, marshes and termite mounds.

Much the same could be said about sulfur dioxide: *Global climate change, prior to the 20th century, appears to have been initiated primarily by major changes in volcanic activity.* Again we could plead innocent.

By 2007, the Intergovernmental Panel on Climate Change grew

> very clear on the following point: observations and measurements unambiguously indicate that the climate system is warming and that humans are primarily responsible for this trend. And the trend has intensified in recent years ... If dangerous impacts of climate change are to be prevented, global warming must be constrained to no more than 2° C in comparison to preindustrial levels ... By 2050, global emissions will have to be reduced by 50–85% in comparison to relevant levels in the year 2000.

As you know, we pushed them the other way. Consider a certain **Table 5: Global per capita consumption of energy, materials, products, and services in 2010, and average annual growth rates for these goods ... :** *The table shows positive growth in every variable, with GDP, energy, and emissions totals growing at ~1.5% per year, but with all industrial measures ahead of this ...*

Carbon Ideologies Defined

When would this thing seem real? Maybe it seemed real in the big cities,
but his worm's-eye view frustrated his curiosity and sense of drama.

C. M. Kornbluth, *Not This August,* 1955

You see, some of us wanted to keep blow-drying our hair, while others enjoyed buying cheap sludge oil and selling it to the neighbors as better stuff, and a few bright actors simply hated to stop making titanium. So what were we supposed to do?

Many of this question's contradictory answers derived from the various ideologies of energy production, which in honor of their most quotidian, useful and ultimately infamous member I will refer to as "carbon ideologies." To be precise, they were "ideologies about carbon." Some of them favored the production of carbon emissions, and the rest opposed it. By my definition, nuclear power, solar energy, coal mining, oil extraction, fracking, tidal electricity and wind turbines could all be named carbon ideologies—for "what to do about carbon" was becoming the question of the day—or at least a question to yawn over, when we were not judging some actress's latest facelift.

Mr. Sam Hewes, Vice-President of the Bank of Oklahoma, once told me:* "I think that we're all ideologues, and we all have a set of beliefs. I have learned that when you in any way attack somebody else's reality, they're either gonna leave you or attack you. I think, because of uncertainty and not really knowing, there are people that believe fully, one side or another side. So I think, the winner's gonna be the people that are gonna convince the others that they are right, although they don't know, either. Those that think that global warming is an unbelievable issue and we're almost at the tipping point where we can't recover, there the model that they have in their head says that we have to do everything that we can to shut down oil. I see this in churches all the time. They make decisions not based on facts and then they call the other side illogical, just because they don't

---

* The rest of his interview begins on II:486.

104

believe the same thing. Well, the other side are not irrational. They're just . . . Science is the new religion."

Or perhaps our religion was *prosperity*.

How should people live?—For a long while most of us had answered: Not by bread alone, but also by electricity.*

Coal, the reliable workhorse, ubiquitous oil and the new ideology of natural gas all promised to keep life easy. Nuclear power offered to solve the matter by producing energy without creating greenhouse gases. Ideologues of presumed good faith advanced their respective agendas by way of the free market. We all bought what they sold.

---

* Even utopia had to have it. Aldous Huxley, 1962: "Lenin used to say that electricity plus socialism equals communism. Our equations are rather different. Electricity minus heavy industry plus birth control equals democracy and plenty. Electricity plus heavy industry minus birth control equals misery, totalitarianism and war." And so the utopians burned carbon.

About Carbon

When they have discovered truth in nature they fling it into a book, where it is in even worse hands.

Georg Christoph Lichtenberg, *Notebook E,* 1775–76

Who would I be to denigrate carbon?—Not *myself*!—for by mass it was the second most common element of my own body, with only oxygen being more conspicuous. (Consider in opposition cesium, whose radioisotopes played such baleful roles in Japan. When we were alive, that element made up two millionths of a percent of a typical human carcass. *Your* body surely contains more.) As one chemist rhapsodized: *No element is more essential to life than carbon, because only carbon forms strong single bonds to itself that are stable enough to resist chemical attack under ambient conditions.* Those bonds, like our economies, could extend almost indefinitely onward. Carbon compounds thus became so numerous* and interesting that we devoted an entire subdiscipline to the carbon family: organic chemistry. And by the way, *combustion to carbon dioxide and water is characteristic of organic compounds.* Prominent among the latter was that large and thriving clan, the hydrocarbons—which is to say, substances composed strictly of carbon and hydrogen atoms—for instance, that dangerous greenhouse gas CH_4, more popularly known as methane.†

An oil geologist explained: *Molecules with five to twenty carbon atoms are liquids: crude oil. Molecules with fewer than five carbon atoms are . . . natural gas.* There you have it.

From our fossil fuels we made new substances almost at pleasure, festooning atoms to our carbon chains. Production procedures could even be altered based on the latest market price of this or that hydrocarbon. Example: One common type of acrylic plastic showed off 15 carbon atoms per molecule. Our chemists could have synthesized it out of either oil or natural gas. Didn't that prove our intelligence?

---

* From a textbook: "The number of compounds that contain carbon is many times greater than the number of compounds that do not contain carbon."

† See "About Methane," II:307.

In our planet's crust, carbon made up a respectable 200 parts per million—another source insisted on 480 parts per million—and in seawater (at least up near the ocean's surface), it showed itself (as carbonic acid*) to the tune of 28 milligrams per liter. But throughout the days of my life, each liter held less carbon than it used to—because the solubility of carbon dioxide lessens in warmer water. As one scientist put it: *Ocean temperature is . . . the primary natural control for atmospheric CO₂ concentration.* In other words, as the temperature of our oceans crept up, they released more greenhouse gases, which further heated the planet, forcing still more carbon dioxide out of solution—a pretty feedback loop.

Accordingly, carbon dioxide rose up out of our water; it thickened in our air. *Humans burning fossil fuels and manufacturing cement are releasing 7.8 gigatons of carbon into the atmosphere yearly.* So I once read, back when I was alive. Later on we released more.

Since fossil fuels occupy most of *Carbon Ideologies,* permit me to leave them alone for now. As for cement, we used to make that stuff in order to thicken our traces upon the earth. In olden times we took silica, and later on we used silica and alumina together, to render lime more adherent and waterproof. Lime comprised the mortar in our cement, and this recipe from 1910 tells how we made it: *When pure chalk or limestone is "burned," i.e., heated in a kiln until its carbonic acid has been driven off, it yields pure lime. This slakes violently with water, giving slaked lime, which can be made into a smooth paste . . . and mixed with sand to form common mortar.* Upgrading sand into amorphous silica, we obtained our uncommon mortar, which was cement. (*Concrete* is a paste of water, aggregate—sand, crushed slag and the like—and cement, which then sets into a hard construction material of high compressive strength.) The liming process was easy if slow, and cheap as long as we declined to count the cost paid by you in the future after the carbonic acid went invisibly upward—0.785 pounds or more for every pound of lime produced.† It was nothing; it was only carbon.

By the way, the CO₂ emitted by roasting limestone did not complete the tally of our carbon expenses in this process—because heating each pound of the stuff into lime cost us 3,800 British Thermal Units, which we could pay in half a pound of dried white pine, or three ounces of heavy grade commercial fuel oil, or five ounces of West Virginia coal, cash on delivery and smoke up the chimney!‡

---

* This term was once used synonymously both for carbon dioxide and for the compound it made when dissolved in water.

† German figure from 2007.

‡ For the heating values of these and other substances, see the table beginning on p. 208.

Between the commencement of the Industrial Revolution and 2011, we offered our planet eight petagrams* of cement-smoke carbon—about 17.6 trillion pounds. But why fret? That was only 1/47 of our total CO_2 emissions.

In 2008, Canada's cement works sent up 6,600 metric kilotons of carbon dioxide—merely 1/83 of her total contribution to atmospheric progress . . .[†]

That year, Portland cement was Oklahoma's second most valuable product of "non-fuel mineral production,"[†] so why would the Sooner State stop making it? Cement-making released more CO_2 than any other American industrial business except for the steelmaking trades. The reason was argue-proof: *Cement continues to be a critical component of the construction industry.*

I myself was an enthusiastic consumer of this stuff. In the final months that I was drafting *Carbon Ideologies,* it happened once again that "good fences make good neighbors," so after his strong and gentle helper had broken regularly spaced deep square holes, partly with a pickaxe and partly with a drill, through the asphalt of my parking lot, my contractor friend towed behind his truck a slowly rotating concrete mixer, and I came out to watch, admire and help place the steel poles, after which the mixer spewed grey slurry into the wheelbarrow, and the helper began to shovel it down around each pole, the contractor pounding and stirring air bubbles out of the stuff, then tamping it just right. Once it had set, he would pound in the polecaps and weld on the downcurving double ribbons of heavy chain, so that no one (I hoped) could dump garbage in my parking lot anymore. Enjoying the company of those strong men who did what I could not, I listened first to a tale of an ancestor who had invented and possibly patented an improved concrete mixture containing ground-up oyster shells, then to a sketch of some early morning alcoholics who refused to guarantee their own cement (at which the contractor demanded, "Well, what *can* you do?"). Then the helper, happy to have gotten this extra work in order to buy Christmas presents for his family, patiently gathered up the excavated dirt, while the contractor drove away to return the mixer.—How much had that operation damaged the planet? (I fall back on the words of Pastor Blevins in West Virginia: "How could manmade equipment put up enough smoke to make a difference?") To begin to calculate that, I would need to know how much cement I had paid for. I asked for the weight; the contractor said he had no idea; anyhow

---

* Eight quadrillion grams.

† Counting only carbon dioxide.

† Number one was crushed stone at $534 million. I wish I could have told you more about Oklahoma cement, but this category had been annotated: "Data withheld to avoid disclosing company proprietary data." In this book we shall often run up against the secrecy of carbon ideologues.

it came by the yard. So I dismissed that issue, preferring to find peace in my new perimeter.

In Japan, the single largest category of industrial carbon dioxide emissions was *limestone in . . . cement production: 40%.** And up went our smoke.

Unfortunately, cement-making was but one element in our highly obscure and heterogenous catalogue of misdemeanors. Hence the length of this "Primer" section. And smoke rose up.

How could we have locked it away again?—By planting more trees.—But we would never have wished to solidify it entirely. As a mid-20th-century soil textbook unnecessarily fretted regarding that contingency: *Without consumers and decomposers to release the fixed carbon, the atmosphere would be depleted of carbon dioxide, life would cease, and the cycle would stop.* Oh, we were releasing it, all right; we took our inspiration from Venus.

The consumers were those organisms that took in various forms of carbon— as carbon dioxide in the case of plants, and as sugars and such for animals. The decomposers were the insects and microorganisms that teemed in the soil, breaking down dead things, thereby liberating their stored carbon. The most prevalent organic compound on our planet (thanks to photosynthesis) was *cellulose.* Like sugar, it consisted exclusively of carbon, hydrogen and oxygen. Its basic formula was $C_6H_{10}O_5$, chained together a great number of times to make each molecule. A hundred pounds of straw contained 40 pounds of carbon. Bury it in dirt, and necrophores would consume it, emitting from their tiny bodies a good 26 pounds of carbon dioxide.[†] The rest stayed underground—for awhile.

Sewage sludge was 50% "degradable organic carbon." Diapers were 24% DOC in Germany and 70% in Japan. (Does this discrepancy say more about diapers or about comparisons?) Japanese plastic bottles were 62.5%. Can you guess where that carbon went?

In 1916, two scientists from Rutgers marveled to learn that every two days or so, provided that they were fed continually, certain bacteria could generate their own dry weight in carbon dioxide.—In short, we hardly needed to worry about depleting the atmosphere of carbon! Again, we should have planted more trees.

On this subject I want to say that *circa* 2015, the forests of the contiguous

---

* 2013 statistics.

[†] A strange parallel, courtesy of an EPA report on American greenhouse gases: "Overall, throughout the time series and across all uses, about 60 percent of the total C consumed for non-energy purposes was stored in [manufactured] products, and not released to the atmosphere; the remaining 40 percent was emitted."

United States sequestered 173 teragrams* of carbon per year, *offsetting 9.7% of C emissions from transportation and energy sources.* That was insufficient; we needed to extend our forests. Hence the following: *We project a gradual decline in the forest C emission sink over the next 25 years . . . to 112 teragrams of carbon per year*—a reduction of 35%. (Our behavior was beautifully consistent.) *Sequestration in eastern regions declines gradually while sequestration in the Rocky Mountain region declines rapidly.*

. . . But what *was* carbon? Does it help to learn that the 1976 *Britannica* called it *a true nonmetal in every sense,* or that its melting point exceeded 6,500° Fahrenheit, by which point platinum and even tungsten had long since puddled?—Two technologists expressed a more magical fact: *Essentially any organic material can be thermally transformed to carbon.*

Table sugar ($C_{12}H_{22}O_{11}$) tastes rather different from anthracite—but pour sulfuric acid over it, and it will begin bubbling and simplifying, all of its hydrogen and oxygen molecules combining into steam (for 22 H's and 11 O's transform perfectly into 11 H_2O's), until only carbon atoms remain. The sucrose has gone gold, then brown, and finally black. The sulfuric acid, having done its catalytic part, lies in a puddle from which rises a ready-to-burn island purer than coal!

Not all such transformations need be thermal. Photosynthesis makes oxygen out of carbon and light; respiration and decomposition produce carbon dioxide.

Ancient decompositions, deep-buried so that their carbon remained trapped, gradually became hydrocarbons. Hence the fact of relevance to this book, again courtesy of the 1976 *Britannica*: *Carbon is widely distributed as coal and in the organic compounds which constitute petroleum, natural gas, and all plant and animal tissue.*

Carbon was ubiquitous, innocuous and sometimes even precious. It could be black or transparent. As the 1911 *Britannica* explained: *It is found native as the diamond . . . , graphite . . . ,*[†] *as a constituent of all animal and vegetable tissues and of coal and petroleum. It also enters (as carbonates) into the compound of many minerals, such as chalk, dolomite, calcite . . . and spathic iron ore.*

The *Britannica* continued: *It is also found to a small extent in the atmosphere.*

A small extent—how could anything be wrong with that? Don't you see why we took our world so easy, back when we were alive?

---

* A teragram is a trillion (10^{12}) grams. Hence the amount of carbon mentioned was 381.4 billion pounds.

† A third native form was later discovered: buckminsterfullerene.

About Agriculture

In a real way the energy for potatoes, beef, and plant produce of intensive agriculture is coming . . . in large part from the fossil fuels rather than from the sun.

H. T. Odum, 1967

My crop of corn is but a field of tares . . .

Chidiock Tichborne, 1586

I n 1971 an ecologist warned: *Man is now unwittingly beginning to speed up decomposition (1) by burning the stored organic matter in fossil fuels and (2) by agricultural practices that increase the decomposition rate of humus.*

Both actions, of course, kept increasing atmospheric carbon dioxide concentrations.

In 1970, our busy human doings sent 8 billion tons of that gas upward. Six billion tons were the combustion products of fossil fuels. The other 2 billion tons derived *from cultivation of the land for agriculture.*

CARBON AND DECOMPOSITION

You see, we used to till the soil. We even supposed that we were thereby "improving" it!* We turned over the top, so that grasses and other plants of no value to us would die and decompose, nutrifying the dirt. We broke up clods, then plowed, because *the more finely divided the soil, the more readily plants grew.* Doesn't that sound harmless? *Pulverization changes the hard soil into a deep mellow seed-bed . . . It enlarges the feeding area of the roots, by placing more plant food and moisture within easy reaching distance . . . It promotes bacterial action in the soil, the fixation of atmospheric nitrogen by bacteria, and the change of plant food from the insoluble to the available form.*

And in those deep mellow seed-beds we planted and we fertilized. A

---

* I remember a sweet old illustration captioned "The Work of the Plow—The Greatest Labor of Mankind."

humus-grubber once explained that *soil is the interface between the living and the dead—where plants combine solar energy and carbon dioxide of the atmosphere with nutrients and water from the soil into living tissue.* We rendered that interface more permeable. More oxygen sank in, which facilitated crop production (plants need to "breathe in" oxygen through their roots—as do their aerobic root-bacteria); and more carbon gases rose out.

The carbon came both from the soil itself—which could contain anywhere from 0.08 to 38.0%—and from the respirations of those plants and bacteria, which "breathed out" carbon dioxide just as we did. Why did they? Let me quote a textbook from 1967:

> The least stable, hence most energy-rich, carbon combinations are the *hydrocarbon* groups* . . . The most stable carbon combination on the contrary is CO_2 . . . In general, therefore, we may predict that usable respiratory energy will result from conversion of hydrocarbons

into "hydrogen-free groupings"[†] such as CO_2. In other words, for exactly the same reason that a powerful hydrocarbon fuel such as gasoline gives up its energy rapidly when burned, meanwhile releasing carbon dioxide and steam, so an energy-filled hydrocarbon such as the fatty tissue of a dead mouse yields hydrogen atoms bit by bit to the ants and other teeming necrophores which it nourishes, gaining oxygen atoms in exchange, until it has been decomposed to pure carbon, carbon dioxide and water. The molecular bonds of carbon dioxide contain less energy than those of any fat; hence carbon dioxide will never turn back into fat of its own accord; only an input of new energy—from the sun, mediated by photosynthesis—can accomplish that.

(When we "consumers" were alive, we, too, stabilized substances by vampirizing their energies—sucking sugars out of lollipops, calories out of avocados, heat out of coal. Up went carbon dioxide.)

Some carbon became vegetable matter. *The first stage in the formation of plant tissue is the union of carbon with oxygen and hydrogen combined as water.* Indeed, where would cellulose be without carbon and its chains? And for as long as the plants were alive, that carbon remained "sequestered," as we had anxiously begun to put it. Then what? We picked up our crops and sent them away, with their carbon inside them.

---

* See p. 106.
[†] These compounds are called anhydrides.

And what about the husks and stalks, moldy oranges and wilted lettuce? *Mature plant residues that provide the raw material for microbial decomposition contain about 50 percent carbon.** Well, we composted them, which decomposed much of their cellulose into carbon dioxide and water.

As for the produce we fed on, we turned it into excrement, which accomplished similar changes. Either way, whatever fixed carbon our crops had preserved reentered the soil (or the water).—Oh, yes, occasionally we did burn them for fuel; up rose their carbon dioxide.

Carbon-rich soils inhabited by "fungal-dominated decomposer communities" and roofed with a diversity of vegetation best retained their carbon. That minimized global warming. Sometimes the carbon mineralized, which safely sequestered it. (Soil organic carbon has been called *the largest terrestrial C sink.*) Or roots might drink it in, transforming it to biomass. That was harmless, while it lasted. After its host died, decomposers would get it if they could—being always drawn to any dead thing: *Most consider that microbial life in soils exists largely under starvation conditions.* We tilled the soil precisely so that they could have at it! In short (I quote the *Encyclopedia of Agriculture and Food Systems*), cultivation *reduces the mean residence time* of carbon *as tillage speeds up decomposition and mineralization.* Accelerated decomposition meant more decomposers turning dead organic matter back into water and carbon dioxide. Sometimes our doings sent half the mineralized carbon straight up into the sky!

According to the Foundation for Deep Ecology: *Our crops . . . were beneficiaries of the energy released as nutrients stored in the carbon compounds in the soil now became available . . . And so it was at this moment that the carbon compounds in the soil were exposed to more rapid oxidation. Carbon dioxide headed for the atmosphere . . .*

Why not send up even more of the stuff? (In our way we were ingenious.) So we fertilized the soil which we had tilled. (Indeed we needed to, because there were getting to be too many of us.) More rapid oxidation meant more rapid metabolic activity, by worms, bacteria and plants. Tilling shuffled around the carbon and oxygen, but that wasn't good enough. In the words of *Garden Manures and Fertilizers* (1949):

> All the chemical and biological processes are going on so much faster in the cultivated than in the uncultivated soil. All our cultural operations are designed to speed up growth and to obtain a high return of fruit,

---

* The same goes for dead animals.

flowers or vegetables . . . This has to be paid for . . . by replenishing the
soil with manures which restore and maintain the organic-matter
status . . .*

So we paid—as did you. Assessment from 1994: *Conservative estimates indicate that about 30% of world food production is directly attributable to fertilizer use.* Well, what were we supposed to do? For God's sake, did you expect us to distribute resources fairly, or practice birth control?

Ready to hand among our repertoire of fertilizers lay, for instance, ammonium
sulfate, which according to *Garden Manures and Fertilizers* "supplies nitrogen"—
one of the three essential substances *most likely to be deficient in the soil.*[†]—There
can be no proteins without nitrogen; indeed, I could have called this book *Nitrogen Ideologies;* and a handbook on sugar beet production explains: *Of all plant*

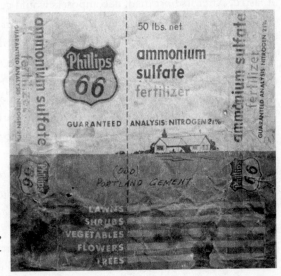

Ammonium sulfate fertilizer,
made by a famous petrochemical firm

---

* What if we didn't pay? A longterm study of Bangladeshi soils concluded: *Generally total carbon status was higher in 1967 than in 1995.* The same went for nitrogen. *Excessive utilization of plant materials as a source of fuel, animal feed, and other uses does not allow for adequate addition of organic matter into these soils.* This meant lower crop yields. Meanwhile, much of the "lost" carbon and nitrogen ascended into some ecosystem somewhere.

† The other two were phosphorus, which we derived mostly from apatite, and potassium, which came from various rocks and brines. "Volumewise, it is the production of N, P, and K fertilizers [in other words, the three just mentioned] that defines the industry." The manufacture of phosphate fertilizers released considerable carbon dioxide; in 2012, Finland emitted 42.4 gigagrams (almost 93.5 million pounds) of CO_2 while making phosphoric acid. Regarding potassium, the first patent in U.S. history, for a process to make the carbonate, burned up to 500 tons of wood to get less than a pound of the stuff. Modern potash-makers did not emit CO_2 directly during the refining process, but they usually employed fossil-fuel-intensive mining machines, centrifuges, etcetera. (For more details, see p. 52 in the sources for this volume.)

nutrients, an application of nitrogen ... has the most spectacular effect on the appearance of the crop. Leaf color changes from pale green or yellow to dark green. All ammonia compounds contain nitrogen. So bring on the ammonium sulfate! In 1973 the United Nations advised Thai farmers to employ the stuff as a top dressing for maize and a basal dressing for peanuts. We didn't stop there.

The nitrogen in ammonium sulfate (or for that matter in the gardener's old standby, bone meal) arrives in the soil as ammonia (NH_3), which soil bacteria transforms into plant-usable nitrates. But one waste product of those microbes is, as usual, carbon dioxide.

The merest tenth of a gram of ammonium sulfate in 100 grams of soil will increase its carbon dioxide production by 500%.* Two early-20th-century American scientists discovered that 1,000 pounds of this chemical applied to an acre of dirt produced far more than its own weight of carbon dioxide: 1,472.4 pounds, to be exact.

Come to think of it, our manufacturing procedure for ammonium sulfate began with the coal we had already roasted in coking ovens[†] preparatory to steel manufacture—and burned coal releases considerable carbon dioxide, not to mention that other greenhouse gas, carbon monoxide.

Another fine fertilizer is sodium nitrate, which *Garden Manures and Fertilizers* assures us can be *assimilated by plants at once.* The two American scientists found that 1,285 pounds of it would release 1,411.5 pounds of carbon dioxide per acre.

Ammonium nitrate, *highly regarded because of the rapid agronomic response,* accomplished comparable atmospheric wonders.

By the late 20th century we had laid pipelines of chilled anhydrous ammonia (82.2% nitrogen) from Louisiana all the way up to Idaho; another ammonia pipeline entered the Corn Belt from Oklahoma and Texas.

In 1991, and other years before and after, we managed to spread 85 million tons of nitrogen-based fertilizers on the fields of our planet. Shall we estimate their carbon dioxide pollution at, say, 119 million tons? And should it be relevant that at least half the fertilizers we deployed got *wasted*—leached and volatilized away?

When appropriately fertilized, tea-bushes emitted 4.7 times more carbon dioxide than did most other Japanese crops—and nearly 10 times more than paddy rice. Should we have stopped drinking tea?—We preferred not to. According to the "Fertilizer Outlook" section of the 2008 *MeisterPro Crop Protection Handbook:*

---

* "Previous drying" of soil releases still more carbon dioxide, so the droughts which global warming now began to bestow on our planet could be relied on to worsen matters.

† For a definition of coke, see p. 556.

Nutrient demand is on the rise worldwide. [Here my heart began to pound with excitement.] The world's population is forecast to increase to . . . 9.4 billion by 2050 . . . The International Fertilizer Association expects world fertilizer demand to increase by 3.1% to 159 million tonnes . . . in 2006–07, and continue to grow to 172 MT by 2010–11,* with most of the increase occurring in Asia.

AGRICULTURAL METHANE

Nor was CO_2 the only climate-changing excretion of our agricultures. Methane, or CH_4, which volume for volume turned out to be a more dangerous greenhouse agent than CO_2,[†] may be generated under anaerobic conditions. It particularly derived from rice.

A *Fertilizer Guide for the Tropics and Subtropics* (1973) proclaims: *Rice has always been the most important grain crop in the world, even more so than wheat.* Or, as the Khmer Rouge used to sloganize in totalitarian Cambodia: *When there is rice, there is everything.*—Like coal in Appalachia, rice in Japan was a cultural talisman. Its tending and consumption perpetuated a very particular carbon ideology. When I asked my Japanese interpreter how life without rice would be, she answered: "We can survive, of course, but we'll miss it. Rice paddies are our *original landscape.* Every Japanese has some kind of landscape that makes you feel nostalgic, and for me for sure, and maybe for many others, that landscape includes rice paddies."—Her "nostalgia" is represented in Saigyo's 12th-century tanka about being lonely in mountain moonlight, lost in a silence disturbed only by clackers scaring away birds from the ricefields; and in Yosa Buson's elegant 18th-century haiku about May rain: *paddy by paddy, it has turned into darkness.*—No doubt those paddies darkened so mysteriously because their farmers had flooded them. Rice thrives best in standing water, in which CO_2-producers drown. (This explains rice's prizeworthy restraint in carbon dioxide emissions.) The winning decomposers are anaerobic bacteria. Nourishing themselves on pure hydrogen—we yearned for our cars and power plants to do the same—they break down organic matter in their own bucolic fashion, exhaling its carbon as methane—maybe 40 billion pounds' worth, or, as might be, 200 billion. Those were 1997 figures.—In 2013, rice farming caused 50% of Japan's methane emissions.

---

* Respectively, 175 and 190 million U.S. tons.

† See "About Methane," II:307.

Do you remember that 100-pound bale of straw with its 40 pounds of inherent carbon?* Call it the poor man's ammonium sulfate. Some rice farmers liked to enrich their fields with rice straw before letting the water in. A report at a Filipino symposium judged that straw's *effect on rice yield has been greater than that of compost*—a happy encomium. As it decayed, straw nutrified the rice-roots with delicious nitrogen salts. This enrichment method was popular in Indonesia, where it increased mean methane emissions by 47% in one type of soil, by 87% in another. From Sumatra's paddies the methane pollution measured eight to 12 times worse, leaving Indonesian researchers to admit: *The amount of emissions may have been markedly underestimated.*

Not wishing to fall behind the rice farmers, dairymen aggrandized their herds wherever that would pay.† Each cow belched up (or farted, depending on which source you prefer) two liters‡ of methane per minute. Not surprisingly, 25% of all carbon-equivalent emissions from milk were enteric methane. Depending on how one computed methane's global warming potential, this might have been another marked underestimation.

NITROUS OXIDE, *or* THE VILLAINIES OF CAULIFLOWER

Meanwhile, from our manure-heaps—not to mention all those volatilizing nitrogen fertilizers—and for that matter even from our dishwater§—rose the greenhouse gas nitrous oxide, NO_2, which despite its lack of carbon atoms managed to trap solar radiation 264 times more efficiently than did CO_2. In fact, it was our third most dangerous greenhouse gas. Fossil fuels gave it off in varying quantities as they burned. From the tailpipes of our cars and trucks it became smog.¶ But I must not badmouth the stuff unreasonably—for my dentist once fed it to me as a "fun" anesthetic.

Food production caused 80% of all NO_2 pollution, thanks to nitrogen fertilizers—whose greediest clients were winter wheat, cauliflower and above all Brussels sprouts. I often sautéed those latter two vegetables in the wok my

---

* See p. 109.

† U.S. Environmental Protection Agency: *Enteric fermentation is the second largest anthropogenic source of CH_4 emissions in the United States . . . This increase in emissions from 1990 to 2014 generally follows the increasing trends in cattle populations.*

‡ Half a gallon.

§ The Germans aphorized: "Nitrous oxide emissions from household wastewater can be roughly determined via the average per-capita protein intake."

¶ Which the Japanese labeled "fuel combustion [stationary sources]." There went 21% of their nitrous oxide emissions.

daughter had given me; and as they sizzled over the ring of blue gas-flames I felt quite happy and healthy, as if I were doing something good.

Cauliflower vendor,
Pakistan

In Germany a round 1% of the nitrogen in their soil became nitrous oxide and got away. In Japan, the single greatest source of NO_2 emissions was "agricultural soils," which released 29%. "Manure management" gave off another 20%. With their customary breeziness, the Americans combined both categories into "agricultural soil management": 78.9% of their entire nitrous oxide contribution for 2014.

Regarding manure, *it is essential for it to be half rotted before it is dug in.* The reason for this admonition (once more, courtesy of *Garden Manures and Fertilizers*) is that manure prematurely plowed under actually reduces the fertility of the soil. Unfortunately, leaving it to half-rot above ground generates more nitrous oxide. In 2008, about 13% of Canada's agriculture-related emissions of nitrous oxide derived from "manure management." That endangered us as much as 10.3 million pounds of carbon dioxide.

Manure not only gave off nitrous oxide; it was also a methane-emitter. In the U.S., *methane emissions from manure management increased by 64.7 percent since 1990 . . . since the general trend in manure management is one of increasing use of liquid systems, which tends to produce greater CH_4 emissions.* Hence the following headline, whose alarmism, founded on both enteric fermentation and on nitrous oxide from manure, might have been hyperbolic:*

Rearing cattle produces more greenhouse gases than driving cars, UN report warns.

"LAND USE CHANGES"

Tilling, flooding and manuring the soil weren't all. We cut down trees at pleasure, to make room for new fields or to sell off timber. (You see, necessity required us to enlarge our populations and economies.) Deforestation has been called *one of the largest anthropogenic sources of emissions to the atmosphere globally*—for where those trees had been, busy necrophores set about altering the soil as usual, turning dead matter into carbon dioxide, water and heat, while the sulfurs went to sulfates, and certain humans hopefully made a living.

We grazed and grubbed 38% of the ice-free land acreage of our planet, then kept at it. *As a result of land-use changes* (so they blandly put it), the Germans sent up more than 13 million tons of carbon dioxide in 2007 alone. *That figure represents the sum of loss of forest biomass . . . and stock losses in organic soils.* (An additional 3.1 million tons *were released from wetlands,* and 15.5 million from grasslands). When I was alive, our greenhouse gas reports bore a line item called "LAND USE, LAND USE CHANGES AND FORESTRY." In 2007, there went 13.24% of Germany's entire greenhouse emissions, not counting methane or nitrous oxide.[†]

In 2014, half the territory of the European Union happened to be farmed. That meant that half of it emitted more greenhouse gases than would uncleared land.

Sometimes we set fires, either to clear new cropland more conveniently, or to nutrify the soil, or, best of all, because it paid. *This sugarcane field in Rhodesia is being burned just prior to harvesting. If not burned, . . . 5 to 10 tons per acre or more . . . of leaf trash would be left, too much to plow under . . . Cutting and cleaning the cane without prior burning is slower and more expensive and also exposes the harvesters to the perils of venomous snakes*—ditto for a field in Venezuela.[‡] In addition to heat-absorbing carbon particles, the smoke contained carbon dioxide, nitrous oxide and methane.

---

* For transportation's share of the blame, see table on pp. 144–45.

† German methane releases are too complex to be considered here. Nitrous releases reached 4,717,844 pounds, or over 490,931 tons of carbon dioxide equivalent.

‡ In the following century, the Nigerian government reported its *two main pollutants and sources* as Saharan dust and "particles from . . . biomass burning that produce huge amount of black and organic carbon. These account for over 70% of air pollution in the country." On the happy side, vegetationless land was paler, so it absorbed less heat.

THE THREE MOST DANGEROUS GREENHOUSE GASES
as of 2011,

their percentage increases since 1750
and their percentages of total national emissions

All percentages [e.g., **+40%**] expressed in carbon dioxide equivalents [at the 100-year global warming potential of 25 for methane, in order to leave source figures as they are], and exclusive of LULUCF (see p. 596). For 20-year methane damage figures, multiply × 3.07. National emissions coded as follows:

> M = Mexico, 2002. Includes LULUCF since this category was not
> separated out in original.
> EU = European Union, 2012
> J = Japan, 2013
> U = U.S.A., 2014

1st

Carbon dioxide: +40% [to 390.5 ± 0.2 ppm]. *Global warming potential: 1 [by definition].*

M: 74%   EU: 82–83%   J: 93.1%   U: 80.9%

2nd

Methane: +150% [to 1,803.2 ± 2.0 ppb]. "The massive increase in the number of ruminants, the emissions from fossil fuel extraction and use, the expansion of rice paddy agriculture and the emissions from landfills and waste are the dominant anthropogenic CH_4 sources. Anthropogenic emissions account for 50 to 65% of total emissions." *GWP: 28 [100-year estimate]. At least 86 [20-year estimate].*

M: 23%   EU: 8–9%   J: 2.6%*   U: 10.6%[†]

3rd [except for Japan]

Nitrous oxide: +20% [to 324.2 ± 0.2 ppm]. "N_2O has overtaken CFC-12 to become the third largest W[ell-] M[ixed] G[reen] H[ouse] G[as] contributor to

---

* As was already noted, rice cultivation emitted 50% of Japan's methane. Enteric fermentation came second, and "solid waste disposal" third at a mere 10%. Unlike the Americans [see next note], the Japanese declined to coyly refer to "petroleum systems." Methane's total "fugitive emissions from fuels" was a decent 2.26%.

[†] In the U.S., the four largest sources of anthropogenic methane were, in decreasing order, "natural gas systems," enteric fermentation from cattle, landfills and "petroleum systems."

R[adiative] F[orcing]."* However, such natural sources as oceans and soils might have equaled or outweighed the anthropogenic ones (as of this writing, our measurements showed a high range of uncertainty). *GWP: 264.* EU: 82–83%

M: 2%   EU: 7.3–7.5%   J: 1.6%   U: 5.9%

A number of other substances were far more dangerous, volume for volume, but not yet abundant. "HFCs, PFCs, and SF_6 [see pp. 177–81] all continue to increase relatively rapidly, but their contributions to RF are less than 1% of the total by well-mixed GHGs." In Japan, however, HFCs now reached 2.25%—a greater abundance than nitrous oxide's.

Sources: *Greenhouse Gas Inventory Mexico,* 2002; IPCC, 2013; EU greenhouse report, 2014; *Greenhouse Gas Inventory Japan,* 2015; U.S. EPA, 2016; with calculations by WTV.

* For definitions of these two terms see pp. 597, 592.

How planet-damaging was agriculture? One good faith reply might have been: Bad, but not terrible.—Between 2002 and 2011, global net* carbon dioxide releases from land use change averaged about ⅑ of our total releases from cement-making and the burning of fossil fuels—about 0.9 metric gigatons, or "only" ¾ of the entire 2011 carbon dioxide emissions of Japan.

(That excluded methane and nitrous oxide, of course).

However, for a century and more our farmers had been burning fossil fuels.

FLIPPING THE SWITCH

Would unmechanized agriculture alone have destabilized our climate? How can I say? That question grew moot about the time that my grandfather was born. Here is a steam plowing contest in Winnipeg in 1908: *Tons of coal and carloads of water are sent into the thin air, and between sunrise of one day and nightfall of the next three hundred and twenty acres of virgin land are doubled in value by breaking.* The coal, of course, became among other substances carbon dioxide, nitrous oxide and methane. The broken prairie grew suitably oxygenated, so as to speed up its own rate of carbon dioxide emissions.—Oh, and by the way, at the competition *influential men from the largest oil corporation in the world* sat

* Every now and then we did the right thing and planted trees, or abandoned fields to nature. These sequestrations got credited against our carbon debits. The net, of course, lay invariably on the bad side.

watching, being *keenly interested in the question of mechanical power on the farm as affecting the market for liquid fuels.*

Such was our progress; and in 1952 a 4-H handbook confessed that *the whole agricultural industry now depends so much upon electrical energy to supplement muscle that many farms would be unable to compete in the present economy without it.*

A treatise from India (1986) elevated our dependence into grandeur: *No progress is possible without an increase in agricultural production. This has meant a steep rise in energy utilisation, to enable the countries to produce not only fertilisers for upgrading production* but also to provide diesel for pump energisation and mechanisation.*

The 2002 *American Electricians' Handbook* drew up a "typical computed load" of 15 kilowatts for a *dairy barn, large size, including milk house and one 7½-hp motor.* The "probable maximum demand" was 17 kilowatts (about 967 BTUs per minute), so any canny farmer would lay in copper wire of appropriate thickness and conductivity for that work.[†] At the merely "typical load," *every hour* of lighting, refrigeration, mechanized milking, pasteurization and hot-water-washing would drink down 51,195 BTUs' worth of current—generated in power plants by 12 pounds of West Virginia coal, or eight pounds of diesel fuel, or (nowadays) six pounds of natural gas!—And don't forget the accompanying "farm residence" at 13 kilowatts (another 11 pounds of coal per hour) with six kilowatts on reserve. Meanwhile, *lighting in the poultry house serves to increase egg production in addition to providing light for seeing. One light should be installed for each 200 ft² of floor area . . . Sunlamps should be installed over feeding troughs.* How many more kilowatts now? We didn't count; electricity was cheap!

As you might have guessed, irrigation, fertilizer application and pest control were ever more often accomplished mechanically. An ecologist warned: *Crops highly selected for industrialized agriculture must be accompanied by the fuel subsidies to which they are adapted!* And in 2005, an oilman who, not foreseeing fracking, unnecessarily fretted that we would run out of his favorite substance, advised us that *80 percent of an Iowa corn farmer's costs is, directly and indirectly, the cost of fuels.*

---

* We paid varying amounts in carbon fuel combustion to make our fertilizers. In 1975 it cost 5,400 BTUs in electric power per pound of elemental phosphorus and 19,500 BTUs for the same quantity of ammonia. Had the currency been West Virginian coal, we would have burned less than two pounds at the generating station in order to make a pound of both together—thereby emitting five pounds of carbon dioxide.

† Watts = volts × amperes. For more detail, see definitions of "ampere," "volt" and "power" on pp. 521, 528, 527.

All of these follies and necessities ran up our carbon bill. One estimate for 2008 pegged our agricultural production at *19–29% of global anthropogenic . . . emissions, releasing 9.8–16.9 Pg [petagrams]* of carbon dioxide equivalent . . . to the atmosphere.*

While we warmed our planet in that way, we simultaneously fouled our nest with various industrial emissions.

---

* About 211.3 to 37.2 trillion pounds.

About Industrial Chemicals

Let us suppose that last night you experienced a very romantic interlude. A candlelit dinner for two preceded by a very dry martini. Soft background music on the stereo and finally, total intimacy for you and your companion. A chemical analysis of your activities would reveal that the candles, the alcohol in your drink, the record you listened to, the clothing you wore, the paint on your walls, the soft carpet underfoot, and even the aspirin you took the following morning were all petroleum derivatives.

Bill D. Berger and Kenneth E. Anderson, *Modern Petroleum*, 1978

Y ou see, we were makers and doers—and proud of it. We synthesized any number of products to sell to each other. Our favorite chemical feedstocks were hydrocarbons.

Here is a picture I took at the Phillips Petroleum Company Museum in Oklahoma. All items depicted are plastic, derived from natural gas.

They could as well have been fashioned out of coal or oil.* Our fossil fuels lay so ready not just for burning but also for reshaping! It was a wonder what they made possible—and you from the future, sourly preparing for the next meter of sea-rise, have reason to condemn us because you've paid more than we for the products that sustained, entertained, nourished, poisoned and defined us.

In case you are wondering what proportion of our fossil fuels became chemicals instead of smoke, here are two statistics from 2014: in France, a mere 14%; in the Netherlands: 49% of all petroleum, and an unspecified *fraction* of natural gas and coal.

I wish that you could have gazed back with me upon our fabrication processes with their vast economies of scale—which, again, reduced our costs and increased yours. (U.S.A., 1978: *A single $15 barrel of crude may be turned into consumer products worth almost $300.*) For that matter, I would have liked to flit from vat to vat, in order to describe for you the strangest acts of creation.— Did you know that we could refine propylene, not to mention detergents, out of crude oil—antifreeze from ethene—rubbing alcohol from propane? Improving petroleum into ethylene gas which we next *bubbled through a solvent,* then catalyzed, we made polyethylene, which was *familiar to most of us as the plastic material of packaging films.*† Kind souls affixed chlorine atoms to that stuff, so that I could have my polyvinyl chloride shower curtain. (It was getting grubby; my guests kept advising me to throw it out and buy another one.) Other wondermongers switched out ethylene's hydrogen atoms for fluorine atoms, and here came Teflon!

These transmutations remained as unseen to me—and most of my neighbors—as the carbon dioxide we exhaled. By the time I wrote *Carbon Ideologies,* industrial security officers tended to frown on public tours, thanks to terrorism, journalism, activism and corporate espionage. So this brief "About" section will retail no vivid descriptions. Instead, let me touch on the greenhouse gas emissions of a few of our favorite petrochemical processes:

---

* For a description of coal's "products tree," see II:10; for natural gas's, II:298; for oil's, II:403.

† Being extremely allergic to poison oak, when I was invited by an activist to spy out a West Virginian mountaintop removal mine (see II:117), I bought the cheapest daypack I could, to spare from contamination the one I generally used. The new pack's substance consisted of Chinese polyethylene. Its zipper was some other kind of plastic. Surprisingly enough, after three years of heavy use (and my daughter's teenaged contempt), this object retained its structural integrity, with the exception of the zipper, which had begun to fail. If I took the pack to a repair establishment I would pay far more than the $20 it had cost. If I threw it away, that would "waste carbon." If I kept using it, sooner or later my laptop would fall out and hit concrete, maybe fatally. Reader, what would you have done?

The oil derrick and "horse head" on this sign near Cushing, Oklahoma, make vivid the fungibility of fossil fuels

When, for instance, we synthesized ammonia for our fertilizers and other techno-necessities, the reaction liberated carbon dioxide: 3,630 pounds of it for every ton of nitrogen we threw in. Thus half a percent of Germany's greenhouse emissions for 2007—only half a percent! (What this "primer" section of *Carbon Ideologies* intends—to shine its wavering light across global warming's dark mountain of hydra-heads—is more necessary than possible. How many half-percents would you find it tolerable, much less useful, for me to picture? While we fretted about power plants, German ammonia manufacturies, hidden behind more frightening monster-heads, exhaled their discreetly baleful half-percent.) They employed both heavy fuel oil and natural gas as their feedstocks. The Austrians, who used natural gas only, emitted less than half as much CO_2 per unit weight of ammonia. The British, who started from hydrogen, were even easier on the atmosphere.—But surely we could make ammonia however we wished!* In Kansas and in various localities of India (*circa* 1979), the high-emissions petroleum route had pencilled out best. Such choices represented what we meant by "the free world."

Another "for instance": Our hydrogen plants incidentally released carbon dioxide—and how could we get by without hydrogen, which was so convenient for making nitric acid?

Had I asked my neighbors what we needed nitric acid for—or ammonia—99% of them would have gone blank. They bought what they wanted when they wanted it. Its origin did not interest them. Why, just the other morning I sat on

---

* Another method, "the easiest route" (now for some reason superseded), was to turn coal into calcium carbide, which we made to react with nitrogen and finally hydrolyzed with water, leaving calcium carbonate and ammonia.

the edge of the bed while a woman rolled on her nylon stockings and neither one of us thought about nitric acid!

In 2007, production of that stinging-smelling liquid donated 0.94% of Germany's nitrous oxide. Since the planet wasn't warm enough to suit us, nor our chemical plants as profitable as in best-case projections, we helped Germany—and looking back on those years, I take inexpressible comfort in the fact that humans kept right on being human. In 2012, Greece emitted 35 times more nitrous oxide per unit volume of nitric acid than did the United Kingdom—in other words, the Greeks were 35 times more wasteful than the British—while we Americans, not quite in the vanguard, were merely 27 times worse. Nobody thought to centralize or rationalize nitric acid production.

Meanwhile, synthetic natural gas proved as practical to make as it was convenient to burn. So we did both, not worrying about a little carbon dioxide along the way.

Sometimes we felt like mixing sulfuric acid with dolomite. (Has that desire ever overcome you?) Up came carbon dioxide.

What next? We put our heads together. One of our favorite corporate slogans was *innovation*.* So through which new means could we excrete carbon gases into the atmosphere?

As our technocrats realized, *carbon dioxide provides soft drinks with a pungent taste, acidic bite, and sparkling fizz.*[†] Well, let's fizz them up, then!—In 1860 the annual American per capita soft drink consumption was 16 ounces; by 1990 we had gotten it up to 47.5 gallons! That was "creating a demand" for you! In our vending machines and supermarkets, hordes of soft drinks waited fizzily for thirsty customers. *The average 2-L soda bottle maintains an internal pressure of roughly . . . 5 atm[ospheres] of CO_2.*[‡] Once some "consumer" unsealed that bottle, out went the carbon dioxide, and up into the atmosphere!

Let a dye chemist in Madras have the last word: *The economics of scale provided the great advantage of reduced cost, which in turn stimulated more applications and demands, thus advancing petrochemical production to ever greater heights.*

---

* Another was *boldness,* which meant pursuing our interests without respecting others. As *Carbon Ideologies* will show, this trait came naturally to fuel extraction industries.

† It "also acts as a preservative against yeast, mold, and bacteria."

‡ "The criterion for shelf-life is the time to 15% loss of carbonation." In other words, even unopened it kept outgassing carbon dioxide. And by the way, each plastic bottle's feedstock derived from fractional crystallization in some petroleum refinery's tower. More hydrocarbons burned in that process.

The Parable of Adipic Acid

But I can't believe that any group which is rooted in the principles of freedom and service can have gone very wrong.

C. M. Kornbluth, *The Syndic*, 1953

Pantyhose, whose semi-ethereal protein-like* substance had the ringing name nylon 6,6, was derived from two chemicals: hexamethylene diamine and adipic acid. Let us briefly consider the latter:—a water-soluble, sour-tasting crystalline white solid with a melting point of 303.8° Fahrenheit. (Reader, I hope that your mean surface temperatures remain below any such figure.) Have you heard of that inconspicuously influential substance? Sometimes we called it hexanedioic acid. Each molecule contained 10 hydrogen atoms, four oxygen atoms and six carbon atoms.

Being a peachy-keen feedstock for synthetic lubricants, urethane foams, plasticizers, you name it, it sprang into *commercial importance* in the 1940s. *Food grade adipic acid is used to provide some foods with a "tangy" flavor,* and justifiably so: After all, its aqueous solution could corrode certain steels.—In the mid-1990s, 3,500 tons a year went for "food acidulant" in American jams, jellies and gelatins.[†] It must have been good for us, being *one of the purest materials produced on a large scale.*

Now, what else could we do with it? (In other words, how could we muck up *some ecosystem somewhere?*) Heating it comprised *the best route to cyclopentane and its derivatives,* whose profitable uses I shall spare you. In that process, clouds of carbon dioxide got "lost"—but only to our atmosphere. As our *decider* President used to say, "Bring it on!"

Most of it was destined for nylon. What a treat to watch adipic acid[†] combining with the hexamethylene diamine so that we could see a pallid syrup forming, settling out into something resembling melted cheddar cheese! Making use of the way that carbon links to itself, and even joins chains to chains, thereby

---

* "Long linear molecules that, stretched, can be made to lie roughly side by side."

† It could well have been the "food acid" listed as an ingredient in the cheap potato crisps on which I sometimes snacked (*Proongles* or *Prankles,* they might have been called).

† Or more correctly, its chloride.

thickening into those long strands which, as pantyhose should, resist oxidation and bacteria, industrial magicians deployed what they called "the nylon rope trick": *Polyamide produced at the interface may be pulled continuously from the open vessel in a startling demonstration of polymerization chemistry.*

Each pound of nylon cost "only" 18,700 BTUs of power or heat—awfully close to the innate energy of a pound of fuel oil—but, as I keep on reciting, our power plants had to burn three pounds to do each pound's worth of thermodynamic work. So, a pound of nylon required three pounds of heavy heating oil, which sent up 10 and a half pounds of carbon dioxide. That was just the beginning.

There was nylon in the hose of my air compressor, nylon in my father's fishing line. The 1956 Sears, Roebuck catalogue couldn't stop pimping it out! Reader, would you like a Cordtex brassiere? *Nylon styles are lined with beautiful nylon sheer and stitched with nylon thread. This lining in underbust offers extra support.* (How lovely!) And for outerwear, why not *Gleaming Rayon Velvet*?

> NOTHING STARS in the firmament of evening fashions like the glowing loveliness of jet black velvet . . . it's the epitome of glamour when topped with an exotic hood and lined in immaculate white—the body of the lining in nylon fleece, the sleeve lining in rayon and acetate velvet.

In short, we had no choice but to continue making adipic acid.

I once met an elegant high-ranking Japanese geisha whose wardrobe largely consisted of her grandmother's handwoven silk kimonos. The fabric looked new; the designs were unfailingly beautiful.—Rarely did nylon "intimates" show any such durability, either of style or of physical integrity. The panties in the photo below were fading more with each wash. Their owner had begun to tire of them. That was as it was supposed to be. As an article in *Resources, Conservation and Recycling* concluded:

> All current economic systems are predicated on growth and . . . it is almost impossible to imagine a different system emerging . . . [B]usiness models in production companies are oriented towards growing sales volumes, so are strongly motivated to . . . build in "planned obsolescence" to product designs. Thus, material efficiency . . . will . . . be opposed by . . . businesses unless they can reclaim value through some other activity.

Undies made in Thailand and sold in U.S.A. (82% nylon)

To be sure, this might have been bad news for our climate—but, reader, don't glare at me like that! It was wonderful news for adipic acid.

We made so much of the stuff that it sometimes got shipped in 200,000-pound hopper cars. (Don't you call that success?) By 1990 a 130,000-ton manufacturing plant had been announced for Western Siberia. Another was up and coming in South Korea. *The continued buildup of capacity in nylon-6,6 intermediates, especially in the Far East, attests to the confidence in continued growth by the major participants.*

Adipic acid was another of those marvel-chemicals which not only offered myriad profit-making applications but could be produced in any number of ways—from cyclohexane, or cyclohexene, phenol, butadiene, whatever, *as dictated by shifts in hydrocarbon markets.* (It also emerged in automobile exhaust, but since we had not yet figured out how to collect and sell that form of it, we simply emitted it, mile after mile, entirely gratis—and up it went. Fortunately, it was invisible.)

In our chemical factories, no matter how or from what we made our adipic acid, there came a moment when we needed to oxidize its hydrocarbon precursor in nitric acid.

As you have read,* producing that *already* generated nitrous oxide. Now the nitric acid must be put to work.—Adipic acid manufacture reduced it to nitrogen dioxide and nitrogen oxide—which could both be captured and reconverted to nitric acid—and to pure nitrogen—which was harmless—and, most relevantly, to what has been called "considerable" nitrous oxide. Up it rose.

---

* From p. 126.

In 2007, adipic acid production emitted (in carbon dioxide equivalents) 0.55% of Germany's greenhouse gases.*

Once again, any reasonable know-nothing might protest that 0.55% of a nation's environmental misdemeanors was so trivial that our attention should have been spent on more conspicuous culprits such as fuel combustion. (In the Appalachian chapter you will hear a lobbyist say: "You can shut all 12 hundred coal-fired power plants down in this country today and you will effectively reduce less than 4% of CO_2 emissions.")[†]—One might argue that since Germany's manufactures of adipic acid had released three times more nitrous oxide in 1990 than in 2007, the latter figure should even be celebrated. At first, even I believed that, back when I was alive. Any decrease in waste, or increase in efficiency, which to my mind comes close to the same thing, is inherently good—and round about 1993, German chemists had invented two new ways of decomposing almost all the nitrous oxide into oxygen and nitrogen. Why not be glad when *anything* got done? To be honest, I could not blame that coal lobbyist for expressing frustration toward government and environmental "activists." "That's where I have a disconnect with the President and the EPA," he told me. "I do not buy the statement that the U.S. must demonstrate leadership. I think *we have been providing leadership throughout my entire adult life!*"[‡]—And so I felt grateful to Germany for "providing leadership" in decreasing adipic acid's nitrous oxide emissions. It had not even been 20 years yet since the most progressive nations (not mine) had signed the Kyoto Protocol, and this fable promised well: In one country, for one category of activity, one kind of greenhouse pollution had been reduced.

Now for the sad part: Considering industrial process emissions as a whole, Germany admitted defeat, as a result of "economic trends." And what could anyone do about those? *To date, a counter-trend has been achieved only in the case of N_2O emissions. It is the result of emissions-reduction measures by adipic acid producers, measures that took effect as of 1997. And yet that trend as well is being increasingly offset by production increases.*

---

* Moreover, it took energy, of course. Some unreacted stuff had to get distilled before sending it through the conversion cycle again, and how could that happen without burning carbon? Meanwhile, "the spent oxidation gas stream must be scrubbed to remove residual cyclohexane, but afterwards will still contain CO, CO_2 and volatile hydrocarbons (especially propane, butane, and pentane)." It could be "catalytically abated," but wiser heads burned it, so that its energy got "recovered," while more carbon dioxide rose to heaven. Other hydrocarbon-laced "waste streams" could be sold to paper manufacturers.

† See II:201.

‡ See II:130.

About Manufacturing

Iron thoughts sail out at evening on iron ships...

Malcolm Lowry, before 1958

And when, instead of reacting them, we *burned* our favorite fossil fuels in the service of industrial production, CO_2 boiled invisibly upward and outward as always. I have already celebrated cement-making's multi-century outgassings.* Other processes accomplished comparable benefits. The titanium dioxide in my toothpaste and the glass in my windows both cost considerable carbon dioxide to make. (*Circa* 2012, our "global energy inputs" went slightly more to titanium than to glass.)

A molten ton of glass cost you somewhere between 120 and 208 pounds of carbon dioxide. One cause of emissions (here glass imitated cement) was lime, which prevented the network of fine oily cracks called "crizzles." Another was the fuel in the kiln: We had to heat the raw materials into the range of 2,400° to 2,750° Fahrenheit before they fused into glass. Strange to say, after an exuberant guest threw a baseball through my kitchen window, we drove to the hardware store, then replaced the pane for pennies!—because *you* were paying the carbon cost.—A store in San Francisco sold glass dildoes; one of them reputedly saved a woman's life when her husband peeked through it and recognized colon cancer, so didn't that justify glass's carbon emissions?—I hope that you have at least inherited a few of our windowpanes. Maybe you pried them out of drowned properties and fitted them into your caves.

Our annual energy expenditure on glassmaking was the merest 30th of that used to produce steel.

In his best-selling *Outline of History* (originally published in 1920), H. G. Wells sang out: *Nothing in the previous practical advances of mankind is comparable in its consequences to the complete mastery over enormous masses of steel and*

---

* In "About Carbon," p. 106.

iron* ... *To-day in the electric furnace one may see tons of incandescent steel swirling about like boiling milk in a saucepan.*

The Germans believed that *the iron and steel industry ... is the second important CO_2-emissions source ... in the area of process combustion.* Like adipic acid, in and of itself it barely added to the burden we laid upon you. For that matter, summing up steel, aluminum and various chemical manufactures all together accounted for less than 10% of all Germany's greenhouse gases. No immediate danger, my friends! Puff by puff, our atmosphere warmed up, even as Gutowski and his colleagues wrote with bleak eloquence:

> Manufacturing is now ... 200 times larger than it was in 1800 on an absolute scale and approximately 30 times larger on a per capita scale ... The enormous material flows ... rival natural geologic flows ... World industry now uses ~190 exajoules [about 180 Q-BTUs] of primary energy, approximately one-third of all energy used globally, and emits ~14 billion tonnes [15.4 billion U.S. tons] ... of CO_2 and 50 M[ega]t[ons] [55,115 U.S. tons] of SO_2, or almost 40% of all global anthropogenic emissions from energy and industrial processes each year ... Furthermore, global manufacturing is growing and increasing its energy use and carbon emissions.

But could these *enormous material flows* be slowed or narrowed through some kind of categorical analysis? In much of *Carbon Ideologies* it becomes my sad task to multiply discouragements and bewilderments: We destroyed ourselves not simply because we burned too much steam coal or heavy commercial fuel oil in our power plants, but also because we followed sound agricultural practices, and because we made useful and beautiful things of all sorts. We composted garbage, which was surely a good deed, and in case of rain we wisely carried our nylon raincoats. We published books, which kept *me* employed. So many customary behaviors imperiled us that the most radical solutions might not have been radical enough. (As Death tells Apollo when the latter tries to save a woman's life: *You may not have all that you should not have.*)—Manufacturing, however, presented us with certain potentially hopeful simplicities:

Gutowski et al. asserted that *most energy is used in industry to make a few key materials, for which energy is a significant driver of cost.* They called those *key materials* the "big five." Steel was one of those.

---

* Well, the poor bugger never lived long enough to admire adipic acid!

Steel wrenches in hardware store window, Sharjah

Reader, wouldn't you like to see the big picture without peering through either flat or window glass? The "big five" *were* the big picture:

ENERGY AND COAL REQUIREMENTS TO MANUFACTURE ONE POUND EACH OF THE "BIG FIVE" MATERIALS, *ca.* 2013,

in multiples of the energy needed for cement

Unbracketed figures express BTUs required to produce 1 lb of the material.

[Bracketed figures] are [lbs of coal burned in a power plant to generate sufficient energy to produce 1 lb of the material].* Assumed inherent energy (HHV) of coal: 12,500 BTUs/lb.

All multiples over 10 rounded to nearest whole digit.

1

Cement: 426 BTUs   [0.1 lb coal]

Making it released 8% of worldwide CO_2 pollution in 2015. Furthermore, when formed into concrete, cement was often reinforced with bars of another "big five" material, steel.

---

* Due to the ⅔ wastages in power plants (see p. 150), this figure reflects 3 × the average innate energy in coal.

23
Paper: 9,888   [2.4]

"Most of the paper we produce is thrown away and not recycled."—Pope Francis, 2015. *The "default half-life" for paper in Japan (after which half of it would be disposed of) was two years. Incinerated or left to rot, it then gave off carbon dioxide. [Meanwhile, we often whitened our paper with titanium dioxide pigment, whose manufacture released more carbon dioxide. In the U.S., production increases of TiO_2 led to emission increases: 47% more in 2014 than in 1990.]*

25
Steel: 10,748   [2.6]

"In terms of climate change, . . . the most important metal is iron . . . because steel is the single largest energy user and carbon emitter in the world materials sector." [Nitrous oxide was another greenhouse gas associated with steelmaking.] . . . "The top seven CO_2-producing manufacturing nations are also the top seven steel producers, and, with the exception of Japan, they are in the same order."*—Timothy G. Gutowski et al., 2013.

32
Plastics: 13,758   [3.3]

"Plastics . . . are produced by the conversion of natural products or by the synthesis from primary chemicals generally coming from oil, natural gas, or coal."—American Chemistry Council, 2016. *In other words, not only were plastics energy hogs when manufactured, they also derived from the three major (non-nuclear) carbon ideologies of resource extraction. Their source materials inevitably emitted carbon dioxide and methane, and sometimes (as we have seen with nylon) nitrous oxide.*

94
Aluminum: 39,983†   [9.6]

The manufacture of this marvelously useful material not only consumed large quantities of energy, but also often released sulfur hexafluoride, which

---

* In 2013, iron and steel combined were Japan's largest single source (48%) of carbon emissions for manufacturing and construction.

† And remember, this is just the *manufacturing energy*. Gutowski writes me that the embodied energy from aluminum "today" (2016) "for all steps: mining to smelting" is 200 MJ/kg, which by my calculation would be 86,188 BTUs/lb.

molecule for molecule was the most potent greenhouse gas known.* *"The increase in total SF$_6$ emissions in recent years is due to use of pure SF$_6$ in aluminum production; in the 1990s, that gas was used solely as an additive."—* German greenhouse gas report, 2009. *Moreover [quoting the same source], "production of primary aluminum continues to be the largest source of PFC emissions in Germany." PFCs, or perfluorocarbons, showed global warming potentials thousands of times worse than carbon dioxide's.*

Sources: as given, with calculations by WTV.

———
* See below, p. 183.

The "big five" released 56% of the world's industrial carbon dioxide. To put it another way, they caused 20% of all *energy and process related* CO_2 emissions— because we had to have them everywhere:

Sharjah Corniche view

After a pleasant morning stroll along the Corniche of Sharjah City, enjoying the view across the Khalid Lagoon—concrete, glass, aluminum and steel incarnated in sunstruck mosques and in skyscrapers whose stacked horizontal rectangles of windows distinguished them with beauty and a similitude of awareness—I sat down on a four-planked steel bench, resting my back against its advertisement which invited me (at an unstated cost in carbon dioxide) **AROUND THE WORLD WITH SATA**, which placed on the same island the Eiffel Tower, the Statue of Liberty, the Dome of the Rock, the Great Pyramid, you name it, while a jet looped around the scene in order to weave a happy contrail—and as two black-robed women* passed by, each carrying her own plastic bag of something, I sat watching the cars and tried to count uncountable construction cranes atop partially risen towers.[†] Without this extravagance of steel, glass, aluminum and plastic, how would our future look? Wishing to tell you all about it, I expended as much paper as I could.

All five of the "big five": paper cartons (enclosing plastic cups), with steel- and aluminum-containing cars and a two-tiered cementscape, in Poza Rica, Veracruz

---

* They wore abayas.

[†] A middle-aged father and a beautiful young mother waved at their little boy and girl as each child whirred by in a miniature electric car, whose tinted passenger windows were only the breadth of my hand. How much current it took to charge the battery I don't know. How much plastic, glass and metal were in each car I cannot tell you.

At first I felt heartened by the disproportionate effects of the "big five." Surely it would be more practical to improve manufacturing efficiencies and reduce unnecessary uses of these few items than to address the emissions of dozens or thousands of other commodities! I trust you will forgive me for this lapse into hopefulness. (Henry James, 1870: *We cry out for a little romance, a particle of poetry, a ray of the ideal.*) Now I promise to wade back into chilly accuracy.

WHY WE HAD TO KEEP MAKING CEMENT

"Best available technology" could have reduced the energy needed for all but one "big five" member by a third, more or less. The exception was cement—which, so it happened, was the only one that could not be efficiently recycled.* So we kept producing it: 1,056 annual pounds for each human alive, with prospective yearly increases of 4.6%.

... AND ALUMINUM

Even though Gutowski singled out steel for its ubiquity, aluminum (as the preceding table shows) was the devouring king, because pound for pound, its manufacture consumed the most energy of the "big five"—94 times more than cement. Whenever we made it,[†] we emitted carbon dioxide (2,374 pounds per ton) and four other greenhouse gases. And yet the sheer *utility* of this strangely beautiful metal was staggering, in everything from spacecraft to bridge decks. Its alloys resisted corrosion marvelously, which enhanced their durability, and thereby saved energy which would have gone for remanufacturing. On the average, an aluminum structure weighed only half as much as its counterpart in structural steel. This reduced energy inputs in transportation, not to mention the gravitational stresses on that streetlamp or industrial roof mobile home. In addition, my friends felt sure that riding their

---

* According to Gutowski. But the *Encyclopedia of Science and Technology* had already thrown down a paragraph on "'green' or sustainable concrete": "To reduce the need for virgin aggregate" in the cement-and-water paste that set into concrete, "recycled concrete is the most promising approach." Aggregate comprised something like 75% of the volume of concrete. Let us ignore the fact that the aggregate, virgin or not, was still added to cement. The good news: Reducing carbonate oxidation in the cement-making process had already lowered carbon dioxide emissions 20% per unit volume relative to the 1980s.

† *Circa* 2007.

aluminum-framed bicycles was "good for the planet." Who was I to nay-say? Surely they did less harm than making or driving a car . . .—Meanwhile, striking a blow for pure aesthetics, the 1956 Sears, Roebuck catalogue invited us to drool over the New Jumbo Carrier Cake Cover of *Lustrous Sunray Aluminum . . . lightweight, rustproof,* and a solid 4 pounds of it (which must have burned 40 pounds of coal) for $4.89—not to mention an automatic playing card shuffler, *made of heavy gauge steel . . . constructed to give many years of dependable service.*

   In the phone book for Abu Dhabi I once saw an advertisement for Widest Steel Industries LLC: *All kinds of Stainless Steel, Aluminum & Carbon Steel structures with 10[,]000 MT* Annual Capacity.* Based on the table above, this company would have needed to burn between 4.5 and 17.1 trillion gallons' worth of medium grade fuel oil to fulfill its stated possibilities.[†]

All-aluminum shell of Sumida
Hokusai Museum in Tokyo

---

* Metric megatons, I presume. More than 22 trillion pounds.
† The first figure would be for 10,000 MT of steel; the second, for the same quantity of aluminum.

Through recycling we could have reduced by a stunning 94% the manufacturing energy requirements of aluminum.* But what you from the future don't realize is that our economies needed to grow, oh, yes; demand kept rising, and how could we disappoint that? Smelling opportunities and profits, we refused to sit around waiting for our aluminum-bearing products to get obsolete so that someone might tear them out of yesterday's bridges; hence there was no good alternative; so please don't tell me that we should have dethroned the king of the "big five"!

...AND STEEL

What about steel? In 1980, the world's recycling rate for that metal was 60%. By 2006 it had sunk to 34%, even as the manufacturing rate kept ascending by 8% a year: We needed more steel! Every year, we produced 440 pounds of it† for each person on Earth.

...AND PLASTICS

As for plastics, why couldn't we recycle, for instance, those 82% nylon feminine underpants depicted in the previous section? Here our difficulty was the antithesis of the one that steel presented us. Plastics wore out quickly; their owners would have been happy to return them to the vat. But they came in so many varieties that separating them out for reprocessing was more trouble than it was worth.‡ Better to start fresh!

---

* The manufacturing energy reduction from recycling was 64% for steel, 48% for paper and 25% for plastics.

† "UK steel consumption is currently around 530 kg [1,166 lbs] steel per person per year and should be reduced to 160 kg per person per year to meet the requirements of the UK Climate Change Act."

‡ Plastics "are victims of their own success: That they can be altered by a wide array of fillers and additives means that there is a great deal of uncertainty concerning their physical properties when they are collected. A second point is that plastics, although used in large volumes, usually are not incorporated as large masses in any given product. This presents a challenge to their collection and recycling."

Plastics in a supermarket, Kyoto

Paper fans for sale, Kyoto

... AND PAPER

Nor could we give up paper, *without which such inexpressible confusion must ensue,* warned John Dickinson back in 1768; we kept fighting the good fight against confusion; by 2013 we needed 127.6 pounds of paper per capita, so that credit card companies could send out junk mail.

THEREFORE

As you see, we were helpless. We had no choice but to wipe out the buffalo herds.

Gutowski and his colleagues considered trends and crunched numbers, then projected the results into half a dozen alternate futures. They concluded: *Under all six scenarios, the allowable accumulated emissions resulting in peak warming of 2°C will have been surpassed by the industrial sector by 2050, with annual emissions continuing to rise for all but the lowest population scenarios beyond that date.*

About Transportation

Transportation is the major culprit of air pollution accounting for over 80% of total air pollutants ... Traffic emissions contribute about 50–80% of NO_2 and CO concentration in developing countries.

European Journal of Scientific Research, 2009

Even the most vibrant experiences become slightly faded over time. What was once new starts to feel programmatic. Which explains why those who are searching for something fresh are heading to Guadalajara.

Travel + Leisure, 2016

We choose to go to the moon because it is profitable!

Bob Richards, CEO of Moon Express, 2017

Surely our entire way of doing things didn't need to be done over! I much preferred to continue living in the way that had always worked so pleasantly for me, for instance taking the bullet train to Hiroshima, the man across the aisle from me reading a snow-white newspaper whose Japanese characters large and small, bolded, bannered or simply laid out within closed rectangles, brought something like satisfaction to his nearsighted squint; meanwhile the train accelerated into a long ear-popping tunnel, pallid blue Kanji crawled across the black information screen, and we all rode at ease in our bright warm world. I let down the plastic tray (stated load limit, 10 kilograms), looked out my oval window as we emerged into a winter morning of concrete, glass, aluminum and steel—Shin-Kobe already!—and after a brief stop we were flying ahead again, the motor's cadence rising far less gratingly than it would have on one of our diesel-powered passenger trains back home; so I settled back, zooming beautifully through the morning.

In the unlikely event that I ever felt called upon to write rank errors simple and large, transportation would have been in the first main category:

THE THREE MOST DANGEROUS SECTORS OF HUMAN ACTIVITY, 2012–14

These categories appeared in formulaic greenhouse inventories.

All percentages calculated excluding LULUCF [see "Agriculture," below, this table]. Since not all activities are listed, percentages =<100%.

G = Germany, 2007
EU = European Union, 2012 [EU-15 and EU-28]
J = Japan, 2013
M = Mexico, 2012
U = U.S.A., 2014

1st

"Energy": Defined as "combustion and fugitive emissions," this flabby category includes electricity production through the use of fossil fuels, most forms of transportation, residential heating, leakage from natural gas pipelines, etc. Were it better broken down, we would see that the primary subcategory was power generation; the second, transportation.

M: 70.4%   G: 80.9%   EU: 79–80%   J: 89.5%   U: 83.6%

2nd (or 3rd)

"Agriculture": As described in its chapter above—but excluding what was called "land use, land use change and forestry"(LULUCF), which referred to the conversion of cropland into settlements, which emitted carbon; and the planting of trees, which sequestered it, and so on.

M: 8.3%   G: 5.38%   EU: 10–10.3%   J: 2.8%   U: 8.5%

3rd (or 2nd)

"Industrial Processes": Manufacturing and chemical engineering, including steel- and cement-making, adipic acid production, etc. *Global per capita CO_2 emissions: 0.99 U.S. tons in 2010, and rising by 2.5% per year.*

M: 9.4%   G: 12.14%   EU: 6.7–7.3%   J: 6.2%*   U: 5.5%

2nd (for Mexico only)

"Waste": Industrial, municipal, and hazardous. Sewage, dumps, etc., etc. M: 11.85%. "Emissions from solid waste disposal and wastewater

---

* In Japan the single largest source was cement-making (40%).

management and treatment underwent significant increases between 1990 and 2002, with 115% and 85%, respectively, brought about by *better management ...*" (my italics).

Sources: Mexican greenhouse report, 2002; German greenhouse report, 2009; Japanese greenhouse report, 2015; IPCC, 2013; EU greenhouse report, 2014; U.S. EPA, 2016; with calculations by WTV.

Having now, as my grandfather used to say after we had admired the "headlights" of TV anchorwomen, "solved the world's problems," I returned to zooming through my easy life, enjoying passing skyscrapers and ricefields.

You see, in our time we took it for granted that as long as we (and you) could pay, we had the right to be carried wherever we pleased. That our non-monetized costs would be payable by you, whose various potential distresses never troubled our sleep, made travel all the better. In this connection my mind turns to Calabar, Nigeria, which from the standpoint of *Carbon Ideologies* was one of this world's most bucolic transportation centers. In 2009 two researchers found that

> the five monitored air pollutants when compared with ... [a]ir quality index ... were in the range of: CO—poor to moderate and moderate to poor in different locations. SO2*—was from very poor to poor, NO2— from very poor to poor, PM10 [small particles] and noise level was poor at all locations ... The study concludes that transport-related pollution in Calabar is indeed significant with possible severe health consequences.

The study did *not* conclude that carbon monoxide, sulfur dioxide and nitrous oxide might incur other possible *consequences.* And with noise elevated into one of the five "pollutants," carbon dioxide got deleted from the list—so there must not have been any. Let us celebrate the miraculous smog of Calabar!

In 2002, "road transportation" comprised 18.81% of all Mexico's emissions; in 2007, 14.13% of all Germany's. These faraway realities could not quite penetrate the wall of boredom with which my egotism defended itself. To be sure, I could remember starting and stopping on a wide highway in the Tijuana smog;

---

* No subscripts in original.

diesel fumes had nauseated me in stalled traffic on the Autobahn . . .—but invisible, near-odorless carbon dioxide stayed joyously inconspicuous in such experiences, and just now, as I sat at my kitchen table in Sacramento, breathing fresh-smelling air and hearing each car breeze past me like a sea-wave, I wished to hurry away from this particularly dreary page of *Carbon Ideologies*.—In 2013, 15.3% of Japan's greenhouse emissions came from rail, air, water or road transportation; of these, "road transportation" made up 89.9%. Tired out by that calculation, I strolled to the refrigerator to eat a handful of blueberries trucked in from Salinas. How else was life supposed to be, with 31% of Canada's energy consumption billed to transportation, while the U.S. Environmental Protection Agency for its part explained: *When electricity-related emissions are distributed to economic end-use sectors, transportation activities accounted for 33.4 percent of U.S. CO_2 emissions* from fossil fuel combustion in 2014?* Most of those "activities" (83.3%) consisted of the doings of cars and trucks. Their emissions kept going up and up.

For a long time my continent had been special. In North America (1980 data), 41.8% of our total refinery output went for motor gasoline. That figure *did not exceed 19 percent . . . in any other region of the world.* But the rest of the world began to imitate us.

In terms of the overall trend, continued the EPA, *from 1990 to 2014, total transportation CO_2 emissions rose by 16 percent* in the U.S., *due, in large part, to increased demand for travel . . .* Well, so what? Climate change might still turn out to be a leftwing figment. And even if it did someday mar some ecosystem somewhere, do remember that travel demand created jobs! Squinting and smiling out at us from the upper righthand corner of a glossy free magazine, the Las Vegas Host Committee Chairman, who used to be Mayor, gloated that come Labor Day weekend, *nearly 307,000 people will travel to the destination. These visitors will generate nearly $206.5 million in direct visitor spending,* and several greenhouse gases.

GERMANY HOLDS THE LINE

Germany, as usual, presented a miracle: After the 1990 benchmark, since growth in miles travelled outweighed improvements in vehicle fuel consumption, her transportation sector could not help gassing out ever thicker greenhouse

---

* Methane and nitrous oxide emissions increased this figure by only a quarter of a percent.

clouds—but between 2000 and 2007, "road transportation's" emissions* actually declined by 4.5%: 30 million tons!

> The likely reasons . . . include reductions in specific fuel consumption, a
> marked shift toward diesel vehicles in new registrations, continual in-
> creases in fuel prices, use of biofuels—and consumers' growing tendency
> to travel to other countries in order to make their fuel purchases.

Setting aside the last factor (which the ever inventive German language called *Tanktourismus,* gas tank tourism), I saw fit to admire one group of humans for making a moderately estimable beginning. Or let me more accurately admit that I longed to exaggerate my belief in them. They had let me down with their adipicacid emissions,[†] but why hold a grudge? Let them be heroes of harm reduction. *Then* I could trust my friends and neighbors when they promised me: "No immediate danger!"

In this spirit I sought out the latest greenhouse inventory available. Five years had passed, and Germany now shared the fortunes of the European Union. Under the category "road transportation" I found the following: *The second largest key source of all categories in the EU-15[,] accounting for 20% of total GHG emissions in 2012. Between 1990 and 2012, CO_2 emissions from road transportation increased by 11% in the EU-15.*

That wasn't pleasant. But where stood Germany in the EU-15?

The Member States Germany, France, Italy, Spain and the United Kingdom contributed most to the CO_2 emissions from this source (77%). All Member States, except for Germany (–3%) and the United Kingdom (–1%), show increased emissions . . .

So Germany still held the line, but what a feeble line it had become! Given our predicament, even minus 99% might not have sufficed.—In 2012, German road transportation released 145,826 gigagrams (call it 321,487,999,60 pounds) of CO_2.—At least minus 3% was still a minus.

"AND CONTINUED TO RISE"

As for sky travel, the most glancing tabulation conveys the same point:

---

* All recorded substances of concern—methane, nitrous oxide, nitrogen oxides, carbon monoxide, volatile organic compounds and sulfur dioxide—showed significant to astonishing decreases.

† See pp. 128–31.

MAXIMUM-RANGE CARBON DIOXIDE EMISSIONS OF SELECTED AIRCRAFT,

in multiples of the "Pampa" Argentine attack jet's (2003)

All levels expressed in [pounds of CO_2 released if entire fuel capacity were burned]. I have plugged in the emission coefficient of 18.4 lbs carbon dioxide per combusted gal of kerosene-based unleaded JET A-1, the "most commonly used aviation gasoline for turbine engines." A different calculation (given in the source notes) yields 25.5 lbs per gal. *Note: Airplane fuel's "total global warming impact is significantly higher [than its volume of use would suggest], due to vapor trails and the other effects of burning fuel at altitude."*

Levels in <angled brackets> = <total fuel capacity, in U.S. gallons>.

1
"Basic jet trainer/attack jet" LMAASA AT-63 Pampa, Argentina, [9,807 lbs] • <533 gal.>.

1.14
"Multirole fighter" Dassault Rafale [English name: "Squall"], **Mark 2, M02,** France, [11,187] • <608>.

96.79
"Wide-bodied airliner" Boeing 777-200LR ["ultra-long-range version]," U.S.A., [949,256] • <51,590>.

Sources: Jane's, 2003; *Plane & Pilot,* 2017; with calculations by WTV.

Air travel showed me that we had transformed parts of our world into many-crystalled excrescences of carbon-emitting "big five" materials. Sometimes I seemed to be gazing down on an immense circuitboard. The awe-inspiring ugliness created by our modes of travel, production and habitation do not directly bear on *Carbon Ideologies*. But as you study the following photograph, I ask you to ask yourself: "What was the work for?"

Knowing that it behooved me to be humble, and maybe or maybe not ashamed, but at least honest with you that in spite of detailing these bad practices, I was no better, and actually worse, than most of the other human beings who lived on Earth when I did, because I, an American, was richer than most, using more steel and flying in jet planes without caring for limits, I still could not

View of Las Vegas freeway from my airplane seat

refrain from telling you this: In 2010, global per capita carbon dioxide from transportation reached 1,804 pounds, and continued to rise by 0.4% per year.*

But all such ill effects, like those from agriculture, were dwarfed by those from electricity generation.

---

* For further discussion, see "About Internal Combustion Engines," II:422.

About Power Plants

The larger the amount of power transmitted, the better on the whole is the commercial outlook.

Encyclopaedia Britannica, 1911

Steam is really just a medium of exchange, like money in an economy.

Encyclopedia of Chemical Technology, 1994

Both nuclear and fossil fuel plants generated steam, whose pressure spun turbines, which in turn drove electrical generators.

Fossil fuel plants created their steam by means of combustion. In other words, they burned oil, coal or natural gas—thereby producing greenhouse agents such as carbon dioxide. Their steam typically rose up to do work at a surprisingly low 100° Fahrenheit—and an outright astonishing 2,400 to 3,500 pounds per square inch: equivalent to pressures from a mile to a mile and a half beneath the ocean!* How our engineers safely, continuously exploited such forces, so that the thermodynamic work they did became not merely quotidian, but *expected* ("consumers" were at a loss whenever their power switches failed to induce instant magic), was a story I would much rather be telling than this one. To increase and enable productivity is, in and of itself, a noble act. But yesterday's forgotten triumphs became today's unregarded perils.† Meanwhile that high-pressure steam whirled the turbine blades at, for instance, 3,600 revolutions per minute, keeping our lights happily on.

As an electricity primer explained: *Efficiency improves when the source temperature is raised. Additionally, turbine blade wear is reduced as the moisture content in the expanding steam is reduced. Both are accomplished by adding heat to the steam* . . . Of course, *adding heat* required additional fuel.

For their part, nuclear plants employed the heat of radioactive fission to make their steam. Boiling water reactors produced steam directly in their cores, at

---

* Some turbines functioned over more extreme conditions—for instance, 1,100° F and 5,000 psi.

† One reason that my generation wasted so little worry on yours (never mind the main cause, natural selfishness) was that we had witnessed the emergence of so many new machines, networks, solutions and defenses that "necessary limits" reminded us of prudes, fascists, grandfathers who *wanted* us to do without! So why wouldn't efficiency improve as needed?

Model of coal-fired power plant, Kentucky Coal Mining Museum. The "smoke" is white cotton.

around 1,000 pounds per square inch. Pressurized water reactors operated at 572° F, with their pressure at 2,000 to 2,500 pounds per square inch.

From a standpoint of *continuous innate energy,* nuclear fuels were superior almost beyond comprehension—whereas fossil fuels simply burned, then burned out. But from a moment-by-moment thermodynamic standpoint, nuclear and fossil fuel plants were comparably inefficient. The majority of their heat was wasted.*

I asked the "Public Information Specialist" at my local utility company why that had to be, to which he replied: **The simple answer is enthalpy of vaporization or latent heat of vaporization. This is the energy (enthalpy) that is absorbed when we convert water to steam. The Rankine Cycle is the classic power plant model to review this thermodynamic cycle which converts heat into mechanical work in a closed loop system.**

In that closed Rankine loop, wet steam, having done work upon the turbines, enters a condenser which chills it back down into saturated water in preparation for recompression and reheating. And here a thermodynamicist relates the sad truth: *No process is possible whose* sole result *is the absorption of heat from a reservoir and the conversion of this heat into work.*

Penny-saving approaches to turbine design, artificially low fuel prices (for instance, exclusion from expense calculations of global warming's cost), diversion

---

* Internal combustion engines are much the same in this respect. See "Comparative Heat Efficiencies of Engines," II:423.

of kinetic energy from useful work, which occurred thanks to exhaust pressure; heat loss in the condenser, antipollution measures, temporary load reductions and other realities took their own various bites out of efficiency; anyhow, there you have it, not quite as concisely as the "Public Information Specialist" told it.—He was a fellow in a hurry, who helped me for no reason that I can see, excepting conscientiousness. And in his hurry he certainly remembered which issues to avoid. Herewith, my favorite of our question-and-answer sequences:

Q. What would you say to people (such as most of the residents I have interviewed in the Appalachian coal country) who disbelieve in human-caused climate change?

A. SMUD*'s vision is to provide our customers and community with innovative solutions to ensure energy affordability and reliability, improve the environment, reduce our region's carbon footprint, and enhance the vitality of our community.

That made me sorry for him. If you from the future have studied the dates of our decline, you might be incredulous or merely contemptuous to learn that his for-attribution utterance went out not in 1906 or 1976, but 2016.

But he was on point about the Rankine Cycle, which kept dragging down SMUD's "vision," not to mention the yearnings of competitor utilities, into our smudgy human realm.—While I was busily kicking lumps of coal down the dirt roads of Appalachia, the U.S. Energy Information Administration compared the energies *available* in coal, oil and natural gas as delivered to the power plants to the energy *actually generated* by them for electricity. Those three sad ratios of loss remained nearly the same:

INNATE ENERGIES *VERSUS* ACTUAL ELECTRIC POWER GENERATED: OIL, COAL AND NATURAL GAS,

in multiples of the energy loss for oil

(U.S.A., 2014)

All levels expressed in [BTUs per pound].

The "innate energies" are what the EIA calls "fuel heat contents (for fuels received by electric power industry in 2014)." These approximate the default yardstick of this

---

* The Sacramento Municipal Utility District.

book: high heating values [HHVs].* As a cross-check, just right of each EIA figure I give an HHV in *italics* for a specific comparable fuel, as copied from the table of Calorific Efficiencies on pp. 208–18.

I have calculated the "actual" electric power figures from an EIA fact sheet called "How much coal, natural gas, or petroleum is used to generate a kilowatthour [*sic*] of electricity?" [One kWh = 3,413 BTUs.]

Let "energy loss" = [1−(actual / innate)] × 100%.

1

Oil
Innate:	EIA figure: 19,857.	*Mexican crude: 18,755.*
Actual:	6,930.	
	Energy loss: 65.10%.	

1.017

Coal
Innate: EIA figure: 9,710, a surprisingly   *West Virginia "average": 12,500.*
low value.

Actual: 3,282.
     Energy loss: 66.20%.

1.032

Natural gas
Innate: EIA figure: 24,718 [BTUs per lb].   *Dry American: 24,646.*
Actual: 8,109.
     Energy loss: 67.19%.

Sources: Mechanical Engineers' Handbook, 1958 (light Mexican crude); U.S. Department of Energy, 1982 (dry American natural gas); West Virginia Coal Association, 2013 (average for coal); and U.S. Energy Information Administration, 2016 (all power figures); with calculations by WTV.

---

* For definition, see p. 534.

(But I defined efficiency in one way, while a power plant's stockholders might choose another. We carbon ideologues loved to deploy prebaked figures, in order to confirm our positions. And the more efficient we were, the less you from the future should blame us.)

The work required of the fuel to ensure that a turbine could do *practical* work continued to be more than we unknowing "consumers" imagined; and as I lay down in the hotel room in Kyoto, chilled from my morning walk through drips of melting snow, it suited me to increase the temperature by a degree or two Celsius, glad that those grey clouds were now on the other side of the window, while the relevant generating station operated somewhere entirely out of mind.

John E. Amos coal-fired power plant, Nitro, West Virginia

And what was the work for?

To manufacture a 2010-model desktop computer we had to burn up to 660 pounds of coal in some power plant. (Of the electric power thus generated, 95% went to make the silicon chip inside the plastic tower.) One of our most characteristic toys, a refrigerator, required somewhere between 1,000 and 1,400 pounds—or we could equally have sizzled away 700 to 1,000 pounds of heavy grade commercial fuel oil. (Of course more than 80% of that refrigerator's substance was steel and plastic—two members of the "big five.")

What if we could have manufactured durable, sufficient-powered, low-emissions refrigerators at a cost of only 50 pounds of coal apiece? Such improvements, made in most or many areas of energy use, might have ameliorated the future in which you find yourself. I suspect that what Gutowski called *best available technology* would not have sufficed. That still more wistful conditional, a necessary beginning, entailed subjecting all methods of energy generation to determined, even obsessive scrutiny. In this connection, please bear with me through another table, which shows that we could and did improve, fivefold:

COMPARATIVE POWER EFFICIENCIES,

in multiples of lowest gas-turbine efficiency as of 1957

Energy required to produce 1 kilowatt-hour of electric power, expressed where possible in [BTUs of combusted fuel].

Since 1 kilowatt-hour = 3,413 BTUs, wherever I could I expressed the corresponding efficiencies as the ratio of BTUs out to BTUs in. Thus for the steam-electric power plant the quotient of 3,413 and 10,500 is 32.50%, which I call the efficiency.

Natural gas efficiency units were not stated. They were probably per gallon or per cubic foot, not per pound. Thus comparisons between natural gas and other fuels are only approximate.

1–1.5

In an American gas-turbine power plant, burning oil "of lower grade" and therefore of half the cost of diesel, at 140,000 BTUs per gallon, "excluding waste-heat recovery," *ca.* 1957 [27,000–18,000 BTUs per kWh generated]. *Efficiency: 12.64–18.96%.*

2.25

In an American diesel-electric power plant, burning 26 API oil, *ca.* 1957 [12,000* BTUs.] *Efficiency: 28.44%*

2.57

In an American steam-electric power plant at "relatively high load factor," burning coal at 13,000 BTUs per lb,[†] *ca.* 1957 [10,500 BTUs]. *Efficiency: 32.50%.*

2.64

In a Japanese nuclear reactor, *ca.* 2011, two-thirds of the heat energy was wasted (not converted into electricity), and simply caused thermal pollution. This figure comes from an anti-nuclear source. *Efficiency: 33.33%.*

< 2.85

In a natural gas power plant, *ca.* 2002. *Efficiency: "Less than 36%."*

---

* Expressed in BTUs per *net* kilowatt-hour.

[†] Assumption in original. Ordinarily *Carbon Ideologies* postulates a coal efficiency of 12,500 BTUs/lb.

3.09

"The overall efficiency of [American] electric power plants consisting of coal-fired boilers and steam turbines has plateaued at about 39%," *ca.* 1993. "The addition of pollutant control equipment has . . . lowered the effective efficiency of the plant." At 12,500 BTUs per lb, the input energy would be [8,751 BTUs].

3.16

"Best" "fossil fuel" power plant," *ca.* 1976.
Efficiency: "About 40%."

3.16–3.56

In a combined-cycle thermal energy generator, *ca.* 1974. "This uses the exhaust heat from a gas turbine to produce steam for a conventional generator, electricity being produced in both stages."
Efficiency: "40–45%."

> 3.32

In a natural gas power plant, *ca.* 2012.
Efficiency: "More than 42%."

3.56–3.96

"Ultra-supercritical and advanced ultra-supercritical plants," *ca.* 2012.
Efficiency: "More than 45–50%."

4.67–4.83

Tokyo Electric Power Company (Tepco)[†] Kawasaki thermal plant, 2016.
Efficiency: 59–61%.[‡]

Sources: Mechanical Engineers' Handbook, 1958; Hirose, 2011; Encyclopedia of Chemical Technology, 1993; Britannica, 1976; Prentiss, Organisation for Economic Co-operation and Development, 1974; West Virginia Coal Association, n.d.; Australian government, 2015; Tokyo Electric Power Company, 2016; with calculations by WTV.

---

[*] It appeared "possible that further development could raise these to 50–60 per cent."

[†] The culprit of the 2011 Fukushima nuclear disaster.

[‡] The Organisation for Economic Co-operation and Development, Paris, came out with even cheerier figures, *ca.* 1974: "Conversion losses, or waste heat, associated with energy generation" (in other words, "electricity conversion losses as percentages of total primary energy") were a mere 12.5% for Japan, 13.8% for North America, 14.2% for "OECD Europe" and 19.0% for Australia and New Zealand. But in one of the annexes the OECD admitted: "Currently, about two-thirds of the energy used to provide electricity are rejected as waste heat . . ."

I heard efficiencies up to 80% being spoken of: *Systems typically produce electricity or mechanical power and remove the waste heat for heating, air conditioning and other applications. Examples are gas turbines with heat recovery units or steam boilers with steam turbines.* Indeed, our "Public Information Specialist" placed his own utility's best effort nearly in that same range:

Q. How efficient is SMUD's power generation? How close is it to the thermodynamic maximum? How much more efficient could it theoretically become?

A. SMUD's most efficient power plant is Cosumnes Power Plant. It is a 2 x 1 Combined Cycle plant and its Net Heat Rate (Btu/kWh HHV*) at full load is 6,719 at 61F and 6,748 at 104F. This would calculate out to a 51% efficiency rate.

I would have liked to view this facility and learn exactly how its efficiency had been calculated—only half of those BTUs wasted!—but the "Public Information Specialist" stood prepared to indulge me only so far: You may see and photograph from the public road the Cosumnes Power Plant . . . I declined myself that pleasure.

---

* *Carbon Ideologies* prefers to express HHVs not in kWh but in BTUs per pound of combusted fuel.

Power and Climate

There was nothing he could do to help her . . . Regret was illogical . . . but
the empty ship still lived for a little while with the presence of the girl who
had not known about the forces that killed with neither hatred nor malice.
It seemed, almost, that she still sat small and bewildered and frightened
on the metal box beside him, her words echoing . . . in the void she had left
behind her: I didn't do anything to die for—I didn't do anything—

Tom Godwin, "The Cold Equations," 1954

I n and of themselves, power plant emissions were not so bad, given other abuses
of the atmosphere. Consider Canada. In 2008, that nation released 603,000
kilotons of carbon dioxide into the all-accepting azure, of which only 130,000
(not quite 22%) came from "electricity and heat generation"*—for as we have
seen, transportation and manufacturing must do their respective parts: 30-odd
percent apiece. Then came the gracious contributions of commercial and residential energy use. That was only the beginning of my inventory, so I stopped looking around me.

POWER GENERATION'S SHARE OF GREENHOUSE GAS EMISSIONS FOR SELECTED COUNTRIES, 2007–14,

in multiples of the 2012 European Union value

Rounded to the nearest 2 digits right of decimal point.

1

European Union, 2012: **25%** (for "public electricity and heat production"
of EU-15).

---

* Meanwhile, electricity generation produced by one measure 37% of my homeland's carbon dioxide
in 2014, making that activity "the single largest source of CO_2 emissions in the United States."

1.20–1.56
U.S.A., 2014: **30–39%**, "the largest portion" ("heat production" not necessarily included.)

1.36
Germany, 2007: **33.89%** (for "public electricity and heat production").

1.61
Japan, 2013: **40.3%** (for "public electricity and heat production").

Sources: EU greenhouse report, 2014; U.S. EPA, 2016; Greenhouse Gas Inventory Germany, 2007; Greenhouse Gas Inventory Japan, 2015; with calculations by WTV.

These figures were problematic enough . . . but had you lived when I was alive, I fancy my tables and numbers would have been as drearily remote to you as they were to me while I patiently, uselessly marshalled them. Even I in my deluded toil can barely imagine that you will trouble to turn over these data, even with your boot, since you are busy enough scouring dead beaches for food. So let me cut to the chase: Throughout the world, power grids grew and grew. They simply had to.

Between 1980 and 2013 American winter peak load capacity increased by 63%, from 572,195,000 to 921,966,000 kilowatt-hours—3.146 trillion BTUs, most of which came from fossil fuels. We lived up to the highest ideals of the Rankine Cycle: To generate our 3 trillion BTUs we wasted 6 trillion more. Burning 674,286,420 pounds of pure carbon* (which is moderately more energy-rich than West Virginia coal) would have produced enough electric current— and offered up 500 million pounds of carbon dioxide . . .

PRIMARY GREENHOUSE GAS AND PRECURSOR EMISSIONS FROM AMERICAN POWER GENERATION, 2014,

in multiples of the value for nitrous oxide

All figures in thousand metric tons [KMT]. 1 metric ton = 1.1023 U.S. tons.

All figures over 10 rounded up to nearest whole digit.

---

* Which contains 14,000 BTUs per pound.

Methane: [?] This must ignore natural gas leaks.

<1

1

Nitrous oxide: 2,178 [KMT]

1.6

Sulfur dioxide: 3,485

992

Carbon dioxide: 2,160,342

Source: Energy Information Administration, 2016, with calculations by WTV.

If you are wondering why this had to be, let the "Public Information Specialist" solve the question:

Q. For what purpose is electricity most essential?

A. Electricity is a basic component of advanced civilization.

About Solar Energy

Look at the bright side always and die in a dream!

Samuel Taylor Coleridge, *Anima Poetae*, entry from 1804

O f course the partially fruitless work of power plant turbines was an ordinary case. We accepted that one-third of everything we demanded got withheld from us—because electric power was *cheap*!—and the most superficial browsings into thermodynamics show that inefficiency cannot be avoided.—From an Olympian vantage point, solar energy might be said to do even less work than coal or nuclear:

SOLAR ENERGY EN ROUTE TO EARTH'S SURFACE

by seasonal angular alterations and by atmospheric absorption and refraction, *ca.* 1957, 1976*

in comparison to the thermal energy [insolation] received in space

All levels expressed in [BTUs per minute per square foot].

———

"The total solar energy reaching the earth is much greater than the energy requirements of all the world's population . . ."

—Farrington Daniels, 1964

---

* Since the carbon content of our atmosphere continued to rise significantly after these years, more heat was trapped and eventually distributed to the planet's landforms by the time I prepared this table in 2015—a red letter year, when average global temperatures reached 1° Celsius above preindustrial levels. However, the warmer air's rising water vapor levels made it cloudier, so that from the 1960s through the time of writing, solar insolation upon Earth actually went down by 3% per decade. Thank God for small mercies—which might have been spurious, for a different scientific team worked out that "downward thermal radiation" had been increasing by as much as 4.25 BTUs per square foot each decade.

One estimate for this total solar energy is 9.9 quadrillion [9,897,816,000,000,000] BTUs per minute, or 174,000 terawatts.

"The sun delivers energy approximately 4,000 times faster than we use it."

—Mara Prentiss, 2015

100%
Just above the atmosphere and normal [perpendicular] to the sun's rays [7.37 BTUs/min-ft^2]. *Loss: 0%.*

67%
"At most, 67 per cent . . . may reach the earth's surface at noon on a clear summer day."—Eugene P. Odum, 1971. [4.94]. *Loss: 32.97%.* But the solar expert who reviewed this chapter in 2017 asserted that a good 77% actually strikes the ground.

23.6%
At ground level in the continental United States in summer, high value [1.74]. *Loss: 76.39%.*

20.7%
Ground level at Stillwater, Oklahoma, June [1.525]. *Loss: 79.31%.*

18.0%
General amount of insolation striking planet's landmass [1.327]. *Loss: 81.99%.*

17.0%
Ground level, continental United States in summer, low value [1.25]. *Loss: 83.04%.*

10.4%
Ground level, continental United States in winter, high value [0.764]. *Loss: 89.63%.*

7.3%
Ground level at Stillwater, Oklahoma, December [0.538]. *Loss: 92.70%.*

4.7%

Ground level, continental United States in winter, low value [0.347]. *Loss: 95.29%.*

Sources: Mechanical Engineers' Handbook, 1958; *Encyclopedia of Agriculture and Food Systems,* 2014; IPCC, 2013; Farrington Daniels, 1964; Darling and Sisterson, 2014; Odum, 1971; Leckie et al., 1975; Dr. Canek Fuentes-Hernandez, 2017; with calculations by WTV.

Nor does the diminution of solar energy end when it strikes a particular surface. Physicists have invented the lovely concept of a *blackbody,* "an ideal substance capable of absorbing all the thermal radiation falling on it." It must also be able to emit all the radiation it absorbs. There is no such thing. Hence further energy losses.

But here I had better quote a senior research scientist at Georgia Tech's Strategic Energy Institute. His name was Dr. Canek Fuentes-Hernandez; I paid him to read this chapter, and it is well that he got something out of it, for he had to clean up several obsolete ideas and amateurish misconstruals—doubtless not the only such to soil *Carbon Ideologies.* Regarding the table he wrote: *The losses you state . . . are somewhat irrelevant since even if year-round 96% of the total solar energy received . . . was lost on its way to earth's surface, we would still get by collecting the remainder for a few months to fulfill all our current yearly needs.*

Could solar power preserve us, then, from our hot dark future? On this subject one might without irony utter the word "progress." High up on my most spiderwebbed bookshelf, an oversize paperbound eco-tract expressing 1975's idea of *self-sufficient living* lamented that *we're going to have to wait a few more years, until solar cells become quite a bit cheaper, before self-sufficiency in electricity will be feasible in the city.* Hence it consigned home solar energy to space heating only. Forty-two years later, as I finished this less than idealistic screed, a newspaper advised me that *millions of Americans now get their electricity, at least in part, from . . . solar panels[,] . . . thanks to their sharply declining cost*—while beneath an awning just outside a "big box" hardware store in Sacramento, two babyfaced salesmen stood pimping out silicon-based solar roof tiles for home electricity (whose efficiency they estimated at "around 20%"). So drink in sunny rays of hope, and postpone your engagement with the darkness! All the same, most everyone I knew, by some coincidence, still intended to *wait a few more years* to solarize. Why rush? There was no immediate danger—nor any immediate salvation; for even if we could have immediately solarized our planet's greenhouse emissions down to zero, existing levels of carbon dioxide ensured continued warming. A U.S. government climatologist wrote me this reminder—and in the chapter about greenhouse gases I shall repeat it: *Over 2000 years, the cumulative amount of heat retained in the Earth system from CO2* emissions is about 8 times larger than if we count only the first 100 years.* Hence only a Manhattan Project to partially decarbonize the atmosphere would have kept your climate as it was when I was alive. Too laborious—we preferred other delusions!

Carbon Ideologies largely neglects solar power, that being associated with decentralization and environmental benignity. Indeed, solar *is* an ideology of hope—not my department. But let me at least introduce it, as a foil to nuclear, oil, natural gas and coal.[†]

A solar collector succeeds to the extent that it approaches a blackbody: It must capture the widest possible spectrum of electromagnetic wavelengths, then re-transmit them into some energy-generation system. The more closely both input and output approach 100% efficiency, the better founded the hope.

Carbon black acrylic paint does an excellent job at 94% absorbance and 83% emittance; while zinc oxide white, as might be expected from what is literally the near opposite of a blackbody, absorbs merely 12 to 18%, although it does emit 88%. So paint your collector black! The substrate might be copper, which offers high thermal conductivity—170.35 times that of ice and 16,042 times that of rigid foam polyurethane—although *aluminum is perhaps the best and cheapest material for direct reflection of sunlight.* Thus our aspirations in pre-photovoltaic days.

In the winter of 1959–60, a certain solar house in Denver received 226.86 million BTUs (mBTU) upon a 600-square-foot collector area. This was equivalent to the inherent energy of 18,149 pounds of average-quality West Virginian coal—but total incidence *when collection cycles operated* was only 161.33 mBTU, while *useful collected heat* came in at 55.72 mBTU, or 24.56% of what had originally fallen on the collectors—worse than the 30-odd percent that a traditional power plant could have burned out of that coal . . . or less than 2% of the original solar radiation reaching the top of our atmosphere. A hairsplitter could thus argue that the Denver house's use of energy was 98% inefficient.[‡]

In another mid-century experiment, English sugar beets, *under intensive cultivation during favorable growing conditions,* got their own solar energy efficiencies measured. It turned out that only 7.7% of the light falling upon them could

---

* No subscript in original.

† See part 7 of the "Definitions, Units and Conversions" section (pp. 572–75) for some technical particulars.

‡ The practical result was that 28.2% of the "house heat load" got supplied by solar, and the rest by natural gas. Well, that was better for the climate than the workings of *my* house.

be converted through photosynthesis. The fraction which became food energy was 5.4%. Hawaiian sugarcane and Israeli maize performed worse.

But why should that waste matter, with so much solar insolation to go around? Two early-20th-century steam plow enthusiasts belted out the good news: *The sun stores up power enough in an acre of plants, in a single season, to plow, sow, and harvest that acre for a century*—if those plants are converted into alcohol fuels (efficiency: half of gasoline's, then burned in a one-cylinder engine. You from the future may see something wrong with *that* picture.*

Decade after decade, our fossil-fueled power plants' ratios of uncaptured to utilized carbon combustion energy ensured an obscenity of greenhouse gas releases, and even when we worked fossil fuels more directly and completely, for instance by burning coal in a stove on a winter night, up went carbon dioxide regardless. Meanwhile, solar energy's dwindlings from searing glory in the near-vacuum of the outer thermosphere down to mildness upon land and sea elegantly powered our ecosystem: our sugar beets and other solar collectors neither brought the sun closer, nor made it burn any hotter.

How efficient a solar collector was our planet itself? From the standpoint of simple energy retention, as my table shows, not very. And thank goodness for that! Otherwise Earth would have been far too hot before the game started.

Where did the heat go? Quantities ricocheted off the dark side of the planet. (On the daylit side, new solar radiation made good the loss.) Meanwhile, Earth kept sunshine busy doing thermodynamic work. Consider one case from Canada's Ottawa River Valley. When I was alive, Perch Lake (46°02' N, 77°22' W) consisted of a modest body of water, about 4,800 square feet in area, and some six and a half feet deep. (For all I know, in your time it has become a desert.†) From the beginning of May through the end of October, Perch Lake used to receive not quite 150,000 BTUs of solar radiation per square foot. In other words, over those six warm months each square foot of the lake took in as much energy as would have been liberated by the complete burning of more than 10 pounds of pure carbon, or not quite 3 pounds of rocket fuel. So why didn't Perch Lake boil away? Two scientists traced what work the sun actually did there.

---

* Dr. Fuentes-Hernandez notes here: "As you point [out], the process of converting solar energy into biological matter is not very efficient. The efficiency of photosynthetic processes converting sun light into chemical energy stored in molecules is indeed around 7%. Furthermore, refining processes are needed to convert biomass into useful biofuels; with some efficiency losses associated with this conversion. Note that fossil fuels, including coal, are indeed sub derivative products of these photosynthetic processes and thus, we may say that from the Olympian point of view you refer to, are considerably less efficient in converting solar energy into electricity (maximum around 4.2%) than for instance the photovoltaic conversion of even the cheapest commercial solar (photovoltaic) cells (> 7%)."

† Or worse. This place figures in the nuclear section (p. 342).

80% provides latent heat of evaporation—the same energy sink that pilfered the useful work of our steam turbine power plants—*and 20% provides sensible heat lost to the atmosphere.*

As went Perch Lake, so went the world—for most of the planetary surface was ocean, which absorbed heat into its water, some of which accordingly evaporated, in the process generating thermal winds which moved air and water about. Seawater distilled itself into clouds; clouds rode the thermals until it came time to rain; thus the dynamic equilibrium which our "civilizations" commenced to alter. To be sure, even without us there had been disconcerting wobbles in the cycle, such as Ice Ages. But from our ephemeral human perspective there was or had been a steady balance, which Fourier in his *Preliminary Discourse* (1807) summarized with pleasing concision:

> The radiation of the sun in which this planet is incessantly plunged, penetrates the air, the earth, and the waters; its elements are divided, change in direction every way, and, penetrating the mass of the globe, would raise its temperature more and more, if the heat acquired were not exactly not balanced by that which escapes in rays from all points of the surface and expands through the sky.

Greenhouse gases, of course, increased energy retention. Had that great power station called the Earth been amenable to human microcontrol, we would have been thrilled to improve its heat into tame electricity, allowing us to acquire new desires and ramp up demand beyond our current greediest extrapolations.

Instead, we locally concentrated the sun's radiation, in hopes of servicing demand.

For a long time I blamed solar's slow progress in the U.S.A. on political inertia, insufficient economies of scale and modest efficiencies. But then the politicians enacted financial credits for renewable energy in any number of states. The cost of solar tiles began to decrease. Best of all, efficiencies increased.

Carbon-blacked copper plates became, oh, so 20th century! Nowadays we aggrandized solar energy onto a bank of molten salts, *to significantly increase their temperature and . . . boost the efficiency of the heat engine (generally steam turbine) used to generate electricity.* Solar heating grew almost quotidian. Better yet, we employed photovoltaic cells, which were, as Dr. Fuentes-Hernandez defined them, *devices that absorb light for its direct conversion into electricity.* Thus the dream, to ban the other four ideologies' harmful byproducts, and with refined passivity to sustain ourselves from sunlight!

Photovoltaics' output *is only limited by the absorption properties of the material[s] used to absorb light (typically these materials do not transform light into heat, but rather into light again;* in other words, *they have very high photoluminescence efficiencies.* Before I turn to their two disadvantages, let me praise their *potential for the smallest energy payback time than any other technology, including wind. This is because, compared to other energy sources, they require little to be produced, maintained, and decommissioned.* Therefore, certain carbon ideologues would hate them.

How these extraordinary devices were made, and even what they were made of, I had better not tell you, for in five or 10 years I would already be out of date. Since this book is already less simple than it aspired to be, why not omit some details? What mattered was the theme, the goal, or, as a self-reverencing executive might say, the "vision": Excite and satisfy demand!

In 1964, Farrington Daniels had hoped for 10% efficiency in solar electricity. In 1972, an automobile engine man declared: *Today's best [solar] cells convert into electricity only about 11 or 12% of the sunlight striking them.* In April 2014, the corporation Okamato fabricated *a silicon solar cell that reached 25.6 percent efficiency, breaking a 15-year-old world record of 25.0 percent.*

Dr. Fuentes-Hernandez now sent me graphs: Thin films of gallium arsenide outperformed gallium arsenide crystals; next came mono-silicon, followed by poly-silicon; those four frontrunners clocked in at not far under 30% efficient. *Remember,* he wrote me proudly, *that efficiencies here are those of directly converting light into electricity.*

That year a scientifically trained anti-nuclear activist in Tokyo had already assured me that *the efficiency of solar power generation is increasing rapidly, and currently the highest is about 20% (module conversion efficiency). And this number is expected to go higher. For industrial use the highest efficiency would reach about 50%. It is considered that 45% has already been reached in the research level.* Why not? According to those graphs, the new Fraunhofer ISE/Soitec multi-junction "research cell" was 46.0% efficient.

Determined to make his field of expertise, so to speak, shine, Dr. Fuentes-Hernandez (who in fact called solar second-best, after nuclear fusion) insisted:

> For a coal fired boiler operating with high pressure steam at a temperature of around 540°C, the maximum theoretical efficiency is 64%. [As *Carbon Ideologies* glumly reiterates, the actual efficiencies of coal power plants have generally been half of that.] Since nuclear uses steam as well, the maximum conversion efficiency is not that much different. The direct conversion of solar energy into electricity in a single photovoltaic cell is

limited to around 34%. However, using multiple cells, the maximum theoretical efficiency is about 69%, not too different from that of coal or nuclear.

And that was not all. Skewering our carbon follies like the zealous solar ideologue he was, he continued:

> If we are willing to claim that the efficiency of fossil-fuels or nuclear is around 60%, and neglect the efficiency of converting solar energy into the biomass that produced these fuels or the efficiency of production of uranium through natural or artificial processes, then we may as well neglect all external losses prior to the absorption of electromagnetic radiation into a solar cell. If we do so, we could claim that the efficiency through which a ... photon ... of absorbed electromagnetic energy is converted into electricity is in fact nearly 100% (or even higher if we consider that physicists have discovered ... singlet exciton fission, whereby 2 electrons can be produced per photon absorbed) ...*

It seemed as if we could triumph over physical necessity itself! If that were only so (and maybe in your time it will be), we could return to the edifying task of deciding how to squander electric power. And if it became harmlessly abundant there would be no more talk of squandering it. In Volume II you will meet a Bank of Oklahoma Vice-President who says of oil: *I'm trying to let you know how cheap this is!*" In that spirit, Dr. Fuentes-Hernandez now addressed the issue of costs and investments:

> ... Solar energy is so plentiful that if ... we could collect [it] for 1000 h (~1/8 of the number of hours in year) we would collect 1 TW/km2.† ... We would need an area of around 120,000 km2 (about the area of North Korea) to fulfill the entire world demand if we had 100% efficient photovoltaic cells, or about the area of Peru (1 285 000 km2) ... [assuming] an efficiency of around 10% (commercial modules are at around 17%) ... Around 2.7% of the world land is already occupied by urban development ... In the US, in 2000 ... about 250 000 km2 [were] covered by urban developments. If ... only half [this] area ... were

---

* "[D]efined as the number of electrons generated per photon absorbed in a photovoltaic cell, [this] is known as internal quantum efficiency ..."

† No superscript in original. Remember that a TW is a terawatt, or 56,884,000,000 BTUs (the inherent energy of 4.55 million pounds of West Virginia coal) *per minute.*

covered with 17% efficient photovoltaic cells[, t]his would supply about 106 TWh/day, or 38,690 TWh in a year, about 3 to 4 times the amount of electric power consumed in the US in 2016 (40 quadrillion Btu = 11 000 TWh...)

Unfortunately, we had not yet learned how to store the sun's energy in batteries the way we could hoard it in heaps, tanks and barrels of fossil fuels. Even Dr. Fuentes-Hernandez had to admit this as *the only compelling reason to keep using non-renewable energy sources that operate 24/7.*

What about that North Korea–sized reserve of photovoltaic cells? A certain young, bespectacled public relations man for an unfortunate, hence infamous Japanese nuclear company* reminded me: "Japan is an island country, unlike other countries operating this [solar grid] system on a larger scale." In other words, where could the Japanese situate their electromagnetic plantation? "And when it rains," he continued, "the solar power will be zero."

Perhaps if the issues of storage and siting were solved we could air-condition ourselves and our crops in solar-powered hives even as global temperatures continued to rise. (Emerson: *Dream delivers us to dream, and there is no end to illusion.*) Better yet, I would make of this book a forgotten error. Solar tiles on every roof and solar-charged battery systems would guarantee a photovoltaic paradise! Unfortunately, as carbon dioxide from fossil fuels increased the heat-retention efficiency of our atmosphere, a bad old story continued to be told, whose gloomy ending you in my future can see all too well.

In my most embittered moments I would never have imagined *a concerted and well-funded lobbying campaign by traditional utilities... to reverse incentives for homeowners to install solar panels.* In 2017, President Trump's new Energy Department decided to "study" the matter—in other words, I suspect, to kill those incentives utterly. The department's appointee happened to be *a former economist at the Institute for Energy Research, a nonprofit funded by* the ultraconservative Koch brothers, who profited (to say the least) from fossil fuels. But for all I know, their onetime employee (a Mr. Travis Fisher) might have been highly objective. It must have been coincidence that he had already *called clean energy policies "the single greatest emerging threat"* to the electric power grid. Perhaps he truly could have explained his intentions and instructions so as to prove that he lacked any animus against renewable energy. Instead, like most of the high-placed carbon ideologues whom you will meet in this book, he refused to comment.

---

* Tepco, which will play a large role in this volume.

About Greenhouse Gases

*The greatest things in the world are brought about by other things which we
count as nothing: little causes we overlook but which at length accumulate.*

Georg Christoph Lichtenberg, *Notebook A,* 1765–70

A greenhouse gas is so defined because it traps solar heat, thereby warming the planet. In our oversimplified carbon ideologies, we often spoke as if there were only one such substance. Specifically, when we expressed our worries about "burning carbon," we tended to mean that ubiquitous carbon dioxide, which may practically be considered a signature not only of respiration, volcanism, fertilization, cement-making and chemical engineering—but most relevantly here, of combustion—for it certainly found its way up from all of our smokestacks, no matter which fossil fuels we burned:

COMPARATIVE CARBON DIOXIDE EMISSIONS OF POWER PLANTS, 2014,

in multiples of those released by natural gas facilities

All levels expressed in [grams of CO_2 given off per kilowatt-hour generated].
$1\ kWh = 56.88\ BTUs.$

The Cirebon figures may display spurious precision, given that the others must be rounded approximation.

1

Natural gas [400 g CO_2 / kWh]. [Another source claims baldly: "Combustion of natural gas emits about half the CO_2 that coal generates at equivalent heat output."]

1.5–1.75

Oil [600–700].

> ### 2.14–2.19
> "New technology" Cirebon coal power plant, Kanci Kulon, Indonesia [856–76].
>
> ### 2.5
> "Old technology" coal [1,000].
>
> *Sources: Charleston Gazette, 2014; Encyclopedia of Chemical Technology, 1994; with calculations by WTV.*

These figures go far to show why coal was not the wisest energy source, back in the days when we were alive.

They also imply that if we had to burn carbon, natural gas would have been our best choice. And indeed, in the ringing words of one syndicated columnist: *Natural gas emits about half as much carbon as coal and can transition us to truly clean power.*—What she wrote might have even been true—in a leak-proof world.

Carbon Ideologies has already mentioned methane and nitrous oxide. Let us call back to mind the first of these.

Like CO_2, methane (CH_4) contains only one carbon atom per molecule. All the same, its ability to effect global warming was 20 times worse than carbon dioxide's—or 21—or 33—or even 86 times worse! (Interested parties uttered whichever numbers suited them.)* It was natural gas's most common component—and our natural gas pipelines were notorious for ineffective seals. On occasion we used to simply burn the methane in "remote" gas wells *because the pipeline gathering systems needed for such gas tend[ed] to be prohibitively expensive.* That was better than letting it leak, but still harmfully careless. As a leading carbon ideologue confessed:

> The flaring of about 1 billion cubic feet per day of natural gas, or about 30% of production, is happening right now in the Bakken Shale play, one in which my company is invested. We're flaring natural gas in certain areas of the Bakken because there's no way to capture the gas through a pipeline, and the Bakken oil can't be produced without flaring that gas. We're wasting that resource . . . It's been estimated that Bakken flaring wastes more than $1 billion per year in natural gas production values.

---

* See II:1, 309, 312.

And to those few of us who did not care about production values (you from the future won't count) he admitted:

> Purposeful venting or flaring of natural gas for lack of market is wasteful and puts huge volumes of methane into the atmosphere. According to the International Energy Agency, flaring and venting of methane amounts to the emission of 1.1 billion metric tons of CO_2-equivalent per year . . .

So that was unnerving. Meanwhile, nitrous oxide caused its own difficulties. Considering the trio *en bloc,* the Intergovernmental Panel on Climate Change concluded:

> Concentrations of CO_2, CH_4, and N_2O now substantially exceed the highest concentrations recorded in ice cores during the past 800,000 years. The mean rates of increase in atmospheric concentrations over the past century are, with *very high confidence,* unprecedented in the last 22,000 years.

But treating them as the root cause of our doom was an only moderately unwarranted simplification.

Yes, carbon dioxide was by volume the worst. In 2007, that gas made up 87.9% of all German greenhouse emissions. In 2013, it comprised 93.1% of Japan's contribution to climate change.* At that period, in the course of each hour it warmed the Earth by half a BTU per square foot.† A dozen burning match tips in a day—who would feel such trivial heat? In the course of a century, to be sure, its steady, inconspicuous power (which climatologists called its *radiative forcing*) might alter our future, but I was pushing 60; I'd soon be dead. As for you, reader, *you* must have begun to sweat . . .

As you remember,‡ carbon dioxide concentrations had been increasing ever since 1750. As they rose, CO_2's radiative forcing power likewise strengthened. By 2100, it might be 3 or 5 times worse than it was in 2011; thus two of several climatologists' scenarios. Of course there was no immediate danger; we carbon

---

* The figures in this paragraph were exclusive of changes in carbon sequestering through land use (a category which we used to abbreviate "LULUCF"; see p. 596).

† In the original units, 1.68 watts per square meter. For conversions, and for more detailed information on GWP and RF, see pp. 591–97.

‡ From p. 96.

ideologues of all stripes went on uttering our jackdaw cries without much inconveniencing each other.

Let this 100-year effect of CO_2, as exerted by a given quantity (a ton, or 100 tons), be called its global warming potential. And let it be quantified as 1.

The global warming potentials of other greenhouse gases might be lesser or greater than 1. As carbon dioxide's absolute forcing ability grew, the GWPs of those rival climate change agents, being relative, proportionately decreased. However, were their absolute forcing strength to increase in consequence of their own rising concentrations or due to chemical interactions, then scientists would adjust their GWPs upward. (Thomas à Kempis, *circa* 1413: *Carefully observe the impulses of nature and grace, for these are opposed to one another . . .*)

Thus inconstant, arbitrarily defined and subject to wide calculation and projection uncertainties, the GWP nonetheless remained a useful simplication that any of us could comprehend, had we bothered to.

When I first encountered the shockingly high GWPs of methane, the halocarbons, nitrogen trifluoride and the rest of them, it seemed as if I had just perceived a legion of new enemies.

Reviewing this chapter, Dr. Pieter Tans at the National Oceanic and Atmospheric Administration wrote me:

> CO2* is easily the most important [greenhouse gas] that needs to be tackled. It is very unfortunate that the I[ntergovernmental] P[anel on] C[limate] C[hange] came up with the 100-year time horizon. [See next table, beginning on p. 176.] They might have done that with the idea that in 100 years future generations would be so technically advanced that they could "undo" the CO2 already emitted. Or perhaps, the responsibility of our generation for the future of our planet and its ecosystems does not go beyond 100 years. The problem is that the residence time of the additional CO2 emitted in the combined "atmosphere plus oceans" system is thousands of years. Therefore there will be plenty of time for slow feedbacks to come into play, such as ice cap melting, permafrost melting in the Arctic, to name a few.

Yes, carbon dioxide was the worst, *so far*. Maybe Dr. Tans would be proven right a century from now, and it would continue to be the worst.—Methane and nitrous oxide were identifiably harmful—but too many of the most dangerous compounds we released into our atmosphere had names, purposes and

---

* No subscripts.

concentrations known only to specialists.* What made their effects still more obscure was that (like all greenhouse gases) they operated one way in the short term and still another over their entire chemical lifetimes. As usual, we took their good qualities on faith, that being less tiring for us and for the entities that made and sold such substances.

Our refrigerators contained chemicals with ominous global warming potentials. They obviously had nothing to do with "burning carbon," so we wasted scant anxiety on them. Besides, they excelled at arresting food decay! A compressor, powered (through cord, outlet, transformers and wires) by fossil fuels or nuclear heat in some inconspicuous generating station, readied these unseen chemicals to absorb heat, after which they were permitted to expand, which automatically cooled them, which sucked away heat from our food, and then it was back to step one again, all courtesy of electricity on demand!

Once upon a time, ammonia used to chill our milk and butter for us. In H. P. Lovecraft's horror tale "Cool Air," it could even keep a dead man alive. But that "environmentally benign refrigerant"[†] stank; besides, it was corrosive and poisonous; so our engineers invented chlorofluorocarbons of all sorts—so many that we distinguished them with hyphenated numbers—for instance, CFC-12, which by the last decade of the 20th century took pride of place in "the vast majority" of refrigerators in the world.[‡] The trade name for the whole group was "Freon," and so until the basic patent ran out they were known as "the Freons."[§] They announced themselves as elegantly as did certain actresses with their custom-blended "signature perfumes"—for thanks to their chlorine atoms, the slightest wisp of them would burn with a deep green flame.—Oh, what excellent work they did, and in such a fine cause! *The Freons are colourless, odourless, nonflammable, noncorrosive gases or liquids of low toxicity,* enthused my 1976 *Britannica,* and one scientific survey lumped them all wistfully together as *an almost perfect industrial chemical.* How prudent, that "almost"!—An illustrated guide to product manufacturing explained that *as gases in the chlorofluorocarbon (CFC) group, which includes freon, waft upward into the stratosphere . . . they gradually decompose, releasing chlorine atoms,* which in turn can destroy 10,000 ozone atoms apiece, drastically decreasing our protection from skin-cancer-causing ultraviolet rays and injuring

---

* And indeed a specialist (an organic chemist) might have been able to predict some of their peculiarities. For instance, carbon-oxygen and carbon-fluorine atomic bonds within a given gas molecule indicate some potential for global warming.

† Whose manufacture produced carbon dioxide. See above, p. 126.

‡ Since they would work for "20 years or more," many of them must have been getting scrapped at the time I was writing *Carbon Ideologies.* The consequence: an uptick in the release of CFC-12 into our atmosphere.

§ My grandfather's *Mechanical Engineers' Handbook* also refers to them as the Genetrons.

food crops. The Americans, who in those days still believed in climate change because they were not called upon to do anything about it, calculated that at the 1976 emissions rate, CFCs would make global warming 10% worse. So all of us together phased out the CFCs, replacing them with hydrochlorofluorocarbons (HFCs), which "have zero O[zone] D[epletion] P[otentials]s and low-to-moderate G[lobal] W[arming] P[otential]s." But some people's definition of "low-to-moderate" was not mine. Please make up your own mind when you see the next table.

(Wouldn't you have liked it if the industry had weighed in? Through an intermediary—just in case I was on some blacklist—I "reached out," as we used to say, to seven large American refrigerator manufacturers, politely inquiring how harmful to "some ecosystem somewhere" their products might be. Six companies never replied; as for the seventh, that industrious intermediary, who was also my friend, contacted its "media company," then reported: *Got her on the phone, she said no, client wouldn't want to participate, 5-27-16.*—Anyhow, wasn't it better not to know? Thus I consoled myself, wishing I weren't writing *Carbon Ideologies*. It was an especially warm July day—108° Fahrenheit, with more of the same scheduled for tomorrow—so I opened the fridge and drank a nice cold beer.)

Refrigerators transported by dhow for sale, Dubai

The CFCs contained chlorine, fluorine and carbon. The HCFCs (and their kindred HFCs) contained hydrogen in addition, which rendered them less susceptible to shedding predatory chlorine atoms. HCFC-22, for instance, did *less than five percent of the damage to the ozone layer that CFC 12 does.*—Molecule for molecule, they were all variously worse than carbon dioxide:

COMPARATIVE ONE-CENTURY GLOBAL WARMING POTENTIALS,

in multiples of carbon dioxide's

(Time Frame: Within 100 Years After Release*)

Each numerical header [such as **1** for carbon dioxide] is the actual GWP. All figures over 10 rounded up to nearest whole digit. After nitrous oxide, all levels rounded to 2 significant digits.[†]

According to the **IPCC**, uncertainties for very long **GWP**s may reach +/− 35%.

• = Targeted [in Kyoto Accords and annexes] for tracking and reduction. The Kyoto Protocol Gases were carbon dioxide, methane, nitrous oxide, HFCs, PFCs and sulfur hexafluoride. Nitrogen trifluoride was later added to the list.[‡]

+ = "Precursors." These substances cause warming more indirectly than directly, by affecting the production or aggregation [in clouds] of other greenhouse gases. Among these is ozone, which I have not tabulated here due to its complexities of generation and altitude dependence.

<1

Ammonia [?].
"The use of ammonia, which is considered an environmentally benign refrigerant, will continue to play an ever increasing role." [Of course CO_2 is released during the manufacture of synthetic ammonia from fossil fuels.] The Intergovernmental Panel on Climate Change asserted that this chemical might actually exert a cooling effect.

> Atmospheric lifetime: <1 year.

1

• **Carbon dioxide.**
The favorite villain of most carbon ideologues.

> Atmospheric lifetime: Once thought to be 120 years. We were dreaming. Dr. Pieter Tans to WTV: "Over 2000 years, the cumulative

---

* Obviously a substance which lingers longer than 100 years in the atmosphere (e.g., carbon dioxide or CFC-13) will exert a GWP worse than the value listed here. But unless we know its rate of decay we cannot exactly quantify the additional harm it does. For instance, CFC-13, which lingers for 400 to 640 years, might show a GWP of nearly [11,700 ×, say, 4] = 46,800 times worse than carbon dioxide's, or it might fade away much faster. In fact its 500-year GWP is 16,400.

† This book's convention for "significant digits" will be "digits to the right of the decimal point."

‡ HCFCs and HFCs were singled out as a group by the Kyoto Accords. The ones in this table are representative but by no means all-inclusive.

amount of heat retained in the Earth system [from CO_2] is about 8 times larger than if we count only the first 100 years . . . I have taken into account that approx. 83% of the current CO2* emissions eventually end up dissolved in the oceans, leaving only 17% in the atmosphere after 2000 years. That number is based on well understood ocean carbonate chemistry. However, the transfer to the oceans occurs very slowly, and I have not estimated how long it takes for dissolution of calcium carbonate minerals to neutralize the added carbonic acid. Estimates are that it may take 3000–7000 years. In other words the factor 8 that I mentioned could still be larger. On the other hand, carbonic acid is another word for dissolved CO2, and neutralization means that the oceans can hold the extra carbon while atmospheric CO2 is back to preindustrial chemical equilibrium values of 280 ppm."

2–3

Water vapor.
Contributes to warming only in the upper atmosphere. "Currently, water vapour has the largest greenhouse effect . . . However, other greenhouse gases, primarily CO_2, are necessary to sustain the presence of water vapour in the atmosphere."
Atmospheric lifetime: 10 days.

3

Propane[†].
A component of natural gas.
Atmospheric lifetime: <1 year.

3

Isobutane[†].
A component of natural gas.
Atmospheric lifetime: <1 year.

3

N-butane[†].
A component of natural gas.
Atmospheric lifetime: <1 year.

---

* No subscript in original.

[†] Re: propane, isobutane and n-butane: "G[lobal] W[arming] P[otential] of the hydrocarbons results almost entirely of the GWP of the CO_2 owing to decomposition."

3–7.6

+ Carbon monoxide.

A common emission from incomplete fossil fuel combustion. Between 1969 and 1980, U.S. exhaust emission standards reduced its per-mile concentrations by nearly 12/13. Many European emissions factors (EFs) also fell. [Average EF of German aircraft kerosene, "all flight phases (LTO and cruising flight), 1992–2015": 1 pound CO given off for every 108.69 pounds of fuel burned.] According to Japan Tobacco Inc., 1 cigarette emits 0.055 grams [1/8,246 lb]. Considered both a greenhouse gas and a "precursor" whose indirect effect is to increase levels of other GHGs. Concentrations vary by continent. GWP range is global average, including direct and indirect aerosol effects.

> Atmospheric lifetime: [?]

21–25 or higher (86 over the first 20 years)

• **Methane**, which is CH_4.

The primary constituent of natural gas. Also an automotive tailpipe pollutant. Released by garbage, flooded ricefields, manure, etc.

> Atmospheric lifetime: 12 years. *Or, if you prefer, "the atmospheric lifetime of methane has increased 25–30% during the past 150 years to a current value of 7.9 years, implying gradually decreasing oxidizing capacity" of our planet, thanks to sulfur dioxide pollution.*

77–93

• **HCFC-123**, which is $CHCl_2CF_3$.

A hydrochlorofluorocarbon used as refrigerant and also as a "blowing agent" to polymerize,* for instance, the foam insulation in refrigerators. A substitute for CFC-11.

> Atmospheric lifetime: 1.3 to 2 years. *But over 20 years its GWP is a horrendous 273.*

80–238. *[In subsequent tables I use the IPCC 2013 value of 159. But the true number might be negative.]*

• + **Nitrogen oxides**, collectively abbreviated "NOx."

A group of chemicals similar to nitrous oxide, and similarly derived from soils, fuel combustion, etc. Considered both greenhouse gases and [especially nitrogen dioxide] "precursors." "NOx ... emission control has both a cooling (through reducing of tropospheric ozone) and a warming effect (due to its impact on methane

---

* A polymer is a large molecule made up of many small molecules. Leather was one such, and cellulose another. When I was alive, the most common polymers we made were plastics.

lifetime and aerosol production)." In other words, leave it to the experts. Concentrations vary by latitude. GWP range is global average, including direct and indirect aerosol effects.

Atmospheric lifetime: [?]

124–140

• HFC-152a, which is CH_3CHF_2.
A hydrofluorocarbon used for refrigeration and for dusting computer keyboards. "Quite toxic."

Atmospheric lifetime: 1.4 to 2 years. *But over 20 years its GWP is 437.*

264–310

• **Nitrous oxide**, which is N_2O. *In subsequent tables I use the IPCC 2013 value of 264. The U.S. Environmental Protection Agency called it "approximately 300 times more powerful than CO_2."*
[Not the same as the various *nitrogen oxides*—see above—which are also greenhouse gases.[*]] "Generated by the action of microbes on nitrogen that leaches or runs off as nitrate from synthetic fertilizers, organic fertilizers, . . . etc." A widespread decay product of aerated manure, sewage, compost, etc. "Increasing mainly as a result of agricultural intensification to meet the food demand for a growing human population." Sometimes used as a commercial refrigerant. Laughing gas. *Ca.* 2007, 10% of global emissions derived from anesthetic use. Released during the production of nitric acid (for fertilizers) and adipic acid (for nylon).[†] Ammonium nitrate explosives give it off. Employed in semiconductor manufacture. Also, as it happens, a component of automotive smog. (In 1979, traveling at 60 mph on "freeways and surface arterials," American cars emitted 3.59 grams of it per mile; American trucks gave off 13.52.) "Responsible for immune system impairment, exacerbation of asthma and chronic respiratory diseases: reduced lung function and cardiovascular disease." "The increase, at least since the early 1950s, is dominated by emissions from soils treated with synthetic and organic (manure) nitrogen fertilizer . . ." Average rate of rise: 0.75 ppb per year.

Atmospheric lifetime: 114 years. *Another source gives 131 years.*

---

[*] In 2009, the German government sadly reported: "Profound changes have occurred in connection with emissions of NITROGEN OXIDES, since efforts to make aircraft engines more fuel-efficient have led to increases in average emission factors . . . Consequently, a mean EF (NOX)," or emission factor for nitrogen oxides, "of about 14.0 g [of nitrogen oxides] /kg [of aviation kerosene] can currently be assumed."—On the bright side, as was proven in the *National Enquirer,* "Nitric Oxide [*sic*] could not only increase your ability to get an erection, it would also work in your brainwaves to stimulate your desire for sex . . . As a doctor, . . . I'm impressed by the way it increases cerebral and penile blood flow."

[†] See "The Parable of Adipic Acid," above, p. 128.

549–1,500

- HCFC-22.

A replacement for the dreadful CFC-12. Not only a refrigerant but also a "feed-stock for . . . synthetic polymers."* *Ca.* 2013 this chemical caused the most warming of any HCFC. In 2014, according to the U.S. EPA, "ozone depleting substance substitute emissions and emissions of HFC-23 during the production of HCFC-22 were the primary contributors to aggregate hydrofluorocarbon . . . emissions." At time of writing, this compound was the main driver of the increase in *all* HCFC emissions.

 Atmospheric lifetime: 12 years. *Over 20 years its GWP is 5,160.*

650–3,800

- 507 series, HFCs 32/125, hydrofluorocarbons.

Sometimes used as a commercial refrigerant.

 Atmospheric lifetime: 4.9 to 29 years. *Over 20 years the GWP of HFC-125 is 6,350.*

1,300–1,430

- HFC-134a.

Another replacement for CFC-12. Widely in use in air conditioners at time of writing, although in Germany it was wisely replaced by isobutane. In Japan, "HFCs," possibly including this one, were "emitted from manufacturing, accidents, and disposals of automatic vending machines." Sometimes found in metered-dose medical inhalers.† Between 2005 and 2011, atmospheric concentrations more than doubled, to 62.7 parts per trillion. At time of writing, this compound caused the most global warming of any HFC. "The largest emissions occur in North America, Europe and East Asia."

 Atmospheric lifetime: 14 years. *Over 20 years its GWP is 3,830.*

1,600–2,310

- HCFC-142b.

A hydrochlorofluorocarbon which "can be used as alternative feedstocks for vinylidene fluoride manufacture."

 Atmospheric lifetime: 17.9 to 19 years. *Over 20 years its GWP is 5,490.*

---

* "Because HCFC-22 depletes stratospheric ozone, its production for non-feedstock uses is scheduled to be phased out by 2020 under the U.S. Clean Air Act. Feedstock production, however, is permitted to continue indefinitely." That was nice.

† "Inhaled HFCs are not broken down in bronchial passages; they are released into the atmosphere, without undergoing any changes, upon exhalation." Well, at least this HFC was better than HCFC-142b.

4,000–4,750*

CFC-11, which is CCl_3F.

This fully halogenated chlorofluorocarbon made up 63% of the 1986 CFC market. Used as a "blowing agent" and refrigerant. "Extremely desirable for use with centrifugal compressors." "Whereas a typical refrigerator/freezer might use 6 oz. (170 g) of CFC-12 as refrigerant, it uses 2 lbs (0.91 kg) of CFC-11 in the blown foam."—Substituted for by HCFC-123.

Atmospheric lifetime: 45 years. *Over 20 years its GWP is 6,730.*

4,800–6,130

CFC-113, which is CCl_2FCClF_2.

A fully halogenated chlorofluorocarbon used as a solvent in dry cleaning, and a low-viscosity cleaning agent in electronics manufacture and metal finishing. Substituted for by HCFC-225ca/cb.

Atmospheric lifetime: 85 to 90 years. *Its 20-year GWP is 6,540.*

6,630

• PFC-14, which is CF_4.

See description of PFCs.

Atmospheric lifetime: 50,000 years.

7,000 to 10,300 [one source lumps them at 7,400]

• Various perfluorocarbons (PFCs).

Used to etch semiconductors. Also appear as uranium enrichment and aluminum smelting byproducts.

Atmospheric lifetime: 1,000 days to 50,000 years. *The 500-year GWP of PFC-318 (life: 3,200 years) is 14,700.*

7,100–10,900

CFC-12, aka R-12, which is CCl_2F_2 *(sometimes called "freon," although "Freon" was the trade name for a group of CFCs).*

In the 1930s, "facing the problem of replacing methyl chloride and ammonia in household refrigerators, researchers found that dichlorodifluoromethane (CFC-12) was the best alternative as a safe, stable gas whose liquefied state had low compressibility. In addition, it was not flammable." It was even odorless. Also used as an industrial refrigerant. Truck and trailer refrigerators employed it. A 1990s scientific survey alleged that an automobile air conditioner "contains about three pounds of freon [= CFC-12], compared to the few ounces used in home refrigerators."—"Traditionally used in medium (−15° C) to high (15° C) temperature

---

* Still another source gives the GWP as 3,400.

refrigeration, while HCFC-22 has been used for temperatures down to −35° C…"
"Reacts vigorously with molten aluminum." At one time CFC-12 was the third
largest contributor to global warming. Replaced by HCFC-22 and HFC-134a,
whose GWPs, as this table shows, were still many times worse than CO_2's.*
Atmospheric lifetime: 100 years.

8,700

• Perfluorocyclobutane, a PFC.
One of several unsalubrious PFCs. *See perfluorocarbons, above.*
Atmospheric lifetime: [?]

8,970

PFPMIE, or perfluoropolymethylisopropyl ether.
One of the PFPEs, "a family of perfluorinated fluids used mainly in industrial applications."
Atmospheric lifetime: At least 800 years. *The 20-year GWP was 4,692;
the 500-year GWP was 31,694.*

11,700–12,400

• HFC-23.
A refrigerant and a semiconductor etching gas. "For process-related reasons, production of HCFC-22 produces up to 3 % HFC-23 as a byproduct … [E]ven when
the HFC-23 is subjected to further processing (for example, to produce refrigerants) or is collected and then broken down into other substances, some HFC-23
is always released into the atmosphere." *Ca.* 2013, most releases came from East
Asia, with "developed countries" emitting "less than 20% of the global total."
Atmospheric lifetime: 222 years.

11,700–14,400

CFC-13, which is $CClF_3$.
A fully halogenated chlorofluorocarbon sometimes used as a commercial refrig-
erant.
Atmospheric lifetime: 400 to 640 years. *The 500-year GWP is 16,400.*

---

* The old name lingered in popular use. In 2017, when I was visiting Las Vegas, my friend Dan
told me that he had paid an acquaintance to fix his truck's air conditioner on the cheap—for
Dan was hardly rich. This time he got what he paid for. As we rolled through the sizzling city
with the windows up, and hardly any coolness inside, Dan decided on Plan B: Buy a cylinder of
coolant (which he called "freon") and make use of it until it had all leaked into the atmosphere,
then repeat. I demurred, but Dan explained the matter: He could not afford a new air conditioner, and "freon" was cheap. How could I blame him?

16,100

Nitrogen trifluoride, which is NF_3.

A byproduct of fluorocarbon production and semiconductor manufacture. Added to the Kyoto Protocol's greenhouse list in 2012. From 1990 to 2013, Japanese emissions of this chemical increased 41 times.

Atmospheric lifetime: 500 years.

22,800–23,900

• Sulfur hexafluoride, or SF_6.

An electrical insulator; common in switching systems [which have a minimum 40-year working life]. "Allows for more compact substations in dense urban areas." Also "blown over molten magnesium metal to induce and stabilize the formation of a protective [oxidized] crust." Sometimes employed to make semiconductors; frequently used in aluminum manufacture. Up through the 1990s it was pumped into high-end auto tires *"for reasons of image"* [my italics]. A cushion in sports shoes (since replaced by nitrogen). In Japan it made a fabulous "insulating medium in the radar ... of ... [the] Airborne Warning and Control System. When the plane ascends, SF_6 is automatically released ... into the atmosphere to maintain the appropriate pressure difference between the system and the outside air. When the plane descends, SF_6 is automatically charged into the system from an SF_6 container on board."—"Used as a tracer gas" in certain British environmental studies since "it is currently the only viable way to measure emissions of methane from ruminant livestock individuals at pasture." SF_6 is another of the perfluorocarbons. "The most potent greenhouse gas the Intergovernmental Panel on Climate Change ... has evaluated." In Germany, "since 1975, SF_6 has been used to enhance the soundproofing properties of multi-pane windows. In such use, the gas is inserted into the spaces between the panes."* This folly was banned in 2007–8, but the Germans increased its application in solar collector manufacture. In 2012, American SF_6 emissions were 8.4 teragrams of CO_2-equivalent, or about 0.13% of the national greenhouse total.

Atmospheric lifetime: 3,200 years. *The 500-year GWP is 32,600.*

Sources: Encyclopedia of Chemical Technology (1994); *Encyclopaedia of Weather and Climate* (2002); *Environmental Science and Technology* (2006); International Panel on Climate Change (2007) and (2013); U.S. Department of Transportation (1979); U.S. Environmental Protection Agency (2014); reviewed by Dr. Pieter Tans, NOPA, and Mr. Ben Coleman, Marshall University

---

* "According to expert-level information ... one-third of the SF_6 used in the process of pumping SF_6 into spaces between windowpanes escapes." One percent of the remainder escaped each year. The rest came out once the window was removed and disposed of.

I am sorry that I could not make my table simple, complete or accurate. There were already many other compounds in the air just then, such as that old war-horse carbon tetrachloride (merely 3,480 times worse than CO_2, and an excellent feedstock for Freon)—indeed, so many substances that I never heard of them all!* So I could only point, not show.

How could the 100-year global warming potentials of a substance vary so much?—Because some computations were newer than others.—As for the confidence levels of plus or minus 35%, such were our best efforts in those days.—But why did CFC-113, whose atmospheric life[†] was about 90 years, do in 20 years more damage relative to CO_2 than it did in a hundred?—Maybe it was how it decayed. By some such magic I explained all these matters to myself.

And why did I omit sulfur dioxide? It certainly deserved to be on the list:

> There have . . . been two dozen times during the past 46,000 years when major volcanic eruptions occurred every year or two or even several times per year for decades. Each of these times was contemporaneous with very rapid global warming. Large volumes of SO_2 erupted frequently to overdrive the oxidizing capacity of the atmosphere . . . These are the times of the greatest mass extinctions.

But since that warming agent deployed its own disputable peculiarities upon our planet, I packed it off to Volume II.[‡] (Volatile organic compounds get their moment on II:352.)

In any event, back when we were alive, most of us rarely read tables such as this. To be sure, the well-meaning expert chemists who in their zeal to profitably please had unleashed CFCs and HCFCs now pored over their latest results, concluding optimistically: *The structure of a molecule affects both its lifetime and radiative forcing. The results present herein may assist with the future design of*

---

* However, the Kyoto Protocol of 2002 was a practical and decisive first approximation, setting reduction targets for six kinds of greenhouse gases: HFCs, PFCs and then the specific compounds carbon dioxide, methane, nitrous oxide and sulfur hexafluoride. (Other direct and indirect greenhouse gases were added to the list in time.) Of course we Americans refused to sign, being the heroes who kept the lights on.

† Atmospheric lifetime was another complex computation. A gas molecule may degrade faster at high altitude, because at the top of the atmosphere the photon flux will be most vigorous. Almost all air is at low altitudes; hence so is the bulk of whatever has contaminated that air. As one technical article points out, "air must cycle through the mesosphere many thousands of times before the entire atmospheric burden would be depleted."

‡ See II:26.

molecules which have shorter lifetimes and lower radiative forcings but which still retain useful function. Most of these substances were unknown to us who had complacently benefitted from their use; the magnitude and duration of their warming seemed peculiar, abstract, improbable; considering such figures bred confusion and despair—and as for their actual *present effects,* in say, 2011, those were still so minute that we could pull our usual trick of making future generations pay.

In their long-lasting ability to harm us, the worst of these substances reminded me of nuclear waste. The 50,000-year activity of PFC-14, with its warming power 525,560 times worse than carbon dioxide's in the first century alone (after that who could say?), how was one supposed to weigh that against the radiation emitted by, for instance, the "uranium legacy site" of Mailuu-Suu, Kyrgyzstan?* We etched semiconductors, sold and used them, then threw them "away," at which point the PFC-14 that had served us had barely begun its climate change. Meanwhile the Soviets enriched their ore at Mailuu-Suu, then went out of business. Two investigators from the International Atomic Energy Agency concluded: *The probability of landslides that potentially can destroy the tailings containment is high . . . Erosion of riverbanks and destruction of tailings containment is only a matter of time[;] consequently fortification of the riverbanks . . . [is] only a short-term solution.* Fortunately, *the site of TSF No 15 has the potential to become (after reconstruction) a safe disposal site sustainable over the time span of 200–1000 years recommended for final disposal of tailings.* Only a thousand years! That would be over before we knew it! *The . . . valley . . . appears . . . sufficiently spacious to be expanded to receive . . . possibly up to 1 million m3†* [35.3 million cubic feet] *. . . of additional tailings.*—Now for the *Strategic Action Plan for Complex Remediation of Uranium Legacy Sites in Mailuu-Suu:*

- *No national strategy for long-term solutions.*
- *National infrastructure lacking—no clear roles and responsibilities.*

But if there was no national strategy, there was also (I am proud to say) no immediate danger. The people of Mailuu-Suu breathed in radioactive dust, and got irradiated from the scrap metal they collected from their "legacy site" . . . while PFC-14 contributed its subtle mite to our atmosphere. *I* never heard anyone complaining!

---

* This place also receives discussion on p. 499 and in II:557.

† No superscript in original.

Atmospheric concentrations of HFCs, PFCs and sulfur hexafluoride were rising steeply, but the entirety of their combined mischief had not yet accomplished 1% of the warming carried out by the so-called "well-mixed greenhouse gases": carbon dioxide, methane, nitrous oxide and the halocarbons.*

Between 1990 and 2012, HFC emissions in the EU-15 nations more than doubled. Between 2005 and 2011, nitrogen trifluoride levels had nearly doubled, but remained under one part per trillion. Sulfur hexafluoride emissions were *at least twice the reported values,* but so what?

By 2013 the HCFCs and CFCs made up 11% of our global warming brew. In Japan, hydrofluorocarbon emissions rose by 99.4% between 1990 and 2013, mostly due to increased deployment of refrigerators and air conditioners. But they still constituted a mere 2.3% of that country's greenhouse gas pollution.

So it was with our human calamities, back when we were alive. Once upon a time, when World War I pushed against the creaking walls of the Austro-Hungarian Empire, a young officer-to-be told his friend: *Each question branches out into more and more questions, and once you start, you get nowhere. One thing I've noticed, though, is that the more narrow-minded a person is, the more easily he'll find a way through this maze. He'll declare confidently, for instance: We'll soon teach the Serbs their lesson, and that'll be that.*

As for our most effective carbon ideologues, how did *they* render the crooked straight?—By omitting.

Excluding agriculture from consideration certainly eased our worries. Forgetting about refrigerants was another path-straightener.[†]

And why fuss over such chemicals as nitrous oxide, which did not have a single atom of carbon in their molecules?—Besides, water vapor contributed more to global warming than other greenhouse gases—so my encyclopaedia said.

---

* These last partially halogenated organic compounds, among which belong CFCs, HFCs and HCFCs. In this group the Freons were actually getting more dilute—a surprisingly pleasant outcome. (See the interview with the retired CEO of Conoco, Mr. Archie Dunham, on II:451.) But they were an exception to this story. Maybe that delayed the misery ahead. Meanwhile four other members (CFC-11, CFC-12, CFC-113 and HCFC-22) caused 85% of all halocarbon-induced warming. "The former three compounds have declining R[adiative] F[orcing] over the last 5 years but are more than compensated for by the increased RF from HCFC-22."

† From a German report on HFC-134a: "The time series show a significant emissions increase since 1995. This increase, which has occurred in spite of decreases in fill amounts, is a direct result of increased use of mobile air-conditioning systems in vehicles." Its atmospheric concentration was 0 in 1990, 14 parts per trillion in 2000 and 63 ppt in 2011.—Well, why *not* forget about that? "Global warming is a major environmental concern . . . ," announced a manual for American air conditioning technicians in 2012. Now for the good news: "No global-warming-based phaseouts are currently in effect for air-conditioning refrigerants in stationary applications." Between 1990 and 2014, HFC emissions in the U.S. increased by 257.9%.

That entry might have been out of date; but even so, why should you from the future blame us for water vapor? Above all, why ever ask what on earth the work was for?* That would have led nowhere, all right!

Carbon Ideologies now promises to bring you on a delightfully easier journey, in which precipices will be screened off behind simplifications, while our puttering narrative vehicle will be cushioned with consolations. For the most part this book will live up to its name, ignoring hordes of lesser known greenhouse gases. In 2014 the European Union put us straight: *The most important energy-related gas is CO_2 that makes up 78% of the total EU-15 GHG emissions. CH_4 and N_2O are each responsible for 1% of the total GHG emissions.* And Dr. Tans of the National Oceanic and Atmospheric Administration insisted in his letter to me:

> [Many of] the other greenhouse gases . . . have much shorter residence times than CO2. My conclusion has to be that climate change is primarily about cumulative emissions of CO2. That has nothing to do with an ideology. We should of course try to lower emissions of non-CO2 GHGs but they are truly secondary on the expected time scales of climate change.

He had spent decades studying just such matters—his official title was "Chief, Carbon Cycle Greenhouse Gases Group"—and for a fact my heart would have felt easier if he'd erased every other greenhouse agent from our ledger of trivial gains and looming losses. However, the main lesson I, a trivially obtuse nonscientist, have taken from the history of climate change is that reality continually surprised us—because we always underestimated! Why shouldn't some half-known gas be the one to actually finish us off? Having said so much, and with that longish preceding table paying off my conscience, I turned thankfully away from, for instance, perfluoropolymethylisopropyl ether. My maze of worries would become a broad, well-lit tunnel once I further limited my concerns—as this book will immediately do. Every now and then I may look over my shoulder at methane or at certain volatile organic compounds—but at least, thank God, this primer is nearly done with.—How shall I leave its wide web of problems? Naturally, with a table!—We knew that burning away rain forests and tilling the soil incurred costs quantifiable as follows:

---

* Regarding essential work, I hereby report that "other relevant products include novelty aerosols (artificial snow, party-streamer sprays, etc.), which emit some 100 t[ons] of HFCs per year."

COMPARATIVE RESPONSIBILITIES FOR GREENHOUSE GAS EMISSIONS, 2007,

in multiples of the figure for food production

1

Food production [14%].

1.21

Deforestation [17%].

4.93

Energy consumption [69%].

Total: 100%.

Source: Encyclopedia of Agriculture and Food Systems, 2014, with calculations by WTV.

Focusing on only 69% of the picture saves time and eases my digestion. There-fore, the remainder of *Carbon Ideologies* simply considers four fuels, as follows:

About Fuels

Fuel—petrol and benzine! That's all we need!

Fyodor Vasilievich Gladkov, Cement, ca. 1920

A fuel is an energy-rich substance which yields up its energy when combusted, giving off energy-poor byproducts. In the case of nuclear fuel, please replace "combusted" with "fissioned" or "irradiating," and note that its byproducts may still contain perilous amounts of energy. So it may be simpler to say: When we were alive, a fuel was a means of enriching the seller.*

Bonhams' modern and contemporary African art auction, initiated in London in 2009, grossed more than [1 million pounds] for the first time in 2013 ... Nigerian collectors placed very high bids; *80 percent are in oil and gas,* remarked the director.

Fuels were so lucrative because they were ubiquitous in our makings and doings. In 2010, fuel combustion produced 4.8 U.S. tons of carbon dioxide for each person on Earth, with an annual increase of 1.7%. In 2012, it caused between 79[†] and 80%[‡] of all CO_2 emissions in the European Union. In 2013, it produced 95.4% of Japan's CO_2 emissions. It would be logical to say that if CO_2 was dangerous, then so were fuels—nuclear fuels once again excepted; those were even safer than swimming with sharks.

---

* Indeed, if you asked me, *What was the work for?* I would answer: *Profit.* One strangely hot September morning in Sacramento (it was going to be 104° Fahrenheit, and the next day would be 106°), I walked my roof with the air conditioning man, and I said: "In your line of work you must know a lot about high temperatures. What do you think about global warming?"—"It's a lot of hooey," he explained. "They talk about *a scientific consensus.* Well, *consensus* is a political term, not a scientific one. What does that tell you? Who's saying it? Follow the money."—In this book I have tried to do so. Consider for instance this aphorism from the Organisation for Economic Co-operation and Development (1974): *The prices on which consumers base their decisions whether or not to conserve energy are, for most energy products, higher than the prices which govern production.* How convenient for the producers!—Another fine rule of thumb: *It has also been shown that the greater the peat consumption, the lower the cost of the power generated* by the burning peat. Of course this proposition was uttered by the utility that sold peat-generated power.

† For "EU-28" countries.

‡ For the remaining "EU-15."

The primary greenhouse gas emitted by human activities in the United States was CO_2, representing approximately 80.9 percent of total greenhouse gas emissions. The largest source of CO_2, and of overall greenhouse gas emissions, was fossil fuel combustion.

U.S. Environmental Protection Agency, 2016

But what did my rival carbon ideologues say? I quote the Heartland Institute:

Affordable, reliable, and plentiful energy enables us to protect the environment while also creating jobs and the goods and services we need. Expensive and unreliable energy, like the kind produced by ethanol and commercial wind and solar companies, destroys jobs and harms the environment. [How did solar power harm the environment? *By being less efficient and more land-intensive than fossil fuels.*]

Fossil fuels are the foundation of economic growth and prosperity. Taxing them or making them more scarce causes economic growth to slow, makes food and other essential goods more expensive, and many of the good things we take for granted are lost.

This last paragraph was true—but not in some "universal" sense; it was true on account of our present sociotechnical organization. We have seen how necessary we made fuels as chemical feedstocks and fertilizers; *many of the good things we take for granted* did depend on them. Ever since Neanderthal times, carbon combustion had decomposed animal proteins for the convenience of our digestions. Then came slash-and-burn agriculture, metallurgy, etcetera. As direct combustibles, fossil fuels appeared irreplaceable in our engines and purpose-built utility plants—but in my time I also heard of nuclear-powered cars, power station redesigns, and even the hypothetical retraining of unemployed coal miners—for in point of fact, "you can do anything. Can you do it at a point where it is *economic* to do?" I am quoting Mr. Sam Hewes, a kind and rather jolly Vice-President of the Bank of Oklahoma (his interview appears in the oil section).* "You can turn gas into electric and sell it as oil where

---

* See II:413 and II:486.

there's tremendous gas reserves," he said, "but how do you get gas from one place to another?"

Carbon Ideologies offers no scenario for replacing nuclear and fossil fuels. (Nor have I any practical idea of how we might have saved the free-roaming buffalo herds.) Whatever "solution" I could have proposed in 2017 would have been found wanting before the oceans rose even one more inch! But even though I rest proudly on my laurels as a nay-sayer, I will at least compare fuels against each other.

RULES FOR FUELS, AND WHY WE BROKE THEM

Other factors being equal (such as cost, abundance and risk), if we had to generate power by creating heat, we might as well employ whichever substances produced the *most* heat per unit of mass or volume—not to mention the least greenhouse emissions for their thermodynamic work.

But those "other factors" *could not* be equal. It would seem preferable when roasting limestone for cement to feed the kiln with commercial packaging waste at 56,854 kilograms of CO_2 per terajoule,* rather than recycled tires at 97,319 kilograms; because by this measure the second fuel threatens nearly twice the atmospheric harm of the first.—Well, what if (as might be the case in a district of auto body shops and wrecking yards) old tires were cheaper and more abundant than packaging waste, so that our competitors were burning those, and offering their cement at prices we in our slavish desire to be good could not match?[†]—Then I would ask how cheap they proved to *you* who must now cope with what all of us left behind.

Here is how another carbon ideologue (who happened to be a physicist) arrived at his particular conclusion:

> Nuclear power is essential . . . The risks of nuclear power are far exceeded by the actuality of fatalities from political instability due to global warming. We are at almost half a million deaths and counting in Syria alone. The civil war there was precipitated in large part by the worsening (due to global warming) of a drought that began in 1998.

---

* In our units, one pound of carbon dioxide for each 7,580 BTUs generated by the first fuel, or for 4,420 BTUs of the second.

† This recalls the hypothetical constraints upon the owner of those inefficient group-driven machine tools. See above, p. 38.

Once again, I must call upon you in my unknown future to say what our nuclear idylls left behind. *Modern reactors,* wrote the just-cited ideologue, *are far safer than those in Chernobyl or Fukushima.* How did he know? What was his prospective gain and loss?—All I can do here in my dead past is to grope at partially informed trial balances.

That fuels vary widely in their heat-making capacities is evident.—But in our day, judging the calorific efficiencies of fuels against each other was not straightforward. They sold us coal by weight; oil, gasoline and diesel by liquid measure; and the various natural and unnatural gases by volume at a given temperature and pressure. Each form of comparison projected its own truth. In 1979, 0.29 gallons of diesel fuel or 0.32 gallons of gasoline had to burn so that an American electric light rail car could roll one mile. So it would seem that gasoline was 10% less efficient than diesel—and indeed, three devout diesel men wrote a book in order to crow that diesel fuel (their loving italics) *contains more heat energy (BTUs or joules) than does gasoline . . .* However, the two liquids varied in density. When the same information was expressed through relative weights, it transpired that to move the car that single mile took 1.29 pounds of gasoline and 2.04 pounds of diesel—which rendered gasoline 158% *more* efficient.*

AVERAGE FUEL CONSUMPTION IN MOVING ONE AMERICAN ELECTRIC LIGHT RAIL CAR ONE MILE, *ca.* 1979,

in multiples of pounds of gasoline required

(Energy needed: About 4.1 kilowatt-hours, or 13,993.3 BTUs.)

1
Gasoline, 1.29 lbs. Converted from 0.32 gal.

1.20
Natural gas, 1.55 lbs. Converted from 37.20 cu ft.

---

* *Encyclopedia of Chemical Technology,* 1994: "As a fuel for internal combustion engines, diesel fuel ranks second only to gasoline."

1.58
Diesel fuel, 2.04 lbs. Converted from 0.29 gal.

2.40
"Manufactured gas," 3.10 lbs. Converted from 74.40 cu ft.

2.47
Coal, 3.18 lbs.

Source: U.S. Department of Transportation, June 1979, with calculations by WTV.

Alcohol *versus* kerosene made a comparable conundrum.

The U.S. Energy Information Administration liked to express the matter in a ratio of BTUs of combusted fuel per kilowatt-hour of potential power consumption. (In that respect, oil, natural gas and coal were even at between ten and eleven thousand.) Someone could have preferred terajoules per virgin weight or of gasified volume. The main thing was to pick a standard, and devoutly mistrust anybody who compared apples and oranges.

Carbon Ideologies will continue to use BTUs per pound.* Professor Gutowski advised me against this; but since the American government had failed decades before to change from pounds to kilograms, and still clung to BTUs, I chose to sink alongside my stubbornly archaic countrymen.

Now to our subject: By combusting equivalent mass-units of all these fuels in equivalent ways, and measuring their proffered heat under similarly controlled conditions,† one might usefully compare the work they can do.

The heating value of a fuel, or calorific efficiency, is the energy it emits when combusted. The higher the heating value of a motor fuel, the greater its fuel economy. More efficient coals and oils cost us less fuel to ship and dispersed smaller volumes of pollutants en route.

Which burns more impressively, a pound of pinewood or a pound of gasoline? The table of Calorific Efficiencies beginning on page 208 expresses this

---

* This is why, however peculiar it may seem, in the table of Calorific Efficiencies I have expressed *all* high heating values, even those of liquids and gases, in terms of combustion energy per pound.

† Subject to the usual limits of human endeavor. Obviously the energies obtained by burning a hunk of peat, blowing up a stick of dynamite, metabolizing a succulent fat or letting loose the radioactive decay of a nuclear fuel rod can be compared only through various conversions.

range of innate energies—a matter of urgent importance to our hungriest category of self-expression, *demand*. Accordingly, when considering the boastings and accusations of various ideologues, I will find steady occasion to cite this table.

If the "average" per-pound calorific efficiency of coal was a 1, then, as the table will show, California crude oil's would have been 1.5, and natural gas's about 1.9—in which case reactor-fissioned uranium-235 was 2.9 million.

To you in the future, if you remain foolish enough to keep burning hydrocarbons for your own short-term gain, I propose the following procedure: Pick the fuel with the highest heating value and the lowest global warming potential. But *our* procedure was: Select for price, convenience and abundance.

Following this method, we continually improved our atmosphere. In 1980, the inhabitants of Planet Earth produced 187.3 Q-BTUs' worth of energy—enough to electroprocess 457 million tons of titanium at standard 30% power plant efficiencies. By 2011 we had nearly literally tripled our efforts, to 518.3 quads—from which, with our customary titanic wastage, we generated 21,080.9 billion kilowatt-hours of electricity—two-thirds of which came from fossil fuels.

> Fossil fuels possess almost ideal properties for humans.
>
> Rolf Peter Sieferle, 1982

As *Time* magazine rhapsodized, not quite two years before Trump became President: *Energy-sector employment is one of the few bright spots for the middle class. Energy jobs pay more than double the average annual wage across all industries in America.* And carbon dioxide levels kept going up.

(In India it was happening still more quickly. But, after all, they needed to develop, didn't they?)

MAGIC CONVERSIONS

Speaking of equivalence, you already know how protean were our carbon-based fuels as chemical feedstocks. Do you remember our proud claim about adipic acid? Out of phenol or butadiene, cyclohexane or cyclohexene we could create it, *as dictated by shifts in hydrocarbon markets.* Meanwhile, from natural gas or oil we could build a solid pair of spectacles! *Carbon Ideologies* too frequently denies its due to productive genius. I hereby interrupt this jeremiad to say *thank you for the magic.*

Industrial processes transformed one fuel into the other according to the following ratios:

A

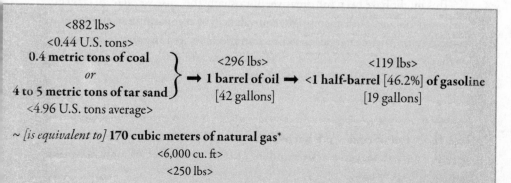

<882 lbs>
<0.44 U.S. tons>
0.4 metric tons of coal
or
4 to 5 metric tons of tar sand
<4.96 U.S. tons average>

} → <296 lbs> **1 barrel of oil** → <119 lbs> <1 half-barrel [46.2%] **of gasoline**
[42 gallons]   [19 gallons]

~ *[is equivalent to]* **170 cubic meters of natural gas***
<6,000 cu. ft>
<250 lbs>

---

* According to one oil geologist, "In energy content, it takes about 6,000 cubic feet of natural gas to equal a barrel of oil." *Note that unlike the other conversions presented here, this one does not inform us of how much oil it takes to make a specified amount of natural gas, or vice versa. It merely reflects innate energy equivalences. Therefore this figure is only very broadly comparable to the others.*

In other words (dividing all these quantities by 296, the poundweight of oil in A):

B

1 lb of oil ~ 0.4 lb of gasoline ~ 3 lbs of coal ~ 33.5 lb of tar sands ~ 0.85 lb of natural gas

Note: Given the widely varying compositions of all fossil fuels, these approximations are +/− 10–15%. For instance, in place of *Carbon Ideologies'* 1 barrel of oil ~ 170 cubic meters of natural gas, another sources gives ~ 150 cubic meters.

From the standpoint of *innate energies* [as lifted from my table of Calorific Efficiencies beginning on page 208], these fuels were considerably more akin:

C

1 lb of oil ~ 0.9 lb of gasoline ~ 1.5 lbs of coal ~ 1.06 lbs of tar sands ["bitumen alone"] ~ 0.77 lb of natural gas

This might have been informative. But usually (because we were all lazy) our carbon ideologues built their bright equivalences by comparing barrels to tons to cubic feet, or innate energies to conversion products. For instance, they advised us that

D

1 short ton of coal ~ 3.8 barrels of crude oil ~ 189 gallons of gasoline ~ 1 cord of seasoned firewood ~ 21,000 cubic feet of natural gas ~ 6,500 kilowatts

Had we divided A by 0.44, in order to match the scale of D, we would have obtained very different ratios:

E [= A / 0.44]

1 short ton of coal ~ 2.3 barrels of crude oil ~ 43 gallons of gasoline ~ 13,600 cubic feet of natural gas

Caveat emptor.

And for precisely this reason, *Carbon Ideologies* will thrust upon you ever so many tables. If you wish to evaluate our fuels in relation to each other, please use your own mind. Have I made errors? No doubt—as have other carbon ideologues . . .—not all of them, but on the other hand perhaps not invariably by accident. And if you'd rather throw up your hands and leave evaluation to others, why, then, that was just what my neighbors and I liked to do! We enjoyed the world we possessed, and deserved the world we left you.

How close could the various equivalences be? Well, our bean-counters tended to agree that

F

1 BTU ~ 1 match tip ~ 1/1,000 cubic feet of natural gas ~ 1/112,500 gallons of gasoline ~ 1/12,500 lb of coal

or

1 metric ton of coal equivalent = 0.7 metric ton of oil equivalent = 27,776,000 BTUs, in which case 1 BTU = 1/12,599 lb of coal

or

according to another source (the National Coal Association): 1 ton of U.S. coal = 22 million BTUs, in which case 1 BTU = 1/11,000 lb of coal

or

according to another source: 1 million BTUs = 90 lbs of coal, in which case 1 BTU = 1/11,111 lb of coal

or

1 quadrillion BTUs = 1 Q-BTU [or quad] = 45 million short tons of coal,* in which case 1 BTU [thank goodness, still] = 1/11,111 lb of coal

These coal-values for 1 BTU differ by no more than 12%.

---

* As usual, the reader should be cautioned against spurious precision. Application of this particular conversion yields 4 quads, while another conversion yields 4.5 quads. As you will find in the table of Calorific Efficiencies, the inherent energies of coals vary.

What should a BTU mean to you? What miser counts his match tips? We lived more grandly than that. As Ayn Rand expressed the matter with her usual compassionate empathy: *Who is enslaved by physical needs: the Hindu who labors from sunrise to sunset at the shafts of a hand-plow for a bowl of rice, or the American who is driving a tractor?* Decade by decade, we Americans enlarged demand—I mean, we sped faster and farther away from slavery. Hence the following datum:

G

100 Q-BTUs = American economy's total energy consumption in 2007

... of which 85% came from fossil fuels.

GLOWING LIKE A BERRY

They all did their collective mite to warm our homes, cook our food and (nuclear again excepted) increase atmospheric carbon levels. Ever since we learned to manage fire, combustion has been associated with some of the sweetest aspects of being human.

> In a white-washed kitchen in the glen the peat-fire glows like a berry, and the cricket—"the cock of the ashes"—sings. And the tangle of Gaelic voices singles out as the Story-teller spreads his fingers for attention . . . For an hour or two the listening farmers and fishermen will forget their bleak existence; the intoxicating talk turns them into kings . . .

A certain dried cut block of that Irish peat would have offered its hearth-audience something like 8,600 BTUs per pound. *Most kinds burn with a red smoky flame, developing a very strong odour, which, however, has its admirers in the same way that wood smoke has.* By the way, it had already released clouds of carbon dioxide, methane and nitrous oxide during its drying. When employed for agriculture, peatlands might emit more than 100 times the CO_2 of other soils. Then the burned peat went right on giving.[*] Well, it was the Story-teller's only fuel, so why blame her? She lived on a windswept island.—One sample of brown peat from Madison, Wisconsin, burned less efficiently than the Irish kind at 7,628 BTUs; the average rating was worse yet: 7,040 BTUs—slightly more energy-poor than lignite. Given its high emissions and low inherent energy, peat should have been reserved for inhabitants of wind-swept places (and in fact, *peat is a significant source of energy for Russia, Finland, and Ireland*)—but how could it be left to them, when we saw possibilities to increase demand? (Come to think of it, I loved to drink heavily peated Scotch whisky.) Proceedings of a resource extraction conference, 1981: *Often neglected in earlier years, solid fuels of low calorific value have more recently seen an upsurge in development. In descending order of heating value, these include sub-bituminous coal, brown coal, lignite, and peat.*

---

[*] *Encyclopedia of Agriculture and Food Systems,* 2014: "Globally, peatlands store more than half of total soil carbon."—*Fertilizer Guide for the Tropics and Subtropics,* 1973: "Practically all cultivated soils in the world are very much in need of nitrogen. The only exception are peat and muck soils where plenty of nitrogen is supplied by decomposition of the organic matter."—Same source: "The use of fire in land clearing, and the increased decomposition due to drainage and fertilization . . . make the organic residues that took thousands of years to accumulate disappear in one or a few decades." In other words, burned peat emitted masses of climate-altering smoke.

HOW SHALL I TELL IT?

Lignite was the most inefficient kind of coal; some was calorifically inferior to peat. Bangladeshis imported it because they couldn't afford the good stuff.* Poles burned it because that was what they had.—Of all the major carbon-based fuels, it took first place (just above firewood) for carbon dioxide emissions, being 112 times worse than natural gas:

CARBON DIOXIDE EMISSIONS OF COMMON FUELS,*

in multiples of natural gas's (2007)

All levels expressed in [pounds of CO_2 released for each 10,000 BTUs released by the combusted fuel].

All multiples rounded to the nearest whole digit.

1
Natural gas [0.0232 lb per 10,000 BTUs].

72
Petrol [gasoline] [1.671].

73
Aircraft fuel [sometimes called "kerosene"] [1.701].

74
Diesel [1.717].

74
Petroleum [1.717].

---

* As the U.S. Environmental Protection Agency quite sensibly explained: "The amount of C[arbon] emitted from the combustion of fossil fuels is dependent upon the C content of the fuel and the fraction of that C that is oxidized. Fossil fuels vary in their average C content, ranging from about 53 M[illion] M[ega]T[ons] CO_2 Eq./QBtu for natural gas to upwards of 95 MMT CO_2 Eq./QBtu for coal and petroleum coke. In general, the C content per unit of energy of fossil fuels is the highest for coal products, followed by petroleum, and then natural gas."

---

* See II:280n.

82

Heavy heating oil [1.890].

94

Hard coal [2.181].

?

Peat [?]. "Higher than the I[mplied] E[mission] F[actor] of hard coal." Possibly worse than lignite.

112

Firewood [2.599].

112

Lignite coal [2.605].

Source: Greenhouse Gas Inventory Germany, 1990–2007, with calculations by WTV.

So a carbon ideologue might tell you that natural gas was good and lignite was bad. But another ideologue could have spun the facts the way I did, in BTUs per pound:

CARBON DIOXIDE EMISSIONS OF COMMON FUELS,

in multiples of lignite's

All levels expressed in [pounds of CO_2 released per pound of the combusted fuel].

[All catch as catch can. In one case, Indian lignite put out 4.8 times less heat per pound than West Virginian anthracite.]

1

Lignite coal [1.396 lbs per lb*].

1.33

Subbituminous coal [1.858].

---

* In case you were wondering, a fuel can release a greater weight of carbon dioxide than its own weight, because its contribution to the CO_2 molecule is only a carbon atom, which combines with two oxygen atoms during combustion. The volume of CO_2 created [and hence the weight] is directly proportional to the number of carbon atoms in the fuel.

1.77
Bituminous coal [2.466].

2.02
Natural gas [2.813]. More usually expressed as 117.10 lbs CO_2/1,000 cu ft.

2.04
Anthracite coal [2.843].

2.19
Leaded [10LL] aviation gasoline for piston engines [3.06]. More usually expressed as 18.4 lbs CO_2/gal.

2.25
Unleaded aviation kerosene [JET A-1] for turbine engines [3.14]. More usually expressed as 21.1 lbs CO_2/gal.

2.27
Kerosene [3.162]. More usually expressed as 21.50 lbs CO_2/gal. But the Germans proposed "a pertinent emission factor of 3,150 g/kg," which comes to [3.0].

2.28
Diesel [3.184]. More usually expressed as 22.40 lbs CO_2/gal.

2.28
Gasoline [3.186]. More usually expressed as 19.60 lbs CO_2/gal.

Sources: U.S. Energy Information Administration, 2016; *Plane & Pilot*, 2017; with calculations by WTV.

Had every drop of the American aviation gas and kerosene-based jet fuel produced in 2016 been burned (and I imagine that it ultimately was), combined emissions would have been more than half a billion pounds of carbon dioxide. What did this "mean"? And for that matter how was a "concerned citizen" such as myself supposed to rank the dangers of fossil fuels—in carbon dioxide per generated BTU or in carbon dioxide per combusted pound?

While I was alive, such ambiguities (not to mention sudden revisions of both measurement and knowledge) deliciously enlarged the possibilities of argument—all in good faith, of course—so that we could not only sustain, but *accelerate* the rise of atmospheric carbon levels, all the while expressing confusion, powerlessness and resentment. Thus the true subject of *Carbon Ideologies.*

No wonder that most "consumers" preferred to leave such matters to the "experts"—and carbon ideologues on both sides obligingly compared apples and oranges.—In 2016, two members of the Safe Climate Campaign warned: *From well to wheels, burning a gallon of gasoline spews 25 pounds of greenhouse gases into the atmosphere.*—Approximately true or not (I could not verify it*), it was but one of several data we would have needed to make any thorough cost-benefit analysis of gasoline. *From well to wheels* implied that the measurement began with the engine-powered extraction of crude oil—some fraction of which probably contained natural gas—continued with its distillation into gasoline at the refinery, and the trucking of that fuel to the service station, and concluded with combustion in an engine of unstated efficiency on some average kind of road. But when was it crude, and when did it become gasoline? Which greenhouse gases escaped along the way; what were their global warming potentials? Finally, what was gasoline's efficiency in relation to other motor fuels?

Subtle, tedious and subject to revisions without end, these calculations repulsed us. Who wouldn't rather just sell the fuel with the greatest temporary financial return?[†] Why not settle for a carboniferous career?—As for the people in my circles, our lives remained easy. We had new choices every year—and on those sadly numerous occasions when we could not gratify demand, it was because we ran out of life, or money; not because we lacked fuel; never because some "future's" imperious protector prohibited us from "consuming" it. (Do you remember the Oak Ridge altruists from 1982? *It is fairly clear that no serious climactic changes will result from fossil fuel use in this century.*) Here were those other fuels:

---

* It is suggestive that when I was finishing *Carbon Ideologies* three of the newest fancy American cars generated an average of 0.88 pounds of CO_2 per mile. At, say, 21 miles per gallon, their emissions per gallon of gasoline would have been around 18.5 pounds.

† From a mutual fund report: *When oil prices plummeted in the second half of 2014, we saw a robust response from businesses, and energy companies worked quickly to halt production activity.* Had they done the same based on global warming potentials, perhaps you from the future could still live in Bangkok or Miami.

VAPOR FROM A MAIDEN'S BREAST

Wood, which pound for pound contained little more than half of peat's warmth (on pages 209–10 you will find a range of sample wood-energies from 4,082 to 5,340 BTUs)—and, as you see, was almost as CO_2-extravagant as lignite—facilitated its own campfire-tales, and often centered a story, as in Jack London's famous "To Build a Fire." You from the future who must worry about heat-stroke and dehydration, please don't forget that on occasion we needed to burn tree-carbon to save ourselves from freezing to death. Meanwhile, here is the happier work accomplished by burning stove-wood in a lyrical Bangladeshi tale: *The kettle's cover teeters and rings as if it wants to shed its husk and fly, as if it is a silver anklet on a maiden who has stepped spellbound onto a field to dance. As if the vapor from that maiden's breast is steaming forth from the kettle's open mouth. And only then is the water ready for making tea.**—How much would those enchantments have warmed our atmosphere? Well, every 519 pounds of combusted wood won us a pound of methane. To release a pound of nitrous oxide we had to burn 76,153 pounds of wood. But as for this book's eponymous gas, each wood-pound we threw in the fire gave off 1.2 pounds of carbon dioxide.

"OUR MOST ABUNDANT ENERGY RESOURCE"

Coal, which might contain twice as much energy for its weight as peat, or three-quarters less, is both cause and setting for the sad horror in D. H. Lawrence's short stories "Odour of Chrysanthemums" and "A Sick Collier." Lawrence was himself a coal miner's son; his coal-haunted tale "Daughters of the Vicar" explores, among other subjects, carbon-based class antagonism:

> But when the pits were sunk, blank rows of dwellings started up beside the high roads, and a new population, skimmed from the floating scum of workmen, was filled in, the cottages and country people almost obliterated.

Coal warms, stains and impels many scenes of this story, as when the heroine *saw the scarlet glow of the kitchen, red firelight falling on the brick floor and on the*

---

* In case too much wood remained unburned, we dreamed up new ways to help the atmosphere. A newsflash from 2017: "And if your phone dies while you are enjoying the wilderness, a tiny stove fed with twigs can convert that heat to electricity and charge it."

bright chintz cushions. It was alive and bright as a peep-show. Later on, in a scene of erotic tension, *"How close it is in this room. You have such intense fires. I will take off my coat," she said.* The rest comes off a page later.—In the face of such magical carbon power, which "made his heart . . . hot and stifled in his breast" and "her lungs full of fire," you may be surprised to learn that a short ton of coal contains no more energy than a cord of firewood! No matter. As the National Coal Association reassured us: *Coal is our most abundant energy resource.*

BETTER YET

Humanity progressed, and thanks to fracking, natural gas soon became even more abundant than coal! We also had ever more oil on hand; in 2015 the stock market declined because that stuff was getting too cheap.

Awaiting fuel combustion in Kazakhstan

When I was alive, fossil fuels made up about 83% of our "overall global energy supply." You can say we loved them. Meanwhile the detractors said:

CRITICISMS OF COMMON FUELS,
1980–2012

Coal

"Compared to oil and gas, coal produces the highest amount of CO_2 per unit of energy generated."

> *George A. Olah, Alain Goeppert and G. K. Surya Prakash,*
> *methanol enthusiasts, 2009*

"Coal is the single greatest threat to civilization and all life on our planet."

> *James Hansen, atmospheric scientist, quoted 2012*

Natural gas

"Conventional natural gas has the lowest greenhouse gas emissions per unit of energy of all the fossil fuels, but since it is used in such high quantities it accounts for over 20 percent of U.S. carbon dioxide emissions."

> *Foundation for Deep Ecology, 2012*

"Natural gas systems were the largest anthropogenic source category of CH_4 emissions in the United States in 2014 . . ."

> *U.S. Environmental Protection Agency, 2016*

Oil

"An evil Power which roams the earth, crippling the bodies of men and women, and luring the nations to destruction by visions of unearned wealth, and the opportunity to enslave and exploit labor."

> *Upton Sinclair, 1926*

"The tyranny of environmental pollution, public health risks, and climate destruction is created at every stage of oil use, from exploration to production, from transport to refining, from consumption to disposal."

> *Antonia Juhasz, 2008*

Nuclear

". . . We are committed to stopping nuclear power before it stops us."

> *Coalition for Direct Action at Seabrook, 1980*

Shrugging off these trivialities, we increased our worldwide carbon dioxide emissions by 106% between 1971 and 2009.* Our inventories of nuclear waste grew meanwhile.—I remember asking Mr. Archie Dunham, the retired CEO of Conoco, whether our planet had a carrying capacity and, if so, whether we were at it. He replied: "How could you even measure it? It would be interesting to know what someone could calculate as the carrying capacity of the planet. Maybe God takes care of that, with plagues and so forth."†

Fortunately, the carbon ideologues addressed all such matters. They said that global warming was or was not occurring. Scientific teams footnoted high-, medium- and low-numbered scenarios for our oncoming tragedy; corporations and non-governmental organizations kindly encouraged low-energy-using nations to "develop." A few earnest "consumers" began to worry that we were burning too much carbon; maybe we should drive less, or set our thermostats closer to external temperatures.—But as we have seen, the combustion of carboniferous fuels was but one cause of our atmosphere's destabilization. Who proposed to abandon agriculture? Was I supposed to stop cleaning my toilet with that inexpensive nylon-bristle brush, or forgo aluminum foil? Chlorofluorocarbons and their cousins leaked happily from our refrigerators, wafting round the planet in unseen aerosols, with busily reactive millennia ahead of them. Most characteristic of all loomed our continued forgetfulness of the question: *What was the work for?*

WHICH FUEL WAS BEST?

This concludes the "primer" section of *Carbon Ideologies*. As you read the "ideology" chapters, you will find more similarities than differences.

Having laid out our basic biological and thermodynamical difficulties, I turned away from the specific methods of resource extraction, and the latest hopes with which we deluded ourselves; those would only go out of date. But the angry vigor with which people defended their own carbon ideologies, not to mention the selfish, blinkered, careerist, arrogant, miserly and callously greedy thinking that caused horrendous accidents, followed by coverups and repetitions of the evil, these are fundamental to humanity. They deserve study. The negligence that led to the deaths of coal miners at Upper Big Branch, West Virginia,‡ was a microcosm of the complacency with which we continued to pollute our future with coal-smoke.

---

* "However," the International Energy Agency consolingly added, "this is 455 million tons less than the record level in 2008."

† See his interview on II:451.

‡ Discussed in detail beginning on II:58.

I sometimes felt out of my depth during the preparation of these early chapters. With some relief I now refer you to people's voices and experiences.

Now for that last introductory ranking of fuels. (We will refer back to it every
now and then—from Bangladesh, for instance; or from Pawhuska, Oklahoma, and
even from Greeley, Colorado.) If energy is the ability to do work, and fuel is something with the ability to release energy, then why not match the fuel with the work?
We loved fuels because we always longed for someone or something else to do our
work for us. And what if we found a magic fuel that could do everything, and at no
cost? Hydrogen-powered fuel cells, photovoltaic electricity, nuclear fusion rendered
somehow "safe," any or all of these might at the eleventh hour validate our smug-
ness. Regarding the four carbon ideologies of this book, every booster pimped his
favorite fuel, touting such allurements as perceived safety, local abundance, profit
and price. Sometimes expediency defined adequacy. (From India, 1986: *In areas
which are heavily forested the direct burning of wood in small power stations has . . .
been resorted to.*) It rarely bore much relation to the good of you in the future.

As you browse through the following spectrum of fuels, try to do better than
we did. To me there is beauty in their rising inherent energies. Perhaps the pattern will show you something that could have saved us. If not, may their vari-
ability entertain. What was the stuff called natural gas? We knew that it consisted
mostly of methane, which contained 23,861 BTUs per pound (or, as another
source had it, 18,056)—so what was "Pittsburgh natural gas," which could offer
up a mere 1,129? Let this be another lesson: When an ideologue defines any fuel's
efficiency with a single number, not with a range of numbers, beware, and ferret
out his vested interest.

This table was arrayed in multiples of the calorific energy of blast furnace gas,
one of the weakest coal-based fuels. When coal is burned in the near-absence of air,
and its tars and volatiles driven off, what remains is *coke*—an important ingredient
for the production of modern steel, not to mention acetylene. (When I was alive,
some of us had to burn two-thirds of a ton of coal to make one ton of steel.)* As
coke in its turn combusts, under certain circumstances it gives off blast furnace gas.
Let this disdained but not entirely useless substance mark the beginning of the
scale. In the midrange you will find gasoline, whose combustion, pound for pound,
produces more total energy than dynamite. For the endpoint consider a comparable
amount of uranium. The last item can accomplish nearly 4 sextillion times more
thermodynamic work than the first[†]—and theoretically could do better still.

---

* But the Japanese were more efficient than the Americans. See II:49.
[†] See p. 218, last entry in table.

COMPARATIVE CALORIFIC* EFFICIENCIES,

in multiples of the thermal energy of blast furnace gas

All levels expressed in [BTUs per pound]. *This can be problematic; see pp. 192–93 and header **663** (pure hydrogen), this table.*

For consistency I have used the high heating value when it is available.[†] For three lower and probably more realistic "working values," see **107** (generic "coal"), **226** ("petroleum") and **264** ("natural gas").

For peat, heating values refer to a water-free, not an air-dried, state. For coal, "(M)" means "mine sample." Otherwise it is a delivery sample. Carbon content is from ultimate analysis unless only the proximate analysis [fixed carbon only] is available, in which case the figure is followed by "(P)." Please note that fixed carbon measures underreport the carbon content.

<D> = Default conversion value throughout *Carbon Ideologies*.

All comparative efficiencies greater than 10 have been rounded to the nearest whole number. After uranium-235, all figures rounded to 2 significant digits.

1

Blast furnace gas, at 60° F and 30 inches mercuric pressure, technology *ca.* 1932 [92 BTUs per lb. A 1960 source gives the spuriously similar value of 105 BTUs per cubic foot.]. *"Blast furnace gas has a very low calorific value and gives a non-luminous flame . . . not hot enough to enable steel ingots. It is, however, very satisfactory for heating ingots, blooms and slabs for rolling."*

12

Pittsburgh natural gas, A.D. 1932, at 60° F and 30 inches mercuric pressure [1,129].

16

Refinery oil gas, at 60° F and 30 inches mercuric pressure [1,468].

22

TNT (dynamite) [1,983]. *A surprisingly low HHV.* This figure must be a conversion of both the kinetic and the heat energy of this explosive into one joint heat energy equivalent, for another source gives the nearly identical 1,924 as the total energy of detonation, of which 6,252 is the heat of combustion. From an assertion in a physics textbook I calculate the comparable 1,942.93 BTUs. Meanwhile, 1 ton of oil

---

* Sometimes called "caloric."

[†] The high heating value of a fuel in [BTUs per pound] = 14,544 × % carbon + 62,028 (% hydrogen − [% oxygen/8]) + 4,050 × % sulfur. See p. 534.

equivalent = 10 tons of TNT [30,965,667 BTUs]. *"TNT is carbon-rich, and releases a relatively large amount of energy during combustion as compared to detonation."*

28

Tanbark, which is "the fibrous portion of ground oak or hemlock bark used in tanning leather" [2,600].

33

PBXN-109 aluminized explosive [3,057].

39

<D> Stated typical international value for lignite coal, A.D. 1974 [3,600]. *At 0.2 mtoe,* this is far lower than the efficiencies of the American lignites in this table. But the Indian, South African and Indonesian lignites burned by the Bangladeshi brickmaker R.C. [see II:280] all fall below this standard.*

44

Cottonwood, burned while still green† [4,082].

47

Carbon monoxide [4,344. Not HHV per se, but heat of combustion, which is thermodynamically equivalent].

52

Black walnut, while still green [4,751].

58

White pine, while still green [5,340]. *A certain energetic fracker asserts that wood gives off "the highest amount of carbon emissions per unit of energy produced—more so than even coal or oil, according to recent research." This may well be true, despite the relatively paltry HHVs of woods shown here; perhaps he was calculating from volume rather than from weight.*

---

* Million tons of oil equivalent.

† The original figures for wood were expressed in terms of 128-cubic-foot cords rather than by 1-lb weight as my scheme requires. In ascending order of energy output they were listed: first white pine (green), then cottonwood (green), white pine at 12% moisture content, black walnut (green), hickory 12% moist. As you see, the order here is quite different.

54–71

Threshed straw [5,000–6,500].

67

"Harvested wood," Japan, *ca.* 2013 [6,177].

72

"Typical" HHV of air-dried peat [7,040]. *Peat "seldom contains more than 60 percent carbon." This substance is rotted organic matter which is slowly being transformed into coal.*

75

North Dakota lignite coal. Williston, unnamed bed. 24.1% carbon [6,580].

83

Fibrous brown water-free peat, Madison, Wisconsin [7,628].

84

Hickory, 12% moisture [7,691].

85–92

Dry shelled corn [7,800–8,500].

87

White pine, dried to 12% moisture content [8,044].

90

Average HHV of dry, resin-free American wood, *ca.* 1958 [8,300].

92

Salt marsh water-free peat, Kittery, Maine [8,462].

96

Dried American city garbage, A.D. 1930 (typically 13% fixed carbon) [8,800].

97

Iowa high-volatile bituminous B coal. Des Moines, No. 3 bed, nut and slack size [1 inch and larger]. 31.3% carbon (P) [8,960].

106

Methanol [9,766]. *An older source gives the HHV of methyl [wood] alcohol at 9,600.*

107

"Coal," per U.S. Energy Information Administration [9,845]. Calculated from the claim that generating 1 kWh requires of 1.04 lbs coal. The resulting calorific efficiency was multiplied × 3 to compensate for the typical ⅔ energy loss in a power plant. As a working "actual" figure, the final 9,845 naturally lies below any best-case HHV.

109

Fibrous brown water-free peat, Hamburg, Michigan [10,026].

110

Colorado subbituminous coal (M). Superior, Gorham bed. 58.8% carbon [10,130].

111–140

"Alcohol," 1911 [10,200–12,900]. See **139.**

113

Illinois high-volatile bituminous B coal. Peru, No. 2 bed. 1¼-inch slack. 36.8% carbon (P) [10,380].

120

Lignite coal, general approximation *ca.* 1978 [11,000].

120

Barapukuria lignite coal ("high volatile"), Parbatipur Upazila, Dinajpur District, Bangladesh, mined beginning 1998. *Discussed on II:264–72, II:282–86.* **48.4% carbon [11,040].**

129

Virginia anthracite coal (M). Pulaski, Langhorn bed. 70.5% carbon [11,850].

129

Wyoming subbituminous coal, general approximation *ca.* 1985 [11,888].

130

Pennsylvania anthracite coal, barley size. 76.1% carbon (P) [11,980].

130

Stated typical international value for "hard" (bituminous and anthracite) coal, A.D. 1974 [12,000]. *0.7 mtoe.*

134

Coke, data from 1962 [12,300].

136

<D> **Calculation based on this claim of the West Virginia Coal Association: "On average, coal contains 25 million BTUs per ton"** [12,500]. *Carbon Ideologies'* **default value for coal. See 147 and 3,961,956,521.**

137

Subbituminous coal, general approximation *ca.* 1978 [12,600].

139

Gasified anthracite, no. 2 buckwheat (rice) size [12,796].

139

Ethyl [grain] alcohol [12,810]. See **111–140.**

143

Japanese charcoal, *ca.* 2013 [13,112]. Charcoal is mostly graphite crystals. See **147** and **152.**

144

Gasified coke, screened 98% through 1 inch, 2% through ½ inch [13,250].

147

One nuclear engineer's general approximate value for the energy in a pound of coal [13,500]. But a certain steelmaker rates it at [12,500]. See **136** and **3,961,956,521.** "Each pound of coal supplies enough electricity to light 10 [old-style incandescent] light bulbs (100 watt) for about an hour." However, that works out to only 3,413 BTUs; consistency with the nuclear engineer's estimate would require that pound of coal to light 40 100-watt bulbs for an hour. Perhaps efficiency losses are considered here.

147
American willow charcoal, *ca* 1958. [13,530].

150
Maryland low-volatile bituminous coal (M). Frostburg, Tyson (Sewickley) bed. 79.0% carbon [13,870].

152
<D> **Combustion of a pound of pure carbon** [14,000].

152
Alabama high-volatile bituminous A coal. Boothton, Gholson & Clark beds. 57.1% carbon (P) [14,000].

153
Colorado anthracite coal (M). Crested Butte, No. 1 bed. 84.5% carbon [14,030].

155
Eastern Kentucky high-volatile bituminous B coal (M). Jenkins, Elkhorn bed. 79.2% carbon [14,290].

158
West Virginia low-volatile bituminous coal (M). Elkhorn, Pocahontas No. 3 bed. 84.0% carbon [14,550].

160
West Virginia low-volatile bituminous coal (M). Big Stick, Beckley bed. 84.7% carbon [14,730].

161
West Virginia low-volatile bituminous coal (M). Skelton, Sewell bed. 84.8% carbon. 14,780 BTUs per lb.

166
Gasified West Virginia bituminous coal, Ethelogan Chiton seam, screened egg size (2–4-inch lump) [15,296].

173

Pitch [16,000].

174–188

Coal tar, fractionally distilled from 932 to 2,012° F [84 to 91% carbon] [16,000–17,300]. *Of course as the temperature increases and the distillation products become less volatile, the BTU yield will go not up but* down *from 17,300 to 16,000.*

176

Chemical energy of fats. *"They are a more concentrated source of energy than proteins and carbohydrates.* Fats provide about 9 kcal/g, compared to about 4 kcal/g for proteins and carbohydrates."* [16,189. Note: This number is not an HHV but thermodynamically equivalent to one.]

184

Average HHV of wood resin [16,900].

188

Japanese "indigenous natural gas," 2013 [17,283]. This is probably the LHV. The HHV might be 20,000 or higher.

193

Oil sands (bitumen alone), *ca.* 1985 [17,768].

201

<D> Commercial fuel oil, heavy grade, *ca.* 1975 [18,450].

201–217

<D> Diesel fuel, grade unstated (85–88% carbon) [18,500–20,000]. [In this book I often take the average of these figures (19,250) as shorthand for the efficiency of "oil."] *"Diesel fuel contains [12%] more heat energy per gallon than gasoline"*—*but only when measured that way, not (as here) per pound. According to the U.S. EPA, "diesel-powered vehicles typically get 30–35% more miles per gallon than comparable vehicles powered by gasoline.*

---

* These organic compounds (consisting of carbon atoms joined to water molecules; hence their name) include cellulose, sugars and starches. According to my college biology textbook: *Without the light-driven conversion* by photosynthesis *of water and CO_2 to carbohydrate, the Earth would rapidly become as sterile as the moon.*

Diesel engines are inherently more energy-efficient, and diesel fuel contains 10% more energy per gallon . . ."

202

Petroleum oil (60%) at specific gravity 10 API [18,540]. *This value is more commonly expressed as 154,600 BTUs per gallon.**

202

<D> Commercial fuel oil, medium grade, *ca.* 1975 [18,570].

203

Commercial fuel oil, light grade, *ca.* 1975 [18,670].

204

Mexican crude petroleum (83.70% carbon) [18,755].

206

California crude petroleum (84% carbon) [18,910].

215

<D> Kerosene (85 to 88% carbon) [19,810]. Another expert proposes [19,400 to 20,200].

223

Light naphtha [20,500].

224

Petroleum oil (60%) at specific gravity 80 API [20,630]. *Again, more commonly expressed as 115,100 BTUs per gallon.†*

226

<D> Automotive gasoline (84.7% carbon), mid-20th-century formulation [20,750]. Other figures: 20,000 to 20,750 and 21,050 BTUs/lb.

---

* A metric ton of crude may vary from 1.5–8% "above the value of 10^7 kilocalories, which is frequently referred to as one ton of oil equivalent." But at the standard conversion of 1 toe = 39,680,000 BTUs and then converting from metric tons to lbs, we obtain: 1 lb [generic] oil = 17, 999 BTUs.

† Note that if these two specific gravities of oil had been ranked by BTUs/gal instead of BTUs/lb, their order would have been reversed.

226

"Petroleum," per U.S. Energy Information Administration [20,789]. See note to "Coal" at **107**.

230

Butane [21,180]. Another source gives [21,200]. *More commonly expressed as 3,261 BTUs/cu ft, or 102,600 BTUs/gal.*

233

<D> Aviation gasoline, mid-20th-century formulation [21,400]. *"Its total global warming impact is significantly higher [than its volume of use would suggest,] due to vapor trails and the other effects of burning fuel at altitude."*

233

Natural gas from Groningen, Holland (81.3% methane), *ca.* 1975 [21,450]. *Calculated from 893 BTUs/cu ft, using the density of methane.*

234

Propane [21,560]. Another source gives [21,500]. *More commonly expressed as 2,522 BTUs/cu ft, or 91,500 BTUs/gal.*

234

Liquefied petroleum gas, A.D. 2011 [21,561 more often seen as 91,410 BTUs per gal].

243

Dry Soviet natural gas, A.D. 1980 [22,388, more often seen as 932 BTUs per cu ft].

258

Liquefied natural gas, A.D. 2011 [23,734, more often seen as 84,820 BTUs per gal].

259

<D> **Methane [23,861*].** Another source gives [18,056.75]. ***Carbon Ideologies' default HHV for natural gas.*** *As you can see from its position here,*

---

* Or, as a contemporary engineer would have expressed it, 50 kilojoules or 12 kilocalories per gram. (Another typical—and comparable—measurement: Natural gas of the "Khuff" type, found in Bahrain *ca.* 1980, was 80.15% methane, with a gross calorific value of 855 BTUs per cubic foot. This = 20,554 BTUs/lb.)

it is a respectable energy source that "reacts readily with oxygen to release heat." (One encomium: "Methane, the almost ideal fuel.") "The single largest source of methane is the flatulence of ruminant farm animals—almost entirely cows and sheep." But leaks from fracking might now have relegated ruminant farts to second place. A dangerously underestimated greenhouse gas.

264

"Natural gas," per U.S. Energy Information Administration [24,328]. See note to "Coal" at **107**.

265

"North American" natural gas, A.D. 1974 [24,381, more often seen as 1,015 BTUs per cu ft]. *Since this substance is mostly methane (immediately above, whose HHV is 2.3% less), the value here may simply reflect differences in measurement.*

268

Dry American natural gas, A.D. 1980 [24,646, more often seen as 1,026 BTUs per cu ft].

560

Rocket fuel: liquid oxygen mixed with liquid hydrogen [51,500]. *For this substance, the HHV was not available, only the "heating value." Hence its comparative efficiency may be significantly higher than listed here.*

663

Pure hydrogen [60,958]. Another source gives [44,711.95]. *A caveat from NASA: "While liquid hydrogen is highly energetic, it has far less energy density than hydrocarbon fuels. Thus, to get an equivalent amount of energy from hydrogen requires a much greater volume."*

820,912

Uranium-235, "thermal energy actually generated" in a reactor [75,523,861].

11.2 million

Upper boundary of following (2012): "Development work is reaching for . . . an increase from about 40 megawatt-hours to over 300 megawatt-hours per pound of [nuclear] fuel . . . This has made it possible to produce power for 18–24 months between extended shutdowns for refueling." [1.024 billion].

93.68 million
Uranium-235, "in theory" [8.62 billion]. Calculation based on this: "Can release around 20 trillion joules . . ."

209.24 million
Plutonium-239, "used in a conventional nuclear reactor" [19.25 billion]. Calculation based on this: "One gram of plutonium used in a conventional nuclear reactor has the potential to release as much energy as a [metric] tonne of oil."

3.96 billion
"If it were possible to convert 1 lb of matter completely to energy, the heat released would be about 3,000,000,000 times the heat of combustion of one lb of coal"—which latter number this nuclear engineer had estimated at 13,500 BTUs rather than the 12,500 of the West Virginia Coal Association [see multipliers **136** and **147**]. I calculate his "if it were possible" product at 40,500,000,000,000 BTUs. However, "in a reactor about 0.09% of the mass of the nuclear fuel is converted." Therefore the HHV-equivalent would be merely 27 million times that of coal, or [364,500,000,000 = 3.64×10^{11} BTUs per pound]. The accuracy of my calculation may be a little spurious, for when this same author worked the numbers himself, describing uranium-235 being fissioned in a reactor with 1958 technology, and employing that figure of 3 million times the energy of the same quantity of coal, he arrived at 3.5 rather than 3.64×10^{10} BTUs per pound. Still another fellow, a physicist, estimated the fissioning of the U-235 in the atom bomb which the Americans dropped on Hiroshima at about 3.45×10^{11} BTUs per pound.

3.79 sextillion [or 3.79×10^{21}]
Uranium-235, completely fissioned (theoretical figure). *Calculated from the statement that a single atomic fission of this isotope yields 200 million electron volts.* [3.49×10^{23}].

Given the immense energy available within nuclear fuel, one can see the attraction of nuclear power—especially since no greenhouse gases are generated. The events at Fukushima demonstrate how wonderfully that mode of power generation has worked out.

NUCLEAR

Fukushima, Japan (2011–14)
Hanford, Washington (2015)

Overleaf: Tepco's infamous Plant No. 1, photographed from about 2 kilometers

Nuclear Ideology

Assertions

UNIQUE OR INTRINSIC BENEFIT

"Wool, a very complex molecule, is improved by small amounts of radiation."

U.S. Atomic Energy Commission, 1965

"A critical source for clean, reliable, affordable energy..."

"Nuclear Matters" newspaper advertisement, 2014

"Does not emit CO_2, so it should contribute to the prevention of global warming."

Tepco public relations official, interviewed in Tokyo, 2014

"The relatively low share of greenhouse gas emissions from energy industries in France can be partly explained by the use of nuclear energy for power generation."

European Union greenhouse report, 2014

"The only energy source that runs 24/7, 365 days a year."

"Nuclear Matters"

"Will free us from the fear that our energy resources will run out within quite a short period of history."

Hans Thirring, 1958

"NUCLEAR POWER... could have a long-run stabilizing effect on energy prices as a whole."

Organisation for Economic Co-operation and Development, Paris, 1974

"Ultimately, if the technical problems of solving nuclear fusion are solved, man will have an essentially unlimited source of energy."

Encyclopaedia Britannica, *1976*

"May help to save the coal-mines from depletion."*

Hans Thirring

"Redesign of the internal combustion engine, removal of sulfur from fuel, and a switch to atomic energy for electric power generation will all hopefully relieve these very serious perturbations of the nitrogen and sulfur cycles."

Eugene Odum, ecologist, 1971

"Almost exhausted oil boy"—captioning a cartoon of a sad-faced oil barrel with legs, sweating as it tries to run. "Today, October 26, is Nuclear Power Day."

Ad placed by Niigata Prefecture, 1985

"At school and at the university we'd been taught that . . . [a nuclear reactor] was a magical factory that made 'energy out of nothing . . .'"

Zoya Bruk, environmental inspector and Chernobyl survivor, after 1986

"A pound of enriched uranium, which is smaller than the size of a baseball, has the energy potential equivalent to approximately a million gallons of gasoline."

Alan M. Herbst and George W. Hopley, 2007

"A thimbleful of nuclear reactor fuel can release as much energy as can be obtained by burning a million times as much coal."

John Tabak, Ph.D., 2009

"The cost of electricity is comparatively low, and the fuel is long lasting, so you can generate energy at a stable price."

Tepco public relations official

"On average, an installed nuclear kilowatt produces nearly twice the annual electricity of a renewable kilowatt."

World Nuclear Industry Status Report 2015 [an anti-nuclear source]

"The densest (in watts per square meter) and safest (in deaths per joule) form of energy known to man."

Eric McFarland, director of the Dow Centre for Sustainable Engineering Innovation, Queensland, Australia, 2015

---

* This pleased the writer because while coal is not irreplaceable as a source of electric power, many coal tar derivatives in commercial and industrial use lack convenient substitutes.

"Besides making nuclear power available, the nuclear reactor provides a copious supply of neutrons which may be used to produce further radioactive elements."

Farrington Daniels and Robert A. Alberty, 1966

"Potential applications . . . include medical and industrial isotope co-production, large-scale radiation-induced chemical synthesis, water treatment, food preservation and other applications that creative thinkers will certainly invent . . . A nuclear reactor might well be the center of an advanced-materials production facility with the electrical power the least valuable of its many products."

*Eric McFarland. These remarks bear comparison with the "coal tree" in the coal ideology section.**

"Studies have shown that radioisotopes such as plutonium-210, curium, thallium or ruthenium could be used in an automotive system. They can be safely contained, can turn out the needed power density, and have a radioactive lifetime of from a few years to as much as 80 years."[†]

Irwin Stambler, 1972

"[A] socialist transition worthy of the name is not possible without nuclear power, assuming that its delivery comes in the most advanced technological forms and under the democratic control of the working class . . ."

Democratic Socialist position paper, 2010

"We therefore conclude that the limit to population set by energy is extremely large, provided that the breeder reactor is developed or that controlled fusion becomes feasible."

Alvin M. Weinberg and R. Philip Hammond, 1971

"I would be afraid to say this in Japan because they'd call me a rightist, but I'm worried about China.[†] If we need to, we can turn nuclear fuel into nuclear weapons to defend our country."

First-generation Japanese-American woman, 2014

---

* See II:10.

[†] However, personal nuclear vehicles would be expensive and "very heavy, with the necessary shielding weighing roughly 2,400 lbs. per car."

[†] From *The Japan Times*, 2016: "China already has the world's fastest-growing nuclear energy program . . . China is an important market for the world's nuclear industry giants. The U.S. last year eased restrictions on its civilian nuclear cooperation with China to allow the reprocessing of fuel from U.S.-designed reactors . . ."

AVAILABILITY AND EXPEDIENCY

"Unlike oil, nuclear fuel is available over much of the world, as in Australia and Canada, where the political situation is stable."

Tepco public relations official, 2014

Embellishments
ECONOMIC APPEAL

"For each 1,000 megawatts of installed capacity, nuclear creates approximately 5,000 jobs during construction, compared with only 1,000 jobs each for natural gas and wind. During operation, a nuclear plant creates approximately 500 jobs, compared with only 60 and 90 jobs for natural gas and wind, respectively."

Tyson Smith, "a lawyer with clients in the nuclear energy industry," 2017.
For a claim about coal and jobs, see II:15.

LOCAL APPEAL

"Nuclear power started at Fukushima. The town was reborn. The symbol of pre-fecture growth . . ."

Fukushima Minyu *newspaper, 1971*

"When we were young, when the Tepco plant was built here, we were very thankful to finally be able to work locally. In this area, when you say, I got a good job, that means, I'm a Tepco employee or a civil servant."

Nuclear evacuee, Iwaki, Japan, 2014

NATIONALISTIC APPEAL

"It is a pity, the current situation of depending on foreign countries" for energy. *And:* "Foreign dependence on electricity: There is already a limit on what corporations can do."

Tohoku Electricity ads, 1976

"Plutonium thermal is needed for the country that lacks natural resources."

Hokkaido Electric ad, 2008

"If we want to keep America working, then we need to keep these plants working."

"Nuclear Matters"

GLOBAL APPEAL

"To Save the Planet, Go Nuclear."

Lamar Alexander and Sheldon Whitehouse, 2016

TAUTOLOGICAL TWADDLE

"With nature power, nuclear power and human power, for the earth's environment."

Kushu Electric ad, 2009

"I don't want to forget the warmth, gentleness and importance of electricity. October 26: Nuclear Power Day. Thank you. Japan's 45 [reactors]. Hamaoka Nuclear Power Plant's fourth reactor was completed"—accompanied by a smiling portrait of the actress Hoshiho Kazuko, 1993.

Chubu Electric, 1993

"From now on, I will turn my face toward nuclear power."*

Kushu Electric, 1997

"Our goal is the nuclear power plant that walks with neighbors."†

Chubu Electric, 2010

"Yotaro is wetting radiation," with a picture of a baby peeing. "We want to talk close to you. Nuclear Power Electric."

Kushu Electric, 1993

"We tried to think about life, about energy with ladies," with a picture of the celebrity Kozu Kanna.

Chubu Electric, 2009

---

* Here the translator sees an implication that not turning one's face toward nuclear power would set one at odds with the community.

† The context is something like: "Try to understand the goal," which of course is nuclear power. "We will not walk away from you. Please walk with us toward the goal."

"Nuclear power is the energy source of life. The town where there is energy is the town where there is [a] smile."

Tohoku Electric, 1996

"Nuclear power makes money."

"Mr. Takagi," mayor of Tsuruga, 1983

"America's nuclear energy plants empower us. Let's keep them running strong."

"Nuclear Matters"

Defenses Against Criticisms

"The radiation exposures in the highly publicized incidents [such as the Three Mile Island nuclear accident] are about equal to the extra radiation you get from spending five days in Colorado."

Bernard L. Cohen, 1983

"The Rasmussen estimate for the frequency of a loss of coolant accident (LOCA) is one every 2000 years of reactor operation."

Bernard L. Cohen. [A LOCA is exactly what occurred at Fukushima, after quite a bit less than 2,000 years.]

"Nuclear power is safer than oil."

Editorial in Fukushima Minyu *newspaper, 1971*

"Watching the safety of [the] nuclear power plant with my big eyes"—captioning a picture of a small boy making pretend-binoculars with his hands.

Fukushima Prefectural Nuclear Power Public Relations Association, 1986, in response to the Chernobyl accident

"Anti-nuclear campaigners have been racking up the figures for deaths and diseases caused by the Chernobyl disaster, and parading deformed babies like a mediaeval circus. They now claim that 985,000 people have been killed . . . Of the workers who tried to contain the emergency . . . , 134 suffered acute radiation syndrome; 29 died soon afterwards. Nineteen others died later, but generally not from diseases associated with radiation. The remaining 87 have suffered other complications, including four cases of solid cancer and two of leukaemia. In the rest of the population, there have been 6,848 cases of thyroid cancer among

young children, arising 'almost entirely' from the Soviet Union's failure to prevent people from drinking milk contaminated with iodine[-]131."

George Monbiot, after 1986

"Safety is our greeting word. October 26 is Nuclear Power Day."

Shikoku Electric, 2009

"Over the years, the safety record in Western-built reactors has been remarkable,
and advanced reactor designs promise to be even safer."

George A. Olah, Alain Goeppert and G. K. Surya Prakash, 2009

"Danger [will] never be a reality, never damage local people."

Japan Nuclear Power Culture Promotion Foundation, 1976

"Reactors are designed to be inherently safe . . . Nuclear power benefits from . . .
the concept of defense in depth . . ."

Encyclopedia of Chemical Technology, *1994*

"If there is [an] unexpected accident, the layers [of containment will] protect."

*Ministry of International Trade, Industry Agency for Natural Resources and Energy,
1996*

"While the trauma caused by this accident [at Fukushima] was great, manifested
largely in the evacuations, no one has died from radiation exposure and no one
is likely to die."

Democratic Socialist position paper

"Japan's 2011 Fukushima nuclear disaster occurred because of the use of out-
dated technology as much as it was the preceding earthquake and tsunami.
Molten-salt and molten metal reactors, for example, . . . could vastly reduce cost
and increase safety."

Eric McFarland

"Nuclear power plants actually have caused fewer fatalities per unit of energy
than other major sources of power."

Xinhua News Agency, China, 2016

About Uranium

*Have you ever heard of the element uranium? It's sometimes talked about in
the news. It's an important fuel in some power plants that make electricity.*

Tyrone Mineo, *Uranium* [a book for children], 2014

The first specimen ever to be accurately identified was carbonaceously
black; but only because we got to it too late. *The silvery luster of freshly
cleaned uranium metal is rapidly converted first to a golden yellow, and then
to black oxide-nitride film within three or four days.* And perhaps the purity left
something to be desired. In due time we learned that this element manifested a
peacock's worth of pretty colors in its compounds: Uranium trifluoride was dark
purple, the tetrafluoride green, the pentafluoride either greyish-white or
yellowish-white, depending. Uranium carbide, which did our thermodynamic
work in sodium- or lead-cooled reactors, was *a dark grey solid with a metallic
luster.* Uranium trioxide, another nuclear fuel, could be anywhere from brick-red
to yellow.

You may recall that carbon made up 200 parts per million in our planet's
crust. Uranium was less conspicuous, at two parts per million. Even so, its busy
irradiations contributed to the warmth of the place we once called Hell. And
whatever it did down there it might as well do up here for us, steaming water in
order to rotate turbines, so that we could keep the lights on.

To be sure, it presented certain difficulties. The *Encyclopedia of Chemical
Technology* called it *a general cellular poison which can potentially affect any organ
or tissue.* It specialized in harming the liver and kidneys.

Since it only came in at number 48 of all the elements for abundance, it rarely
did us mischief. (As that children's book explained: *Everyone is exposed to a small
amount of radiation every day. This isn't harmful.*) Volume for volume, "natural
uranium" was 1/1,000 as "hot" as radium-226—which, by the way, was *literally*
hot, emitting 180 BTUs per hour for every pound. (As for plutonium, that *feels
hot, like a live rabbit.*)—Uranium rock did not even feel warm.*—We ate a mi-
crogram or two of uranium every day in our food crops. If it got into our blood,

---

* In case you are wondering why in that case uranium instead of radium became the subject of this
chapter, here is one answer: To extract one pound of radium required between 2.85 and 6.65 million
pounds of uranium ore.

it ended up in our skeletons. Otherwise we mostly pissed it out.—Inhaled uranium dust, of course, was not so good.

Following human nature, we set out to concentrate the nasty stuff.* We had already been unknowingly doing so, thanks to agriculture. Uranium's natural occurrence in soil was 0.7 to 11 parts per million. The phosphate fertilizers we spread not only helped warm the atmosphere,[†] they also boosted cropland's uranium levels to 15 parts per million. And indeed, one way to get uranium was to soak those phosphates in a hydrocarbon solvent: organic esters dissolved in that traditional oil well product, kerosene. Thus even this very first manufacturing stage begins to reveal the ironic fact that nuclear fuel, touted as the anti-carbon ideology, required carbon-based feedstocks—and considerable carbon-combustion-derived electric power.

(Come to think of it, our fossil fuels frequently contained *relatively high levels of uranium as impurities.* Thus we added value to the fields, roads, back yards and streams near power plants.)

But mostly we left those phosphates to agrochemists, mining the conglomerate rocks and sandstone in which 90% of uranium could be found,[†] as a mix of two isotopes[§] of significance: U-235 and U-238. U-235 is capable of continuous energy release through fission; U-238 is not, at least not without help. In fact, U-238 inhibits the fission of U-235. In 1997, U-235 comprised only 0.72% of all uranium; and since it decayed more rapidly than U-238, in your time it will be slightly harder to come by.

Our task became to "enrich" uranium rock until it held an increased proportion of U-235.—Time to burn more fossil fuels!

First we had to dig up the pitchblende, uranite and other ores which contained decent concentrations of what we were after. I hate to pooh-pooh the good old days when any solvent American could buy a pocket-sized Geiger counter

---

* Before radioactivity was discovered, the Austrians were already coloring ceramics with it—or rather with its salt, sodium uranate, which made a fine yellow-green glaze. Uranyl sulfide could be oxidized into a pigment called uranium red, which I hope was never an ingredient in lipstick. Uranyl nitrate, which my 1911 *Britannica* considered "the most important uranium salt," existed as yellow prisms. In those days photographers still fooled with it, although commercial uranium printing papers got discontinued *ca.* 1898. One fellow was still at it in the early 21st century. I have seen reproductions of his prints: murky, grainy, reddish-yellow—and radioactive, I suppose. When I was alive, platinum printers might employ it to tone their images blue, olive or red. "The tones obtained with uranium are not stable and in time may change."

† See "About Agriculture," p. 111.

† In your time, with the oceans surely dead from acidification, you might have no objection to straining the seas for uranium, in the unlikely event that you still enjoy large-scale electric power. At three parts per billion (that is, before the Antarctic ice cap melted), marine uranium added up to more than 11 trillion pounds.

§ For nuclear terminology, see pp. 536ff.

from Sears and Roebuck, *with earphone head set, radio-active material for testing and AEC booklet "Prospecting for Uranium"*—but in fact what we were about to undertake required heavy machinery with internal combustion engines: Now came pulverization, which we preferred not to accomplish with hammers and human slaves. *Some ores are highly refractory and require intensive processing, while others break down between the mine and the mill.* We fed the *refractory* rock into jaw crushers, at an undisclosed cost in BTUs per hour, some of which we paid up front in coal or oil, while the rest we left to you zero-interest beings of the future. Meanwhile the miners paid in their own way, for these operations emitted the gas radon-222 (half-life: four days), which is *highly unstable and highly dangerous because of its intense radioactivity.* For that we established a brilliant work-around: Tailings got *placed in large piles and covered to prevent a local health problem.*

We ground up our coarse heap in rod mills, hammer mills, whatever it took. Next came the *oxidizing roast,* which required heat inputs (again provided by coal or oil, of course). Having solubilized the powder, we leached it in an oxide-laced bath of acid or alkali. *Most mills use acid leaching, which completely extracts uranium. Because of its low cost, sulfuric acid is preferred* . . . (To make sulfuric acid, which might have been our favorite chemical,* we often roasted sulfur, with fossil fuels for heat inputs.)

Our treasure remained no good to us until we could precipitate it out of solution. One method was solvent extraction. It takes energy and usually petrochemicals to make a solvent, which then adds to global warming by emitting volatile organic compounds.†

The other way involved ion exchange, with the help of ammonia. (You may remember that in 2012, carbon dioxide from ammonia production caused 0.4% of all greenhouse emissions from the EU-15 countries.) Or, instead of ammonia, we could aim for a higher uranium concentration by adding the powdered white oxide called magnesia, whose manufacture generally required high heat, and therefore, considerable energy, very likely from fossil fuels.

But enough on distracting subjects; at last we had our yellowcake—which was not necessarily yellow—and at the bottom of a corrugated round tunnel we now glimpse, courtesy of a photo in that same children's book (which unlike any other source absolutely guarantees the following: *METAL MANIA! All the uranium on earth is the result of a large star exploding more than 5 billion years ago*), yes, that powdered yellowcake, while a bespectacled technician rests his chin on

---

* See "About Sulfur," II:26.

† These get their eponymous section on II:352.

his hand, peering in at us through what one trusts is a heavy glass window.—It was refining time.

So we dissolved the yellowcake in nitric acid (whose manufacture had already cost our climate in nitrous oxide emissions*), then extracted it back again with tributyl phosphate—*in a kerosene or hexane diluent,* which must have originated either in a coal mine or an oil well—and after re-dissolution, evaporation, pyrolysis and reduction, all of which I shall spare you, we had uranium dioxide (*brown to copper-coloured*), one of the most common reactor fuels.

Oh, but it was far from ready yet! It had not yet been "enriched." As that children's book reminds us: *It takes a lot of energy to create enriched uranium. The uranium isotope needed for nuclear energy makes up less than 1 percent of all uranium!*

So through two or three more stages we converted it into uranium hexafluoride. Unlike the prettier incarnations of uranium, whose hues might appeal to many a child, "hex" was *an extremely corrosive, colorless, crystalline solid, which sublimes with ease at room temperature.*

Permitting it to sublime, we diffused it through a membrane of silvered zinc. The U-235 and U-238 within the invisible gas had very slightly different molecular weights. After the "hex" had gasified, we sent half of it through the membrane, passed half of that through another membrane, etcetera, each time pumping the undiffused half back to go through again; and of course at each sifting a smidgeon more of the lighter isotope came through. In case you are wondering how many repetitions were required, well, it depended on the enrichment. The first time, at Oak Ridge, when the purpose was to spread destruction and terror to Japan, thereby not only wrapping up the war but also cowing our Russian allies, the slide rule crew worked out that raising the proportion of U-235 atoms from 0.72 to 99% would take 4,000 diffusion cells. We later settled on a 96–97% enrichment for weapons-grade fuel, and 2 to 5% for reactor fuel.[†] Even the latter (whose "cascade" was modest at *more than a thousand separation stages*) cost considerable thermodynamic work. *Gaseous diffusion units are enormous in size, often covering hundreds of acres, and requiring huge amounts of electric power to operate.* At Oak Ridge's enrichment facility, K-25, the coal-fired power plant *designed to generate electricity to run the barrier diffusers* for an atomic bomb was rated at 238,000 kilowatts—more than 800 million BTUs an hour. If it ran at capacity, during that hour it would have burned 97 tons of coal!

Eventually we learned that centrifugal separation could be *more efficient,*

---

* See "About Industrial Chemicals," p. 124.

† At any time weapon fuel could be de-enriched into reactor fuel.

although the results were *highly corrosive*. Moreover, efficient or not, those gas centrifuges must have drawn considerable power in their own right, with their *high velocity . . . attained under temperature and pressure conditions that favor evaporation.**

At any rate, that was how we enriched uranium. Then we reconverted the enriched hexafluoride back into dioxide, or one of the other nuclear fuels: pure uranium metal, or *sulfates, silicides, nitrates, carbides, and molten salts.* What quantity of climate-altering emissions we made along the way I cannot begin to reckon. Here is one cryptically suggestive passage from a European Union greenhouse report, pertaining to France—the only mention of uranium in the whole document:

> Uranium tetrafluoride: N_2O emissions data is taken directly from annual statements of pollutant emissions since 1990 and emissions are derived from continuous measurements since the 2012 submission.

By 1943 we had even discovered how to irradiate that most common and supposedly unfissionable isotope, U-238, feeding slow neutrons to its greedy nuclei. The resulting "capture reaction" transformed it into U-239, which speedily decayed into neptunium-239—which expeditiously became plutonium-239. This isotope possessed a magnificent shelf life. After 24,360 years, half of its atoms would still be Pu-239. While that might not have been a desirable feature in fuel waste, it was awfully convenient from a military point of view. Furthermore, Pu-239 readily fissions, as was so empirically proved at Nagasaki.

The milder plutonium-238 powered certain machines on the Apollo 14 spacecraft; while Pu-239 presently entered use in the "mixed-oxide" (MOX) reactors containing plutonium and uranium in a ratio of about 1:17. Wasn't thermal efficiency the ideal in our steam turbines? As the World Nuclear Association promised us: *A single recycle of plutonium in the form of MOX fuel increases the energy derived from the original uranium by some 12%, and if the uranium is also recycled this becomes about 22%.* Well, viva MOX reactors!—One of those exploded at Fukushima, but plutonium contamination remained almost indetectable throughout the red zones, so I would call the World Nuclear Association right on the money, wouldn't you?[†]

---

* We could also choose electromagnetic separation, "an incredibly labor-intensive process." And there was another kind of magic (whose energy costs, were unknown to me), which was accomplished with "laser radiation of precise energy to sever the uranium-fluorine bonds of the lighter isotope but not of the heavier one."

† On June 6, 2017, five unfortunates at the Plutonium Fuel Research Facility in Ibaraki Prefecture happened to be "taking stock of a radioactive substance in an old container." The container was stainless steel, inside of which lay 300 double-bagged grams of powdered uranium and plutonium oxides

Back when we were alive, we liked to sell our inventories of everything as rapidly as possible, so that we could sell some more. MOX possessed a very agreeable quality in this regard: The more quickly somebody put it to work, the better!

> Plutonium from reprocessed fuel is usually fabricated into MOX as soon as possible to avoid problems with the decay of short-lived plutonium isotopes. In particular, Pu-241 (half-life 14 years) decays to Am-241 which is a strong gamma emitter, giving rise to a potential occupational health hazard if separated plutonium over five years old is used in a normal MOX plant (where radiation levels are very low). The Am-241 level in stored plutonium increases about 0.5% per year . . .

As for the "hex," someday we might even address *the problem of the disposition of 500,000 [metric] tons of depleted UF$_6$ stored at Paducah, Portsmouth, and Oak Ridge.* We could always blend in plutonium oxide, and whip up a fresh batch of MOX. Or we could do something else, sometime, in some ecosystem somewhere. But why not put all that off, just as we did with such other trivialities as global warming? Let the future worry, we said to ourselves. Why not live in the present? For *Carbon Ideologies* now delights to inform you that unlike its three main rival fuels, nuclear could be *fun!* The daughter of a uranium processing worker once assured me: "During the early nuclear testing times in Nevada the Sands Hotel used to pack lunches for guests to enjoy with their families while out watching the atomic testing in the desert outside of town. "The Chamber of Commerce in Las Vegas would print when and where the tests were going to be, along with the best viewing site on their yearly calendars."

Back to "atoms for peace": How efficient was plutonium? Well, from a thermodynamic point of view it fully justified the extravagant energies poured into

---

"used in past experiments." In other words, the substance must have been MOX, or something like it. The worker on the spot had prepared himself as well as any Boy Scout: suited, gloved and masked. No immediate danger, colleagues! But as soon as he withdrew the six sealing bolts, out rushed black powder— the result, perhaps, of a buildup of helium from the plutonium's radiodecay. "Although masks were covering the workers' noses and mouths, radioactive material was detected inside the noses of three" of the men; "internal . . . exposure was detected in four" and "suspected" in the fifth. The employee who had opened the cylinder got "up to 22,000 becquerels . . . in his lungs," or something like "360,000 becquerels overall . . . Under current labor standards," which they may or may not have been obeying, "that translates into 1.2 sieverts a year, and perhaps 12 sieverts over 50 years . . ." The two figures in sieverts respectively translate into 136.986 microsieverts per hour (comparable to Okuma Town's "more than 100" micros after the nuclear accident) and 27 microsieverts per hour—equivalent to certain gratings and drains I measured in Okuma and Tomioka. To make better sense of those numbers, see table of Comparative Measured Radiation Levels on p. 244. As for those five workers at the Plutonium Fuel Research Facility, well, why worry about a touch of prospective bone cancer?

its production. One pound of it would have carried out every last BTU of the hourly work for which that coal plant at Oak Ridge was rated—non-stop for almost three years.

CALORIFIC EFFICIENCIES OF COAL, OIL, NATURAL GAS, URANIUM-235 AND PLUTONIUM-239,

in multiples of the thermal energy of coal

as simplified from the table of Calorific Efficiencies (p. 208)

All levels expressed in [BTUs per pound].

All comparative efficiencies greater than 10 have been rounded to the nearest whole number. All figures after U-235 rounded to 2 significant digits.

1
Coal, based on West Virginia Coal Association average [12,500 BTUs/lb].

1.54
"Oil," calculated from average for diesel fuel [19,250].

1.95
Natural gas, dry American, A.D. 1980 [24,381].

689,503
Uranium-235, "in theory" [8,618,790,909].

1.48 million
Estimate of uranium burned in a 1-million-kWh Japanese reactor [18.60 billion].

1.54 million
Plutonium-239, "used in a conventional nuclear reactor" [19.25 billion].

To complete the preparation of nuclear fuel we shaped our uranium dioxide, MOX, or whatever else pleased our fancy, into spheres and pellets, furnace-fired them, then dropped them into long* zirconium-jacketed rods.—For marine reactors we actually alloyed the uranium with zirconium. This latter element

---

* At Fukushima Daiichi these ran 13 feet.

resembled uranium in its ductile silveriness, but resisted corrosion and disdained to absorb neutrons, which create radioactive byproducts.—Those built up all the same.

One might imagine that once a reactor began to run, the cost of the energy it generated would fall to near nothing. What did anyone need to do then but let those long-lived fuels go right on fissioning?—Strange to say, nuclear power cost about the same as coal.

The lifetime of a fuel rod was two to five years. After that, the rod grew so contaminated with those byproducts (or, if you'd rather, "daughter nuclides") as to impede neutron travel: No neutrons, no fission.

I liked the way they put it: *The used or spent fuel contains a large inventory of fission products.* They sounded like salespeople laying out their wares. But they were more generous than that. Wherever possible, they would give their *fission products* away—to you.

We had dreamed up three methods of waste disposal, depending on the danger: *dilute and disperse,* wait out short-term decay and *concentrate and contain.*

Here was a typical to-do list:

1. Submerge spent fuel in a water pool for a decade.
2. Transfer it to a *dry storage facility.*
3. Sweetly bundle it in titanium or stainless steel.
4. Lovingly jacket it in concrete or steel.
5. *The fundamental safety criterion for the permanent repository is that the engineered package retain complete integrity for at least 1000 yr.* (In Japan, where everything was better and brighter, a shorter interval sufficed, at least for "low level nuclear waste," which *must be kept apart from residential settlement for up to 400 years.*)

The daughter nuclides cesium-137 and strontium-90 *provide most of the radioactivity, radiation and heat during the early years of a disposal facility.* Another fission product was iodine-131—more short-lived than the other two, but particularly dangerous to children. We will meet all three of them at Fukushima.

Depleted uranium (0.2% U-235) offered twice the density of steel. It was 19 times denser than water, 11 times denser than lead! So it made for efficient ballast in ships. And the military loved it, because DPU shells could *self-sharpen during armor penetration by failure along adiabatic shear bands.**

---

* Another source: "And D[P]U has the added advantage of catching fire on impact."

About Nuclear Reactors

The true romance which the world exists to realize, will be the transformation of genius into practical power.

Emerson, before 1844

Now, why did we even need fission reactors? Why couldn't the reliable irradiation-heat of nuclear fuels suffice?

You have seen* that a pound of pure carbon, completely combusted, will give off something like 14,000 BTUs. In the more modern and practical metric system preferred by physicists (and I will respect it in the three following Japan chapters, because when I was alive that country reported its experiences in kilometers and kilograms), the high heating value of carbon would be 30.87 BTUs per gram.[†] Of course it can supply that heat only once.

A gram of plutonium-239 gives off 1.824 plus or minus 0.18 BTUs *every second*. Thus in only 17 seconds that isotope would have released as much energy as the same volume of carbon. After one half-life, 24,100 years, it will still be emitting almost a BTU per second.

A gram of plutonium-238 offers a robust 432 BTUs per second. However, the half-life of that stuff is merely 87.7 years.

These qualities failed to satisfy us. Let me quote a sad epitaph from 1991: *Pu-238-based radioisotope thermal generator systems delivered 7 W[atts]/kg[‡] and cost $120,000/W.*

These devices reached their niche market. In the far-travelling *Voyager* space probes they produced electricity for more than half a century. Thus they satisfied demand. But we deserved to get more than that out of plutonium!

Fortunately for us and our power to do evil, more energy can be released through nuclear fission, which entails the splitting of an atomic nucleus into two pieces by collision with another particle, most practically a neutron. "Enriched"

---

* Above, p. 213.

[†] The nuclear energy units of the *Encyclopedia of Chemical Technology*. In the full metric employed by Japanese and most everyone else, 32,564 kilojoules per kilogram.

[‡] Or 0.1806 BTU per minute per pound. Twelve and a half pounds would be needed to give as much power as the plug-in vibrator on p. 68.

uranium falls susceptible to chain reactions whereby the broken nuclei of an atom throw out their own neutrons, which strike and fission other nuclei. A single fissioned atom of U-235 liberates a respectable 3.04×10^{-14} BTUs. In the process, it becomes first neptunium, then plutonium, which also fissions nicely.

U-238, the isotope we started with, still made up most of our fuel, since "enrichment" of even 5% remained perfectly adequate for our peace-loving purposes. And in case you are fretting about wasted fuel mass, please allow the World Nuclear Association to emit some reassurance: Even U-238 *can become plutonium-239 and by successive neutron capture Pu-240, Pu-241 and Pu-242 as well as other transuranic isotopes . . . Pu-239 and Pu-241 are fissile, like U-235.*

When each fission gives rise to several others, the result is a nuclear bomb. In a nuclear reactor, the number of fissions remains constant.

To achieve that steady state, we mixed the fissile substances in our fuels with "moderators" such as graphite or water, which slowed down neutrons. In addition, we transpierced each reactor core with so-called "control rods," whose material (often boron, in the form of boric acid) *absorbed* neutrons. As our mechanisms withdrew the control rods from the core, the number of fissions increased. When we fully re-inserted them, the reactor shut down.

And so the control rods retracted, at which neutrons and flying "fragment nuclei" obeyed us, slamming into others and halting dead, at which point their kinetic energy began doing thermodynamic work—far, far more of it than could burning coal or oil! And, yes, irradiation still gave off its own heat—about 5% of what the fission accomplished.

In 2011, the year of the Fukushima accident, nuclear power generated 2,517.7 billion kilowatt-hours (8,592,910,100,000,000 BTUs, or 8.6 quads)—nearly 12% of all electricity on Earth.

We had various nuclear batteries by then, and *when shielding is not a critical problem, as in unmanned satellites, they are extremely convenient, being reliable and light.* Some unshielded batteries could produce a continuous 320 watt-hours per pound—six times better than chemical ones. But since demand kept rising, they were hardly going to solve our so-called "energy crisis."

So we built several types of reactors, for several purposes. *Carbon Ideologies* concerns itself only with reactors intended to produce municipal electric power. Concerning the boutique reactors that made medical radioisotopes, and the outer space reactors which cooled their cores with liquid hydrogen or liquid sodium, I have nothing to tell you. And of the electricity-generating sort, let us ignore heavy water reactors, graphite reactors and fast breeder reactors. Two commercial light water designs will be relevant (to my country as well as to

Japan): pressurized water reactors and boiling water reactors. Both made steam, which then directly or indirectly did mechanical work upon turbines, turning them to create electricity. Both often although not exclusively took uranium dioxide as their fuel.

PWRs were of the indirect kind; they used heat exchangers. For whatever reason, they were prevalent in western Japan. Or, to put the matter more exactly, they employed *a closed circuit of high pressure, high-temperature water* to *transfer . . . heat from the reactor core to . . . U-tube steam generators.* (The nuclear icebreaker *Lenin,* and one American merchant ship, the *N.S. Savannah* ("not competitive economically" with her oil-powered sisters) were both powered by PWRs.)

BWRs brought water straight into the core, then irradiated it into steam, which spun the turbines without any intermediaries. These electricity-makers were common in eastern Japan. Unlike PWRs, they got by without boron.

As to how efficient they were, well, what can I tell you? The Rankine Cycle* stole perhaps two-thirds of their steam-heating BTUs, just as with fossil fuel plants. (Two practice problems in a physics textbook from 1971 estimated the conversion success of the energy in fission fragments to electrical energy at a flat 25%.)—All the same, if only poor old Josephine Cochran could have lived to see the magic! Her steam-powered automatic dishwasher, which could wash and dry 200 dishes in two minutes, was *the sensation of the 1893 Columbian Exposition.* But it didn't catch on—too expensive. A nuclear-powered dishwasher could have autoclaved those dishes into outright purity, and for cheap—didn't they promise us that nuclear power was cheap?

In contrast to power plants using fossil fuel, nuclear reactor plants emit no compounds of carbon, nitrogen or sulfur, and thus do not contribute to . . . global warming, glowed the *Encyclopedia of Chemical Technology.*† Then it mentioned certain aspects of "normal operations":

---

* See "About Power Plants," p. 150.

† Takashi Hirose, 2011: "Before saying that 'nuclear power supplies one-third of the demand for electricity,' it needs to be said that 'twice as much energy as the energy they [the reactors] produce is used to heat up the sea.' I want to ask, what kind of global warming debate is it that never discusses this fact?" He claimed that Japan's nuclear plants were cooling themselves by emitting 100 million kW [= 5,688,400,000 BTUs per minute] "in the form of heat" into the ocean. Back in 1971, one ecologist was already worrying that *"thermal pollution* will become an increasingly serious problem" from nuclear power. "The shift from fossil fuel to atomic power reduces air pollution but increases water pollution . . . There is no agreement as to its ultimate effect on the global heat balance."—Given the immensity of that heat sink, I myself am skeptical that our nuclear thermal pollution perceptibly warmed the ocean. But I always hoped for the best, back when I was alive.

These emissions do include radionuclides of the noble gases xenon and krypton,* which readily diffuse throughout the atmosphere. Small quantities of soluble radionuclides are released into lakes or streams that provide very large dilution factors.

In short, we pulled our usual trick and threw those "away."

To shield the operators, we had to build in four-foot thicknesses of high-density concrete—which of course required that "big five" material, cement, at an unmentioned production cost in carbon dioxide—reinforced with bars of another "big five" substance, steel.

Like our fuel rods, the reactors themselves grew contaminated with fission products. Radiation could turn rubber inelastic, reduce the notch-impact strength of carbon steel (although for other metals, *most show appreciable increase in yield strength*), and in high doses affect aluminum alloys: *ductility reduced but not greatly impaired*. One anti-nuclear source asserts that *the lifespan of each part is around 30 years at maximum*. I lack knowledge on the subject. But parts of the reactors did become waste, which had to be stored somewhere for a very long time. It was typical that we invented various temporary solutions, leaving the consequences, like our carbon dioxide, as our personal gift to you of the future.

Our cost-benefit analysis of nuclear power and its competing carbon ideologies went as follows: Get subsidies wherever possible, calculate the ratio of immediate income *versus* immediate costs, such as employee compensation, and expel from the equation such future costs as global warming, air pollution and nuclear waste disposal. Again, we figured that *you* could pay for all that.

Bit by bit, each reactor's coolant water grew "hot" with nickel-63, niobium-94 and nickel-59, whose respective half-lives ranged from 100 to 760,000 years. Outside the core, cobalt nuclides released the most radiation. Their predominant member was cobalt-60, which could be inhibited with zinc, although an occasional result of that cure was more radioactivity, from zinc-65 "hot spots."

It might take four to six years before a PWR got significantly contaminated with cobalt-58 and cobalt-60. Then what? I don't exactly know. The procedure must have depended on how "hot" they were, for how long. A reactor could operate for decades.

One radiocontaminant especially pertaining to BWRs was nitrogen-16, whose half-life was only 7.1 seconds, but since it kept being created, and often

---

* The half-life of is Xe-133 is five days; of Kr-85, 11 years.

existed as radioactive nitric acid, it ate away at whatever encased it. Mixed with hydrogen, it formed corrosive radioactive ammonia.

The six reactors at the Fukushima Daiichi power plant (four of which went critical as a result of what became officially called the Great Tohoku Disaster), and at least two of the four at the larger-capacity Tokai Daini* 12 kilometers to the south, were all BWRs. Daiichi's Reactor No. 1 had been delivering electricity for two weeks short of 40 years. Daini's eponymous Reactor No. 1 was only 29 years old. Their fuel consisted of uranium oxide, with one exception: Daiichi's 34-year-old Reactor No. 3 had been converted to "pluthermal" in 2010. In other words, its steam now came from MOX. Every global-warming-fearing type should have been thrilled about that, because *one gram of plutonium used in a nuclear reactor has the potential to release as much energy as a tonne of oil.*

And in case of emergencies, both plants expected to save themselves with fossil fuels. As my country's Nuclear Regulatory Commission reminded us (a year after the accident): *Nuclear power plants need electrical power 24 hours per day, even when the nuclear reactors are shut down, to run equipment that cools the reactor core and spent nuclear fuel.* At Plant No. 2, only one out of three emergency diesel generators would fail. At No. 1, matters fell out differently.

Their continuous inflow of water kept the core from overheating, and operators deployed various other coolants. (Have I mentioned that as contaminated fuel got "hotter," it gave out all the more literal heat?) *A loss of coolant is conceivable,* admitted the 1976 *Britannica,* which otherwise expressed a moderately pro-nuclear attitude. *The loss would immediately stop the chain reaction, but the heat residue could melt the fuel, releasing a large volume of radioactivity. Coolant systems are designed to make the probability of such an accident extremely remote, and [even more so with] . . . an emergency coolant in the form of automatically sprayed borated water. In the very remote possibility of both coolant and emergency-coolant systems failing, further engineering safeguards must be provided.*

As so often in human endeavor, those safeguards were inadequate.

---

* Daiichi essentially meant "No. 1"; Daini, "No. 2," and so we will call them. No. 1 would appropriately be the star of the show.

1

Lower Than for Real Estate Agents

For the Fukushima disaster of 2011, the consensual estimate is a 1% increase in cancer for employees who worked at the site and an undetectable increase for the plant's neighbors . . . even after the catastrophe in Japan, the likelihood of work-related death and injury for nuclear plant workers is lower than for real estate agents . . .

Craig Nelson, "A Radiation Reality Check: From bananas to bricks, radioactivity is everywhere—but it's nothing to be afraid of" (2014)

The following table is meant to orient the reader to comparative radiation levels. It is expressed in multiples of the lowest level in my studio, which happened to be my darkroom coating counter.

Nearly adjacent spots can read wildly differently. For example, in this table a certain drainpipe appears twice. When I frisked it at 1 foot above ground level, it was 200 times more radioactive than my coating counter. As I bent down and continued measuring, the level quickly rose. At ground level it had more than doubled, to 533 times the benchmark value. The moral: If you enter a hot zone, please do not stay very long unless you know which side of the street to stand on.*

Please note that nearly all the Japanese readings in this table were made in 2014. In 2011 many of them would have been much higher. For instance, on March 16, 2011, one spot in the town of Iitate reportedly read more than 95 microsieverts an hour. When I measured this same place on October 23, 2014, the level was less than 5 micros.

Do not conclude from this that the nightmare of Fukushima was approaching its end. Cesium-137, a major pollutant, has a half-life of 30.2 years. As you will see, 10 half-lives, or 302 years, must pass to reduce the concentration of this isotope to 1/1,000 of its peak level.—Why then did the measured radiation decline so rapidly?—Because the cesium had been washed a few centimeters underground, or into rivers, or took up residence in mushrooms and trees.

In the abandoned town of Tomioka I got the chance to measure the drainpipe of a certain house which had been under decontamination eight months before. On that first occasion it had measured 22.1 micros per hour (evidently the crew had not yet gotten to it). On my return, with the yellow tape around it removed and the crew long gone, it measured 32 micros. How would you like to live in that house?

---

* For a different picture of the same raw data, see p. 425, where I have averaged my measurements by locality. That table provides an approximate (and very superficial) answer to such questions as: "Which is more radioactive, San Francisco or Tokyo?"

COMPARATIVE MEASURED RADIATION LEVELS, 2014–15* (with Hiroshima readings from 2017),

in multiples of lowest Sacramento interior reading

(from pancake frisker† data)

All levels expressed in [microsieverts per hour]. Benchmarks interpolated in **boldface**.

"<R>" means "reported"; in other words, this value was not measured by me.

"Average frisk" is the arithmetical average of several instantaneous readings in NORMAL mode. "Short frisk" is a steady NORMAL reading measured over 1 to 10 seconds. Dental X-rays have been read in MAX mode. All others measured in 1-minute timed counts.

The full data from which these readings were selected appears in the end matter.

1 sievert (Sv) = 1,000 millisieverts or "millis" (mSv) = 1,000,000 microsieverts or "micros" (microSv). 1 rem = 10 mSv. For definitions, see pp. 545ff. All figures over 1 million rounded to 2 significant digits.

<1–7.61

<R> **0.04 to 4 mSv per year. Worldwide variations in natural background dose.** [0.00457 to 0.457 microSv/hr].‡

1

My studio darkroom coating counter, Sacramento. The lowest measurement of which the frisker was capable [0.001 micros per second].§ A full

---

* Dental X-rays measured in January 2015. Dunsmuir, California, and Portland, Oregon, readings also taken in 2015. The Hiroshima readings were made in January 2017. All other frisker readings taken in 2014. Some benchmarks from other sources were published earlier, and here so stated.

† This device is introduced on p. 397.

‡ *CRC Handbook of Chemistry and Physics,* 87th edition (2006): "The natural background from all sources in most parts of the world leads to an equivalent dose . . . of about 0.04 to 4 mSv per year for the average person." [To convert from millisieverts per year to microsieverts per hour, multiply by 0.114.]

§ Here Richard Crownover, M.D., Ph.D., remarks: "It is an odd quantitative choice to scale everything to the tiniest measured value. It's the least precise measurement and any slight change in that number would throw the scale about wildly."—My response: (1) Since 0.001 microSv/sec. is not only the tiniest measurement but also the tiniest measurable value for this device, it is a logical endpoint to my scale, and maintains consistency with most other tables in *Carbon Ideologies.* (2) I made this particular measurement many times, and always got the same reading. It is, of course, possible that my darkroom reading, or any of these other lowest measurements, was wrong, but I presume that most or all of them were comparable; getting the lowest possible year later, having frisked many other places, I had duplicated this lovely reading in only a few places, so I shall list them all here: Dunsmuir, California (in a forest overlooking the Sacramento River), some surprisingly urban sites in Portland, Oregon; Subway L3 in Barcelona (well underground); a breeze in the middle of the Neva River in Saint Petersburg [Russia], and the marble floor of the cafeteria in the Tretyakov Gallery in Moscow [0.06 microSv/hr]. *This equals 0.5256 millis/yr.* On a per-minute basis, the cumulative per capita fallout dose in the Northern Hemisphere from 1945 through 1999 was 1/27 of this amount.

1.9

<R> **1 mSv per year. Maximum dose advised for ordinary citizens, per the International Commission on Radiological Protection.**[*] Obviously this conservative standard generally goes unmet. *See entry for multiple 4.6.* It approximates the total body irradiation one receives from natural sources at sea level "in most parts of the world." For every increase in altitude of 1,500 meters, an additional 0.28 millis a year [0.03 micros a minute] may accrue. For professional workers, the ICRP maximum was 20 mSv. [0.11416].

2

The Darjeeling tea palace, Ginza, Tokyo. Coal House (built almost entirely of coal), Williamson, West Virginia, U.S.A. My studio's kitchen counter in Sacramento. Kievsky metro station, Moscow (deep underground). Shore of Lake Geneva, at Cully, Switzerland (near Lausanne). Air dose in La Rambla, Barcelona. Also typical of downtown Poza Rica, Veracruz, Mexico. [0.12].

3

Air dose in National Orchid Garden, Singapore. Concrete office interior, sex workers' organization, Dhaka, Bangladesh. Ironwork railing of Gogol's grave, Moscow. Sewer grating in Passeig de Sant Joan, Barcelona. Lump of coal on ground, mountaintop removal mine, Cook Mountain, West Virginia. Air dose near Barapukuria Coal Mine Company conveyor, Bangladesh. Stone step of El Tajín archaeological site (Totonac pyramids), near Poza Rica.

---

reading deep in the Barcelona subway, for instance, is highly plausible. (3) The frisker was calibrated (see p. 401), and in Japan it read in close conformance with other scintillation meters. At 2.0 micros (the manufacturer's lowest testing benchmark), it read 2.04 micros. At the manufacturer's tested 80 micros (my own highest field reading was 41.5 micros), it read 77.0 micros.

[*] But the ICRP's earlier standard was four times less conservative: "The ICRP recommended exposure limit to man-made sources of ionizing radiation . . . [1990] is 20 mSv/yr averaged over 5 years, with the dose in any year not to exceed 50 mSv."

Front of railroad station, Fukushima City, Japan (same reading for 3 consecutive days). **Air dose [at waist level] on bridge between Atomic Dome [= Ground Zero, 1945] and Peace Park, Hiroshima.** [0.18].

3.67

<R> Decontamination target for households in Iwaki, Japan* [0.22].

3.8

<R> **2 mSv per year. Japanese national target air dose ("1 additional milli").** Based on the same assumptions as in the footnote to **3.67**. [0.22832].

4

Mary Cook's headstone, Cook Mountain, West Virginia. Carrea Sant Anna, Barcelona, after a rain. **Air dose by Atomic Dome, Hiroshima.** Fallen leaf on gravel walkway, Tsurugajo Castle, Aizu-Wakamatsu, Japan. Interior of taxi van, Hirono, Japan. Interior wall of Enrique's Restaurant, Poza Rica. The last 3 readings were the highest in their respective locales. [0.24].

4.6

<R> **2.4 mSv per year. Average worldwide annual dose, according to Mr. Kida Shoichi, Nuclear Hazard Countermeasure Division, Iwaki.** [0.27397].

5

Deserted playground, Naraha, Japan [0.296]. Granite wall of Jardin de Veant, Lausanne. Bricks in Bangladeshi brickyard, 60 km north of Dhaka. Interior of Taylor's Books, Charleston, West Virginia [0.3]. On a different day, Taylor's Books measured only 0.18.[†]

5

Granite countertops in hotel suites in Charleston, West Virginia, and in Portland, Oregon [0.3].

---

* The "long-term goal in the Fukushima cleanup": An air dose of below 0.23, calculated from 1 milliSv per year, "assuming that an individual spends eight hours outdoors and 16 hours indoors."

† Dr. Crownover remarks: "Discussion of Taylor's [B]ooks. Release of [r]adon gas from the ground can vary significantly by weather conditions, particularly temperature [whether it is] raining or not. It can also be different if windows are open or closed. I first encountered this in an undergraduate Modern Physics lab where on a warm day with the windows open [Oregon] background radiation measurements were increased by a factor of 8. Radon levels can also increase in a closed space with brick/stone walls. Background measurements are not stable."

5.7

<R> "An average radiation background dose for a human being," according to James A. Mahaffey, Ph.D. (3 mSv per year). An identical value: "The average resident of the United States is routinely exposed to about 300 millirem . . . every year from the soil beneath their [*sic*] feet and the sun shining overhead." [0.34248].

6

Air dose on stone curb by Atomic Dome, Hiroshima. Ocean vantage point, Naraha [0.36]. The highest reading found in both locales.

6.83

<R> Average American yearly dose, mid-1990s [0.41].

6.85

<R> **3.6 milliSv per year. Alleged average worldwide dose.** [0.41095].

7

Koriyama Station [Japan], outdoor plaza [0.42].

7.61

<R> **4 milliSv per year. Old recommendation of International Committee on Radiological Protection, re: ceiling dose.** [0.4566].

8

Ten Commandments courthouse tablet, Pineville, West Virginia [0.48].

9

Granite curb of Molotov's grave, Moscow. Sidewalk by air dose marker, Fukushima City. Ocean beach in Okuma red zone, 3.5 km from Reactor No. 1. This last was the lowest reading found in that red zone. [0.54].

9.51

<R> **5 milliSv per year. "For an individual steadily receiving 500 millirads [= 5 milliSv] per year, the chance of dying from cancer or leukemia is increased by 30 percent."** That claim (from 1971) is much disputed; see header **190**. **"A read of 5 millisieverts is one of the thresholds for whether [Japanese] nuclear plant workers suffering from leukemia can be eligible for compensation . . ."** [0.570776].

10

Granite pedestrian zone of Anichkov Bridge, Saint Petersburg, Russia [0.60].

13

Air dose upon entering greater Iitate, Japan [0.78].

16.67

<R> **0.1 millirem/hour. Per** *First Responder's Guide to Radiation Incidents* **(2006), the ceiling for safe background dose. [1.0].**

19

Decontaminated driveway, Iitate red zone [1.14].

23

Cherry trees not exposed to ocean wind, Iitate red zone [1.38].

29

<R> **More than 15 milliSv per year. "Living on the atoll and consuming local food," Bikini, 1997—a half-century after U.S. nuclear bomb tests. [1.712].**

29

Roadside 10 m from bags of contamination waste, Iitate red zone [1.74].

34

A moment at cruising altitude (elevation unstated), Tokyo to Singapore [2.04].*

38

<R> **20 milliSv per year. Upper limit of Japanese green zone (this color meant "decontaminate with priority," and here the government hoped soon to lift any ban). Lower limit of Japanese yellow ["residence restriction," or daytime use only] zone designation. [In 2014,**

---

* Dr. Crownover inserts: "Readings in airplanes with the pancake detector would be sensitive to position (e.g., aisle or window seat, but more importantly the detector would not respond to the full spectrum of radiation in the air which is very different than on the ground . . . Also, radiation exposure on aircraft is sensitive to the 11 year solar cycle and also to solar particle events or 'space weather' . . . In Europe there is, or was, a phone number that sensitive travelers can call to determine whether they should delay flying for a few hours (particularly pregnant women)."

central downtown Tomioka was one such zone.] Legal ceiling for a nuclear worker in non-emergency conditions. The ICRP maximum for professional workers, including radiologists, dental technicians, etc. [2.283].

39

Edge of red zone, downtown Tomioka, Japan [2.34].

41

A sampled moment at airplane cruising altitude (36,000 ft), Washington, D.C., to San Francisco [2.46].

42

A sampled moment at cruising altitude (37,013 ft), San Francisco to Los Angeles [2.52].

49

First torus of White Bird Shrine, Iitate red zone [2.94].

54

Average frisk, roadside hydrangea flowers, Iitate red zone [3.25].

55

Airplane reading, maximum in five-minute interval, 35,988 feet over North Pole [3.32].

61

Two steps past no-go zone warning sign [i.e., for red zone] on Hometown Appreciation Road, Tomioka [3.66].

68

Ten steps farther into red zone in same place as previous. Identical reading standing in center of cherry tree boulevard, Yonomori, Tomioka yellow zone. [4.08].

71

Average frisk, deserted crossroads beside radiation dose digital sign, Iitate red zone [4.25].

83

Short frisk, vegetation in abandoned downtown, Okuma red zone [5].

83

<R> "A passenger in a plane flying at 12,000 meters [39,360 feet] receives 5 microSv/hr from cosmic rays . . ." [5]. This figure seems high to me; see **41** and **55**.

87

"WELCOME TO OKHUMA,"* downtown by JR station, Okuma red zone [5.19].

95

<R> **50 milliSv per year. Upper limit of yellow zone designation; lower limit of red [no-go] zone. [5.708].**

99

Goldenrod by wrecked houses, Okuma red zone [5.94].

100

Cherry trees exposed to ocean wind, Iitate red zone, average frisk [6]. Interior of closed vehicle at this spot measured 3.5.

115

Downtown street between former town hall and the tsunami-damaged Ono train station, Okuma red zone [6.9].

123

Fence of abandoned old age home in Okuma red zone, with clear line of sight to Reactor No. 1 (at about 2.5 km away) [7.38].

133

Average frisk, interior of Ono station, Okuma red zone [8].

142

Destroyed shrine, Okuma red zone [8.52]. The road two steps past this spot measured 20 and higher (which encouraged me to make a short frisk).

---

* A rare transliteration. Hence I have preferred "Okuma."

181

Air dose in front of fish hatchery, Okuma red zone [10.86].

183

Air on side street in Tomioka green zone, just off Highway 6 (which was probably contaminated by the trucks of decontamination workers), frisked at waist level [11].

190

<R> **100 milliSv per year. 0.5 percent increase in probability of fatal cancer, if this dose is received for the entire year.*** [11.416]. *This was the average dose received by the "liquidators" who entombed the reactor at Chernobyl.*

200

Drainpipe in Tomioka green zone, frisked at 1 ft from ground level [12].

> 217

<R> Hot spots in Aiikuen Orphanage, Fukushima City, March 2013 [> 13].

228

<R> 120 millis per year. Alleged annual dose in an unnamed Brazilian village. [This would be awfuly high.† The Brazilian town of Guarapari‡ is said to receive 10 millis per year]. [13.699].

---

* International Committee on Radiological Protection assessment. Whether each additional year of exposure increases the risk by another 0.5% was not clear to me.

† Dr. Crownover writes: "It is possible that an unnamed Brazilian village is this high due to unfortunate geology. For instance, there is a spot in Africa that is a naturally occurring 'reactor.' Another possibility is that industry has blessed the village with enriched waste of some sort. I sat next to a person on a plane many years ago who described himself as a 'historical geologist.' His career was devoted to predicting where the toxic wastes from past mining/smelting operations had ended up; for instance where a slow river bend a hundred miles downstream in the watershed might have collected high concentrations."—One might hope that there is indeed such a village. Then any nuclear ideologue with sufficient pep could pooh-pooh my 41.1-micro reading in Okuma as being merely three times "normal background."

‡ In this place, according to the Atomic Energy Commission (1968), "Radiation dosimeters were given to selected individuals ... To induce these people to wear these packets continuously, they were disguised as religious medallions ..."

333

Drainage grating by fish hatchery in Okuma red zone, short-frisked at waist level [20]. At 3 inches, this same grating measured 30. I chose to reach no closer.

350

The same side street in Tomioka green zone that measured 11 micros at waist level, now frisked 3 inches from pavement [21].

365

Another drainage grating in Okuma red zone [21.9].

492

Pavement in Okuma red zone [29.5].

533

Same Tomioka green zone drainpipe that measured 12 micros when frisked at 1 ft [see multiple **200**], now frisked at ground level [32].

692

Grass near hatchery in Okuma red zone [41.5].

———

*This last is the highest reading I personally measured in the field. Of course it is disturbing. The municipal officials accompanying me were unimpressed. They said that in 2011 this area read 100 micros and more. Exactly how dangerous was that grassy spot when I frisked it? If sustained over a year, its radiation level would produce a dose of 363.54 millis. Needless to say, nobody was planning to camp there for a year. And so I must reiterate that so long as the level is, in and of itself, subacute, the peril at that level depends on the duration of the exposure. Even in Okuma I probably increased my lethal cancer risk only negligibly, assuming that I managed not to inhale too many alpha or beta particles. I remained in Okuma for all of four hours, and much of that in a closed vehicle. Had I idled away 40 hours by that grassy place of the 41.5 micros, I would still have been considered safe enough by 1958 standards [see **1,250**].*

761–1,712

<R> Exposure in outer space (400 to 900 mSv per year) [45.662–102.7397].

880

<R> Mission to Mars (1.14 Sv over 30 months) [52.778].

1,000 if single dose, or 3.6 million sustained over 1 hr

<R> [0.6 millis] Japanese chest X-ray, *ca.* 2007 [60 micros, received in about 1 sec. Hence this same dose sustained over an hour would be about 216,000 micros or 216 mSv.].

1,250

<R> [0.3 rem per week or 7.5 mrem per hour for a 40-hr workweek.] Maximum permissible external dose for an American nuclear worker in 1958. [75].

> 1,667

<R> Stated level in Okuma immediately after the nuclear accident [> 100].

Reader, how brave do you feel when you go to get your teeth checked? Back home in Sacramento my kindly dentist agreed to receive the pancake frisker as a special patient. I laid it down in the chair, and the beautiful Amanda trained her X-ray camera on its head. The results were thought-provoking:

2,520 if single dose

Dental X-ray camera, fired through protective lead apron, intensity at 0.20, which was the setting Amanda used for most teeth [151.2]. *My first thought was that pregnant women had better ask for two lead aprons. But Dr. Richard Crownover reassured me: "The shielding for dental X-rays is not placed in the direct beam. It is designed to be adequate shielding for peripheral leakage and scatter outside of the imaging field. If your numbers refer to an apron in the direct beam then you will be criticized for alarmist advice to pregnant women."*

4,117 if single dose

Dental X-ray camera, without lead apron, intensity at 0.20 [247].

4,183 if single dose
Dental X-ray camera, without lead apron, intensity at 0.23, which is used for molars [251].

———————

Because I laid a dosimeter next to the frisker for each of those three readings, the results were even more enlightening. The lead apron blocked out 1 of the 2 micros. Without the apron, a full 2 micros fell upon my instruments. Increasing the setting from 0.20 to 0.23 did not much raise the radiation level, but it must have increased either the duration or the width of the beam, because instead of another 2 micros, the dosimeter registered 37. 9—about the same as if one had stood at that grassy place in Okuma for a whole hour.[*]

9,921
<R> **100 milliSv per week. Emergency exposure limit for nuclear workers in Japan, 2014 (susceptible to upward revision, which the government was considering). [595.238].**

16,250 if single dose
Security X-ray for carry-on baggage, Yaeger Airport, Charleston, West Virginia. (Frisker sent through screening machine.) [975 over about 1 min]

16,667 if single dose
<R> Full-body CT [computerized tomography] scan: 10 millis. [Assuming the duration of the scan is an hour, this would be 1,000 micros an hour.]

17,100
Security X-ray, Los Angeles International Airport [1,026]. One hopes for the workers' sake that this device is shielded.[†]

———————

[*] Dr. Crownover again: "If the dosimeter was placed near the pancake detector head and moved even slightly it could be subject to highly variable scatter from the detector itself." Fortunately it was not moved.

[†] The comparable X-ray in the Mexico City airport did not seem to be. I was forbidden to use the frisker, but I did have a dosimeter in my pocket, and *after passing through security,* in the course of a five-minute dash to the departure gate that device accrued 11 micros. The security workers were mostly young women. I worry that their ovaries were getting cooked.

833,333

<R> **50 milliSv per hour.*** [5 rems]. **Maximum dose for an American "first responder" not saving lives or valuable property.** Estimated level in collapsed plutonium storage tunnel during 2017 nuclear accident in Hanford, Washington.[†] **[50,000 micros/hr].**

2.5 million

<R> 25 mSv received in 10 minutes by the crewmen who sought to release steam from Reactor No. 1 before it exploded. Over an hour this would have added up to 150 millis. [150,000].

8.33 million

<R> 500 milliSv, over any time period [50 rems]. [500,000 micros/interval]. "A total dose of 50 rem of gamma radiation, largely independent of the rate of administration, causes approximately the same number of mutations as occur spontaneously over a lifetime."* This author (writing in 1958) believed that such a dose received in a few hours and over the entire body would probably cause radiation sickness. This may be too pessimistic.

11.67 million

<R> 0.7 Sv per hour* [70 rems]. Possible onset of radiation sickness. [700,000]. 16,868 times higher than my highest measurement in Okuma.

11.67 million

<R> 1 Sv per hour* [100 rems]. Definite onset of radiation sickness. [1 million].

21.67 million

<R> 1.3 Sv per hour, most of which was received almost at once by a female victim of the Hiroshima A-bomb. She was bedridden with "acute radiation syndrome" for 3 months. [1.3 million].

---

* Or per "incident," per short-term dose. The exact amount of time is unspecified. The same applies to the entries for 1 and 50 Sv/hr. We are told that the first responders at Chernobyl took in "up to 20,000 millisieverts," or up to 20 Sv, over what must have been several days—a very dangerous and perhaps lethal amount of radiation.

† See p. 416.

33.33–50 million

<R> 2–3 Sv per hour.* "Permanent female sterility is possible." [2–3 million].

50–66.67 million

<R> 3–4 Sv per hour.* "Uncontrollable bleeding in the mouth, under the skin and in the kidneys after the latent period"; 50% will die within 30 days. [3–4 million].

75 million

<R> 4.5 Sv, instantly. "At Hiroshima a dose equal to or greater than 450 rads (4.5 grays) extended to almost 1 mi. . . ." [4.5 million, instantly].

100 million

<R> 6 Sv per hour. Probably lethal within 1–3 weeks. [6 million].

279.89 million

<R> 16.793 Sv per hour.* Dose taken by the helicopter pilot Colonel Vodo-lazhsky [and probably sustained for only a few minutes] in one of his flights over the Chernobyl reactor, 1986. [He died, but not right away, and possibly not just from this.] [16,793,394.598]

166.67–833.33 million

<R> 10–50 Sv per hour.* "100 percent fatal after seven days"—but see slightly different claim for 50 Sv/hr, below. "Conditions for this exposure have depended on touching proximity to a supercritical sphere of naked plutonium." [10–50 million].

171.67 million

<R> 10.3 Sv per hour. Measured value in Reactor No. 1's suppression chamber, June 2012 ("likely to be fatal if a person stays there for an hour"). [10,300,000].

833.33 million

<R> 50 Sv per hour* [5,000 rems]. All exposed persons die within 48 hours. [50 million].

---

* Or per "incident," per short-term dose. The exact amount of time is unspecified. The same applies to the entries for 1 and 50 Sv/hr. We are told that the first responders at Chernobyl took in "up to 20,000 millisieverts," or up to 20 Sv, over what must have been several days—a very dangerous and perhaps lethal amount of radiation.

March 2011

WHEN THE WIND BLOWS FROM THE SOUTH

0: Picaresque Wanderings of a Dosimeter

The golden rule of journalism—keep one's dental appointments—I had now neglected for a couple of years, but in obedience to the current practicalities of Japan I made haste to cultivate my hygienist, who pressed the X-ray camera's snout against patients' cheekbones and therefore wore a dosimeter badge clipped to her pinkish smock. Thanks to her, I grew acquainted with the phone number of Carol (on subsequent diallings I got Ginger), who connected me with a salesman named Ray*, who allowed that he did still have one Geiger counter in stock—or, more precisely, a post-Geiger-Müller sort of gadget which, said Ray (who had not actually inspected it but seemed to be interpolating from some data screen), resembled an electronic calculator. Current and cumulative exposure, X-ray and gamma, a programmable exposure alarm—oh, delicious! Never mind its inability to detect alpha or beta particles;† wouldn't those be approximately innocuous so long as I refrained from ingesting them? (Within the body, remarked my radiation incident guide, *alpha and beta emitters are the most hazardous* since they *can transfer ionizing radiation to surrounding tissue, damaging DNA or other cellular material.*) Five hundred dollars plus shipping, credit card only; thus spake Ray, who must have known he was sitting pretty, for the

---

* His alias, as required by my publisher's lawyer.

† An alpha particle is a helium nucleus ejected from an excited atom. It can be stopped by paper. A beta particle (which may also be considered a wave) is an electron thrown out of an excited atom. It can be stopped by flesh and clothes. Both of these particles may prove permanently harmful if taken into the body. Consider plutonium-239's alpha particles, which will keep radiating for thousands of years. A certain Rick Anderson, we are told, kept *a plutonium paperweight on his desk. It's slightly warm to the touch and harmless as its alpha radiation cannot penetrate the skin.* I hope he had fun. Gamma waves possess for more penetrative power than alpha or beta. A thick lead shield is a prudent defense. Most accrued dosimeters, such as the one Ray sold me, measured only gamma. [Neutron rays, somewhat similar to gamma, will not be considered here.] For recapitulation and elaboration on all such matters, including units and conversions, see pp. 536ff.

other companies I had contacted accepted only back orders, two weeks having already lapsed since the reactor accident. At this stage in my nuclear career I could not even tell apart a Geiger counter, which was what I thought I had purchased, from a dosimeter, which was what Ray so helpfully sold me. Well, in a hot zone dosimeters are more crucial anyhow. Moreover, I had heard (possibly from Ray) that in Japan just now one couldn't buy any dosimeters at all. As we neared the end of our negotiations, I wondered aloud whether Ray's product came with a probe to stick into my sashimi, because that, I proposed, might be *fun*. Eliding the issue of fun, Ray (who informed me that he had had a long hard week) assured me that I could hold the machine six inches away from, say, a glass of drinking water, after which I would really know something. With potable water unavailable in much of the disaster area, and Tokyo's water supply spiced up, so rumor had it, with variable radioactivity, I considered myself canny to have pounced upon this capacity for pinpoint monitoring.

At this point in our transaction, any self-respecting used car buyer would have kicked a tire while nodding wisely; I accomplished the equivalent by inquiring as to which units of measurement his product employed. Millisieverts or millirems, replied Ray. I confess that this answer left me feeling long in the tooth, for in my day it had all been roentgens. When I phoned my friend the retired radiologist, he came down solidly in my camp, announcing: "I'm too old to care about millisieverts." Vaguely remembering from my university education that 400 roentgens was a lethal dose, I dusted off my copy of *Physical Chemistry* (1966), wherein I learned that *the development of the nuclear reactor, which promises to be a very important power source, has led to many challenging chemical problems.* That's nice. Actually, *a lethal whole-body single dose of radiation for man is about 500 roentgens;* now it was all coming back, except that I'd better learn how to convert to sieverts.* Perhaps it wasn't hard and fast, for my college physics textbook (1974) advised that 600 roentgens would finish me off, while the merest hundred might cause radiation sickness. Even 25 to 50 roentgens could bring about "blood-cell destruction." Dear reader, how do you vote? And should I have asked Ray's opinion?

My lean old Swedish neighbor offered me the loan of his Geiger counter, a souvenir of business dealings in the Jewish Autonomous Oblast of the Soviet Union, but I could not guarantee my ability to return it. At my request, he took a reading right there in my parking lot, a good 10 days after the Japanese

---

* One millisievert is 100 millirems. A sievert is a "unit of equivalent dose . . . defined as the absorbed dose multiplied by a weighting factor that expresses the long-term biological risk . . ."

radioactive plume had officially reached our city. The dial slept at zero. Well, hadn't the newspaper promised us that the levels were practically indetectable? (As a hardboiled American dosimetrist remarked to me: "We would pull the amount of air you would breathe in a month over a paper filter until we could see lines on a spectrograph, because that's what we do.") Sighing, my neighbor wondered whether his toy was still calibrated. He slipped it back into the worn leather case, and because I could see that he loved it very much, I knew I had done right to leave it in his keeping.

My new dosimeter (whose brand and model number I decline to insert here unless Ray pays me) resembled a calculator far less than a pager. It was a fat, blue, boring little plastic thing with a pocket clip. After an hour or so I figured out how to turn it on. And what do you know? The number in the narrow window was zero. The device's compactness pleased me, and its ugliness reassured me, but how could I ascertain whether it even functioned? To be sure, a hand-signed certificate of calibration came enclosed; and Sacramento's background radiation ought to be negligible even now, so the fact that the window displayed zero and only zero was no cause for suspicion; all the same, I preferred not to trust my life to a calibration certificate. Hence another kindhearted neighbor, who had connections in the local fire department, carried into my living room a padded envelope hand-labeled **DANGER—RADIATION**. My 12-year-old grimaced in horror; she preferred to stay in the kitchen. Out came a plastic box containing a radioactive point source of unknown magnitude, which my neighbor's bare hand now conveyed from her side of the sofa to mine. A trifle uneasily, I do admit, I inquired about the safety of these proceedings. My neighbor assured me that the source was weak, which could also be said of her surprise bonus: one of her Great-Aunt Lou's orange-glazed dinner plates, which had become a rarity as soon as our government sequestered that orange glaze for the Manhattan Project. Well, I clinked my dosimeter against each of these items for a good five seconds, and the display continued steady at zero! My Swedish neighbor's Geiger counter would have been more appropriate, assuming that it still functioned—but I continued ignorant of the difference between a scintillation counter like his and a dosimeter like mine. In due course I would realize that Ray had misrepresented the machine's capabilities; all it could read, except in extreme cases, was *accrued* incident radiation. I was going to have to take Tokyo's drinking water on faith. Not yet comprehending this reality, I sat glumly beside my neighbor, still wondering whether the toy had arrived broken, and halfway imagining that my leg was receiving a radiation sunburn. Why wouldn't she at least move that plate away? I asked her how positively she knew that the point source's box was not in fact empty. She brought the ghastly thing up to her eyes, thereby nearly interpret-ing a pencilled notation in what could have been microcuries. I thanked and embraced her; she meant well. Then I went to bed.

Well, I left my dosimeter at my bedside in the measurement mode all night, and in the morning it was still at zero. But then the Radiation Gods saw fit to bestow upon me a sign. Since my best friend's brain tumor had begun to grow again (or perhaps it wasn't growing at all, depending on which doctors said what), it came time for his gamma knife surgery. His wife and I accompanied him to the chamber while he got strapped down. Then we returned to the waiting room to worry about him. During that 10-minute interval, the dosimeter registered 0.1 millirem. Did that mean that the gamma knife had just then slightly irradi-ated us through the wall, or simply that the dosimeter's lowest increment was 0.1 millirem? Either way, life was looking up. I decided to forget millisieverts and dwell in millirems.

My radiation incident guide (a gift from the neighbor with the orange plate) informed me that 0.05 millirems or less per hour falls within the bounds of normal background exposure, while even 0.1 millirem can be considered unex-ceptional.—Well, what *should* be exceptional? Could we but safely trust to one true figure, my investigations at Fukushima would have been less troubling. I felt outright alarm about carbon dioxide levels because a benchmark gave me grounds for comparison: 315.98 parts per million in 1959. Thus the 407.42 parts per mil-lion of 2016 did not exist all by itself; it meant something *by reference.*

What value, then, should I fix for a "natural" (tentative equivalent: "accept-able") radiation level? Once I knew that, I could tell you how dangerous were the silent towns and waving goldenrods of Fukushima.

According to that radiation incident guide, our global average dose was 360 millirems per year. I went goggle-eyed, for 365 of my Sacramento or San Fran-cisco days at 0.1 millirem apiece would do me a trifling 36.5 millirems' worth of damage. How could these two figures vary by a factor of 10? Was my dosimeter inaccurate?—In later years, while gathering my bouquet of statistics, I discovered another purported average of 300 millirems, and a third (courtesy of a Japanese official) of 240 millirems. Meanwhile, a standard scientific reference pegged our exposure at anywhere from 4 to 400 millirems! (In your time, reader, the average might be 500 millirems or considerably more, depending on the requirements of progress and demand.) All right; that helped me.—The incident guide's author did or did not belong to the pro-nuclear camp—and for all I can tell, she added in a few chest X-rays, airplane rides and slumber parties in stone castles full of radon gas—but her approximations remained within the fold.

Next question: How much radiation was "safe"?

The International Commission on Radiological Protection, which had formerly expressed satisfaction with 400 millirems, now advised ordinary citizens, perhaps unrealistically, not to exceed 100 millirems.* That would be 0.011 millirem per hour. Ten times more devil-may-care, the incident guide remained at peace in any field up to 0.1 millirem per hour. That suited me.

My readings in San Francisco and Sacramento remained at 0.1 millirem per *day*—a plausible, even typical dose for the U.S. Pacific Coast; thanks to its high altitude and inland location, Denver received triple this amount. Thus the dosimeter eased my mind.

As I read over this chapter six years later it seems quaint to me that we Californians worried about Japanese fallout, whose presence in our state approached negligibility—but in those days how could we know what might happen? (It took our minds off climate change.) I visited a health food store to buy iodine tablets for my journey, only to find that my fellow citizens had cleared out the shelf! Meanwhile, the dosimeter went on accruing a soothing 0.1 millirem per day.—This made me happy, for my daughter's sake.

The radiation incident guide advised that if I were a "responder" of the best official type I should limit my dose at any one occasion to 5 rems; hence that would be my ceiling for Japan. Five rems divided by 10 days would be 500 millirems a day, or 5,000 times what I was getting in San Francisco. In case you are interested, I will tell you that the U.S. Environmental Protection Agency's maximum dose for *protecting valuable property* is 10 rems, and for the saving of human lives a toasty 25 rems. Certain mild symptoms may appear at 30 rems. Radiation sickness manifests itself at 70 to 100 rems. Above 350 rems, any recovery is likely to be followed by a relapse. Two hundred and fifty to 500 rems constitutes the lethal dose for 50% of humans within 60 days.—Do rems sound a trifle like roentgens? Well, "rem" is an abbreviation for "roentgen equivalent man."—At 5,000 rems (or, if you like, 50 sieverts), all patients will die within 48 hours.

In Japan the authorities danced fluently between millisieverts per hour for air and becquerels for drinking water. The former is a unit of biological damage; the latter has to do with atomic disintegrations per second. Nobody I met over there could keep them straight.[†]

---

* All these numbers, including those for background values, appear on pp. 244ff., headers (<**1–7.61**: [0.04 to 4 mSv = 4 millirems to 400 millirems per year]), **1.9, 3.8, 4.6, 5.7, 6.85** [3.6 millisieverts = 360 millirems].

[†] For units and conversions, see pp. 545–52.

My friend Dave Golden, who had a finger in every pie, somehow managed to make me an appointment with Dr. Jean Pouliot, Vice-Chair of the radiation oncology department at Mount Zion Hospital in San Francisco. Dr. Pouliot was a pleasant man of middle age. With him came a quietly competent, pretty young physicist named Josephine Chen. Dr. Pouliot unlocked the door of a windowless room, picked up a meter the size of a small laptop and approached a certain metal cabinet whose front bore a radiation warning. The meter did not respond. Nor did my dosimeter. Sighing, he unlocked the cabinet, pulled aside a nest of lead bricks and withdrew a cylindrical object. Still his meter showed nothing; evidently the battery had died. Josephine brought my dosimeter close to the object, and the alarm sounded. I felt pleased. In the quarter-hour we spent in that room, God regaled us with 0.6 heavenly millirems!*

"Well, it's a little high," said Dr. Pouliot. "Maybe we should put it away."

His meter being dead, I could not calibrate my dosimeter against it. And given my experience with the neighbor's orange plate, I wondered whether my dosimeter was insensitive to, or inaccurate in, low ranges.† But at least it was doing something. My homework might not be well done, but I hoped to earn an "A" for diligent intentions.

Dr. Pouliot considered my 5-rem ceiling dose a trifle dangerous. When I showed him the very page in my incident guide where the EPA recommended it, that tolerant man said that after all, they ought to know what they were doing. The veteran dosimetrist whom I have mentioned before (he was, indeed, a reactor emeritus) later flattered me thus: "Well, for a member of the general public you've got guts. I don't say you're crazy. If I were taking food and water to those people at the reactor I could go for five rems . . ."

The next day I flew off to Japan, accruing 1.2 millirems (about an eighth of a chest X-ray‡) in 11 and a half hours.

1: Not Known as of Sunday Afternoon

On March 11, 2011, a nine-magnitude temblor struck the eastern coast of Japan's main island. A tsunami followed. The day before I departed Tokyo for the

---

* This corresponded to one of the higher (but not highest) levels I would find in 2014, in the radioactive town of Tomioka; it was about half of my maximum reading in the Okuma red zone.

† Now I understand that the orange plate must have simply been a mild irradiator.

‡ A chest X-ray is 0.0001 sievert—although other sources peg it at 6 and 7 times that. [See header for **1,000** on p. 253.] One sievert is 100 rems, or 100,000 millirems. Hence a chest X-ray is 100 microsieverts. In *Carbon Ideologies* I sometimes use another stated value for this procedure: 1.0 microsieverts, or 6 millirems.

disaster zone, the casualties had been totted up as follows: Killed, 12,175; missing, 15,489; injured, 2,858.* In the affected area there happened to be a pair of nuclear power plants owned by the Tokyo Electric Power Company, or, in English-language parlance, Tepco. The six-reactor Fukushima Nuclear Plant No. 1 endured the catastrophe with more cracks and leaks than its counterpart 12 kilometers to the south.† Once the power failed, the fuel rods began to overheat. Although Plant No. 1's brave workers, struggling almost sleeplessly under terrifying conditions, did mitigate the result,‡ three reactors melted down. Reactor 1 exploded on the twelfth, Reactor 3 on the fourteenth and Reactor 4 on the fifteenth. The fallout's main plume apparently originated in Reactor 1, although useless efforts to depressurize the containment vessels by venting contaminated steam made their own contribution. On the fourteenth, the American naval base more than 300 kilometers to the south was already registering 1.5 millirems an hour. What must be happening in the Japanese towns around the plant remained mostly unreported. To prevent the sizzling cores from vomiting up more radionuclides, the only cure was water, tons of it, applied desperately from fire engines, helicopters, and, eventually, crane trucks. By the twenty-sixth of the month, water in Reactor 2 was emitting at least a sievert per hour of radiation. At this rate, a person would receive that 5-rem dose in about three minutes.

The situation seemed unpromising, all the more so since I was not the only ignoramus in Japan:

March 27:

Q: *Where did this radioactive water come from?*

A: *Plant officials and government regulators say they don't know.*

April 3:

How much water has leaked and for how long was not known as of Saturday afternoon.

---

* The final death figure was ~18,500.

† Of these six, the infamous Reactor 1 was a General Electric, Mark I, whose construction commenced in 1971. Reactors 2, 3, 4 and 5 were "more advanced B[oiling] W[ater] R[eactor] models but also had Mark I containments."

‡ Unappeased by heroics, Takashi Hirose said: "An electric power company that lost control of its nuclear power plants because of an electric failure—it's a pathetic story."

Peter Bradford, formerly a member of the Nuclear Regulatory Commission and now serving on the Board of Trustees of the Union of Concerned Scientists, said to me: "I'm getting increasingly concerned about the failure of the Japanese public to get accurate information. In the first week I thought the Japanese government was being cautious for good reason. In the third week, there are more and more symptoms that details are being held back. Just now there's first of all that one extremely high radiation reading, which was declared to be a mistake, and secondly the discovery of iodine-131,* which has a very short half-life and would only be present if there's some recriticality, and they said that's also a mistake. That's two mistakes."

"What would the worst case be?"

"If one of the cores was able to go critical to produce even a small scale nuclear explosion."

"How much of Japan would become uninhabitable?"

"It's hard to say. It depends a lot on the wind. So far the Japanese have been lucky with the winds blowing west to east, out to sea."

WHY THIS CHAPTER IS SHORT ON STATISTICS

Although my letter of press accreditation informed those very few Japanese who were interested that my duties involved *interviewing individuals and officials on behalf of our publication,* I did not see it as my duty to obtain figures on casualties, radiation levels, etcetera, which might well be lies and would certainly be superseded. (The stunning capacity of the Japanese official to say absolutely nothing is matched only by the absurd degree of trust which his public places in him; while the cynical suspicion of the American electorate finds its perfect mate in American officials' complacent and sometimes even blustering dishonesty.) Nor could I imagine that "experts" had any more to say about the profoundest questions raised by this continuing tragedy than those who suffered by it, comprehendingly or not. Finally, I could see no benefit in seeking out the people in greatest emotional pain. As you read this account, you will see that my interviewees were, for materially devastated individuals, relatively "lucky." Only the family in Ishinomaki had lost one of its members—so far. This selection was less the fruit of my deliberate policy than the consequence of the fact that those not grieving the death of a relative felt more inclined to open their hearts to a stranger. God knows, their circumstances were sad enough.

---

* From a chemical primer: "Another element released by uranium fission is iodine-131, which is ... worrying. This was widely scattered over the North of England in 1957 by an accident at the Windscale ... nuclear power plant, and over Europe by ... Chernobyl ... in 1986."

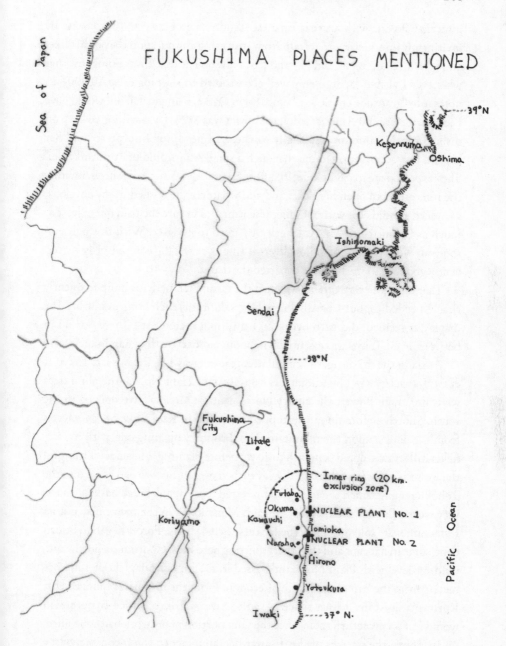

FUKUSHIMA PLACES MENTIONED

However conservatively considerate I imagined this approach to be, it scarcely put me in the clear. My interpreter, to whom I had been close for many years, was sluggish and irritable as I had never seen her; she admitted to being depressed, not to mention enraged at Tepco and her government. Her cousin, who had not met me, expected that I would do harm, and therefore admonished me (a) to interview no one without that Japanese standby, a go-between; (b) to begin by inviting my interviewees to refrain from answering any question they didn't like; and (c) above all, to pay and pay and pay. I always felt that I was doing just that whenever I visited Japan, being well accustomed to slipping crisp 10,000-yen notes into "gratitude envelopes." Once upon a time, a single such bill would have been a trifle over 80 American dollars; now it was $125. I was willing to keep on disbursing this amount, especially to those in need; my interpreter and her cousin, however, informed me that such a small sum would be "unthinkable." They expected me to pay at least 40,000 or 50,000 a go. I dug in my heels, inviting the interpreter to open her heart and add whatever she wished to my envelope, as indeed she did, not without quiet resentment; I'm sure she paid out at least as much as I remitted to her. At length we agreed to disagree. With this ugly episode our work began, the yellow-liveried employee waving compulsorily yet fervently goodbye to the bus as it departed the depot.

That day and every other I watched the dosimeter, perhaps more frequently than I needed to, but I hardly knew how salubrious each hour might be. The display, it seemed, did turn over in 0.1 millirem increments. There was no in between. And Tokyo was essentially no more radioactive than San Francisco.

At six in the morning, the cumulative reading was 1.5. The bus left Tokyo at eight. I was, let's say, 230 kilometers from Reactor Plant No. 1.* The plum trees were already in flower; the cherry blossoms must already have opened in the south. Shortly before noon we stopped for lunch in Koriyama, 58 kilometers from the danger spot, the mountain-ringed country opening out, with its ricefields still straw-colored (a month to go yet before planting) and snow shining on the western peaks; just then the display turned over to 1.6. We had come into the Tohoku region, which the interpreter referred to as Japan's bread basket, "so I'm very worried about the future," she said. Many items in the convenience store restaurant were sold out. Here the Japanese Self-Defense Forces began to be evident, some in flat caps and the others sporting hard hats. Continuing northward, we drew level with Plant No. 1 and passed it, reaching Sendai (about 100 kilometers from the bad place) in midafternoon. From the dosimeter I judged that Koriyama must be at least twice as radioactive as Tokyo, which hypothesis I would test on my return there, once the safer portion of my work had concluded.

In Tokyo the stresses of the disaster had approached the inconspicuous: a blackout here and there, shortages of diapers and sanitary wipes, which people

---

* All distances given here are to the poisonous Plant No. 1. Tokyo distance is measured somewhat randomly from the large and central Setagawa Ward. The (spuriously) exact distances are 232 km to Plant No. 1 and 222 to Plant No. 2.

were sending to their relatives in the stricken zone. As for Sendai, that city could be described as "recovering"; for although the airport had not reopened, heating-gas remained unavailable, and milk, yogurt, eggs and cigarettes were still in short supply, at least the two-hour waits in petrol stations had come to an end and the electricity was back on. The downtown appeared untouched, aside from warning signs posted on this or that building.

I hired a taxi to take me down into the Wakabayashi district, which had been harder hit.

"I was on duty in the car," said the driver, whose name was Mr. Sato Masayoshi.* "There were no passengers. I heard the earthquake alert on the radio. I looked for a wide open place to park, since the buildings were shaking. You couldn't stand! I was sitting on the median strip. It lasted a good two minutes, moving between east and south, laterally.† When the tremors stopped, I got out of the cab, tried my cell phone, which did not connect, and used a public phone to call my family. It rang and rang but nobody answered. So I drove to the office, received permission to stop working and hurried home. The traffic jam was terrible, but everybody was okay. We had no electricity for three days. My grandchildren enjoyed it."

He pointed. "Over there, there's the restaurant that shook so much. And you see this gas station! The ceiling dropped . . ."

"Did the tsunami come here?"

"No, this is all earthquake."

"What was your opinion when you first heard about the reactor accident?"

"Sendai is 80 or 90 kilos from the power plant,‡ so I'm not really worried about it. The wind in this season blows from the land to the sea. If it blows from the south, that will be a problem. The highly contaminated water needs to be released, they say . . ."

That was the word I so often heard: *contaminated*. It sounded less frightening than *radioactive*.

"How contaminated is the sea around Sendai?"

"I don't think they've measured it yet."

Gazing down at the dosimeter in my shirt pocket, I was pleased to see it still at 1.6. We came to a shed which had been uprooted and left in a tilt on the shoulder of the road. I photographed it, and then the driver remarked, a trifle

---

* Following the Japanese convention, in this book I give family names first.

† In other words, the ground moved not up and down but from side to side—perfect conditions for a tsunami.

‡ As indicated, it was actually slightly farther away.

indignantly, I thought: "Today a fishing boat in Choshi Port,* even without in-spection their catch was refused!"

I wondered aloud if fish, eels and other such foods might be getting danger-ous. Not caring to pursue those implications, or perhaps simply wishing to re-turn to business, the driver announced like a tour guide: "And now we're making a right turn to the place where the houses are gone. Here to the left there's a highway. In some places the highway blocked the water. Some of the people who ran up on top of it survived."

"Are you worried about the next earthquake?"

"Since the Miyagi Coast Earthquake in 1978 it's been a long time. This latest one was not the one which the experts were discussing. People are talking about the next one; yes, there may be another . . . Here the water came," he continued, gesturing at some mud-fields decorated with fallen trees and stumps. "On ac-count of the salt water, you won't be able to grow anything here for five or six years. They were growing soybeans."

A crossroads in Sendai

A fallen pine, cables, heaps of mud, bent pipes, metal grilles, fallen poles as thick around as my shoulder, with their fat plastic veins exposed, these sad and ugly objects varied themselves monotonously all the way to the mud horizon. On one side of the road the former fields were flooded with seawater. On the other, on the edge of streaming tidal flats which used to be ricefields, a two-storey

---

* In Chiba Prefecture, near Tokyo.

concrete house, windowless but seemingly intact, supported a second home which had smashed up against it, embracing it drunkenly, the roof twisted like sections of ruined armor, both structures choked with wooden rubbish, a tree hugged up against them, while on the ground lay a woman's empty purse; and a detachment of goggled, web-belted, booted, camouflage-uniformed Self-Defense Forces from Hokkaido were dissecting the two houses in search of bodies. The slogan one often saw on their helmets was: "*Let's cheer up, Sendai!*" Their crane's claw closed upon a wad of white documents—for a human being, a massive arm-ful, as was proved when a soldier took these treasures and began to pull them apart, seeking identification papers, I suppose. A cool breeze blew from the sea; I wondered if it were poisonous with beta particles. In any event, my dosimeter remained comfortingly at 1.6. On and on in the house-lots, sad heaps of trash which used to be houses hid their secrets. In this prefecture alone, the current figure was more than 7,800 deaths. Here came a civilian cyclist, stern and skinny, riding up the dirt road and passing us, continuing down among the house-stumps; perhaps he was looking in upon his home. Slowly, slowly, while the sol-diers stood around, the crane-claw opened and closed, pulling up a heap of crackling tree stumps. A young soldier informed me that his bunch had found no corpses yet. When I photographed him, he pulled himself up tall and straight. He said that he was not worried about radiation; the likelihood of its coming here was low. I shook his hand, which was hot and sweaty from its sojourn in the dark glove.

Self-Defense Forces soldier on
corpse detail

It is hard to describe to you the littered flatness, everything pulverized into irrelevance, some foundations still visible, a few pines far away stripped of their lower branches (but this last, the driver said, had been a pre-quake phenomenon). One of the driver's colleagues had lived here. Now he was staying at his son's. Every now and then a lone figure in a hard hat and coveralls stood at the side of the road, accomplishing I know not what. Here came three houses screened by pines, which must have shielded them, for they appeared nearly intact, although one of them lurched. The neighborhoods of Okada, Gamo, Shiratori and Ara-hama were gone now. The former geriatric home was full of rubble and trees. By now the trees had already started to decompose, so that when they formed up the sides of houses, they infiltrated them like subtly woven rattan, perfectly fitted by the weaver-upholster called death. Very occasionally the empty doors and windows of better-off buildings were protected by blue tarps taped into place. We drove slowly south through the smell of tidal flats, toward the Natori River, passing blue and grey stretches of rippling water, and a sign: Seaside Park Adventure Field.

"What are you feeling?" I asked the driver.

"I have no words. We often drive this road."

Buddha in destroyed house

Here came mud and muck and shining water, a car in water up to the snout, a policeman in a hard hat, more fallen trees, a red sports car turned onto its side, the light now pretty on the ricefields. In one place, the road had been licked away underneath, the asphalt looking silly as it stretched through the air.

"Were most people drowned or crushed?"

"I think they drowned. Some of the cars were in a traffic jam. I know of one person who climbed up a pine tree to survive. His decision to give up the car was good."

The cool air was dust-prickly in my throat. The driver and the interpreter were both wearing masks. I wondered some more about beta particles but decided to rely on the inverse-square law. The Natori bridge had been closed off with a checkerboarded barrel. A man with a lightstick-baton and hard hat stood demoralized beside a flashing police car. Behind him, a boat had been pounded sideways into the muck.

"Mr. Sato, do you think that nuclear power is wise or unwise?"

"There are three nuclear plants in this prefecture. They are on higher ground than Tepco's, so I think that is good."

"So you approve of nuclear power?"

"Well. Due to the greenhouse effect, oil and coal are not clean, so as long as they secure the safety, I think that nuclear power is good."

An old woman in baggy clothes and a flapping shawl staggered down the road. Here came a small cemetery, the steles all upright but the mud churned up between them disgustingly. In the port, the trade show palace appeared in good health from the outside. A glittering stack of Toyotas awaiting export had been crushed. It was strange to see new paint jobs on pancaked cars.

"So what will happen in the other season when the wind blows in from the south?"

"Well, we don't have it like that so often."

"It might only take one time," I said.

"I agree!" he said with a meaningless laugh.

GOURMANDIZING

Due to the lack of natural gas and the hordes of soldiers and volunteers in Sendai (the Metropolitan Hotel had been entirely turned over to relief workers), my accommodation consisted of a hot springs more than an hour's bus ride out of town. Here various hard-pressed employees of the Osaka Gas Company were staying, and in the morning one sometimes saw a truckload of Self-Defense Forces outside. It was a half-empty, second-rate place where the sashimi came wrapped in plastic, although one could only admire the fervency of their many rules (*we firmly refuse your request to enter the baths when you are drunk or if you have tattoos on your body*).* The waitress proudly assured me that the food was local insofar as possible, so while I was eating it I grew angry again at Ray the salesman, who had promised that the dosimeter could make pinpoint measurements of local radioactivities, and of course at Tepco; for how could I have any idea how carcinogenic the fish might be, not to mention these slightly less than fresh greens accompanying them, or the crab claw in my soup? I was not unmindful of the fact that I could eat while so many others went hungry; nor was I so concerned on my own account, for a man in his 50s has already won a victory of sorts; but what about the pregnant women, the young children, the people who should have had decades to look forward to? In the words of yesterday's paper: **Govt. holding radiation data back: IAEA gets information, but public doesn't**. In the body of the article, an unnamed Meteorological Agency official explained that the Japanese government made its own forecasts—never mind that they had been released only once, because, explained an official named Shioya Seiji, "we can't do it since accuracy is low." The unnamed official then remarked: "If the government releases two different sets of data, it might cause disorder in the society." Was that why the official statistics offered varying units of measurement, so that in Koriyama the drinking water at the bus station was proclaimed safe on account of its radioactivity being less than 100 becquerels, I think per liter; while the newspaper reported the radioactivity of this or that city in millisieverts per hour? Nobody I met knew what these numbers meant. How convenient!—As I look back on this moment, having learned more about the difficulties of converting from becquerels to such units of biological dose as millirems and millisieverts, it occurs to me that my suspicions were arguably unfair to the Japanese government. But surely some approximate conversion could have been made; more could have been explained. Not knowing how much I might be harming myself, I chopsticked another previously frozen tidbit of horse mackerel into my mouth.

PRESENT INTEREST

Since this is *Carbon Ideologies,* I considered this matter of the reactor to be the real story. Sad as the rest of it was, the damage had been done, the people killed

---

* "Additionally, in open air bath, we set a rule saying that is mixed bathing and strictly refuse you from wrapping bath towel around yourself."

and property ruined; and now recovery could continue until the next quake. But this other horror wrapped up in becquerels, sieverts and millirems was just beginning, and nobody knew how bad it might be.

(I had asked Peter Bradford: "Could it happen here in the States? I understand we have some reactors of the Japanese type."

"I don't think the likelihood is driven so much by reactors of that kind as by the fact that we're just about as vulnerable as the Japanese to complacency about what used to be called a Class Nine accident. I don't think we're any less vulnerable than the Japanese.")

About the earthquake-tsunami and the concomitant reactor disaster it may be apposite to cite the words of Buddha: *Nothing in the world is permanent or lasting; everything is changing and momentary and unpredictable. But people are ignorant and selfish, and concerned only with the desires and sufferings of the present moment. They do not listen to the good teachings; nor do they try to understand them; and simply give themselves up to the present interest, to wealth and lust,—to,* for instance, the tax credits awarded those who dwell near a nuclear reactor, not to mention what the reactor enables and impels. In Tokyo the subway car darkens for a stop or two, thanks no doubt to power shortages. The information screen by the lefthand door informs us that one line happens to be down due to "blackout," while two bullet trains have been cancelled due to "earthquake."* Classifying the power inputs and outputs of Japan's electric grid (capacity: 290 gigawatts, of which Fukushima Daiichi once contributed 1.6%) would be vexing, and from the standpoint of this essay needless. Is, for instance, a bullet train nourished on atomic power? According to a Japan Railways telephone operator, it is not, although "for security purposes" she declined to specify which kind of energy it does employ to flit so luxuriously past the blossoming cherry trees. The latest earthquake, a small one, has delayed our departure from Shinagawa Station by the merest quarter-hour. The pallid young salaryman across the aisle peers down over his white dust mask at his shining laptop; the stylized yellow man and woman glow side by side in their black square in order to inform us that both lavatories are occupied; and we fly onward over houses and gardens. From Buddha's point of view, it scarcely matters whether all this ease derives from uranium pellets, solar cells or perpetual motion; in any case, our complacency alone protects the lovely roofs and trees of this present instant from becoming the rubble into which the very next moment might in fact cast them. But

---

* In 2010, Japan consumed 21.9 Q-BTUs (171.3 million BTUs per capita). In 2011 the respective figures were 21.0 Q-BTUs and 164.4 million BTUs. This approximately 5% decline surely resulted from the accident.

how many of us (excepting monks) can live and hope—in other words, chase our present interests—without disregarding our inevitable ends? I say we are "better off" pretending that the bullet train won't derail, that nuclear power is safe and global warming is a hoax. The peril is remote; probably we will die from something else.—When the peril comes nearer, present interest advises against disregard. (That is why you in my future do believe in climate change.) The more present the interest, the less present or apparently present the danger, the more irresistible the disregard. Hence the following parable, courtesy of the paterfamilias of the family who hosted me on Oshima Island. Refilling my sake glass as we sat in the dark and chilly mud-stained dining room, he remarked that following an infamous tsunami back in the Meiji Era,* numerous oceanfront plots here and elsewhere were banned from resettlement, but "somehow," he jocularly continued, people forgot or set the edict aside. Of course, even had they complied, this latest terror would have carried off ever so many, since it rolled in higher than any wave seen by the people of Meiji. Who can blame Oshima's inhabitants for not predicting that?

However, the corporate engineers and presidents, the prefectural governors, the authorities whose task it should have been to maximize public safety, these super-actors on the civic stage, who made policies affecting and even determining the fates of others, must be held accountable should they abandon themselves to *their* own present interests. The reason that I unalterably opposed nuclear power was so obvious (to me) that all my life I remained astounded that everybody on Earth was not likewise against it: Dangerously radioactive nuclear wastes had to be stored and guarded for periods insanely in excess of any civilization's frame of reference. Were it possible to render the fuel rods harmless in, say, five years, even then I'd worry about carelessness and greed, but at least I would be willing to consider the use-value of nuclear power. Having reached that point, I would, of course, remain among the smug ignoramuses against whom Buddha's warning was directed. Tepco's complaint-apology—how could we have been expected to foresee so high a tsunami?—approached legitimacy, but fell short. "The cooling facilities survived the earthquake, at least partially," remarked my interpreter. "The disaster occurred because the cooling facility was totally destroyed by the tsunami. *The cooling facility was located lower than the reactor itself.* Their assumption was a 5.7-meter tsunami, when the wave was actually 14."—Well, should Tepco have been expected to prepare for a 14-meter tsunami?

---

* A.D. 1868–1912.

Whatever your answer may be, please consider Buddha's admonition an instant longer. *They do not listen to the good teachings; nor do they try to understand them; and simply give themselves up to the present interest, to wealth and lust.* If the present interest required us to consume more and ever more energy, then dangerous forms of energy generation might become accepted as necessary. So it proved, in the time when I was alive. In 2010, the year before the great disaster, nuclear power did work all over the world, to the tune of 2,600 terawatt-hours, or 8.873 quadrillion BTUs—the equivalent of burning a billion metric tons of coal, and without any carbon emissions! You may recall that two-thirds of the energy given off by a reactor's fuel rods (as by coal, oil and natural gas) was waste heat*—but never mind the 17.7 lost quads; hadn't those 8,872,448,000,000,000 BTUs that the nuclear turbines actually bestowed upon us sufficed to at least begin fulfilling our present interest? And, as some nuclear ideologues might say, weren't a few meltdowns preferable to the continuing upcreep of CO_2 in the atmosphere?

CARBON DIOXIDE EMISSIONS FROM FOSSIL FUELS: JAPAN, U.S. AND WORLD,
1980 and 2011,
in multiples of the 1980 Japanese value

All levels expressed in [million metric tons].

All percentages rounded to nearest whole digit.

Japan, 1980	1 [947.0]	
Japan, 2011	1.25 [1,180.6]	*Japanese increase: 25%*
U.S., 1980	5.04 [4,775.8]	

---

* Hence the International Energy Agency's axiom that "the primary energy equivalent of nuclear electricity is calculated from the gross generation by assuming a 33% efficiency." And see "About Power Plants," p. 150.

	5.80	
U.S., 2011	[5,490.6]	*American increase: 15%*
	19.47	
World, 1980	[18,433.2]	
	34.40	
World, 2011	[32,578.6]	*Total increase: 77%*

Source: U.S. Energy Information Administration, 2014, with calculations by WTV.

NONE OF US ARE PARTICULARLY CONCERNED

Three blocks away from the pedestrian mall where on this sunny breezy afternoon members of the group called Antiatom proffered petitions against nuclear weapons and reactors, in an almost undamaged quarter one of whose steep-roofed houses had been blue-tarped along the ridgeline, stood the Sendai City War Reconstruction Memorial Hall, which presently served as a temporary evacuation center for 31 voluntary evacuees from Fukushima; 120 people had so far been cared for. One entered through the back door, the earthquake having now rendered the lobby's ceiling ducts liable to collapse. In Japan, as we shall often see, neighborhood attachments ran deep enough that communities tended to relocate as coherent entities. Hence the Memorial Hall housed people from a specific place: the northern sector of the radiation-poisoned zone.

Rather than seek out some bureaucrat who might have denied me entry privileges (I had already been refused permission to sleep in several evacuation shelters), I waylaid the first non-uniformed individual who seemed in no hurry—in this case, a bespectacled woman about 25 years old who had fled the Haramachi-ku ward of Minami Soma City.* Officialdom had drawn two rings around Plant No. 1. The inner one, 20 kilometers in radius, constituted an area of involuntary evacuation. Residents of the outer ring were merely advised to leave, at their own expense; if they wished, they could remain at home, keeping indoors as much as practicable. The woman, whose name was Hotsuki Minako, had lived in the outer ring.

---

* Also transliterated Minamisoma. For a brief description of this place in 2014, see p. 462.

She said: "On Friday there was an earthquake. On Sunday or Monday, on the news they said to stay inside. We tried to wait and see, but since we have kids, just my two kids and I came to Sendai with my husband. In a couple of days, my husband's parents also came here."

"So now your home is empty?"

"Yes."

"Could you please tell me more about how you left Minami Soma?"

"After we saw the video of the reactor explosion, we immediately moved. Even after the explosion we thought we could come back . . ."

"Did you feel or hear it?"

"No. We only saw the television image. There were three explosions, I think"—holding her fist to her mouth in thought. "And because we had kids, we were concerned. Otherwise we would have simply stayed inside."

"If your children could be safe there, would you ever go back?"

"Life here is just fine, so we are not too concerned to go back."

She had a very oval, girlish face; her bangs spilled over her thick eyebrows. Her hooded blue sweatshirt seemed too large for her.

Soon, she believed, her family and neighbors would be moved again, to a hotel, "so that the community itself will continue." They had already stayed first at a relative's, then at an elementary school. She supposed that after the hotel they would be moved still again.

"Do you think you'll be returning home anytime soon?"

"I have a feeling it will take a year or more."

"Are you worried about leaving your house uncared for?"

"Damage from the earthquake was not that significant. Only some roof tiles fell off. I'm a little anxious about rain coming in . . ."

She had described the earthquake matter-of-factly, but when I inquired into the emotions of that time, she said: "The first tremor was big, and that night came very big aftershocks. So I was unable to sleep from fear."

"When you think about radiation, what comes into your mind?"

"I worked as a clerk for a Tepco subsidiary. So I've heard about the danger of radiation and about controlling it, but I hear it's not *that* scary. But now, when I hear on television that it can affect your blood and so forth, well, I didn't know that."

"Was Tepco a good employer?"

"Those who worked at the nuclear site seemed to enjoy their job, but I only saw them once a month. I was in a clerical department . . ."

"How are you managing for expenses nowadays?"

"We are using our savings. I heard the city would pay some 50,000* per household, but I was unable to attend the registration for that. The city office is not really functioning. I've lost my job, but I don't know whether I can register for unemployment in this prefecture."

"Should we try to find out for you at the prefectural office?"

"My company has not finished the clerical procedures related to our termination, so I cannot ask. The prefecture can't tell me . . ."

She had two children, ages seven and five. Just now they were at the park with her husband. I asked how they were managing, and she replied: "They've regressed to a younger state. At home they can do everything themselves. Here, I don't know whether it's from staying so long and living like this, they say: *I can't do this* . . ."

I requested to see how her family lived. She hesitated. "My husband's mother is a bit depressed, so . . ."—At length I prevailed on her to at least ask the older woman, who kindly allowed the interpreter and me upstairs and inside the long, almost empty room, over whose floor stretched many long, narrow tatami mats, very bright and clean, a few bags of belongings in a neat row against the wall. Sheets and blankets had been folded into neat squares and stacked. One child was here after all, a boy, next to whom his mother quietly went, sitting on her heels, with her fingers spread on her knees; he smiled shyly but cautiously behind her. They had nice light, and a window view of a wall with other windows; surely it was better here than in Ishinomaki, where, I was told,† one tatami sufficed for three people. The family with whom they shared the room had lived beyond the outer evacuation ring, but they chose to leave after the supermarkets and stores in their district closed. Now they had returned home, with what consequences to their health I certainly cannot say. Twenty other people from Minami Soma remained downstairs.

Ms. Hotsuki Keiko, the mother-in-law, was lying down. She sat up when we came in, smiling politely, lowering her eyes, discreetly half-stretching; perhaps she had been sleeping. She appeared to be not much older than her daughter-in-law. Bowing as respectfully as I could, I inquired how the quake had expressed itself to her.

"At that time I was at home. I rushed out of the house, where there was a big plum tree. I held it for a long time. During this time, furniture fell, and plates fell out of the closet."

Since Minami Soma lies some distance from the ocean, the tsunami caused

---

* At the time of the interview, this would have amounted to about U.S. $625.

† By Mr. Kawanami Shugoro. See below.

Members of the Hotsuki family in temporary housing

her no personal fear. But her aunt and uncle had drowned in their car. Fortunately, she said, the family could recover their bodies. Unfortunately, the cemetery had washed away.

"We were allowed within 30 kilometers. The recommendation was to stay inside. The city mayor told us to evacuate 'on your responsibility,' so some are still living there."

"What is your opinion of the reactor accident?"

"Everyone has always said that nuclear power is safe ..."

"Mrs. Hotsuki, here is a question that baffles me. As a citizen of the country that dropped atomic bombs on Japan, I wonder how this could have happened in your country twice. First you were our victims, and then, it seems, you did it again to yourselves."

"We don't know much about the nuclear bomb," explained the older woman. "They're pretty far from here, Hiroshima and Nagasaki, and we just heard from our parents that some plane came over and so forth. They didn't talk about it."

"Why didn't they?"

"Unless you go to that area and see that atomic site, then maybe you have no interest in it." Trying to give me what I appeared to expect, Mrs. Hotsuki gleaned through her memories, then presently grew animated, gesturing and almost grimacing as if she were close to tears, nodding her head as she said: "Once I saw a display in Chiba Prefecture, all about the kamikazes. I was so moved I couldn't stop crying."

Less moved by the kamikazes than perhaps I should have been, I resurrected the matter of Hiroshima and Nagasaki. It turned out that both of the Hotsuki women believed the atomic bomb to have been worse than the reactor accident, because "at least we evacuated." Minako, the young daughter-in-law, explained that "if you just brush it off, said the prefectural office, it should be okay, and you don't even have to take the radiation screening. So we felt better."—As Orwell would say, ignorance is strength.—Or was the prefectural office correct? Alpha particles were nearly safe outside one's body; beta particles once washed or brushed away could do no further harm. While I essayed to formulate why that procedure might be inadequate, a pretty girl wearing a red armband bowed herself in, announcing that the child-minders were here again, today with candy; she also wished to inquire whether anyone might be sick. So perhaps it was all perfect; no matter how politely I pleaded, neither of my two interviewees would accept a 10,000-yen note, not even for the children's sake; wouldn't you rather believe that they lacked for nothing?

Having scored my interview, I dared to risk an encounter with officialdom, and so met a certain bespectacled, pimpled and narrow-faced young man named Mr. Maeda, who identified himself as "just an employee of this facility. If you put this in your article, you must contact the city office. That's what we have been told." (I most inexcusably neglected to follow his instructions, but, reader, if you wish to do so, the telephone number is 022-214-1148.) He photocopied my letter of press accreditation most alertly; fortunately, my interpreter had always reminded me to keep it neatly folded, in homage to its pretense of importance.— "In your opinion," I inquired, "how dangerous is the radiation?"—Mr. Maeda replied: "None of us are particularly concerned."

AN OLD MAN PLANTING SEEDS

Ishinomaki, they said, looked now the way that Sendai had two weeks ago. In Sendai some people stayed for two days on their roofs until the water subsided. In Ishinomaki there were those who were trapped on their roofs for a week.

On the other hand, Ishinomaki was better off than Rikuzentakada. Isn't there always a worse place?

The 50-kilometer drive in the veterinary science professor's car would ordinarily have taken an hour. Ever since the quake, there were traffic jams. It took nearly two hours to reach Ishinomaki; and, indeed, in the course of researching this chapter I had almost daily recourse to the $400 or $600 creeping taxi ride or the much cheaper half-day stalled bus ride (the region's railways being broken), on this highway or that expressway, frozen in traffic or not, so many kilometers toward or away from Reactor No. 1, the long windshield wipers sometimes dancing in rain of unknown salubriousness, the radio news on low, the cab creeping and stopping between other cars in a like situation, the driver occasionally misplacing his Japanese patience.

In Ishinomaki the first storey of the supermarket was open and newly gleaming. Most goods were present in pre-quake abundance. Only one yogurt was allowed per customer, several shelves were bare, and others held milk brands from Kyushu and Hokkaido, which were not normally sold here. The brand new washing machines were all sold out, the tsunami having ruined ever so many; and automobile dealerships were booming for much the same reason.

The professor's name was Morimoto Motoko. She lived in Sendai. During the tsunami, her two teenaged children had overnighted (foodless) in care of their teachers; now they were staying in Osaka with relatives. She was making this drive to bring supplies to one of her students, a young man named Utsumi Takehiro, who now bowed to us; so did his mother Yoshie. They got into their car, and we followed them home, Ishinomaki being less easy to navigate than before. "If you go beyond the number 45 road," said Takehiro drily, "the scenery changes."

Passing the vegetable market which was now a temporary morgue, we rounded a corner, and I saw many grooves cut deep in the smooth tan earth, with a line of cars and people perpendicular to them on the far side, and white coffins down in the farthest of those open trenches. Takehiro's grandmother was buried here. The tsunami had caught her. From the way that he spoke about her, I came to believe that he had loved her very much.

"I didn't see the body, actually," he said. "My parents saw 100 bodies every day. They finally found her. Now it takes a year or two before they're cremated. First, we temporarily bury them. Then they're disinterred. There are only a few crematoria,* so we have to take turns."

"I'm sorry," I said.

"Our dog was also killed, because he was chained. We took his body to Niigata, where my father was working, and cremated him. But you need a special vehicle to transport a human body, and those are in shortage."

---

* His mother said that facilities permitted only 20 bodies per day to be burned in the entire city.

Mr. Utsumi Takehiro

Now came heaps of mud, canted trailers, gouged walls, crumpled cars, the crazy skeleton of a shed barely supporting its intact roof, many relief workers and blue-clawed cranes, sunken muddy acres which reminded me of a slum I'd once visited in Thailand, half-smashed houses on a muddy ugly plain with wet trenches tunneled through it, man-high mounds of debris on the roadside; and so we came into the Tsukiyama district (the clouds like sheets of white slate, the sun in the pinetops and the dust in my throat). The oil slicks were particularly nasty. Several large oil tanks had exploded, setting off numerous fires. We rolled past the wreckage of the paper mill, whose round bales of product lay dripping everywhere, sometimes oozing down like Spanish moss. Paper was now in short supply, remarked Mrs. Utsumi.

"My uncle was rescued by helicopter, and he appeared on TV," said Takehiro proudly.

An American battleship lay on the horizon of the pale blue sea. Here came a long mild wave, its crest so clean. One of its predecessors, the tsunami, had dragged a giant fuel tank onto what remained of the dike. More heaps of mud

framed our scenic drive, accompanied by fuel tanks thrust against and through
roofs, cars leaning against trees, and every so often the ground storeys gnawed
out of buildings, especially sheet metal ones—in short, block after block of ugliness (unlike war-damaged cities, Ishinomaki offered few scorchmarks or round
projectile-holes), plastic streamers hanging in trees; and presently we turned
down a street of new-made junk-lots and Takehiro said: "This is my house."

His next door neighbor, Mr. Kawanami Shugoro, made us black coffee on a
butane-powered hot plate on the dust-choked rickety table inside his blighted
house, which appeared intact on the outside. He wore his cap, presumably for
warmth. Fat hunks of ceiling dangled down from the rafters, the drywall torn
like cardboard. Everything in the living room cabinet was in place, but the cabinet itself tilted at about 30 degrees.

Mr. Kawanami Shugoro

Mr. Kawanami said: "When the earthquake came I was at home. My office had some meeting, so I was trying to change into a suit. At that time there was not much damage, so I changed back into my work clothes and drove the clerical worker to her home near the supermarket. Then I headed toward the office. Then there was a traffic jam, and they said that the tsunami was coming, so I made a U-turn, meaning to come home again. I saw water coming out of the canal by the senior high school and vehicles were floating; so, since that direction was no good, I made another U-turn and took a higher direction. At the river's edge, the fire department personnel told me not to go that way, but it didn't look bad. All the same, the water level seemed a bit higher, and then I saw it come over the dike. So I fled. I had to sleep on the ground for four days. I went to Yamato to confirm that my grandchildren were all right. Then there was a gas shortage, and it was so cold. I found a garbage bag to keep me warm—so cold, so cold! It was snowing. I tried to find someplace warm; I took more and more garbage bags for a shirt . . ."

"What color was the tsunami?"

He laughed. "I don't remember. It was black, they said, with oil."

He was a cheery, rugged, white-bearded old man in a cap—66 years old, with the face of a workingman. At the shipyard he was in charge of safety and hygiene. The neighbors stood around us. Cans of juice were on the filthy table. His wife had led some panicked Chinese girls up onto the second storey of a parking lot, and they all survived. "Everyone went to the roofs," he said. The second or third tsunami-wave had been in his opinion the bad one, people floating in their cars and calling out for help until they sank. Mr. Kawanami said: "These images were in my brain yesterday, and I got depressed and confused . . ."

A couple of his acquaintances had fled; once they reached high ground the wife returned home for their valuables, because she was a strong swimmer. Fortunately, they recovered her corpse, which still gripped one of the two bags of precious things.

"When you think of all you've suffered," I asked him, "do you think the reactor accident might be better or worse than that?"

"What shall I say? I can't even imagine. This area is where elderly people are residing. It requires money to rebuild a house, and many people are scared. My wife says that if everybody leaves, then only we will stay. We think that since we are old, we can stay until we die, since *this*"—he must have meant the tsunami—"happens only on a thousand-year scale. I planned to retire this year and live a nice relaxed life. But the money for my future will have to be spent on repairs. Moreover, the people at the nuclear plant, they are talking about some nuclear explosion. Our governor is so proud of our nuclear plant, compared with Fukushima.

"I'm directed to have my hair cut," he laughed. "Is the barber back yet in the center of town?"

"Yes," said a neighbor.

In the filth, muddy dishes were neatly stacked. Fresh water was still too rare for washing just yet.

At my request, Mr. Kawanami took me upstairs to admire the sand and silt. He said: "When the wave came, each tatami mat struggled like *this*!" and his arms writhed.

Thanking him and departing with my best bow, I was next introduced to Mrs. Ito Yukiko, age 66, who received me narrow-eyed, with shoulders drawn in and fists on her lap as she sat on the edge of her chipped, cracked concrete porch, wearing orange windpants, a dingy sweater, a white-striped wool cap pulled down over her eyebrows. The toes of her slippers touched the mucky, rubbly ground, which happened to be decorated with broken dishes. Here as everywhere else in that neighborhood the smell of diesel was nauseating. Her two young granddaughters, wearing galoshes, played at sweeping the doorway, then settled down to read what might have been comic books. They were very shy; I left them alone. I also did not ask, since no one mentioned them, where their parents might be.

Mrs. Ito Yukiko

A neighbor of Mrs. Ito's

"I was born in the beach area," she said. "I have experienced the Chilean tsunami, and also another one in this prefecture. So I knew well that when an earthquake comes, you have to take care in case of a tsunami. But *this* one," she said grimacing (and stopped to pick up a spoon which one of the little girls had dropped), "*this* one was different."

Well aware that quake-deformed doors might trap people behind them, she carefully opened the house-door in advance, then rode out the temblor just within, for fear of getting brained by her roof tiles. Unlocking the safe, she removed "the memory of the ancestors," evidently their Buddhist memorial tablets, and then, believing she still had time, searched for a cotton *furoshiki* cloth in which to wrap them. One of the granddaughters then suggested that she might wish to take the cell phone. And so they fled in the car. Reaching a safe place and sending the two girls ahead, she then drove back in hopes of retrieving their dog and her wallet. Here her hands began twisting tighter and tighter together in her lap, and when she said, "I took a narrow way, and then I saw the tsunami in the middle of the road," I found the horror in her round reddish face nearly unendurable.

Ishinomaki tsunamiscape

"The first wave took all the belongings away from me, so then I ran to where the wave was lower. I know that a human cannot escape the tsunami once she is caught in it, so I removed my shoes and climbed a wall, and first it was unstable, but I found a stable place, clinging with my toes. The water was up to my waist, and then it was up to my chest; I was holding onto the roof so that I couldn't be carried away; I was screaming *help, help, help!* to the spirit of my late husband... Then it came."

On the porch the grandchildren went on reading comic books in that fishy, diesely wind (and since it might for all I knew be blowing from the reactor, I inspected my dosimeter, which at six in the morning had been reading 1.9 accrued millirems and just now after three hours in Ishinomaki turned over to 2.0, which signified that the radioactivity here was once again at least twice that of Tokyo's—not bad; never mind those hypothetical beta particles riding on the wind); and a crow cawed; there was a heap of tires; in a glassless window, sodden futons hung out to dry.

"I was on the rooftop, so I was rescued at the end, before it got dark. I didn't see my granddaughters for two days, but their teacher told me they were all right..."

Behind a leaning grate, her old neighbor was picking up clinking things from the mud of his former yard.

"How relevant is the nuclear accident to you?"

"The power plant may be necessary, but they ought to publish the facts. It seems to be stopped all right, but is it really? They've said that in some fishery products the concentration is low, but accumulation will be bad..."

Gazing down at the sand by her feet, I saw a small fish-mummy, convulsed.

Dead fish by Mrs. Ito's house

In a narrow zone of sand between two ruined houses, an old man was plant-ing seeds. Streams of plastic twitched in the tsunami-pollarded trees. A twisted cypress, still green, rested against the patio wall of a house which had been smashed open on its eave-end. I bowed goodbye to Mrs. Ito, who slowly crept into her house.

CONCERNING A *KOTO* NOW UNDER REPAIR

How many such stories would you care to hear? I collected a number; they are much the same in that one quality which causes journalists to seek them out, just as are the grimacing, often swollen, frequently forehead-bruised corpses whose images faced us on the fluttering blue tarp-wall of that temporary morgue at Ishinomaki; their expression much depends on the angle of the head. The survi-vors who view them keep calm, in the best Japanese manner, gesturing each other forward with a polite *"hai, domo!,"* offering one another the best views of those horrid faces, whose eyes are usually closed.

One woman was explaining to another: "I came here to look for my mother-in-law, but since the faces are swollen it's difficult, and the number I specified was wrong; that's why I couldn't identify her right away . . .'"

On the other side of the long rectangle of sunlight, a priest was ringing a bell, and a photograph gazed down upon a bed of donated flowers.* Relatives bowed over the ritual bowl; candles flickered. The priest bowed. My throat ached with dust.

Thinking to learn more, I asked a policeman for information, so he referred me to his chief, who could do nothing without the big boss, who when I asked how many people had died in Ishinomaki rewarded me with a perfect answer: "Our policy is not to answer regarding the individual numbers." Bowing and thanking him, I said that in that case I had no further questions; reddening, he bowed deeply and apologized for keeping me waiting.

So let us at least momentarily leave the narratives of loss at rest with their eyes closed (the bulldozers clearing more long narrow corpse-trenches in the dirt, 20 bodies per line, three temporary cemeteries in Ishinomaki, and a long green line of Self-Defense Forces dividing into two detachments to break open build-ings in search of bodies), while we consider what *meaning,* if any, we can find in them. In this context let me now re-introduce you to Takehiro's mother, Mrs. Utsumi Yoshie.

---

* All the flowers here were donated, I was told.

A temporary grave in Ishinomaki

Searching for bodies in Ishinomaki

"What lesson if any do you see in this event?" I asked her.

"Since March eleventh, something is finished. I feel that something different has begun. We have never had the experience of losing everything to this extreme. The good lesson," she laughed, "is to keep valuable things on the second floor!"

"Will your lives be worse?"

"Of course I believe they're going to be better," she replied, sitting with me in the dirty wreckage inside her house, with smashed things everywhere.

Ishinomaki street

"Why?"

"Well, I don't know why. The passage* of daily life will create another sense of value. Unless you think that way, you cannot advance."

I told her how brave I considered her and all of them to be, at which she remarked that for some time she had been taking lessons in playing the *koto,* a traditional stringed instrument whose notes I have sometimes been graced to hear in Kyoto's and Kanazawa's secluded teahouses: slow, quiet and (to me) melancholy notes reassembling some blurred ghost-face out of the melodies of olden times. I hope never to forget how it was for me in that small chamber in Gion when the lovely old geisha Kofumi-san danced the "Black Hair Song," to which

---

* The interpreter said "accumulation," but this must be the meaning.

Kawabata and Tanizaki allude in their greatest novels.* It pleased me that Mrs. Utsumi also knew and indeed had mastered this tune, whose simple mention made her faintly smile. She wished me to meet her *koto* teacher, who might help me understand more about Japanese attitudes; unfortunately, that person could not be reached on the phone. But for a fluttering instant the two of us had lived again in the Japan of March tenth, 2011—the day before Ishinomaki became newsworthy.

I wondered whether she had time to play for me on her *koto,* but the instrument had been submerged in the filthy wave. Right now it was under repair. Very softly she said: "A *koto* is like a living thing to me, so I was very sad. We lost our dog, but when I saw the *koto* so dirty with mud I felt so sad . . ."

I asked her sons which possessions they themselves had been the most distressed to lose.—"All!" they laughed cheerfully. Since not enough chairs remained, they stood around us inside that dark and chilly ruin with its bitter stench of dust.

And what did they all think about the reactor accident?

"I think the fact that it occurred cannot be helped," the mother said. "I want Tepco to work hard and the vegetables to be grown again. I would like to buy the vegetables from Fukushima, but . . ."

"Do you think it will become contaminated here?"

"Yes," she said steadily.

"What will you do?"

"We have nowhere else to go."

"Have you suffered any nightmares?" I asked her then.

"In my case, I didn't see the tsunami with my own eyes. I wasn't able to return home for two days, so I didn't experience that, but the fire, well, we could see it so vividly from our hill† that I was almost afraid it might come to us; at three in the morning there was an announcement that the flame might come to us, so we evacuated."

"But no nightmares?"

"No. The fire burned two days . . ."

"I cannot live in our home," Mrs. Utsumi said later. "It's too scary. I cannot live there again, even if we have to build a new house . . ."

Not knowing what else to say, I repeated that she was brave, and she said: "I

---

* See the "Jewels in the Darkness" chapter of my *Kissing the Mask* (New York: Ecco, 2010).

† Another translation comment: In Japanese the pronouns may sometimes be left implied. The literal translation here was the sentence fragment "the fire so vividly seen from my hill." Here and throughout I have emended the raw translations in such ways as this. My interpreter has seen and approved.

think that if we live decently, that will give my mother-in-law peace of mind. She would not have wanted an expensive coffin for her temporary burial; she would have preferred to have the money go to her grandchildren."

I nodded. The dust ached in my chest.

"To console my dead mother-in-law, I would like my two sons to work hard to rebuild this city."

One saw small germinations of this stoically and at times quixotically resurgent impulse here and there in Ishinomaki; I remember a gang of workmen with white rags around their foreheads, dragging soaked tatamis out of a warehouse, needing a wheelbarrow and a man on each end, such was the water-weight; and then the car rolled down past a wide hill of garbage, many men in rubber boots standing in a puddled courtyard, waiting for I know not what; but less than a month had passed since the tidal wave, so that more conspicuous than these were the antediluvian survivals: for example, the age- or diesel-blackened torus of a small shrine standing alone and out of true over sandy mud struck me with déjà vu, and later I remembered an image made by the great photographer Yamahata Yosuke in Nagasaki, 1945: not dissimilar to Ishinomaki, 2011, except that in the former case the wreckage around the torus appeared to be almost exclusively wooden; moreover, one wheeled fragment, evidently deriving from a cart, seemed more slender than any modern counterpart, and the backdrop was all white-veined grey smoke.

Before we returned to Sendai's stopped ferris wheel and rectangular dirt-fields speckled with pale trash, the elder of the two boys, whose name was Yuya, said to me: "I would like to eat food from this area to help the farmers."

"You mean, you wish to eat produce grown near the nuclear plant?"

He nodded with a calm smile.

Professor Morimoto having already gone home, they drove us to the bus station. I told them that there was no reason for them to wait for us to board our bus, but Mrs. Utsumi assured me that they had nothing better to do.

WHEN THE WIND COMES FROM THE SOUTH

In the night there was a tremor at the hot spring, then a moderate quake, and there came a swaying and a shaking as I lay on my tatami, knowing that I could do nothing but relax, being on the fifth floor. Fortunately, the room contained little furniture (people had told me how televisions and books could literally fly off the wall to do execution). As the blue-white dawn glanced through the rush blinds, the dosimeter still pleasantly at 2.0, the new taxi driver called to report that the road had been "broken," so an early start would be best. The power was out again in Sendai, it seemed, and when we stopped to pick up Professor Morimoto, now on another mercy-errand to a student of hers on Oshima Island, we found her shaken and discouraged. The elevator was dead, of course, so the driver and I carried her suitcases of batteries and other provisions down six flights of stairs; then we sped along the crisp road, the stiff tan grasses soon shrugging off the frost as we descended. By now the dosimeter read 2.1. Laughing, the driver, a strong man in late middle age, remarked that he and his wife had just finished clearing up earthquake damage from their home, and now their crockery lay smashed on the floor again! The railroad station at Sendai was now leaking through the roof, he remarked, so it might have become unusable. Meanwhile the interpreter looked up from the newspaper to report that restrictions on the fisheries in Miyagi Prefecture might be put into place for two months, which I imagined could possibly turn into 20 or 50 years. Early plum blossoms and very occasional palm trees kept us company as we passed the straw-colored ricefields; a seagull overflew us. The radio announced that 916,000 households had been "powered down" by last night's event. Here came another hour-long traffic jam because the trains were stopped, and after awhile we entered the diesel-flavored ugliness of Furukawa, which the creeping of all vehicles allowed us to inspect to our hearts' content: small banks, billboards, automobile carriers, undistinguished houses behind hedges, pachinko parlors surrounded by empty parking lots, carwashes, a tombstone shop laid out on blacktop overlooking a dirty concrete-lined canal. We stopped in a dark convenience store so that the two women could relieve themselves; the washroom was out of order. Half an hour later, their experience repeated itself in an establishment whose dim shelves were partially bare, a single clerk receiving a long line of customers who appeared to be buying mostly bottled liquids. His cash register was asleep, of course. Everybody acted very patient. Soon afterward we began to see the long, long crack in the asphalt, running parallel to the white line; sometimes there were fragments of pavement sticking up like bedraggled roosters' combs. At one place, two yellow-uniformed road workers shared a long gauging-pole between them, inserting it into a fissure in the highway as if they were fascinated.

I asked the interpreter how often she had experienced a damaging quake.

"They come all the time in Tokyo, but there's usually no damage. A moderate one, every two or three years, I guess . . ."

The cracks gradually grew more impressive. They were at their worst at the edges of bridges. The driver sighed and shook his head; the two women were silent. Then the highway improved again.

In Kurihara the power likewise appeared to be gone, as in Ichinoseki with its dark McDonald's and its empty parking lot around uninviting New Kouraku Power Entertainment Space. After a longish time we came down into Kesennuma, 172 kilometers from Nuclear Plant No. 1, meeting ever larger heaps of broken lumber, then ruined buildings, mounds of metal and masonry, upturned cars, but everything already wrapped neatly out of the streets. The driver groaned, *"Awh! Ehh!"*

I never knew how Kesennuma used to look; all I know of the place is street after street of rainy trash, wrecked cars, burned cars, trash in puddles, trash-hills with sludgy pools between them, a bad-tasting rain (and for all I know, the most dangerous thing I did on that entire trip was to hold the interpreter's dripping umbrella for her while she went to the washroom). Sometimes filth-darkened fibers, cables and splinters hung down in doorways like teeth in a monster's maw. The dirt roads had on occasion come to resemble dikes between rectangular ruin-fields* heaped with garbage and filled to the brim with stinking water. Many houses resembled auto wrecking yards. Here came a green army truck with a red cross on it, rolling slowly. On higher ground, where it was less watery, the former neighborhoods were simply mucky like bombed, looted and vandalized construction sites. And in one puddly, muddly stretch, another vermilion shrine-torus stood alone above junk and filth, just as at Ishinomaki.

That muck, that particle-laden slime had to be gotten rid of, and the Japanese were on the case. It would take years. Between 2011 and 2013 alone, they disposed of 40 metric kilotons of the stuff, whose official name was "tsunami sediment." To be precise, they dried it out and interred it in landfills. There was so much of it that its methane-emitting organic matter† remained a tiny but still measurable and reportable fraction of Japan's greenhouse gas inventory for 2013, by which year the cleanup was not over. It must have taken much effort to shovel and sweep and scrape all that nastiness out of Kesennuma.

"Kesennuma," they say, means or derives from an Ainu word meaning "bay." Across the street from the harbor whose street sign was buckled and torn and whose power wires were having a bad hair day, the flooded parking garage smelled like the sea and rain spanked down onto the sidewalk. A gaunt cyclist in grubby grey pedaled past, his dust mask down around his neck. The milky-green-grey sea did not seem foul. The rain made the air less dusty, although possibly

---

* A more exact term would be "ruined house foundations which now resembled broken-walled fields."

† It was 4.5% carbon—slightly more than a tenth of the comparable percentage for human waste. Its "decomposition half-life," or the time required for half of it to decay in a landfill, was estimated at 36 years—the same as wood's. The greenhouse gas emission factor of that nasty substance was 15 kilograms of methane per ton, or in our units 15 pounds per half-ton.

more radioactive; I never forgot that the dosimeter couldn't tell the difference. After hauling Professor Morimoto's boxes of batteries over to the ferry landing, I stood gazing uphill across the concrete chunks and through the rebar, over the crushed-matchbox houses, chairs, futons; here was a house whose upper storey looked virginal but whose first floor had entirely disappeared except for one wall, so that the ruin resembled a sick letter P. The rubble led my gaze up to two red

View of Kesennuma

roofs and somebody's pine, which had been manicured into cloudlike lobes of green in traditional Japanese style.

The unevenly humming ferry bore crates of apple juice and other supplies. A longhaired adolescent boy whose shirt said HAVE A NICE YEAR 2009 was one of the many who wore dust masks, probably for hay fever; the dosimeter was steady at 2.1. Three pre-teen girls with teddybear chains on their backpacks stood at the bow. A tiny girl in a pink windbreaker sat in her mother's lap, playing with a toy pistol, laughing gleefully, reaching out uncomprehendingly at a horizon of broken ships. Lumber floated here and there, and another boat lay sunk as if by enemy aircraft. Fingers and claws of wreckage protruded from the chilly sea. After half an hour of filthy slicks, scraps of foam rubber and styrofoam, a row of multicolored garbage-flecks, a seagull flying very low, a lost bamboo pole, the orange prow of a ship sticking out of the water upside down like the bill of a dead porpoise, we landed at Oshima Island (165 kilometers from Plant No. 1; population about 3,000), where Professor Morimoto's student, Mr. Murakami Takuto, awaited us.

Mr. Murakami Takuto

The Murakami family's is the last tsunami story that I will tell. They were of old stock, their ancestors being "marine soldiers" who fought on the side of the Heike in that 12th-century civil war about which so much great literature has been written. *The Tale of the Heike* opens in a way not devoid of reference to the events of this essay: *The bell of the Gion Temple tolls into every man's house to warn him that all is vanity and evanescence. The faded flowers of the sala trees by the Buddha's deathbed bear witness to the truth that all who flourish are destined to decay.*

The first storey of their house had been half submerged. The second floor was fine. Almost all of their electrical appliances must now be replaced, from the rice cooker to the new television to the heating system, which unfortunately and uncharacteristically had not employed natural gas.

In the dining room, which now needed work, Grandmother Fumiko (born in 1933) said, speaking very slowly, tilting her wide, handsome face: "On that day I was in the garden when the earthquake broke out. When it stopped, I came in; there wasn't much damage, just some glasses and candlesticks. Then I heard the tsunami alert: someone from the fire department calling on the loudspeaker. I cannot run like others. Then I saw the wave: lots of bubbles, so it was white. It was low. And I saw another big wave coming behind it, and so I tried to run. I ran to a higher place. Had I taken the big road I would have been drowned. I took the narrower, higher road. I looked back; the neighbor's house was floating. After that, I picked up a bamboo stick and used it as a cane. In this city an elementary school is used as an evacuation center. I still live there. I just came here to welcome you.

"In the beginning we couldn't communicate with anyone. After five days the parents came and I learned that the three grandchildren had survived. It was so scary that I trembled and couldn't stop. I couldn't sleep. Friends offered me clothes, riceballs* and a futon, so I'm doing fine."

"Does the government provide funding?"

"Not yet."

She then said: "For 350 years our family have been living here, and our ancestors' saying is that in the Meiji Era the big tsunami could not come up to here; therefore this house is safe. If I believed the saying of the ancestors, I wouldn't be alive."

"Are you concerned about the accident in Fukushima?"

"The radiation, when it rains, they tell us not to get wet . . ."

(Her grandson later told me: "About radiation people on this island don't know anything.")

I made my usual remark that after Hiroshima and Nagasaki it seemed particularly sad to me that the Japanese were once again suffering from radiation, to which the old lady replied, clasping her hands: "I just want them to be careful."

"Do you believe that the contamination will come here, or will the situation improve in a year or two?"

"We get vegetables from different places, so we will be affected."

"The pines are all fallen and gone," said the grandmother, stretching out her left hand toward where they used to be, out across the clouds and then the broken trees and sand, and over the sea toward the former location of the great rock which the two grandsons used to call their "target" when they swam together. "From here we used to be able to see the sun rising through the pines. We were so proud of that. Now the ocean seems closer. That's a little scary."

In the garden she had grown corn, rapeseed, spinach, pumpkins and white radishes.† She said: "I feel so lonely now that I have nothing left to work on."

Takuto, the interpreter and I went for a walk. Down on the futile breakwater of the wrecked beach we found a dripping Chinese book for boys and girls—the property of his late grandfather. "But we never read it anyway," he laughed, leaving it for others. I found a field now prettily sown with scallop shells, a bamboo grove hung with garbage.

---

* *Onigiri.*

† Daikon.

The topmost plate in the stack on the lowest shelf shows what level is. This china cabinet leaned alone in a field; the house was gone.

Searching for bodies in Oshima

We met a fisherman in an orange jacket; he said that a third of the island's inhabitants had died on March eleventh. He said: "First they ran, then they returned to fetch something important; they didn't survive."

"Radiation?" he cried. "No, that's Fukushima. We have nothing to do with that."

Walking past him to the end of the concrete jetty was nearly pleasant, the gulls calling from their low islet (lower than before, it seemed; much of Oshima had subsided one or even two meters, which was why as we approached toward the ferry terminal on the following day we found the tide coming in over the main street), the sea-wind smelling so delicious that I could not make myself wear a mask, my dosimeter still at 2.1. The setting sun cast a white trail on the water, and a helicopter, probably from the Self-Defense Forces, hummed out behind a cloud. As the day failed, the sad tokens of the tsunami withdrew into the shadow, until Oshima appeared nearly whole. Clouds and dust touched the low piney coast.

Takuto said to me: "I would like to do everything I can for this island. I would like to grow up and be a human being and help."

Although our clothes were now getting quite dirty (since we expected to discard them after entering the hot zone), that kind and hospitable family refused to let us use our sleeping bags. Father and son laid out futons for the two women, and a bed for me in the adjacent chamber. That meant that the rest of them slept downstairs in those chilly rooms which stank of muck. Our host's flashlight wavered slow and white around his belly, Professor Morimoto's cell phone glowing as she and her student giggled over some stupid display, the interpreter switching on her headlamp, which illuminated her face, and I writing these notes with the aid of my American flashlight which was more yellow than anyone else's. Now it was nearly dark, my two lady companions fading into the wall across the room. Although the Murakamis accepted half a dozen cans of American food (the interpreter's share of our provisions; on the next day I divided with her what remained), they insisted on cooking us dinner. Ashamed and grateful, we came downstairs to the table, where Mr. Murakami's stubbled, moustached face gleamed in the light of the Coleman lantern. He was one of the head teachers of an elementary school. After the earthquake, he had permitted some students to depart in the care of their parents. I could tell that he felt guilty about what could have transpired; as it happened, however, they survived the tidal wave. He pressed on me a satellite-photograph disaster map of Oshima, based on that infamous great wave from the Meiji Era. With his spectacles high on his forehead, he showed me the family home on the map. He said: "Far too optimistic."

His mother, Mrs. Murakami Kaoru, in her checked apron, stayed nearly always on her feet, her pale arms and cheekbones shining, the other grandmother slowly nodding her heavy head at the two of her three grandsons who were present, while bananas and aluminum foil shone softly in the dark. Mrs. Murakami invariably bowed to the grandmother when offering her food, with a polite *"hai, dozo."* On the following morning she could not be dissuaded from driving us to the ferry in the family's car; it would only have been a 15-minute walk; I don't know where they found the petrol; the last I saw of her, she was in the doorway of a dark and ruined waterfront cafe, smiling and waving to us with both hands as the boat pulled away. Given the absence of refrigeration I cannot imagine how she managed so well to make that ad hoc stew, many of whose ingredients were perishable. Light rested on the shoulders of the riceballs and the bright white pickled shallots. Mr. Murakami said: "For the first five days, we got only one riceball per day, so I became thinner."

An hour before dinner he had already been promising me treasure: a bottle of sake rescued from the first floor after the ocean departed. The sodden label was nearly invisible in the darkness. Again and again he filled my water glass to the brim, meanwhile offering it around to the other guests. Embarrassed to take so much from him, I finally pleaded tipsiness, at which he happily continued to fill his own glass, not least, he remarked, because it was Saturday night. He kept saying to his wife in English: *"I love you."* She smiled with pleasure. I am happy to report that on the following drizzly sober morning he said it again.

In the midst of dinner the electricity came back on, and they happily shouted: *"Surprise!,"* the grandsons grinningly illuminated; I assured our hostesses that they were even more beautiful by electric light, and the grandmother clapped her hand over her laughing mouth.

Whenever I mentioned Hiroshima the whole family grew sad and silent, so I hated to bring up the matter; but it seemed my duty to raise it once more with the patriarch, which I did while we were still eating in the dark. The whites of his eyes seemed to flare. "Because Fukushima is prosperous on account of their fishery," he said, "I fear their decline." To me this seemed so Japanese, to worry about others first! He went on: "Atomic power is very dangerous. To me, it's so dangerous. To me, it's like war."

That afternoon I had asked Takuto how he imagined the worst might be, and he replied, not quite a week before the Japanese government admitted that the reactor accident was a Level Seven like Chernobyl's:* "Like Chernobyl. Oshima could be contaminated. In the summer the wind comes from the south."

---

* A year later, *The Japan Times* asserted that "in terms of soil contamination" the accident was in fact only one-eighth as severe as Chernobyl's. For a slightly lower 2014 estimate, see table on p. 469.

2: They Did It for the Nation

I won't deny my selfish relief at leaving the stinking ugliness of Kesennuma and Oshima, not to mention my anticipation of getting safely home where such things never happened (Sacramento, my home city, was second only to New Orleans for flood risk in America). As well as I could determine from my dosimeter, the radiation in Kesennuma and Oshima seemed to be the double of Tokyo's: 0.1 millirem every 12 hours. Now it had come time to return to Koriyama, and from there to make a foray or two into the evacuation zone.

The power seemed to have been restored in all the cities along the way.

At five-thirty in the morning in Oshima, the meter read 2.2 millirems, having turned over once since dusk. At nine-thirty on the Tohoku Expressway, after two hours of moderate rain and just north of the Ohira exit, it reached 2.3. Shortly before one in the afternoon, just as we came into Koriyama, it showed 2.4. By eight that night, thanks no doubt to a certain pleasure-drive which I will shortly recount, it was at 2.5, and before six the following morning it had achieved a glorious 2.6 millirems: four times the Tokyo baseline, in short. According to the newspaper,* the actual level was closer to 44 times Tokyo's; but, optimist that I am, I'll keep faith with the toy Ray sold me, for Ray would never lie.

Here is an appropriate place to say that my dosimeter's accruals for Koriyama, the highest of any 24-hour interval excepting the days of my two international flights, were not alarming: 146 millirems per year—barely 40% of the emergency guide's supposed world average. And that American dosimetrist (who inhabited the nuclear industry, and preferred to be called "Source ML") opined that as much as half of Koriyama's new energy field might consist of pre-disaster "natural" background radiation. To reach my danger threshold of 5 rems per single "incident," I would have had to hang around Koriyama for more than 34 years. All the same, one might not wish to marry and raise children in Koriyama.†

THE WIND THAT COMES FROM THE SEA

Regarding that day's pleasure-drive, I will tell you that shortly after five in the evening, once the nasty and potentially hazardous rain had ceased, the interpreter and I hired a taxi to convey us to Komatsu Shrine, of which none of us, including the

---

* The *Yomiuri Shimbun*. See source note. This is the source referred to in the next footnote.

† If the *Yomiuri Shimbun*'s figure is correct and consistent, then three years would suffice to reach that same nasty 5-rem dose—which of course would be almost safe amortized over such a long period: 0.19 millirems per hour, or not quite twice the incident guide's ceiling of innocuous normality.

driver, had ever heard; the interpreter and I had chosen it after a glance at her map. The driver was a bald old man who insisted on his financial rights, grimacing and arguing, so that for awhile it appeared that we could not negotiate. His stubbornness had nothing to do with the danger; the question was whether to pay by the hour, the meter or the job. Finally we compromised on all three. The driver then said that this journey might not even be allowed, because it seemed as if Komatsu Shrine might even lie (here I expressed innocent surprise) within the forced evacuation ring. He radioed his boss, who gave us his blessing, and off we went. Needless to say, I watched the dosimeter, expecting radiation to rise in proportion to the inverse-square law; but anyhow I have already spoiled the suspense of that business.

The driver had picked up only two fares that day, both of them insurance company operatives who were verifying earthquake damage. Now to Sendai the highway was open, he continued, so that made it smooth. He was a good talker, and I had already begun to like him. He had driven a Korean reporter toward Sendai but a landslide blocked the way. I asked which shrine in Koriyama was the most beautiful, and he mentioned one which was somehow related to the famous temple at Ise; the dosimeter remained at 2.4 and the evening was clear, the bare trees on the verge of appearing springlike. Here came more lovely silver-white plum blossoms in the dusk. The driver said that he used to party in Kesennuma; he was an angler; the fish were so good there. I did not have the heart to tell him what the fisherman in the orange windbreaker had told me in Oshima—that all the fishing was ruined there for years to come;* the driver must know that anyway.

He said that Koriyama was very quiet at night now.

Rolling up the ever emptier road into the grassier hills, we saw here and there a long straight line of cabbages in the grass, and jade-colored onions growing high. We were all wearing masks on this occasion, the driver for the dust, and the interpreter and I for the—dust. The taxi was hot; my mask made me a trifle nauseous; but I thought it best not to roll down the window. The driver said that after 40 years in service he now drove only part time; he grew vegetables himself, depending on the season; that was his enjoyment.—Did he worry about the contamination? ("Contamination" was the word they all used; oh, it sounded so much better than "radiation"!)—Not at all, he chuckled. "My wife," he laughed, "she told me since it was raining not to go out, but I don't care at all! The government always says: no *immediate* effect on your health! Ha, ha, ha! Every day they announce the level of radiation within the prefecture. Compared with an X-ray check, which is 600 sieverts, that doesn't sound scary at all!"

"Was that sieverts or millisieverts?" I inquired.

---

* The most common fish was (or had been) flatfish.

"I thought it was 600," he said vaguely, "but however strong the contamination is, it's not comparable. We have never thought about these things."*

The yellow-orange sun suddenly winked on our left, sending down silvergold rays upon the dry ricefields. The driver's bald head was as pale yellow as the bamboo leaves in sunset, and he seemed rather happy, the local road winding us under bluish-purple clouds, ostensibly toward Funehiki, and then as the driver mentioned a fine cherry-blossom-viewing spot (the season was still too early), we turned onto National Highway 288.

Because he had been born in 1941, I asked him how he compared the dropping of the two atom bombs with the reactor accident, and he said: "There was a movie I saw as a child, a black-and-white movie, so vividly showing the ruins of Hiroshima and Bikini Island. Now, Koriyama is farther than the 30-kilometer ring; in fact, it's 60; but if there's a hydrogen explosion, I'm afraid that Koriyama may be in the forced evacuation zone. Having reached the age of 70, if we're told to evacuate, I said to my wife, where should we go, Sado or where? When I see the situation of evacuees, they don't seem to be in a good condition! In Koriyama there are three evacuation centers, mainly from the forced evacuation ring around the reactor. Most people are in a hall called Big Palette† which can accommodate 3,000 people. There's also a baseball stadium which holds three or four hundred, and ..."

Night-globes glowed in a roadside restaurant, and then at a gas station. We rolled across the Inasokamatsi River, then down through a cut which was tasseled with golden grass, a grey-green volcanic-shaped mountain projecting itself ahead upon the pale cloudy sky. An old man in a russet robe toiled slowly uphill toward his house, swaying from side to side. Presently we entered the municipality of Funehiki, which began narrow and opened wide with high mossy walls. A convenience store glowed with electric power, as did a traffic light's red eye; and the driver said: "You know Chernobyl? I watched on the news; there was a 90-year-old woman who grows her own vegetables, lives alone and gets sick occasionally but is basically fine."—How inspiring, I thought.—Passing a grocery store and a geisha's glowing likeness, the driver now said: "All these conflicting reports, they make me so anxious I can hardly eat; and when the earthquake came, the ground made the sound *grrrr,* and my wife is really scared now. When the first big shock came, I told my wife to go outside; I was holding a big TV which was not repaired; we lost only that teacup and four tiles in the bathroom."

---

* In fact, 600 sieverts would be about a hundred times more than the lethal dose. One figure for a chest X-ray was 60 microsieverts (= 6 millirems). So the driver was off by a figure of exactly 10 million. Is it a misrepresentation to say that this misunderstanding victimized him? He supposed himself safer than he was.

† My interpreter assured me that this name had no meaning and was therefore representative of the Japanese craze for English words.

Turning away from the route which led to the famous limestone caves, he said: "Where we are now is about 40 kilometers from the plant. If you stay on this road, you will go straight there," at which the back of my neck prickled slightly. "If you pass that point up there, you approach the mountain and then you go down again . . ."

"Have you ever been there?"

"Once. It was a tour. At that time we never thought this could happen."

That afternoon while we were waiting for the rain to stop, the interpreter and I had swallowed our Cold War era potassium iodide tablets, courtesy of my friend Dave Golden, who once purchased a bottle at some gun show. The bright yellow-green, crumbling pills were to be taken only in the event of fallout, said the label. (My tongue tingled for days and I got a rash; the interpreter remained unaffected.) In retrospect I am ashamed that we did not think to bring one more tablet along for our driver. Fortunately, the meter remained at 2.4. A young woman stood bareheaded at the side of the road, her long hair wide at her shoulders as she checked her watch, perhaps waiting for someone to meet her. Would it matter to her health if she did or did not wash the dust out of her hair tonight? Leaving her behind, we entered the town of Tokiwa and stopped at the shrine. The mask had fogged up my glasses so badly that I pulled it off, gratefully inhaling the chilly air. The interpreter and I ascended the stone stairs. Above the wooden-slatted offering box, the immense corn-hued tassel, the size and proportions of a girl-child's skirt, barely swayed in the breeze; it was ferruled (if that is the word) by a tall hexagon engraved with the name of the person who had dedicated it. Climbing the last wooden steps in stockinged feet as tradition required, I peered into the windows of the place and, as usual, saw mostly darkness, interrupted by the reflection of that tassel behind me and by indistinct golden gleamings deep within. My heart revolted at slipping the mask back on, but I did, descending toward the pines and clouds and down to the steep edge of this high place, down the decrepit stone steps, which might have been damaged by the earthquake. I could smell the pines.

Informing the driver that the dosimeter still indicated a safe amount of radiation, I asked whether he would be willing to take us farther.

"Sure," he laughed. "I'll take you to the point where you can't go anymore."

"The radio announces it every day," he remarked. "For the past few days, this area has had a very low level."

So we drove up the road toward Futaba,* the pallid roofs of houses fading into the low oaks. "This area is close to the plant, but they say it is not just distance

---

* Also called Futabacho. Plant No. 1 was situated between the towns of Futaba and Okuma (or Okhumamachi). I never visited Futaba, but my observations of Okuma in 2014 begin on p. 466.

but also geography," he explained. Suddenly we reached a yellow signboard, no more impressive than any sidewalk restaurant's, whose red letters warned: **DANGER: ENTRY IS RESTRICTED 10 KILOMETERS FROM HERE**. We had entered the voluntary evacuation zone. From this point on, the road was quite empty. The next village was virtually lightless, the main exception being the yellow windows of three sidewalk vending machines from which no one had yet pulled the plugs. Another sign for Futaba and an indication that we were still on Highway 288 separated us from the following village. Now it had become nearly pitch dark, although I could infrequently make out the silhouettes of forest ridges.

I told the driver that this excursion was very interesting. He chuckled: "I am ready to cooperate as much as possible, because anyway I don't have much longer to live."

There came another winding stretch. Soon we would arrive at Miyako Oji (20 kilometers from the reactor), beyond which rose a mountain, which would, I hoped, protect us from beta particles and gamma rays. I inquired whether there might be any legend relative to Miyako Oji; the driver replied that there was not. "What's popular in this vicinity is beetles for the kids. They produce them here." As a matter of fact, the interpreter had once purchased a pet beetle for her sons (although whether or not it came from Miyako Oji she did not know); it failed to thrive, I am sorry to say. When we turned right at the junction for Inaki, it was quite dark.—"Most of them are gone, I guess," said the driver.

Warning signs at the edge of the outer ring

And so we came to the inner ring, where the tall signs with their black and red letters interrupted the road, the prefecture proclaiming that further travel was forbidden while the police merely announced that it was restricted. In the darkness beyond lurked a police riot bus, empty or not. With nothing to see and the hazards (including perhaps arrest) unknown, I could not in good conscience ask the driver and the interpreter to go any farther that night, brave though they were. We did take a spin through Miyako Oji, whose houses seemed intact but very, very dark. A lonely white dog trotted up to the taxi, looking up at us hopefully; as we continued on our way it darted crazily back and forth. The driver remarked that the evacuation centers such as Big Palette did not allow any pets, which therefore had to be left behind. Should I have tried to take it to some animal shelter in Koriyama, if there were such a place? Who knew how contaminated the creature was?

"It's the first time that I've seen this here," said the driver. "It's like a ghost town. About 20 years ago in Koriyama, on Christmas Eve a cable fell, and there was a blackout, so black! And this is the first time since then."

As we began to drive away from the reactor I pulled my mask off, and instantly tasted dust in my throat, which made me anxious, because what if the dosimeter might be lying? It was all I had to go on, really.

The old driver said: "What's the most scary around here is the wind that comes from the sea. That's when the radiation comes. In summer, that's when it comes."

A quarter-hour later, at eight o'clock, the dosimeter clocked 2.5 millirems.

THEY DID IT FOR THE NATION

On the following morning, a Sunday, which began chilly and breezy while the dosimeter displayed its predicted 2.6 millirems and the hotel television explained that the overflow trench for the contaminated No. 1 reactor would soon itself overflow, we set out for the danger zone once more. As before, I was wearing my ball cap (a convenient resting place for airborne beta particles), my housepaint-spotted old raincoat (quite inappropriate for appearance-conscious Japan) whose virtues were its hood and its expendability; this magnificent accessory was intended to accompany a disposable poncho whose sleeveless armholes I would seal with masking tape around the raincoat's sleeves; beneath this was an optional sweater of polyester fleece, for the tsunami zone had been chilly; then came my 15-year-old long-sleeved shirt (just broken in; a shame to lose it, but anyhow I had already employed it in chemical experiments), and in the breast pocket of this lived my dosimeter; beneath this shirt I wore another lighter one; and the idea was to pull on my yellow kitchen gloves at the last moment, taping them around the cuffs with masking tape; then came my old bluejeans and underpants, my grubby socks still soggy with tsunami scum, my late father's old shoes, disposable shoe covers at the ready—and, of course, my respirator, guaranteed to filter out 99.97% of all particulate matter, although, since I had bought it at an American hardware store, the label advised me that misuse might cause injury or death. I had brought a second set of all the exotic items for the interpreter (who in due course would inherit the dosimeter). Needless to say, the poncho, gloves, masking tape and shoe covers had not graced my person so far on this adventure; all the other clothes I had been wearing unstintingly, day after day, since I had to suppose that everything I did not store in Tokyo might become contaminated, so why throw away more than I had to? Although I succeeded in showering every day except in Oshima, I doubt that I made for professional success; and the notebook I carried, a scarlet-spined yellow affair emblazoned with a pink-tutu'd ballerina who curtseyed from beneath a cloud of multicolored butterflies, might have been what tipped the scales, causing policemen to snicker softly the instant that my back was turned. Never mind; even in former years, when I was younger and slimmer and needed to dress up for interviews in my one and only business suit, my best achievement had been a look of mild surprise on the interpreter's face, accompanied by this encomium: "You look almost handsome!"

On the previous night's drive to Miyako Oji we had brought our yellow kitchen gloves, respirators, etcetera, but the dosimeter persuaded us not to use them. Moreover, we both would have felt ashamed to protect ourselves so ostentatiously without doing likewise for the driver. In my American imaginings of this final visit to the hot zone I had envisioned a walk of some sort, probably on my own; any taxi driver would have stayed inside the vehicle, with the windows cranked up against beta particles. Just in case someone accompanied me, I had brought double everything. By the time I had packed canned food and waterbags for two, not to mention that pair of disposable backpacks donated to the cause by my friend Jay, my suitcase and duffel bag were full; the two respirators, purchased at the last minute in order to augment and at the crucial moment supplant all the others, rendered the final luggage-zip nearly preposterous. Now of course this does not excuse me from having forgotten the safety of any hypothetical third party; never mind the fact that a sane person might well decline to drive anywhere that such accoutrements were advisable; in short, last-minute logic (and decency) prohibited the interpreter and me from setting forth in any such dress, although we did bring them with us just in case.

And so we were each wearing a medium-quality surgical mask, purchased at a nursing supply store in San Francisco; we offered our new driver, whom I will introduce in a moment, a fresh mask of his own, but he was satisfied with the one he had. Then, as I said, I wore my hat, raincoat (unzipped so long as we were in the car), heavy shirt, light shirt, underwear, jeans, socks and shoes. Upon our return to Koriyama the yellow gloves would go on, the shoes wiped down with a damp cloth before permanent removal to plastic bag (the interpreter insisted on keeping her footwear), and then pretty much everything else listed in the previous sentence—as well as the gloves—also got disposed of in a suitable place—contaminated presents for a contaminated town. The fancy respirators, the backpacks and all the other usable items we gave away to an evacuee at Big Palette that night, before proceeding to the gymnasium for our radiation screening.

As it proved, this day's accrued dose would be no higher than the previous two Koriyama days: 0.4 millirems in 24 hours.* I flatter myself that prudence played a part in this result; I paid attention to wind direction as well as distance, and consulted the dosimeter every few minutes; still, we seem to have had two very lucky days (a statement I intend to retract should I come down with some cesium-characteristic cancer within the next few years). The interpreter later informed me that in the newspapers she had read that the maximum recorded radioactivity suffered by any inhabited place fell 40-odd kilometers north of the reactor: 16,020 microsieverts over 21 days, which worked out to 7 millirems a day; at that rate it would have taken only 66 days to achieve my ceiling of 5 rems.†

First we went to Big Palette. I hoped to find an evacuee who knew how to enter the inner ring without police interference. En route, the driver explained that Koriyama was "the Oriental Vienna," an appellation I never would have imagined. My tongue was still tingling and stinging from the potassium iodide. The driver said: "Well, we have no direct impact from the reactor, but I don't like the rumors."

As soon as we stepped out of the taxi we spied people passing in and out of Big Palette. I stopped a youngish-looking woman who was carrying her granddaughter against her chest. The child and her mother were from Okuma, five kilometers from the reactor; the grandmother hailed from Kawauchi Village, right on the edge of the 20-kilometer inner ring. It was to Kawauchi that we would go today.

---

* In the short run, at least, Koriyama's readings held consistent. Half of the following day was spent in reaching Tokyo; hence, as arithmetic might predict, that day gave me a mere doubling of the Tokyo baseline, or 0.2 millirems.

† This was Iitate Village, which I would visit three years later. A different (one-time) statistic gave Iitate's dose as 9.13 microsieverts [0.913 millirems] per hour—9 times over the safe dose, 21.9 millirems per day, so that the 5 rems would be reached in 228 days. When I went there I finally had a scintillation counter; the highest level it measured was 8.7 micros.

The grandmother said: "We had been helping the victims since the twelfth," the day after the earthquake-tsunami. "On the sixteenth, we ourselves were made to evacuate. It's like I'm seeing a dream. The life is hard. My daughters are all living very close to the reactor, so they lost everything."

She did not want to visit Kawauchi just then, and a man who was going there today preferred to organize his things first, so we hired the driver at the head of the long line of taxis which waited there; he asked other drivers for directions, and they looked at us, and then the driver said we must visit his company's office before setting out. I spoke against this, expecting as usual to be quashed by some higher-up, but there was nothing for it; his boss came out and inspected us, after which he and the driver worked out a price. I required a receipt; the driver prepared it, and it got stamped right there. I said that the journey might possibly take longer than arranged for in our agreement, in which case I could pay more. The driver, unassuming to the point of shyness, appeared disinterested in these details.

I asked about Big Palette. "I hear there are about 2,000 people," he replied. "On the floor they live. It's terrible, their situation," he said as we crossed a bridge, headed toward Iwaki. "Yesterday the television showed the situation. Norovirus* has spread among the evacuees in Big Palette . . ."

Typical living quarters at Big Palette

---

* A kind of stomach ailment.

That night I saw the place, with its long sprawl of cardboard partitions, and apathetic families lying on the floor; it reminded me of a certain homeless camp I once visited, where people lived under a freeway. Big Palette held the inhabitants of eight towns and villages. One man said: "There's nothing to do there. We just have three meals which we're supplied . . ." There was no order here at all, I was told; people grabbed whatever space they could. The young woman who showed me where she slept (in a relatively warm and private place, courtesy of a friend who had gotten there first and bequeathed the spot to her) had worked for Tepco in a clerical capacity. She said that her employers had explained that if a contaminated truck happened to roll by in its progress from Plant No. 1 to Plant No. 2, all she needed to do was slip plastic bags over her shoes.

 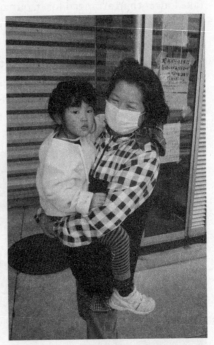

Nuclear evacuees at Big Palette. These people arrived earlier, so they had more living space than was the rule (see previous photo).

Evacuees just outside Big Palette

We were still in central Koriyama when the meter went to 2.7. *"Ehh!"* cried the interpreter anxiously. Feeling a trifle nervous myself, I put on my second best mask, the surgical one which I had worn last night; so that in that department I now approximately resembled my two companions as we took Highway 288 toward Okuma.

When I inquired whether he imagined that Tepco could control the situation, the driver smiled and laughed. "I don't know," he replied. "Well, they seem to be working hard."

We twisted up through the yellow-green hills toward Ono, the bamboos shining in the sun, a man working the soil; that wasn't yet prohibited here, as it already was in Iitate Village, which lay 40 kilometers to the northwest of the plant and hence outside both evacuation zones; it was said that the inhabitants of Iitate would soon be required to evacuate. A young man in long sleeves walked his poodle at the side of the road. I found myself checking the dosimeter more often than usual. The driver was silent. My upper lip sweated within the mask. Coming down into Ono we saw some broken stone along the road-edge which might have had nothing to do with the earthquake, and a few specks of snow on the mountainside. It seemed like such a beautiful place to go hill-wandering. The driver pointed out some *nara* trees: good for growing mushrooms, he said; and a few days later, the news screen on the bullet train from Hiroshima to Tokyo announced that mushrooms in a certain zone near the reactor could no longer be harvested, having exceeded the legal radiation limit.* Nearly everywhere I looked in Ono there were small square garden plots whose vegetables were coming in, in neat rows, young and green; were they poisonous? The sun was strangely warm on my wrists, or perhaps they were tingling from the potassium iodide. "They are farmers," the driver remarked with satisfaction; and I knew that he too must have had rural origins.

We swung onto Highway 349, then left onto 36, toward Tomioka, which had been subject to forced evacuation. That place-name meant nothing to me in 2011; now it has become my personal reification of the Fukushima nightmare; closing my eyes, I can see its weedy half-wrecked central business district and I can hear the creaking and clattering of its shutters in the wind; but of course on this drive to Kawauchi just after the accident the appearance of Tomioka, which lay in the no-go zone, could not yet have deteriorated into something more sinister than the tsunami-earthquake relics which I have already described.

In the center of Tamura City (a valley paved with tiled houses) were many lovingly manicured pines, and behind people's hedges sometimes rose the irregularly phallic boulders so beloved of Japanese gardeners. Above an ancient stele stood a willow not yet in leaf. The convenience stores had not closed. We left Tamura, which was, the driver informed us, a new conglomeration of small villages regathered for certain administrative benefits; I wondered whether the

---

* Around Chernobyl the comparable forbidden country food, which one survivor's grandmother especially missed, was sorrel, "the thing that absorbed the most radiation."

place would remain inhabited. In fact the residents would soon be expelled; none of them would sleep in their homes until 2014. A police car slowly rolled up the hill ahead of us on the quake-cracked road, the cruiser's lights flashing. Then it turned around.—"Maybe he's too close to the radiation!" laughed the driver, and who can say he was not correct? For this is a story about things in which we can scarcely believe, let alone understand.

We stopped to chat with an old man in boots and waders and a fishing cap; across his shoulders he bore one of those long poles from which harvested rice is hung to dry.—"Sorry," said the interpreter; "I cannot understand his dialect." She made out that the ricefields across the road were his; he owned a largish acreage of four *tang,* or, if you like, 1,200 *tsubo.* He said that the farmers could not sell their products now.

"Is it safe here?" I inquired.

"They don't tell us that it's safe."*

Bowing and thanking him, we returned to the taxi.

Old man near Tamura City

---

* This reminds me of the words of a water company president in West Virginia after a chemical spill (II:173): "We don't know that the water's not safe. But I can't say that it is safe."

Here came a vehicle on the otherwise empty road; our driver asked the old lady at the wheel if we could get to Kawauchi. Politely covering her mouth all the while, she said: "You can go." The meter remained at 2.7 millirems.

Now we paralleled the river, beyond whose far edge grew many slender-trunked *nara* trees; apparently they were Japanese oaks. I asked the driver to stop. Greenness was welling up in what had been until not long ago a winter forest. I had a strange, not quite eerie feeling. So beautiful, the green lichens on the boulders! In the cool shade of the cedar trees the ground was so thick with needle-leaves that my steps grew soundless. Sunlight came in low and green on the sides of the trees. An unknown bird whistled its two-toned call over and over. I would have liked to picnic sitting on one of these low fat boulders. Delighting in the cool wind at my back, whose degree of particulate contamination was of course unknown, I strolled across a little bridge toward the pinkish-grey *nara* trees, beyond which rose another wall of cedars. A stand of young green bamboos was growing beside me. Looking down into the jade stream with its white fans and ribbons of foam emerging from each mossy boulder-islet, I forgot where I was, and for a moment removed my mask, which might have been useless anyhow.

We rolled along, and at the side of the road, not long after we had seen some wooden boxes which the driver said were employed for the collection of wild bees, a sandwichboard-type sign unimpressively announced: **ENTRY RESTRICTED BY POLICE**. And so we came into Kawauchi Village, 10 kilometers from the inner ring. The houses were silent. The driver said: "They may have evacuated. This is no good."

On the hillside just off the road rose a pleasant wooden house. Seeing an old man in wading boots performing some chore, I asked the driver to stop again, and the interpreter and I went out to introduce ourselves to Mr. Sato Yoshimi, who said: "I went to Koriyama to evacuate, but just returned today."

"Why did you return?"

"I've been there at Big Palette for about a month, and I just had to look at my house. I'll go back to Big Palette today."

"What made you choose to move there?"

"People here were told: If you're within 20 kilometers, you have to evacuate. If you're within 30 kilometers, you might want to. So, to be on the safe side, this village was specified for evacuation."

I did not completely understand this; but who precisely had specified the evacuation, and how voluntary it was, might not be something which the broken-toothed old man cared to spell out. His white mask hung down between chin and neck.

"How did you experience the earthquake?"

"I was at the site," he answered. "I was working at the Number Four Turbine. I've been working at the reactor for more than 30 years."

"Was it a good job?"

"Well, before the accident I enjoyed it. You never imagined . . ."

"And then what happened?"

"It was about 2:30. Within the building the tremor was terrible, and the lights started to fall. Lots of sand and dust—you couldn't see where you were going. I was within the controlled area, where you have to wear protective gear specified by Tepco, and you have your own dosimeter."

"Do you still have it?"

"I left it in the reactor building."

"Did you see the tsunami?"

"I immediately evacuated before it came. From the Number Four building I went on foot with my colleagues. There was lots of water leaking from pipes since the ground had sunk. You work in a team—six people. All of us evacuated together. There is an office four kilometers from there. We checked in there. When everyone had arrived, we were told to go wherever on our own responsibility."

His employer was a certain subcontractor of Tepco, called Nito Resin. They were still paying him, he said; he had received last month's salary.

"How long do you suppose you'll be living at Big Palette?"

"I don't know. It depends on the radiation here. Unless the restriction is lifted, I don't think I can come back. Here it's pretty low, 0.5 or 0.6 millisieverts.* My daughter is within the 20-kilometer limit. So she and her mother went to see her home. I think they can enter for a short time."

A brown creek flowed beneath the cypresses at the edge of his steep lot. Across the road lay his garden: daikon, green onion, cabbage, long beans. I wish I could tell you whether anyone ought to have eaten what he was growing.

Bowing our goodbyes to him, we continued on down the road, while he, bending painfully there in the driveway, slowly returned to watering a plant with his turquoise plastic pitcher, while a young child wailed and whined inside his house, my dosimeter still pleasantly at 2.7. When we reached the fork we took the

---

* Per hour. I suspect that he meant *micro*sieverts, or 0.05–0.06 millirems. (This would be about 1.4 millirems per day, or about 4 times what my dosimeter was reporting for Koriyama. At this rate, a resident of Kawauchi would have accrued 5 rems in about 9 years and 9 months.) Had he actually been accruing 0.5 millisieverts an hour, every two hours he would have received the International Committee on Radiological Protection's maximum recommended *yearly* dose.

righthand turning as he had advised, while the driver said: "Normally, reactor workers die at an early age, so I'm surprised, frankly, that he's still alive. One of my friends was working there, and he wanted to retire. He opened a noodle shop and died very soon."

"How old was he?"

"Forty something."

"Was it cancer?"

"I don't know the details."

This anecdote said more about the driver than it did about the reactor, or nuclear power. Anyhow, one person is a small sample. Passing more dry ricefields, my forehead burning and itching, perhaps from an insect bite, we passed two dogs running loose outside the Kawauchi municipal office and reached the inner ring where a line of police stood in their blue vests with the reflective yellow stripes on them, their white masks covering them from their chins to the bridges of their noses, and their white hats firm and straight over their eyes, their white-gloved hands open at their sides and their boots shining. They prohibited us from going farther, so I had the taxi driver turn right and park a block away, on a street where locals drove in and out of an unmanned checkpoint as they pleased, lifting the flimsy barrier aside. These people were always in a hurry—a great hurry, in fact. Whenever the interpreter and I waved them down, even the outbound ones would always say, in violation of their famous Tohoku politeness: "No time!" Invariably, they were headed for Big Palette. I strolled into the forbidden zone, just so that I could say I had done it. The interpreter took a cautious step or two behind me, then stopped. The driver sat with the windows rolled up. Every time I looked at him, he anxiously started his engine. Should I have insisted that he also enter? My dosimeter had not registered any recent increase; regarding gamma rays the situation seemed safe enough, and perhaps this story would have been more dramatic had I been more pushy, but then again perhaps not, for what would we have seen but more empty houses, and then quake and tsunami damage, and finally the reactor plant, which from drone photographs in the newspaper resembled any number of muddy construction sites? I think that the driver would have done it if I had asked; as for my loyal, courageous interpreter, she said simply: "I will follow you." Perhaps she and I should have suited up with respirators, yellow kitchen gloves and all the rest of it, and then walked toward Plant No. 1. Honestly, I lacked the ruthlessness to ask it of her. Or I could have set out by myself, leaving the two of them to wait for me there. Why didn't I do that? Perhaps I was afraid and didn't admit it to myself; but I believe that I simply couldn't see the point.

Inner ring checkpoint in Kawauchi

The birds were singing, the plants were growing and the trees were coming into flower. It was very warm now. Moss grew on a wall, and in the deserted houses the curtains were all drawn. If you can, try to see those curtained houses and the shadows on their silver-ringed roof tiles, the blue flowers in someone's back yard, which like the other lawns there still appeared decently trimmed, probably due to the coolness of that season.—In due course I will tell you how Kawauchi looked three years after this.—At the side of another house, a few potted houseplants had begun to wither, but the others still looked perfect. Perhaps more people returned home from Big Palette than was generally imagined.

Behind an outer door, an inner sliding partition was wide open. We called and called, but no one answered. I informed the checkpoint police, since last night's taxi driver had remarked that burglars had begun to take advantage of the evacuation.

In the shade of an old wooden house, several bicycles leaned neatly beside clean shovels. A line of sandbags, perhaps tsunami protection, followed the house around the corner.

What is there to say about this place, morning-shadowed and bird-songed, the meter at 2.7 millirems, the shadows of power wires gently dancing across the ribbed concrete facade of a workshop, a small black beetle crawling on a sandbag?

Here is another house for you: a wheelbarrow neatly tilted on its scoop-edge, handles high against the wall, wheels outward, then some empty gardening pots, a shovel leaning in parallel with the wheelbarrow, and clean sticks in parallel with them, then a small broom already fallen into the moss.

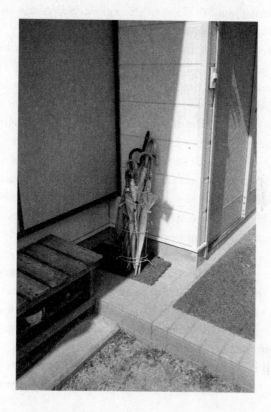

From the main checkpoint a bus emerged, then a truck, then three cars, the police waving them all through with their white-gloved hands before they re-closed the barrier, and these vehicles all headed back in the direction of Kori-yama. Then a man on a motorcycle approached them from our side.

"Unless your purpose is very strong I cannot allow you," the policeman said.

"But I have a brother inside. Can I go another way?"

"You may be able to advance and go a little farther," said the policeman.

So the motorcyclist proceeded to the unmanned checkpoint which the interpreter and I had breached. Later the taxi driver, who had spoken with him, remarked that this fellow had complained of a burning, tingling sensation, which of course is one of the first symptoms of massive radiation exposure. Most likely it was psychosomatic, or he had some sort of allergic reaction; nobody we asked in Koriyama, even at Big Palette, had heard of anybody getting radiation sickness.

Then the driver summoned us; he had discovered an actual inhabitant: bearded and greying, with a very red workingman's face, in a blue slicker and cap; he must have been about 50. He wore green gloves and a mask and green boots. The metal grating of the Showa Shell service station was only half raised. He stooped just outside, hosing down a patch of pavement. He worked unceasingly while he spoke with us; nor would he allow us into his house next door, in whose second-storey window the drapes parted for an instant and a lovely feminine hand flickered, folding a towel over the curtainrod; his wife or daughter must be doing laundry indoors. The drapes closed again. The workingman said: "This is the stay-at-home area. This is my area. We'll leave soon. There's a cat we left to stay inside, and we felt so sad not to let it out. I'm the head of the fire department here, so I come back every day, and every day I check the radiation level at the village office. Today is 0.38 millisieverts.* On the seventeenth, everyone evacuated . . ." He worked on, never stopping until I made him a present of my best respirator, at which point he did pause to bow deeply, then hurried back to work.

We saw one or two more carloads of people, and they always roared off toward Big Palette, answering all questions with the remark that they were too busy. The driver said: "They believe it's too dangerous here, I think."

With the cool wind still at my back and the sound of the brook louder than almost anything else, I inspected a very young bird on the grass, rust on a guardrail, manicured pines, then gazed over blank empty pavement. I went up one more driveway and rang the bell. The chimes echoed on and on; the door was locked.

For some reason, what I remember most is the bicycles leaning neatly up against the empty houses which shaded them.

Every time I looked at him, the driver eagerly started the engine. He reminded me of the forlorn boy who must stand on the snowy sidewalk just outside the hot springs near Sendai, ready to bow if any guest comes in. Finally I asked how he was.—"I'm not really concerned," he said, "but somehow I feel uneasy."

"What makes you the most uneasy?"

"I see the cars but no people."

Taking pity on him, I told him to commence our return, driving very slowly down the smooth pavement to the fork of highways 399 and 36, and then as the

---

* Per hour, presumably. And once again he must have meant microsieverts. This would be 0.038 millirems per hour—such a low figure that this man should have been moving his family back home instead of fleeing to Big Palette.

road began to rise back up into the hills, but long before we reached Mr. Sato's home, I made the driver stop again, for I now perceived one more chance to accomplish my journalistic vulture-swoop, for here again was human life—namely, a middle-aged couple wearing those nearly useless paper masks over their mouths, and striding out of their house and down the gravel driveway to their separate cars. I rushed to halt them, and the interpreter bowed with her best politeness, requesting the favor of five minutes, just five minutes, but the wife said: "We don't have time. This is the first occasion that we have checked our home since we evacuated to Tochigi."—"How long ago was that?"—Shedding all remnants of that celebrated Tohoku patience and politeness, she cried: "We don't have time; we don't have time!"—at which they ran to their cars without bowing goodbye, the man sweating around his mask, and off they drove at nearly reckless speed, up Highway 399 toward Koriyama and then Tochigi.

The driver remarked that they seemed afraid.

Reascending Highway 399 past terraces and plum blossoms, my wrists stinging strangely, no doubt simply from sunburn or that potassium iodide, we proceeded toward Koriyama; now we were descending the mountainside, a brown stream glinting white in the sun, at which time the dosimeter reading increased to 2.8 millirems. I said nothing. Looking into the car mirror, I saw the sad bewildered fear in the driver's eyes.

"My eyes have been pretty watery for the last two or three days," he said. "Is it related to the radiation?"

This gentle, stolid rule-follower, who had been born in a traditional thatched-roof house and who was proud of his 86-year-old mother's health, who had prepared that receipt for me in advance and therefore firmly refused payment for the extra two hours which my loitering had required—never mind the hazardous duty bonus I tried to give him (he did take a fraction of it), struck me as one of those innocents so useful to authority everywhere. I asked him whether he knew what radiation was, and he said: "I don't know. Does it evaporate? Is it a liquid?"

"Should Tepco be punished?" I inquired.

"It was the government's policy," he said loyally. "They did it for the nation."

3: Cherry Blossoms

On the day that I arrived in Hiroshima, the Nuclear and Industrial Safety Agency classified the reactor accident as a Level Seven—comparable to Chernobyl. One agency said 370,000 terabecquerels had been released so far. Another said 630,000 terabecquerels. I figured that nobody knew and everybody lied.

I asked the wide-faced taxi driver which he considered worse, the reactor accident or the detonation over his city of the atom bomb, and he replied: "Of course the nuclear bomb! It instantly killed more than 200,000 people!"*

I mentioned that the Tohoku people appeared to know or care little about what happened at Hiroshima, to which he replied in his reedy voice: "Of course. More foreigners visit the museum than Japanese. At that time, I was three years old. One day before, we were ordered to evacuate because Hiroshima was a military capital and in danger." He laughed, quite cynically and bitterly I thought. His mother took him to the countryside, but on the day after Little Boy was dropped she returned to Hiroshima to look after relatives, which was why she—lucky woman!—became eligible for atomic victim status. "She showed no symptoms, but when I was 14 she got recognition. Myself, if you have that *hibakusha* health book, if you're a victim, that means that no one wants to marry you, so I didn't want to get one."

He went on: "Those who used to live close to the dome"—which is to say, the hypocenter† of the explosion—"they hide it, since they are discriminated against."

"How many years did it take for the radioactivity to go away?"

"For that I'm not sure, but in 1945 they said that for 50 years no plants would grow, and soon weeds came."

"Do you think that the reactor accident at Fukushima could affect people here?"

"I think it's rather irrelevant to me. I won't be affected.

"You're going to the museum, and you'll see that the atomic bomb gives you burns and hair loss," he explained wisely, and so we pulled up there, on the edge of the Hongawa River, among the pinkish-white clouds of cherry blossoms.

I visited the Hiroshima Peace Memorial Museum, and my heart grew as brown-grey as a salinized ricefield. The torn, stained rags of the summer uniform and chemise which the 13-year-old schoolgirl, Oshita Nobuko, had sewn herself, oh, yes, these flattened, faded bloody tokens, weren't those enough to see? Just as one can find on display at that museum a lightbulb painted black except for a neat ring of transparency at the base, in order to decrease, however slightly, the probability that the Allied bombers could locate nighttime targets, so I could see of various radioactive issues, matters and agendas what little there was to see; and I have told you what I saw. What did I see? What did I know?

At Hiroshima my dosimeter registered 0.2 millirem per 24 hours—twice

---

* A wall display at the museum stated that 140,000 had died by the end of 1945. A book published in my country, lumping Hiroshima and Nagasaki together, placed the total at 220,000 souls.

† Little Boy detonated in midair. This spot was the epicenter. The hypocenter was the place on the ground immediately beneath.

Tokyo's background.* At first I thought that I had found some artifact of the bomb, but that American dosimetrist later remarked that this reading was probably within the pre-atomic norm for that area. Six years later, returning with a spot-measuring device,† I found the levels at Ground Zero to be lower than the supposed American average.‡

Across the river from the Fuel Hall

On the bench across from the ruined Fuel Hall§ with its Atomic Dome within, a pigtailed child sat upon her young mother's lap, giggling and rubbing noses with her, the blue sky glaring through the blank window-holes in the brick.

Then the petals began to rain down, losing themselves upon the whiteness of the granite flagstones, floating down onto the long dark hair of two young women who sat drinking coffee together, turning their faces toward each other.

---

* New dosimeter readings in 2017 were the same. I also measured the levels in Kyoto, which again were twice Tokyo's. According to Dr. Ed Lyman of the Union of Concerned Scientists, variations of 2 or 3× are common; the cause can be change in altitude, bedrock type or local choice of construction materials.

† The "pancake frisker" scintillation meter described on p. 397.

‡ I refer to a mid-1990s figure for the average American yearly dose, which works out to 0.41 microsieverts an hour. My readings around the Atomic Dome varied from 0.18 to 0.36 micros. See the table on p. 245 (header **3**) and p. 247 (**6**).

§ So labeled in the English-language sign. Accordingly, I resist the interpreter's emendation of "Industrial Promotion Hall."

February 2014

HARMFUL RUMORS

1: Safe Behind a Wall of Ice

For the past three years my dosimeter had sat silently on a narrow shelf just inside the door of a house in Tokyo, upticking its final digit every 24 hours by one or two, its increase never failing—for radiation is an attribute of time. Wherever we are, radiation finds and damages us, imperceptibly at best, ever varying within the bounds of natural fluctuation, measuring error and other such expressions of inconsequentiality. Sometimes people age suddenly, oh, yes; and every now and then come nuclear disasters. You might not call *those* imperceptible, but before the end of 2011, my American neighbors, who tended toward surprise at the notion that this or that forgotten crisis might not have been solved, lost sight of the accident in Fukushima. A tsunami had killed hundreds or thousands; yes, they remembered that; several also recollected the accompanying earthquake, but as for the containment breach* at Nuclear Plant No. 1, that must have been fixed, they were pretty sure—because its effluents no longer shone forth from our national news. In 2013, two out of the nine evacuated towns had formally reopened (when I saw them in 2014, they still looked abandoned). Meanwhile my dosimeter accrued its figure, one or two digits per day, more or less as it would have done in San Francisco—actually, a trifle more. And in Tokyo as in San Francisco, people went about their business, except on Friday nights, when the stretch between Kasumigaseki and Kokkai-Gijido-mae subway stations—half a dozen blocks of sidewalk, commencing at the anti-nuclear tent which had already remained on this spot for more than 900 days, and ending at

---

* In the words of Mr. Takashi Hirose, the author of *Fukushima Meltdown:* "There had been hydrogen explosions in the concrete building where #1 and #3 reactors were housed, another explosion inside #2 reactor's containment vessel, an explosion in the spent pool of #4 reactor, and the situation in general was in shambles—all things the reader well knows."

the Prime Minister's lair—became a dim and feeble carnival of pamphleteers, Fukushima refugees peddling handicrafts, etcetera, their half of sidewalk demarcated by police barrier railings, and in the street a line of officers wearing reflective white belts and double white reflective stripes resembling suspenders on the backs of their dark uniforms as they slouched at ease, some with yellow battery-powered bullhorns hanging from their necks; and at the very end, where the National Diet glowed white and strange behind other buildings, a policeman began setting up a microphone on a monopod, then deploying a small video camera in the direction of the muscular young people in their Drums Against Fascists jackets who now at six-thirty sharp began drumming and chanting: *"We don't need nuclear energy! Stop nuclear power plants! Stop them, stop them, stop them! No restart! No restart!,"* at which the police assumed a stiffer stance, with their hands on their hips or at their sides, the drumming and chanting almost uncomfortably loud, commuters hurrying past us along that open half of sidewalk between police and protesters while a plainclothesman with a white armband departed the videographer's presence, then strolled moodily behind the other officers in order to convey some private matter to the police captain; and another file of people, evidently all disgorged from a single train or subway car, walked past, staring straight ahead, one man covering his ears. Finally a fellow in a shabby sweater appeared, and murmured along with the chants as he too passed round the corner. He was the only one who appeared to sympathize; hardly any others reacted at all. The drummers were banging away as if for their very lives, swaying like dancers, staring forward like competitive athletes on the home stretch, raising clenched fists, looking blissed out and endlessly determined. Their musical, exuberant defiance slightly uplifted me. After awhile, I retraced my steps, walking the open stretch of sidewalk toward my subway station, and was astounded to see how that listless scattering of the half-seen had become a close-packed disciplined crowd. There must have been 300 and more. They chanted and raised their hand-lettered placards. It was the last night of February 2014. Perhaps in another three years' worth of Fridays they would still be congregating here to express their dissent,* which after what had happened at Plant No. 1 must be considered pure sanity itself; all the same, they were hurried past, overlooked, and left to chant in darkness while the dosimeter accrued another digit. Another uptick, another grey hair—so what? With radiation as with time, moment by moment there may indeed be nothing to worry about.

I was happy to see that dosimeter again. By the time I returned to Japan, the

---

* They were. On, for instance, March 9, 2017, between six and seven hundred turned out.

interpreter had recalibrated it from millirems to millisieverts,* the latter being her country's more customary unit of poisonousness, so that the number of interest was now preceded (so long as the radiation remained subacute) by a decimal point and zeroes. Why not? The thing was hers now; I had given it to her, because she had to live here and I didn't. It was the best I could do for her. I still deplored its inability to detect beta waves, given that by July 2013, and very likely sooner, Tepco had found *high levels of strontium and other radioactive substances that are emitting beta rays in groundwater from a well at the port* of Plant No. 1.† (The concentration of these poisons was a hundred times greater than the legal maximum. And by the way, the well was only six meters away from the ocean, but why be a pessimist? The ocean can swallow anything.) Fortunately, the dosimeter could stay occupied in measuring such gamma ray emitters as cesium-134 and cesium-137, the accident's commonest soil pollutants, which in that same happy month were *both about 90 times the levels found Friday.*

For the rest of that year the news kept getting better and better. *JDC Corp. discharged 340 tons of radioactive water into the Iizaki River . . . during government sponsored decontamination work.* Well, we all make mistakes. You see, *the company had not been aware that water from the river would be used for agricultural purposes, the sources added.* Ten days later, *Tepco now admits radioactive water entering the sea at Fukushima No. 1 . . . fueling fears that marine life is being poisoned.* Come to think of it, the dangerous hydrogen isotope called tritium† had already been detected in the ocean back in June, at twice the permitted levels and climbing. Tepco would fix that, too, no doubt.

In August, the Nuclear Regulation Authority, which thus far had treated the leaking as a Level One, in other words an "anomaly," now prepared to recategorize

---

* A millisievert ("milli") is a thousandth of a sievert. A microsievert is a thousandth of a milli, or a millionth of a sievert. My customary daily dose in Tokyo or Sacramento was about one micro. One millirem equals 10 microsieverts.

† In 2010 an anti-nuclear organization asserted that a "reactor site may have anywhere from two to 20 miles of buried and underground pipes intertwined beneath the power plant property." The pipes at Daiichi were probably compromised by the earthquake; in addition, groundwater appeared to be flowing into the lower levels of the reactor buildings, and then back out again into the water table and thus to the sea.

† "The necessary ingredient of a boosted fission bomb." Hence this substance was manufactured at the Savannah River Plant in South Carolina for 35 years. Written T or 3H. A hydrogen isotope whose atomic weight is about 3 instead of 1, hence the name. "Easily permeates most kinds of materials including concrete and many grades of steel . . . Clinically shown to be more effective at damaging and destroying living cells even than gamma rays." This last was disputed by pro-nuclear sources. For instance, a certain American nuclear scientist said: "The good news about tritium is that even if you inhale or ingest an awful lot, it is going to flush out of your body . . . Just have a few beers and you're done." Meanwhile, my old ecology textbook, copyright 1971, advised that given sufficient reliance on nuclear power, tritium "could contaminate the entire global hydrological cycle." The half-life of 3H is 12.5 years.

it as a Level Three, "a serious accident." Soon *The Japan Times* was calling the situation "alarming" and speaking of trillions of becquerels of radioactive matter, which is to say *about 100 times more than what Tepco has been allowing to enter the sea each year before the crisis.* Tepco now estimated that 20 to 40 trillion becquerels of tritium *may have flowed into the Pacific Ocean since May 2011.* Fortunately, *the size of the release is roughly in line with the allowed range of 22 trillion becquerels a year,* so why worry?

To prevent No. 1 from exploding again, and maybe melting down, Tepco thought only to cool it with water and more water, which then went into holding tanks, which, like all human aspirations, eventually leaked. As one anti-nuclear killjoy remarked, *Of course, a reactor running out of control will not be put back in order just by dumping sea water over it . . . What [this] achieved was to destroy all the reactors, and to flush . . . the radioactive substances which were inside the reactors out . . . This is the worst case.*—No doubt Tepco meant well, and I for one believe that the worst case could have been even worse.

By September, South Korea had banned the importation of fish from eight Japanese prefectures. And in February 2014, when I set out for Japan, the cesium concentration in one sampling well was more than twice as high as it had been the previous July. A day after this report was issued, the cesium figure had to be more than doubled yet again. As for strontium levels, Tepco confessed that it had somehow underreported those; they were five and a half times worse than previously stated.

Three hundred tons of radioactive water now entered the ocean every day. Cesium concentrations on the surface appeared to be greatest (41.5 becquerels per kilogram) from latitudes 36 to 39 inclusive, where the Kuroshio and the Oyashio currents kissed, while the most cesium-tainted planktons were found at latitude 25. There was a scientific explanation, which the newspaper headlined as follows: **Cesium levels in water, plankton baffle scientists**.

In short, the marine environment was being poisoned to an unknown and almost certainly underreported extent. As for the land, it may suffice to tally the number of nuclear refugees: 150,000.

(Did you ever wonder why they couldn't just shut down their problem? I quote a Ph.D.: *The products of the fission reaction continue to generate thermal energy as a result of radioactive decay. There is no way to stop this process.*—By the way, he assured us: *The safety of [pressurized water reactors] is comparable to that of [boiling water reactors]. Both have excellent safety records.* As for a loss-of-coolant accident, which is what had happened at Plant No. 1, *the prospect that several independent systems would simultaneously fail is highly unlikely.*)

Another hilarious little anecdote: Nearly 2,000 of the poor souls who toiled for Tepco in the hideous environs of No. 1 had somehow taken a dose of more than 100 millisieverts. (Bear in mind that a maximum of *one* millisievert per year for ordinary citizens is the general standard determined by the International Commission on Radiological Protection. A hundred millis per year increases one's fatal cancer risk by 0.5%. A hundred millis in a week was the emergency exposure limit for nuclear workers in Japan.)* But the tale had a happy ending: *The workers will be allowed to undergo annual ultrasonic thyroid examinations free of charge.*

The total cost of the disaster now approached 100 billion yen.

As the Japanese government used to often advise us, there was "no immediate danger."† And important plans were being drawn up to save the world. *Tepco plans to isolate the area by injecting liquid glass into the soil and building water-proof walls . . . by early October.* I never heard more about that. The next step would be to build an electric-powered wall of ice around Plant No. 1, so that no more groundwater could flow in and get contaminated; this magic defense would only cost 30 or 40 billion yen, and hopefully no other earthquake or tsunami would strike that place until the radioactivity had subsided.

How long might that be?

As you are probably aware, the half-life of a radioactive element is the time required for half of its atoms to decay into something else. An area polluted by radiation will thus remain dangerous for not one but several half-lives, until the poison has mellowed into something approximating the unpoisoned state. One Japanese official† proposed to me that when radioactivity has fallen to a thousandth of its initial strength, it may be considered a safe equivalent to normal background radiation.

Dr. Edwin Lyman of the Union of Concerned Scientists sounded less than thrilled with any such standard, saying: "Is this adequate for purpose of rehabilitation? One of the issues is, if you're talking about returning land to unrestricted use, do you clean it all the way to background level or not? If not, you're increasing people's risk. So it's a value judgment. I would say offhand that it's not reasonable to have a different standard." In short, reduction to one one-thousandth remained unacceptable to him. But since I liked the Japanese official, who gave me many multicolored printouts, let's call a thousandth good enough.

---

* The authorities in their infinite compassion were now considering raising this ceiling.

† This comment was frequently uttered in press conferences shortly after the accident. In 2013 the interpreter and I used to laugh about it. Takashi Hirose wondered aloud: "The stock phrases used by both Edano [the Chief Cabinet Secretary] and the TV 'experts'—'no immediate effects to health'—mean what? That radiation effects start one year or 10 years later?"

† Mr. Kida Shoichi, in Iwaki. His interview is discussed below.

Very well then. Take 1,000 and divide it in half. Now you are at 500. Divide this in half again, and repeat the procedure until you arrive near 1. You will find that 10 such halvings are required to reach a quotient of about 1.17. So 10 half-lives must pass to reduce a thousand radioactive atoms to one.

The half-life of tritium is 12 and a half years. Therefore, 10 half-lives will take 125 years. If the reactor sites somehow stopped polluting the ocean on the day I wrote this, it would be well over a century before any tritium absorbed in, say, a clamshell had become harmless.

Tritium is, unfortunately, one of the shorter-lived poisons in question.*

In point of fact the radiation at Plant No. 1 had only to decay to a level at which robots and protective-suited humans could handle the fuel rods, which would be conveyed to some storage dump. One former Tepco worker[†] opined that the ice wall might need to operate for merely 10 or 20 years. Mr. Yamasaki Hisataka, a roundfaced, delicate, slightly corpulent man with bangs, was one of the original members of an NGO called No Nukes Plaza, which dated from 1987, the year after Chernobyl. On my arrival in Tokyo I asked him: "At this point is the Fukushima situation the same, better or worse?"

"The same. I wouldn't say it's worse."

I said nothing. Looking at me, he then said, "The radiation travelled through the Pacific. As a result, it has reached the American west coast. There's no direct danger yet, but in the future who knows what might happen?"

"What do you think should be done?"

"Not to allow them to discharge anymore. Unless we stop it now, it's going to get worse and worse."

"Do you think this ice wall is practical?"

He laughed. "No. Concrete or some other permanent barrier would be better."

Did I mention that Japan's reactor-studded islands were entering one of their cyclical periods of earthquake-proneness? One doomsayer wrote: *There is simply no doubt that the Great Tokai Earthquake will come in the near future*, and the nuclear disaster which that might cause at the Hamaoka Nuclear Plant could ensure that *the very center of Japanese society would be annihilated*.

This is why Tepco's wall of ice resembled the fairytale of Sleeping Beauty, in which it was possible to wall off time with something permeable only to the brave.

---

* See Table 1, p. 537.

† Whose interview begins on p. 366.

THE CONVENIENCES OF IWAKI

A Tokyo coffee shop girl told me that she had worried about fallout just after Plant No. 1 exploded, but she did not worry anymore. She must have been safe, because she thought she was.—Would she think likewise if she lived 40 kilometers from Plant No. 1?

The latter distance haunted my mind because in 2011, when the authorities had drawn that pair of rings centered at No. 1, the 40-kilometer circle demarcated the voluntary evacuation zone. The 20-kilometer circle marked the restricted or "forbidden" zone. Only one-third of the original area remained restricted; could it be that my pronouncements about 10 half-lives were too pessimistic? As the work of decontamination progressed, and wind patterns grew better known, the two circles were abolished, and replaced by specific areas of irregular shape.—Once I decided to base my explorations near the 40-kilometer mark, the most convenient place appeared to Iwaki.

According to the encyclopaedia, this was a *city in southeastern Fukushima Prefecture . . . created in 1966 with the merger of . . . five cities,* including the original Iwaki. *The complex of cities prospered with the opening of the Joban Coalfield in 1883. Mining ceased in 1976 . . . The Onahama Port district is Iwaki's industrial center . . . Pop: 355,812.* Well, that encyclopaedia was 20 years old. Ten years ago, so I was told, Iwaki had been by area the largest urban entity in Japan. Now it was still the third largest, so there ought to be plenty of hotels.

On the map of Tohoku, Iwaki looked to be isolated (an injustice to its sprawl), and almost but not quite on the coast, hence perhaps moderately safer from tsunamis than the localities I had seen in 2011—although the tidal wave had killed more than 300 Iwakians. The cartographers rendered it in larger type than the row of villages to the north: Tomioka, where I would soon experience several radioactive idylls; Okuma, where that fall I would actually look upon Nuclear Plant No. 1; not to mention Miyaoji and Namie.

Being closer to the 50- than to the 40-kilometer circle centered at No. 1, Iwaki never got annexed into the exclusion zone, although the mayor had issued a voluntary evacuation notice for the districts of Shidamyo and Ogi in Shimo Okeuri, Kawamae. Now all that could be repressed behind a wall of ice.

Embarking on the Super Limited Hitachi Express, which was also known as the Super Hitachi 23 Limited Express, I traveled northward from the icy snowpatches on the walkways of whitish-skyed Tokyo, where many householders had their flowerpots out, and the blooms seemed none the worse. Brown ponds reflected the ricefields. There came a few patches of snow in shady alleys, then Ishioka's recapitulation of the Japanese architectural monoculture—no sign yet of anything strange or broken as had been the case last time, but of course even in those days I had not noticed anything sinister until Koriyama. We stopped in Mita, with the dosimeter steady at 0.271 accrued millisieverts. I saw a wall of bamboo just before Katsuta, a farmer leaning on his staff just before Tokai, a man bent over a long irrigation pipe in a brown ricefield, and then the complex agro-industrial metal skeletons of Hitachi. After our next stop I got my first glimpse of the sea, which was ultramarine across a straw-colored rectangle of grass; the pines stood outspreading; no tsunami had uprooted them, and the radioactivity must be negligible. White cranes were wading in a grey-green river over which our express rushed, and a raptor wheeled low over the water, almost hovering.

Here came Nakoso, still 70 kilometers south of No. 1. Well before Izumi we drew away again from the sea. We paused at Yumoto with its eponymous spa, set within low wooded hills. And presently I arrived at Iwaki Station, three years after what is now called the Great East Japan Earthquake of 2011, and six months after the newspaper proclaimed: **Openings of Iwaki beaches offer semblance of normalcy**.

2: Harmful Rumors Defined and Combatted

At 3:36 in the afternoon of March 12, 2011, when its exposed fuel rods reacted with the air's hydrogen, Reactor No. 1 had exploded. In Iwaki the radiation achieved its maximum at 4:00 in the morning of the fifteenth: 23.72 microsieverts per hour, or about 380 times Tokyo's average background exposure. The municipal authorities of Iwaki distributed iodine tablets to residents under 40 years old or pregnant. For the others, after all, there was "no immediate danger." Radiation levels rapidly dropped to a mild 0.2 microsieverts per hour. During my stay in Iwaki, the dosimeter almost invariably accrued its single daily microsievert, which is to say a mean 0.042 micros each hour.

Disembarking from the Super Hitachi Limited Express on that chilly afternoon, I asked the girl at the tourist information office which local restaurants she would recommend, and she replied that Iwakians had always been proud of their seafood, but due to the nuclear accident, unfortunately, fish now had to be imported.

From one of the maps she gave me, I saw that the international port was a few kilometers south along the winding Route 6. Several beaches lay east, and then Route 6 followed the shoreline north toward the reactors. I could not tell where the current exclusion zone began.

As it turned out, the hotels were nearly always full; one of them smelled of cigarette smoke and was only for decontamination workers; I asked the receptionist at another why she was so busy and she laughed, "Tepco!"

In spite of those reminders of No. 1's nearness, Iwaki strove to present itself as safe, and maybe it was. Contrary claims were considered to be "harmful rumors." *Even after one week from the earthquake, harmful rumours on the nuclear problem kept blocking the distribution of goods to Iwaki* . . . In other words, for that week certain fuel and grocery trucks refrained from making deliveries here, for fear of contamination. Moreover, *harmful rumours due to the accidents at nuclear power stations degraded the status of Iwaki local products tremendously. To regain its reputation, Iwaki City has participated in more than 50 events held in Tokyo metropolitan area* . . .

Thanks to such efforts, Iwaki had nearly defeated those harmful rumors. Most people's fears, if they had not yet shed them, seemed to have been buried behind a wall of ice.

THE BLACK BAGS

Consider my first taxi driver, who was roundfaced, ingenuous and patient, smiling often, wrinkling up his cheeks. Although his hair was still black, he had entered middle age.

"How has life been here since the accident?"

"No change!" he laughed. "I'm not a person who's very sensitive. I'm kind of dull. I just think this area is safe. Please forgive me for that."

"What was your experience when the tsunami struck Iwaki?"

"There was no tsunami exactly here. Of course it was terrible on the seashore, but I had to be at a different place for my work, so I really couldn't imagine."

"And what did you think when you heard the radiation warning?"

"Even in Iwaki most people tried to flee, but I didn't mind.* And I have an elderly person at home, so I couldn't leave anyhow!" he laughed.

I thought he would be a perfect conductor to the exclusion zone.

"How far north can one go nowadays?"

"A little more than an hour. Then you'll see police officers stationed in the road."

"In your opinion, is anyone to blame for what happened, or was it just bad luck?"

---

* Such was his cheerfully understated Tohoku stoicism. As a matter of fact, his house had been damaged.

"Just bad luck."

"How do you feel about nuclear power now?"

"Ah! That's difficult! Since we use electricity . . ."

First I asked him to drive south, to the international port of Onahama, so we rolled through the sprawl infested by France Bed, Slumberland, Lawson, Daiyu 8, Big Boy, Rabbit and ABC Mart. On the way he indicated some barrackslike temporary housing for nuclear evacuees. In general, he explained, he felt cheerful and hopeful, "because the taxi business is now thriving due to the nuclear power situation. Those who work for Tepco use taxis. If I were a fisherman I would be suffering a lot, but that's not the case for me."

"How do you define radiation?"

Smiling, he replied: "My image is that it causes thyroid cancer," and he helpfully indicated his neck.

"What's your favorite thing about Iwaki?"

"Commodity prices are low, since pay is low.

"Here the tsunami was terrible," he now remarked. "Some ships came onto the land." But when we arrived at Onahama, he said, "Maybe just two big ships were on the land. The vessels are all repaired. There was some souvenir shop that has not yet been repaired. And this is under construction now," he said, pointing leftward at a fence. "It used to be . . . I can't remember."

"Do people here still eat local fish?"

He chopped the air with his hand. "No, because they're not catching them anymore."

In the international port, which I had half expected to find deserted, hordes of immaculate metal-towered ships still rode in the harbor, with a line of seabirds like buoys before them. It felt pleasant to stroll among many gulls on the great dock, with the white-laddered ships humming, diesel-perfumed. I checked the dosimeter, and a gull picked at a dead fish; all was right with the world. Far and high above us on one great ship, a dozen sailors in blue coveralls and white hard hats were unwinding a fishing net from a great spool. They were bound two hours southward, where the fishing was still considered safe. But all the smaller boats rocked empty on the water.

A patrol boat captain told me that fishing for personal consumption was not restricted, and that once a week there was "test fishing."

"Is it the same as before? No mutant fish?"

"The same. At the market they test the fish, and if it's all right they sell it, at a low price."

"How are they surviving?"

"Guaranteed money from Tepco."

He declined to let me take his photograph, and during my time in Iwaki, more people than not did the same. This fact contrasted weirdly with the vulnerable openness of all those Tohoku people back in 2011, with muck and brokenness undeniable in the tsunami zone, and the radiation danger still so new that harmful rumors were just getting hatched.

I asked him whether any unemployed fishermen might be amenable to an ocean cruise in the direction of No. 1, but this was not Mexico, where wildcatters will do any unauthorized deed for money, so he directed me up the concrete stairs to the darkened offices of the Fishermen's Union, where a man in the hallway said: "You cannot. The Coast Guard is very strict here. It's prohibited."

Returning to the taxi, I asked my driver: "What do you imagine that the exclusion zone looks like?"

He thought a long time. "Just no people. You can't enter without permission, so . . ."

"Do you suppose there are any animals there?"

"The rumor goes that pigs and wild boars have interbred."

"How far can we drive?"

"Until Tomioka."

"Let's go."

So we motored up the coast, taking Highway 15. On this lovely morning the pastel sea was very still. Now we were crossing the flats whose tsunami-destroyed homes had been deconstructed into foundations, concrete squares upgrown with grass. What I remembered from 2011 was the raw black stinking filthiness of everything that the tidal wave had reached, and surely Iwaki's affected coastline must have been similarly death-slimed, but these pits looked merely mellow and "historic," like archaeological sites, at least until we came to a boarded-up black house which had been wrecked by the tsunami. The driver said that the city would build a dike on the landward side of it and prohibit construction on this side.

"Do you believe in global warming?" I inquired.

"I think it's true. Coal, electricity and all this are being used; people's activities are continuing 24 hours a day."

Leaving the lighthouse behind us, we reached a gleaming skeleton of half-built apartments for people who had lost their houses, and the driver now remarked: "Beside the incinerator they put some irradiated ash. There's no other place to put it."

"Where did the ash come from?"

"When they did decontamination in Iwaki, there were lots of radioactive debris."

As we began to wind up into the forest, then reentered Highway 15, he asked whether I would like to see this waste, and I assented. Ahead we could see snow on Mizu Ishi Mountain, which resembled a peak less than a blue-grey ridge.

So we arrived at the Northern Iwaki Rubbish Disposal Center, whose monument is its own big smokestack. And that is how I first saw the disgusting bags of Fukushima.

"This is debris they burned in Iwaki, not fallout." He laughed. "They don't know where to put it."

I got out of the taxi. Down a 45-degree slope, behind an aluminum or stainless steel wall, stood a neat close-packed crowd of dark-wrapped bags containing what he kept saying was ash from the decontamination of Iwaki. In this he was mistaken; the city did not burn radioactive matter. But in saying that they did not know where to put it he uttered a truth, or, if you like, a harmful rumor. On the metal wall was a radiation symbol. Not knowing how dangerous these bags might be, I strode slowly toward the edge of the grass where the slope began. This was close enough, I thought. The dosimeter remained at 0.272 accrued millisieverts, where it had been all day. I liked that.

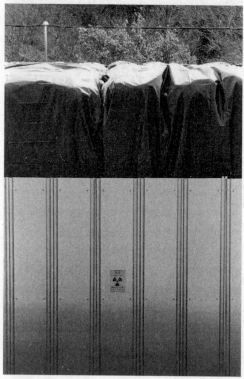

Black bags at the Northern Iwaki Rubbish Disposal Center

Within a day or two, I hardly noticed those evil bags unless they happened to be many together. The closer to No. 1 one drew, the more one saw. By early 2016 there would be nine million of them.

Black bags near Naraha

MY FIRST TOMIOKA IDYLL

Flashing through Mizushina Village, which demarcated itself first by a line of snow at the border of forest and fields, and then by small fields of cabbages, we passed a white crane in a concrete-lined stream, patches of snow in the road, a lovely brown-green pond in the forest below.

We were now, I was informed, around 30 minutes south of Tomioka.

"Is Tomioka in the voluntary or mandatory evacuation zone?"

"We are not directly involved with that, so it is not my business."

Leaving greater Iwaki at last, coming into Hirono, which had only recently been declared safe, we passed a square reservoir for the ricefields, and the snow thickened a trifle. Who would have known there was poison in this land with its many bright green pines, its wall of bright bamboo, and now its ricefields rectangularly outlined in snow?

Some workers in blue uniforms and masks were digging up someone's yard.— "Decontamination," explained the driver.

Other laborers were decontaminating a ditch.

Turning left onto the Sendai road, glimpsing the ocean not far away, we presently entered Naraha, just south of the Sports Park where more great neat blocks of waste from decontamination lay wrapped in greenish-black tarps down in two sunken fields.

The driver said: "Continue down this mountain and you will see Plant No. 2."*

"Is it dangerous?"

"There is no problem from that. It's No. 1 that exploded. I think they are trying to decide if they should abandon No. 2."

"What do you think they should do?"

"I don't need it!" he laughed. "Even though we do use electricity, we are not the ones who need this nuclear plant. The big cities use it."

Down we went, and he said: "You can see No. 2 to your right," and at first I couldn't, on account of a certain snowy grassy hill, but then I saw the stack alone, cradled with scaffolding. That meant we were 12 kilometers south of No. 1.

"Before three-one-one,† I entered it," he said, almost bemused. "I was surprised how strict the security was. They were concerned about terrorism."

Pretty soon after that we came into Tomioka.

On that first occasion the place did not look so bad; I noticed only a smashed house.—For me Tomioka, whose pre-accident population must have been between ten and sixteen thousand (nobody lived there anymore),‡ resembled the contested Iraqi city of Kirkuk in that each time I returned I felt less safe, because I came to know more and see more. Even this time, I am proud to say, I understood enough not to follow the sign that directed me to the Tomioka Town Office.

I wrote in my notebook, *a partially abandoned feeling.*

It seemed quiet enough, to be sure, but an American could easily set this aside because in many of our small towns the pedestrian had become a rare sight. And there was plentiful vehicle traffic just ahead—Tepco workers, I supposed,

---

* That is, the four reactors of Fukushima Daini.

† Just as Americans used "nine-eleven" as shorthand for the September eleventh terrorist attacks, Japanese sometimes said "three-one-one" or "three-eleven" to refer to the tsunami of March 11, 2011. Occasionally one heard "eleven-eleven."

‡ In addition to the radiation problem, there was no water. But I later did hear of a Mr. Matsumura Naoto, said to be "the sole remaining resident of the radiation-tainted town of Tomioka." "After signing some papers for the police, he comes and goes as he pleases, and the officers look the other way. Unauthorized entry to the zone is punishable by a fine of up to 100,000 yen or detention." Although he "spends six to seven hours a day" outdoors "feeding pets and livestock," his reported cumulative dosage was insanely low. Maybe he left his dosimeter inside. In Tomioka there might also have been a woman who stayed to care for a paralyzed relative.

queuing up at the police checkpoint where the exclusion zone began. Not wishing to be noticed and possibly asked to leave, I had the taxi pull over on the first side street. Leaving the driver for awhile, the interpreter and I got out. Snow lay against the Night Pub Sepia.

Just off this road lay a building whose reliefed facade spelled **TSUBA**, and between the **T** and the **S** the epidermal layer had been torn open. This place had once been a pachinko parlor. It would be my landmark on other visits.

Ruined pachinko parlor, Tomioka

There was a dead weed, tall and half-frozen, outspreading its lace-tipped finger-stalks, projecting its complex and lovely shadow upon the white wall of a silent house whose windows were all curtained. I remembered how Kawauchi had been three years earlier, with the residents freshly fled from their homes. Tomioka was not like that. Tall grasses rose up before the side of another home, the tallest one reaching halfway up the window, whose outer pane was so canted that only two corners now kept it in place. There were reflections and shadows of other weeds in the glass, behind which the blinds were drawn. In a third house, a dead weed elegantly leaned before a sliding glass door that had been papered over. In a fourth, golden weeds reached and bent before crumpled blinds.

I gazed up a power pole, and found its wires wrapped with creepers.

Grass was growing up toward the snow-patched roof tiles of another silent home whose curtains had been neatly drawn sometime between a day ago and three years ago—most likely not the former, since the front door was utterly overgrown.

For the first time I now suddenly heard that amplified recorded voice, evidently a young woman's, but distorted and metallic, so that the interpreter and I came to refer to her as the *robot girl*. This time she was reminding the workers not to spread radioactivity; they should dispose of their protective gear at the screening place on their way home. It must have been lunchtime, or the end of a shift. But where were they? We walked past closed garages, a shuttered sliding door, curtained windows, then down the street to a house where an all-male crew in blue uniforms and light paper masks were decontaminating: digging with shovels, pausing, then dragging picks and rakes across gravel. I asked one of them if the radiation concerned him and he replied that he felt no worry. He refused to let me photograph him.

Now I caught first sight of the forbidden zone's present boundary. Partly occluding a side street was a tall narrow signboard, which said, as rendered in my interpreter's beautiful English:

> **TRANSPORTATION IS NOW BEING
> RESTRICTED.
> AHEAD OF HERE IS
> A "DIFFICULT TO RETURN ZONE."
> SO ROAD CLOSED.
> —NUCLEAR DISASTER FIELD MEASURE HQ.***

Tomioka was warmish and still, the patches of grubby snow beginning to sweat, and a crow cawed far away; then for a moment the only sound was the scraping of those decontamination workers' shovels.

It was strange to look south up the rise of highway we had descended in the taxi; somehow it seemed that up there it was safe, and here the wrongness began, when really it must have been just as dangerous all the way up the hill. The weeds were motionless, and now the *robot girl* fell silent. As the interpreter and I walked among those forsaken houses we saw more windows covered with paper, and sometimes uncurtained windows through which we could look into furnished rooms. Every now and then, but not too often in this district, a window was broken. Behind a fence, metal was banging on metal. Perhaps it was torn siding. Was a breeze blowing over there? The banging went on and on like a telegraph.

There was an asphalt path, perhaps for cyclists, and after it crossed a deserted road, weeds began to grow out of it. I came into this vegetation, and it began to touch me.

Still I thought that Tomioka appeared only a little shabby, not really, as the interpreter opined, abandoned—weeds could have done much more in three years. A flutter of wings, and geese rose screaming from the tall grass. Startled, I watched them go, then checked the dosimeter, which had already registered another digit, so that my chest tightened. As another high stalk wrapped around

---

* After the accident, Tomioka had of course been evacuated. In 2013 the town was partitioned into zones of various radioactivities. Southern Tomioka, where the annual levels were a pleasant 20 millisieverts and below, became a green sector, and hence relatively easy to visit. Fifteen hundred people used to live there. They were hardly in evidence now. The central district, formerly inhabited by 10,000, was now a yellow or "residence restriction zone, with annual doses estimated between 20 . . . and 50 millisieverts. Residents can visit this zone in the daytime, but cannot stay overnight." Then, of course, there were the red zones; what would Fukushima be without them? Their levels exceeded 50 millis. To enforce all these delineations, the government had erected 128 barricades in Tomioka. I did not understand any of this at the time.

my ankle on that overgrown sidewalk, I wondered what particles might adhere to it.

A silent house of broken roof tiles hunched beside another whose window was shattered. Red shards lay bright on the empty street.

To the left and the right were those vertical rectangular free-standing signs, flanked by traffic cones, with knee-high metal railings as if for bridges deployed behind them, so that scofflaws such as I must step over them to come into the forbidden zone. That is what I did, I confess, but only for a moment or two. Of course the dosimeter refrained from another immediate increase, and then we turned north, passing the taxi and walking along the highway toward the checkpoint, which might have been the most dangerous action of that day since many construction and decontamination vehicles were passing in and out, emitting dust. We had brought no masks, because three years ago I had found that drivers tended not to wear them and the interpreter experienced difficulty in speaking with a mask on; as for me, I declined to use protection when the people in my employ wouldn't or couldn't. So I read the dosimeter, and choked in more truck-dust, facing away from the road whenever I could. That was my way of standing up to harmful rumors. Then we arrived at the central government's checkpoint.

I asked the nearest police officer why some houses looked worse than others,

Nuclear checkpoint on Highway 6

and he said he didn't know, because he was from Tochigi Prefecture. No doubt it was better for him not to be from there. When I began to inquire about the radiation levels here he said: "Sorry; I'm busy with my duties."

Across the road stood two other officers, badge numbers TI 562 and 558. I asked if anyone still lived here in Tomioka and they said that almost all the people were gone. As to why some houses looked better kept than others, one policeman replied: "I have heard that some workers," evidently meaning decontamination laborers, "are weeding."—They each had white dosimeters in their breast pockets. I asked how much radiation they expected to accrue, and they replied that it was their first day here. Whether the government rotated them briefly in and out to keep them safe or to keep them ignorant was a question I might have spent all my money to solve, so I left it, and again that loudspeaker voice, feminine yet brassy, echoed through the hills.

As we entered the taxi and began to drive away, I pulled the dosimeter from my shirt pocket and felt a kind of pleasure, which presently became uneasiness, to see that it had already upticked by another digit. Within that single hour it had registered three times as much radiation as it had in the 24 hours from Tokyo to Iwaki. In other words, the dosage in Tomioka was potentially 72 times greater than in Tokyo.[*]

On the way I kept the dosimeter near my ankles, where the weeds had touched, and since it stayed at 0.275 millisieverts for the rest of the day, I (being rather ignorant in my way) decided that my bluejeans and shoes must not be overly contaminated, at least not with gamma emitters, and I had not yet been capable of measuring alpha or beta anyway. The driver drove cheerfully back toward Iwaki, past the sand-striped beaches where as he explained one could not swim due to those leaks of radioactive water. The pastel ocean-line looked lovely, and Iwaki was innocuous. It felt relaxing to be once again on the safe side of that imaginary wall of ice.

"I EAT WHAT IS IN FRONT OF ME"

If you walked east of the train station for 15 minutes and then ascended a steep little hill, you would reach the campus of Higashi Nippon International University, where I met the fresh-minted graduate Mr. Takamitsu Endo, who was smiling, lightly moustached and pale. He had majored in information technology.

---

[*] Projecting from an hour to a year is a pretty speculative game. However, on two out of my other three visits the readings were comparable. So, for what it is worth, I note that $72 \times 365 = 26{,}280$ micros per year, or 26.28 millis. Recall that the international standard for a maximum recommended yearly dose is 1 milli.

The interpreter characterized him as "gentle and timid." He wished to become a social worker here in his home prefecture, "to be with my parents."

He had been an exchange student in Ohio when the earthquake struck.

"How were you feeling when you heard about the disaster?"

"I was anxious about my parents."

"And how are you feeling about it now?"

He hesitated, hissed between his teeth, covered his mouth with his hand and finally said: "It's hard to tell. I think about my parents, my friends."

"How long do you expect this area to be poisoned?"

"For a solution, well, I believe it takes time," he replied, smiling and opening his hands. "The government is more focused on the Tokyo Olympics than on this."

"How dangerous is it here in Iwaki?"

"I don't know. In the beginning, people were talking about it. Nowadays there's nothing said. It seems that no one is worried."

"In the long run, will the situation here be better, worse or the same?"

"If another earthquake comes, it will be dangerous. But currently the prospect of another Kanto Earthquake is greater."

"Do you worry about eating local food?"

"I don't think twice about it."

"And local seafood?"

"I feel a little resistance to eating seafood."

"When you go to a sushi restaurant, do you wonder where the fish came from?"

"I eat what is in front of me."

"Would you go swimming here?"

"Well, I haven't swum in 10 years."

So he had it all figured out.

In the vocational office next door, an embarrassed young clerk who because this interview had not been previously cleared with his superiors preferred to be called "a company employee" assured me in contradistinction to the first taxi driver that no one in Iwaki had run away when No. 1 exploded; he himself did not feel especially concerned, although perhaps he might now begin to consider food safety now that he had a six-month-old child—but with the child still unready for solid food, this issue could be postponed. No one was really concerned about radiation, he repeated, for after all there had been no effect. I mentioned the suffering of farmers and fishermen, and he remarked that none of them were his friends; anyhow, the restaurants of Iwaki no longer carried local produce or seafood, so everything was safe.

Come to think of it, Iwaki *was* safe—or at least almost as safe as Tokyo. No doctor or hospital employee in the city would agree to meet me, even though I had promised to ask only one question: Have you seen any health effects of radiation in Iwaki?—Perhaps I gave up too easily, or was losing my touch. Maybe the whole city's bill of health was glowingly perfect, although it did disconcert me to read in the newspaper that the Fukushima Prefecture Dental Association would now with family consent examine extracted teeth of children aged five to 15, first checking *for cesium or other isotopes* and then, if those were present, for strontium, which, by the way, had long since set a proud culinary benchmark: When two of the Manhattan Project's leading scientists considered poisoning the German food supply with a radioisotope, strontium *appeared to offer the highest promise.* Perhaps they were hoping to cause bone cancers:

RELATIVE STRONTIUM-90 CONCENTRATIONS IN PERCH LAKE* (CANADA), 1963

(absolute values not given; lake water = 1)

Perch flesh	5
Bottom sediment	200
Clam tissue	750
Mink bone	1,000
Perch bone	3,000
Muskrat bone	3,900

For Sr-90 concentrations at the Hanford Nuclear Reserve, see table on p. 417.
Source: Odum, 1971, from Ophel, 1963.

---

* This was the same Perch Lake whose receipts of solar radiation were discussed on p. 165. In case you wonder just how it happened that the place became so convenient for testing strontium, I quote the following scene-setter from a scientific paper: "Perch Lake is situated near the southern edge of the Canadian Shield in the Ottawa River valley on the property of the Chalk River Nuclear Laboratories."

Well, but a test is not a result. Maybe there was no cesium in any child's teeth. As early as 2012, the Ministry of Health had announced that most sorts of vegetables that had been prone to absorb perilous levels of cesium in 2011 now registered non-detect for that element. In 2013 the news got even better: *Doses of up to 15 millisieverts of cesium-134 and -137 were found in adult residents, much lower than the 100-millisievert threshold considered to increase the risk of solid tumors.*—And who could blame the doctors who refused me? They were busy, I was a nobody, and so long as they declined my companionship I could hardly use their names in the process of spreading harmful rumors.

Thanks to a taxi driver who worked out with him at the local gym, I now met the farmer Mr. Hamamatsu Koichi, not to mention his industrious wife who kept carefully trimming and cleaning green onions during the interview, smiling and sometimes even sweetly laughing at my corny radiation jokes; he spoke for both of them and did not give her name. They might have been around 60. Mr. Hamamatsu wore an apron over a light down vest, and elastic wrist gaiters over the sleeves of his striped shirt. He had a round face and cropped grey hair. His eyebrows rose and curved like caterpillars. His wife wore trousers and a plaid overshirt whose feminine touch, which perhaps she had fashioned herself, was a scooped collar of the same fabric. Her lush bun of hair remained mostly black. On their urban farm, whose area was 600 *tsubo*,* they raised cucumbers, eggplants, broccoli, cabbages, lettuce, *hakusai*[†] for kimchi, and of course those green onions, whose fresh smell and enticing jade color excelled any counterpart I had seen in an American supermarket. I could not help but perceive their establishment as good and healthy. Perhaps it truly was, and after all there was "no immediate danger."

"When the accident occurred," said Mr. Hamamatsu, "the earthquake damaged the roof of this greenhouse, and we placed the plastic sheet here. On account of that damage, we could not evacuate."

"What were you feeling when you learned about the radiation?"

"The explosion was shown on the television. I heard that the American embassy reported radiation danger all the way to 60 and 80 kilometers from Plant No. 1. Iwaki is less than 40 kilometers from there, but we didn't have fuel for our vehicle, so the idea of escaping didn't enter our heads."

"And how do you feel about radiation now?"

"Because it's invisible . . . ," he said hesitantly, "and something happened in Russia;[‡] there I heard people evacuated, but people here, well, it must be okay. The TV news said it went even to Shizuoka[§] and damaged their tea, so for sure this place must be contaminated, but we have no place to go.

---

* A *tsubo* was originally a little enclosed garden. I often heard Japanese farmers use this term as a unit of measurement. 1 *tsubo* = the area occupied by two ordinary tatami mats side by side = 3.306 square meters.

† The meaning is napa cabbage.

‡ Chernobyl.

§ A prefecture whose center is roughly 150 kilometers southwest of the center of Tokyo, or about 375 kilometers southwest of No. 1. If the television news was accurate, then the radiation probably also affected Tokyo. In October 2014, when I was touring the Okuma red zone, an official originally from Shizuoka insisted that he had never heard of any fallout in the latter place (see p. 504).

Mr. Hamamatsu Koichi

"In the beginning we were not allowed to sell. We had to take our produce to the Agricultural Union and measure the radiation per kilo. Because it was too radioactive, we could not sell it, and in the second year, even when the radiation was low, we had to sell at a very low price because customers were afraid. Now the radioactivity is low enough that all the vegetables are fine. When we introduce it to the market, we must attach the certificate . . ."

In other words, that "company employee" at Higashi Nippon International University must have been misinformed: Local produce was still allowed, provided that it tested well at the Agricultural Union. And if the tests were accurate, why not? At the train station I bought a kilogram of Iwaki tomatoes and sent them off to No Nukes Plaza in Tokyo for analysis. They tested absolutely nondetect for cesium.

"Which crop shows the greatest radioactivity?" I asked Mr. Hamamatsu.

"Rape flower, they say."

"And what do you think about the fish?"

He smiled. "Ah, I don't think we can eat it, because the leak is going to the sea, and surely some information is being hidden. It's not at all fine yet."

"In 10 years will you be able to go right up to Plant No. 1 if you wish?"

Folding his hands across his apron, he smiled at this new joke, and his wife also laughed, returning to her onions. "There may be some spot where you can never return."

A MOTIVATIONAL PASSAGE

That last was a blunt statement. Ordinarily I had trouble in getting Iwaki people to converse with me about Tomioka and other such places.

I kept asking them how safe they felt; mostly their replies came out calm and bland.

The Japanese-born Korean mother and daughter who ran a restaurant three or four blocks west of the railroad station were more openly worried, but had no idea what anyone ought to do. (I felt the same way.) The tsunami had destroyed their previous establishment, but with that customary Tohoku stoicism they assured me that it had been an old building anyhow. In their opinion, the people within the 30-kilometer zone had been compensated more than enough for the nuclear accident, while Iwaki here in the 40-kilometer zone received nothing. When I met them again in 2017, their position on the evacuees had fossilized. That pair worked hard; one night the mother was feverish, but she kept right on cooking, serving, taking orders, while the daughter rushed around trying to help her. Because only their toil preserved them, they could not but disapprove of able-bodied people who lived at public expense. Perhaps they believed that sturdy perseverance could negate the loss of home, community and possessions. And for all I know, reader, whenever you stash a few roots and tubers in your larder, you will feel comparable resentment for the less fortunate denizens of your hot dark future. We would all prefer to subscribe to Mr. Hamamatsu's assertion: "People here, well, it must be okay." And if we are managing, then those who cannot manage must be deficient. There are no inescapable horrors—only harmful rumors.

3: "Something Will Come In"

The next day's driver was a very active bald old man who wore a paper mask over his mouth, as do so many Japanese when they have colds or wish to protect themselves from germs. I requested his description of the disaster and he very practically replied: "In the beginning we were busy. Now that the bus services have recovered, it's quieter for taxis."

"How do you feel about radiation?"

"Since it's not close to here, it has no reality. If someone around here got cancer, I might feel something, but it's invisible, so I don't feel anything."

I felt proud of him. He too had stood up against harmful rumors.

On occasion he had entered the prohibited area with the Red Cross. There had been three or four checkpoints. "Now we regularly take nurses from the Toyoko Inn," a chain hotel, "to the No. 2 power plant. We leave at seven in the morning. They serve those who temporarily return to their houses."

Temporary housing near Hisanohama

He drove me half an hour north of central Iwaki, back to Hisanohama* by the other ruined Fishermen's Union, then turned westward off the highway, not very far. Asphalt and gravel channels ran straight and far between the long barracks-like units where some of Fukushima's 150,000 nuclear refugees were housed, with patches of snow enduring between many of those silent buildings, which were besieged by clouds and wind and adorned by rectangular shadows. A tiny black cat watched from between a windowpane and a white curtain. Everything was still and raw—a different kind of still from Tomioka. Walking up and down the place, I listened in vain for the sound of a radio or television. At last I finally met one person, a former Hirono resident whom I will here call "Michiko," because a few hours after happily flirting with me she telephoned the interpreter in a panic to ask that her name be changed; her relatives had admonished her to have nothing to do with this project. She spoke some English, and whenever she

---

* *Hama* means "beach."

had trouble the interpreter helped her. Here then is a combination of their words: "We left after midnight of the tsunami day, March eleventh. I put the dog in the car and went to higher places. It must have been 2:30 in the morning. There were 40 or 50 of us. We arrived at about 3:00 a.m. We couldn't sleep. The dog had to stay in the car. Everyone watched the TV. It showed a horrible picture. And soon it was morning. I took a walk with the dog and there was another explosion, and we were told to evacuate and we were so surprised: What explosion? We took Highway 36 to Iwaki. Normally this takes 30 minutes, but there were so many cars because Highway 6 was closed. We were told not to open the window, because if you open, *something will come in*. We were thirsty and didn't know what was going on. I arrived at Iwaki at 10:30 in the evening, left the car and took another car to Niigata. We arrived at 4:30 in the morning without any sleep. All the roads were humped and cracked. We stayed in Niigata for awhile, then went to Tokyo, left our dog at a kennel because he wasn't allowed at the soccer stadium where we were sleeping, and the dog was crying, so miserable! I told him, we'll come get you soon. We stayed at the stadium for some time. Then we applied for this housing and we got it. My house is still all right, but if there's another explosion it might not be safe. So I'm afraid to go back there."

Such was the tale, or harmful rumor as some might say ("since it's not close to here, it has no reality"), of this displaced woman in the plaid red vest, which is to say this high-cheekboned older woman of the pretty white teeth and wavy shoulder-length black hair and the rising laughlines at the corners of her eyes, with her hands folded across her grey skirt as she looked up at me, cocking her head as she sat on the narrow porch of her temporary home, which for a fact was better than any of the cardboard-partitioned spaces and mattress on the floor which I had seen three years ago in the hallway of Big Palette in Koriyama, where a skinny man was lying on his side, and on the next mattress a scared woman with a white mask over her mouth sat on her heels, cradling a little boy whose dark eyes stared up at the ceiling; just as the tsunami-slimed rubble of 2011 had given way to grassy hollows in the ground, so her exile had faded into something routine; there was no immediate danger. Leaving her alone behind her wall of ice, we drove back to Iwaki.

"BECAUSE IT WAS MY HOME TOWN!"

Southeast of the city center lay a smaller older hillier section of tightpacked little houses, after which the taxi ascended to the edge of the forest, then onto a plateau of fields walled on both sides by low forest ridges. In 20 minutes we had reached the district called Takaku, where another 90 households of Tomioka people lived in a barrackslike complex like Michiko's.* They paid for electricity and utilities, not for rent. Some of them worked for Tepco.

They were shy and hasty and busy, but finally one girl, just to be rid of us, urged us to talk to a certain Mrs. Yoshida, "who knows everything" and whom we caught, poor soul, in front of her flower boxes. She was an old lady with red hair, wearing *tabi* socks, and as we talked with her she kept folding bluejeans in the chilly wind, having evidently taken in the rest of her laundry.

Her former home lay in the restricted area of Tomioka.† She said: "There is no decontamination and the grass is growing. The whole radiation level is high. Some civil engineers will remove large debris. It looks terrible. Even this area here is better," gesturing around her.

Born in Kawauchi, she had lived in Tomioka for 53 years. Her son died in the earthquake. Her grandchildren moved far away.

"How was your house?"

"It was nice. Only two years before, we remodeled it. The scenery was pretty."

"And the neighbors?"

"They're all here."

"Will you be able to move back someday?"

"No. We must demolish the house. The government will pay, since it's more than half collapsed."

Although she had been gazing sorrowfully downward, when I asked her what she had most loved about Tomioka, she began to smile, and said: "Because it was my home town!"

This made me very sad.

"IF IT IS NATURAL, IT MUST BE ALL RIGHT"

Ms. Kuwahara Akiyo, a slightly weary-looking woman with blue eyeshadow and tiny sparkly earrings, was a part time contract employee of the Tomioka Life Recovery Support Center and Iwaki Taira Interchange Salon. When I asked her to describe her home, she said: "There is a fire station. Behind the fire station there is a checkpoint. A little past the checkpoint is my house. I can return only

---

* Mr. Endo Kazuhiro, the protagonist of this chapter's final Tomioka idyll, explained to me: "In that temporary housing there are 90 households, and of these 80 sometimes participate in the residents' association. Some live elsewhere. It is easy for those people to meet their rent because they get forty to sixty thousand yen per month from Tepco."

† She must have meant the yellow sector, whose annual radioactivity was between 20 and 50 millis.

Ms. Kuwahara Akiyo

once a month. I used to work for a day nursing service in Tomioka. I've been living there for 12 or 13 years.

"In Tomioka there's not much improvement. In fact it's even worse! There are animals: mice and rats; also those pigs interbred with wild boar. The authorities are trying to kill them, because they damage the houses, and poop inside, and bite the pillars. They don't know humans, so they are fearless. Sometimes you can see them on the street. And there are so many rats! Whenever we go to Tomioka, we first apply to the national call center, and then there is a station where your radiation is measured. Wherever we go there, they give us some chemicals to kill the rats. Other than the rats, there is no damage to my house.

"I know somebody whose lock was broken. Then they took the car key and stole the car.

"Yesterday I went there. I did cleaning, aired out the house and picked up some stuff."

"Do you ever feel anxiety about radiation?"

"A little bit anxious, but we have a dosimeter, and we check it. I don't stay there so long. Yesterday I got one microsievert in two or three hours. In Okuma and Futaba they get five or six, or even 20 or 30 microsieverts in one or two hours.* A friend of mine in Futaba said the atmosphere there was like that."

"So when you think about radiation, what do you think?"

"Invisible," she said smiling, "and that is the cause of the anxiety. You don't feel any pain or itching; it just comes into your body. But it is not as strong as a medical X-ray. If you want to stay, it's of course impossible. But for an hour or two, it's okay. In Okuma and Futaba, you'd better not stay all night, but some people, the nuclear power workers who are trying to close the plant, they do stay, and to them we feel so grateful."

"Is it completely safe here in Iwaki?"

"Even natural radiation exists, and if it is natural, it must be all right. I think it's safe to live in Iwaki. There are nuclear plants all over the world. You can't flee anywhere."

"What is your feeling about nuclear power?"

"My husband's work is related to nuclear plants. Until the accident we were told that it was totally safe. Nuclear plants we cannot live without." Then she said: "This was a manmade thing, but we made something so evil, so fearful, so dangerous."

"When will Tomioka be safe again?"

"It could be a hundred years. And whether the hospitals or shopping will ever come back, I don't know."

What an optimist she was! She imagined that her town would be re-inhabited in only a hundred years!—And what a pessimist I was! Three years later, I revisited Tomioka and found economic demand newly allured by opportunity, in the form of a solitary convenience store.† Moreover, certain decontaminated districts would soon be cleared for resettlement. Meanwhile, back in 2014, I did my best to resist harmful rumors. In short, I inquired at City Hall.

GOALS AND MEASUREMENTS

Mr. Kida Shoichi was a decontamination specialist in the Nuclear Hazard Countermeasure Division.‡ He was a patient man, not young, with kind eyes. I

---

* When I visited Okuma in October, my scintillation meter measured contamination as high as 41.5 micros per hour. (Had this level been sustained, it would work out to 363.54 millis per year.) Fortunately, my dosimeter readings in this red zone averaged only 3.033 micros an hour, which would be 26.57 millis a year. Most of the time I was shielded in a vehicle, which surely reduced my dose.

† See II:620 for some remarks on that triumph.

‡ The title on his business card could be gloriously translated: "Administrative Management Department, Risk Control Division, Nuclear Hazard Countermeasure Division, Decontamination Specialist."

remember him with gratitude, for although I had interrupted his day without notice, he took a half-hour to answer all my questions, showing knowledge and honesty in the process. He considered his city to be not at all badly off, because as he explained, "from Daiichi, the northwest direction* is high in radiation, so Iwaki is low."

Activating the laptop on his desk, he showed me the 475 monitoring posts in Iwaki. Then he zoomed in.—"Here is the city office where we are right now. And here is the monitoring post. It's 11:05 a.m., and we're receiving 0.121 microsieverts per hour."[†] Had the winds stayed consistent for 24 hours, the dosimeter should have accrued 2.9 microsieverts—three times as much as it did in Tokyo—but on that day as on almost all the others I spent in that city, only a single microsievert damaged my body. Of course I passed much of that time indoors—although, come to think of it, the dosimeter mostly lay within the interpreter's house, in a basket just past her front door. Wishing to cross-check that device, I told Mr. Kida that by my measurement Tomioka had registered as 72 times more radioactive than Iwaki. Clicking on Tomioka, he corrected me: "No, the highest is about three microsieverts per hour, so that's only 26.8 times more than here.—No, here's a place that measures four. And here's a five, so that's 45 times higher . . ."[‡]

Since he now looked a little sad, I refrained from asking him to continue scrolling through the remaining Tomioka readings. Besides, my would-be comparison was once again apples and oranges, since the dosimeter could not measure local radiation moment by moment.

Mr. Kida next zoomed in on a monitoring site at Futaba Town: 13.61 microsieverts per hour at present.[§]—The year previous to the accident, Futaba's air dose had averaged 0.043 micros an hour. An anti-nuclear ideologue might accordingly denounce that place as 317 times more radioactive than before. But one station at one moment, compared to a townwide average, should not have caused any stir. Besides, had I opened my mouth about the matter, some harmful rumor might have come out! So why not simply reflect that if those 13.61 micros kept up until tomorrow, all over Futaba, then Futaba might be maybe 327 or possibly only 113 times more dangerous than Iwaki?[¶] Better yet, why not assume that in due course

---

* That is, toward Iitate, whose red zone is described beginning p. 435.

† About ⅒ of my nuclear "first responder" guide's safe ceiling of 0.1 millirem.

‡ As you will see, in the red zones I found dramatically "hotter" spots than that.

§ That would be 1,192.24 millisieverts per year, or nearly 1,200 times the ICRP maximum recommended annual dose. Later that year, returning to Tomioka with that "pancake frisker" scintillation counter, I measured similar and higher radiation fields.

¶ The first figure would be [13.61/0.041, my rough dosimeter average for Iwaki], the second [13.61/0.121, the reading at the Iwaki monitoring post].

everything would get wonderful? Thus Tepco planned for tsunamis and American politicians prepared for climate change. We humans were much the same.

"Okuma and Futaba are still prohibited," Mr. Kida remarked, which seemed sensible enough.

"When will Plant No. 1 be safe?"

"You have to wait until natural reduction occurs."

"Five hundred years?"

"I believe so."

The next day he telephoned me with a correction: In only 300 years, barring further accidents, the radiation would have decayed to one-thousandth of its original strength. If I peeked out of my grave in three centuries, I might well find him a perfect prophet, given my 10 half-lives calculation in Table 1* for the two most slowly decaying isotopes, strontium-90 and cesium-137.

But I received a more pessimistic answer when I asked Ed Lyman, the Union of Concerned Scientists physicist: "If there's no further emission from the plant, is 300 years long enough?"

"Well, there's radioactive decay plus transfer to soil; it will go deeper and deeper and get into the water and so forth. Then there's a weathering coefficient.

"To reduce by a factor of a thousand, it would take hundreds of years. I'd hesitate to give a blanket answer."†

One study of cesium from decades before the Fukushima catastrophe determined that *losses from soil to river sediments* were *50 percent greater in the year after fallout than in subsequent years*—a heartening finding, perhaps, to those whose land does not abut downstream bends of that river . . . but absent accurately local quantifications of the *losses* one could scarcely guess how much cesium might leach away over any given time, and therefore when the area might be safe. In the Columbia River Basin of the United States,‡ only 0.263% of the plutonium contamination from the Hanford reactors had eroded in 20 years.

Well, anyway I had my blanket answer from Mr. Kida: 300 years.

The basic strategy for decontamination, he had explained, was to make an area give off less than one millisievert per year. Thirty percent of the cesium reduction around Iwaki had already happened naturally, and he gave me some multicolored information sheets to prove it. In November 2011, almost every

---

* See pp. 537–38.

† *The Japan Times* was more optimistic, and even prepared to be pinned down. In its best-case estimate, decontaminating No. 1 "will take nearly a half-century." Successful decontamination remained undefined.

‡ A place which will be described below, pp. 406ff.

part of the area had been shaded yellow, meaning "decontamination required." By December 2012, the yellow had melted into a crescent along north and east, with a few strange patches to the south.

When I passed on this good news to Ed Lyman, he remarked: "My impression is that the cesium-134 is decaying away but the cesium-137 levels are not being reduced that much." (The latter isotope, as you will see on page 538, will take nearly a century longer than the former to reach one one-thousandth strength.)

Weren't these fine tidings just the same? Iwaki was becoming safe much more rapidly than a simple half-life model would have predicted. But my thoughts rose like those weeds at Tomioka which kept growing up through parking lots, filling in for eyelashes in the dark broken window-sockets of damaged buildings. Ed Lyman had contaminated my equanimity. I could not stop wondering what was happening to the cesium. Did it sink deeper and deeper into the earth—or stop at clay—or leach into storm drains and rivers . . . ? I had read that it sometimes got concentrated in roots. No doubt the radioactivity of the ground must vary just as the ground itself did. Presumably every square meter of Iwaki required measurement. The city's report (the one that blasted harmful rumors) explained that topsoil in schoolyards was being removed if it measured more than 0.3 microsieverts per hour—which multiplies out to 2.63 millisieverts a year, more than twice the ICRP standard; but on the other hand, one Bangladeshi brickyard, and the window-ledge of a certain bookstore in West Virginia, showed comparable values*—and if Iwaki's children were affected only when they played outside at recess—which might or might not be a reasonable assumption to make about a gamma emitter present in low concentrations—then they would accrue a mere 60 microsieverts per year.[†] Oh, those radiation arithmetic games! *The removal of topsoil was completed in 101 out of 131 facilities by the end of December [2012], which resulted in the maximum of 80% of reduction in radiation levels, and 50–70% in most facilities.*

A 50 to 70% improvement could not be adequate—although what if it simply *had* to be? On this subject, Ed Lyman told me that 50%, not 70%, was the general outcome so far as he knew, in which case the time to bring down the radiation to one one-thousandth of original strength would be shortened by only one out of 10 half-lives—for cesium-137, to 270 from 300 years. But again, why spread harmful rumors?

By Mr. Kida's estimation, the average annual natural radioactivity in the world

---

* See "Comparative Measured Radiation Levels," header 5, p. 246.

[†] Assuming a 1-hour per day exposure in the schoolyard for 5 days a week, with 3 months off, so (52 minus 12 weeks), I calculate an exposure of 200 × 0.3, or only 60 microsieverts per year: an extra two months of Iwaki radiation.

was already 2.4 millisieverts,* and so I suppose that a pro-nuclear ideologue could label that schoolyard topsoil unremarkable. But Japanese areas unaffected by the accident received about 1.4 millis a year. The goal for Iwaki was to wait out a few half-lives and decontaminate where necessary until the radiation reached 0.23 micros per hour, or slightly above 2 millisieverts per year.—Mr. Kida now remarked that he would be satisfied with a target of even 0.5 micros an hour. I was surprised; that would be 4.4 millis a year, which exceeded the old ICRP standard.[†]

"Are any decontamination measures possible for water?"

"It's difficult. We do measure. No part of Iwaki shows a high level of radioactivity. Ten kinds of fish are prohibited; those on the sea bottom[†] are most radioactive. If you eat a little, it doesn't matter. If you eat one kilo every day, it's dangerous."

The maximum safe value for fish was 100 becquerels per kilo; some catches measured 200 or 300. But he assured me that there had been many "no detects" in 2012 and after.

Indeed, as I write this, I have before me the very long printout he gave me that appears to list many or all of the fish testing results from March 2011 until February 2014. Iodine-131, whose half-life is short, had mostly dropped to "no detect" by June 2011. In November that contaminant reappeared, at very low levels, then vanished again by March 2012, after which its placeholder stayed reassuringly blank. (For that matter, by 2013 the iodine in Fukushima's land environment was below *the critical 50-millisievert level . . . However, . . . [the U.N. Scientific Committee on the Effects of Atomic Radiation] also noted cases of children who had been exposed to as much as 66 millisieverts.*)[§]

As for the two cesium radioisotopes of concern, these had intermittently announced themselves in marine tests from the time of the accident until now, but although their initial concentrations occasionally (as in samples number 11 and 13) reached such measurements as 12,500 or 14,400 becquerels per kilo, even in that bad time right after the three reactors exploded they mostly ranged from the low hundreds to around 2,000. With the exception of sample number 43 (9,900 becquerels per kilo), they quickly stopped exceeding the low hundreds at all, and early in 2012 they reached the middle double digits—where they now remained.

I said: "We have heard of many new releases of radioactive water, so would you expect that these numbers will now increase?"

---

* A middle-of-the-road figure. See the table of Comparative Measured Radiation Levels, headers 4–6, pp. 246–47.

† See p. 245 fn.

† I.e., bottom-feeders.

§ "Quick evacuation . . . reduced radiation doses by a factor of 10."

"In Iwaki, what they are catching right now is at some distance from the plant, so . . . I don't know."

Who could blame him for not knowing? His department was Iwaki, not Futaba or Okuma. And, after all, what might be the true purpose of this testing? Perhaps it had never been intended to inform the public how dangerous it might be to consume the fish that swam just offshore from Tomioka—only to fight harmful rumors about the seafood now for legal sale in Fukushima. But this is merely my cynical conjecture. I could have optimistically hypothesized that the tritium from Nuclear Plant No. 1 would dilute to an indetectable level as soon as it entered the ocean.

Continuing on this subject, how could anyone even decide how much radiation was safe in food? According to that Tokyo anti-nuclear activist Mr. Yamasaki, cesium fallout due to nuclear testing had once (I did not ask when or for how long)* gifted Japanese rice to the tune of 10 becquerels per kilo. In Fukushima Prefecture the current reading was twice as high. "If you measure the soil," he said, "you can tell. If you measure the rice contamination, you can see the soil contamination."

"What is the risk from food to the average Japanese who is uninformed and does not live in this area of contamination?"

"That is a very difficult question. The rice currently in circulation is 10 or 20 becquerels per kilo maximum. If you eat this every day and the cesium accumulates internally, 4,000 becquerels will be in the whole body after one year."

"And in your opinion, what is the maximum safe limit in the body?"

Mr. Yamasaki laughed. "There is no safe limit. When we lived in Toyoma Prefecture as children, thanks to nuclear tests we constantly had about 400 becquerels in our bodies."

Meanwhile, when I asked Mr. Kida whether one could live out one's entire life in safety as a citizen of Iwaki, he nodded. "Up to Naraha the radiation level is low," he said.†

As for Tomioka, if those two nuclear evacuees, "Michiko" and Ms. Kuwahara, were here before me right now, what "blanket answer" would I present to them? The time to one-thousandth of the 2011 contamination value, which you have just heard Ed Lyman calling inadequate, would still be three centuries. Assuming that my projected annual level for Tomioka of 26.28 millisieverts was a

---

* This would likely have been in the 1960s.

† But a certain labor activist, Mr. Kawama Tatsuhiko of the National Railway Mito Motive Power Union, claimed that radiation might actually be twice as high in Iwaki as the sensors reported, "because they replace them in specially decontaminated areas. It's way too low," he insisted. My own dosimeter and pancake frisker measurements did not confirm this.

useful approximation,* and that decontamination was ineffective, then between four and five† half-lives (151 to 181.2 years for cesium-137) must achieve the tolerable target of one milli. If decontamination succeeded, then subtract 30 years. That was my first approximation of an answer.

"BECAUSE IT'S ENDLESS"

Remembering a mention of gamma cameras being used in Kawauchi, I asked Mr. Kida whether I could see or use one. He had nothing quite like that available in the office, but promised me an excursion with a scintillation meter; so on the following day his deputy came to collect me. He was a pleasant roundish young man named Mr. Kanari Takahiro.

"In Iwaki," he said, "we have never felt the imminent desperate situation. In fact my parents returned home in only two weeks. I myself couldn't flee since I am a civil servant."

"Around here, do people tell any jokes about radiation?"

"The radiation level is not high enough to create any joke."

All the same, he too might have been a little cynical, for he referred to nuclear countermeasures as "casual countermeasures."

"Are you sad not to eat the local fish?"

"I don't care. Whatever fish, anywhere it comes from . . ."

The decontamination procedure he described was this: Measure, then scrape up surface soil and cut tree branches as needed; place in a bag.

The scintillation counter (which is to say an Aloka TCS 172B gamma survey meter) cost about 100,000 yen.† He called the operation of it "not difficult." As the technical name implied, it measured only gamma waves, like my dosimeter.

We met a "superior" and a construction boss. One was tall and the other short. They wore hard hats, work jackets with white belts and suspenders, and baggy pants tucked into calf-length rubber boots. The short one wore a mask over his mouth and the tall one didn't. I told them they were heroes. Reducing radioactivity by a factor of two was better than nothing, assuming that they didn't smear around the contamination too much. Anyhow they were endangering themselves.

The work was all paid for by the central government. I inquired into whether it was unionized.—No, they curtly said.

They said that by and large the decontamination workers received about

---

* In fact the yearly dose based on an average of my own scintillation measurements for that town in October 2014 was 38.56 millis. While my few readings cannot be any better than indicative, still they do indicate something. It is an ugly picture.

† Since 26.28 halved five times is 0.82125.

† About U.S. $1,250.

4.5 micros a day, which works out to 1.63 millis a year. (Frankly, I suspect that those numbers were underestimations.) They also said that the poor souls who were trying to ice away No. 1 were getting 40 millis a year.

Hereabouts, they told me, they took in less than 0.23 micros an hour—which is to say, 5.52 micros a day—2.02 millis a year if they never took time off. Again, I am skeptical. Their supposed quarter-of-an-hour micro was only twice the 0.121 micros that Mr. Kida had said was Iwaki's air dose. How could they be digging, bagging and breathing radiocontaminants all day without much higher exposures? Both of the cesium isotopes were beta emitters; as for the strontium-90, scientists called that *one of the best long-lived high-energy beta emitters known.*[*] On occasion these men must surely inhale beta particles—and from a dosimetric point of view, beta was invisible.—To answer my question they first considered, then estimated their doses; they never checked their dosimeters. "The radiation level is so low here that we are not worried."

But as a rule, I pursued, what maximum did they set themselves? It was very hard to say, they replied; perhaps there was no rule. They hardly seemed to know or care. "Maybe two years ago or even last year you might have trouble, but now in normal work you never have to worry."

So maybe the Iwaki people were perfectly right to eat whatever was set before them. I myself have always loved happy news.

Decontamination "superior" and construction boss at a radioactive waste site near Iwaki

---

[*] See Table 1, p. 537.

Rolling up Highway 245, then turning east, we arrived back in Hisanohama[*] near where the wrecked old Fishermen's Union used to be. A traditional Japanese-style house was being decontaminated. It looked as if a gardening crew were on the premises, except that all the gardeners were masked. Other than that, they eschewed special defenses. They wore heavy work clothes, gloves and hats; that was all.

I had Mr. Kanari begin to test things.

In the yard, the bag of waste that the nearest worker was filling emitted 0.3 microsieverts per hour. Another bag was 0.26. The target for decontamination was 0.22.

The "superior" said: "Dead leaves are an especial concern; the plants are up to the homeowner."

"What happens to these bags?"

"We remove them and put them in temporary storage for three years."

Temporary storage turned out to mean, for instance, those fields in Tomioka where so many black bags sat.

"Then what?"

He smiled, laughed, shrugged.

But Mr. Kanari told me: "First we remove the soil, then put it in a bag, then it has to stay on the homeowner's property until a formal decontamination begins."

The "superior" explained: "Unlike the United States where there are rules, here we make up the rules."[†]

I had Mr. Kanari test a potted houseplant: 0.14 micros; the construction boss's boot, 0.11; a dirt berm behind the house, 0.33; a rain channel in the sloping concrete driveway, a surprisingly low 0.16; then the drainpipe: 0.49. That last reading was not so good. It worked out to 4.29 millis per year.

Pointing to the pipe, I asked him: "How can you decontaminate it?"

"Structurally speaking, unless you break it, it's rather difficult. We must see what the property owner says."

Here were two black bags side by side, at 0.35 and 0.38 micros; and there was the next door neighbor's drainpipe at 0.17.

The two decontamination men agreed that radiation generally went about five centimeters into the ground.[‡]

---

[*] About 16 kilometers north of the city center.

[†] In October a more rule-bound and very ancient decontaminator named Joko (who declined to be photographed) asserted that there were three successive forms of storage: a "laydown area" at the decontamination site, then a "laydown yard" followed by intermediate storage facilities "until final waste treatment is decided."

[‡] I heard the same thing that autumn in Iitate, and my single excavation with my own scintillation counter accorded with it.

Across the highway lay a field of vegetables. I asked Mr. Kanari to read the crops for me, but he said: "The scintillation meter measures only the air, not the thing." All the same, he did as I asked: 0.13 away from the dirt, 0.18 close to the dirt, where a napa cabbage was.

We drove to another place on a hill; it was called Ohisa. There were some heavy blackish-green tarps, and then a blue tarp "as an afterthought," they said. There was a DO NOT ENTER sign. The air read 0.17 about 2 meters from the sign. Mr. Kanari entered the site and placed his meter almost on a bag, which read 0.19.

"What do you think of these readings?" I later asked Ed Lyman.

"Not very impressive. One to five times background. In my apartment" in Washington, D.C., "I register 0.2 microsieverts per hour."—I did the math: 4.8 microsieverts a day, or 1.74 millis a year. My studio's kitchen gave off just over a milli in the same period. How glad I was to live on the west coast!

Mr. Kanari was very reluctant to measure the forest, since the municipality's only concern was to decontaminate where people actually lived, but I finally coaxed him into entering a snowy bamboo grove at the mountain's edge behind the field of Chinese cabbage: 0.43 micros per hour, or nearly four times Mr. Kida's stated goal of a single annual milli. Mr. Kida, of course, had said that 0.5 microsieverts per hour was acceptable, so he might have felt quite contented here. Maybe even Ed Lyman would have been at peace.

Once upon a time in Kyoto I used to sit in a temple garden called the Shoren-In, whose bamboo grove at the end of the winding path afforded a quiet view of a carp pond the rock border of which achieved asymmetrical grace. Camphor trees shaded my dreams at the Shoren-In, but what I most loved was the bamboo grove itself, whose jade towers decorated the sky. Now I remembered that place while Mr. Kanari stood beside me, holding his scintillation meter as we all listened to the sounds of those lovely bamboos, which resembled birdsongs or windy hollow gurglings, and I almost felt happy.

"How dangerous is it?" I asked the decontamination workers.

They replied that no one should stay here more than half a month.

"What should the authorities do about this?"

"Because it's endless." Mr. Kanari sadly smiled. "And this is not the place where people often go."*

He was probably right, and anyhow I'm not the sort of cad who would want to spread harmful rumors. The silver lining in that radioactive cloud was that if

---

* One Ministry of the Environment typescript proffered a nice color photo of masked workers "decontaminating a forest (by removing fallen leaves)."

everything continued perfectly from here on out, a mere two half-lives (60 years for cesium-137) would bring the levels below a milli. With respect to Japan, arithmetic had become one of my favorite distractions.

For instance: Radiation sickness—this statistic I have already expressed in the dosimeter's previous unit, millirems—may begin to appear after a dose of 0.7 sieverts, which you would get from a thousand chest X-rays, more or less, but the situation comprehended by this warning is a *rapidly delivered dose,* such as a nearby H-bomb detonation or a brief foray to the core of Reactor No. 1. Meanwhile, 35 millisieverts a year delivered over 20 years, which works out to the same total dose, may cause no perceptible harm.* Employing a comparable example, the International Commission on Radiological Protection concludes: *The resulting whole-body activity at the end of the period is significantly different* between a quick dose and a slow one. As I researched this matter I grew surprised at how little is surely known about the perils of extended subacute radiation exposure. The so-called "linear hypothesis" would have it that the multiple of any given dose is the multiple of the danger. *All responsible bodies involved in considering radiation hazards have agreed to use the linear theory . . . ,* warns an anti-nuclear tract from 1971, but a blandly, moderately pro-nuclear book from 2009 qualifies the "linear hypothesis" as follows: *A dose of 2x mSv [= millisieverts] of radiation, where x is greater than 100, produces double the response,* and by "response" he means nothing pleasant, *of a dose of x mSv of radiation. The effects of very low doses . . . are less well understood . . .*

MY SECOND TOMIOKA IDYLL

Thanking the decontamination men for their helpfulness, I said goodbye and bowed. No doubt they were happy to be rid of me. As for poor Mr. Kanari, I now so far imposed on him as to wangle another pleasure-drive to Tomioka. I will never know whether he liked being out of the office or whether he was one of those unfortunates who cannot say no. He might have been curious. After all, he had not visited there since before the accident.

So we rolled north, soon arriving in Hirono, where the nuclear refugee I have called Michiko once lived. Mr. Kanari thought the radiation levels here to be comparable to Iwaki's. I would have asked him to stop and take readings, but

---

* It could, of course, cause cancer, but since nearly half of all Japanese die from cancer anyway, the radiation will be a mere suspect, not a definite culprit. If, say, iodine-131 contamination precedes a significant increase in the number of childhood thyroid cancers, radiation may be plausibly blamed—but the cause of any one thyroid cancer, however likely, is not certain; whereas the symptoms of radiation sickness do afford casual certainty, for all the good that does the patient.

sooner or later the welcome would wear thin, and before that happened I wished to be sure of my Tomioka idyll. Before we knew it, we were in Naraha, former population 7,500; here was J Village on the right, where many Tepco workers were based; whenever we set out for Tomioka, the interpreter, who eschewed going behind radioactive bushes, liked to stop here to have a good pee. It was a very corporate sort of place. Bright-eyed desk sentries ensured that I could neither wander around nor interview any decontamination laborers. In the vicinity of J Village we began to see those black bags in the middle of fields.—"Temporary storage," remarked Mr. Kanari. "Once the intermediate disposal site is ready, they'll move it there."

"What will it be?"

"Well, we hope there will be such a place, but none of us wants it near us."

Cruising up Highway 6, we now came into Tomioka Township. I had to give him directions. From his expression, he might have been having second thoughts.

So we came home again, where the dead grass rose tall on the far side of signs for the exclusion zone. Much of the snow had melted. Ahead was the checkpoint; there stood the police, monitoring that double lane of traffic cones. When I asked Mr. Kanari to pull over by the pachinko parlor called **TSUBA**, he kept the engine running. Assuming that he had done so on purpose, I said nothing, but the interpreter mentioned it, so he went back to the car and shut it off. Did the radiation make him anxious or did he simply wish to get back to work?

So far our day had consisted of levels well below the radiation incident guide's warning mark of 0.1 millirems, or 1 micro, per hour, so that Mr. Kanari could perfectly well engage in his "casual countermeasures." But now, right by **TSUBA** his scintillation meter read 4.2 microsieverts per hour*—comparable to an airplane measurement I took above the North Pole[†]—or, if you would rather, approximately 10 times the value of that mildly dangerous drainpipe in Hisanohama. The three of us strolled down the now familiar side street, past the Night Pub Sepia (3.5 micros), where decontamination workers were still scraping away at the same house, and the echoing, crackling voice of the *robot girl* continued to roar from the loudspeaker. Not another human was in sight.

At a nearby house with yellow danger tape around it, the base of a drainpipe read 22.1 microsieverts.[‡] That was a little perilous, I would say.

---

* Or 36.79 millis a year.

† Page 249, header **55**.

‡ The daily dose would be 530.4 micros—in other words (based on my dosimeter readings), this drainpipe was about 530 times more radioactive than the various spots I encountered in a day in Sacramento. According to my own scintillation counter measurements made later that year, it had become 368 times more radioactive. Had I been chained to the drainpipe for a year in punishment for my sins, I would have accrued 193.6 millis.

The grassy field on which the interpreter and I had walked was a cool 7.5 micros, while the main highway on which the decontamination trucks kept raising and leaving dust was only 3.72.

"Those are pretty elevated readings," said Lyman.

After 15 minutes, my dosimeter had accrued a new digit, and Mr. Kanari definitely wanted to get out of there. I asked him to drive us through the checkpoint, but he was halfhearted about trying, and the police turned us back.

"Based on your measurements," I inquired, "would it be okay to stay overnight in Tomioka?"

Hissing thoughtfully, he finally said: "Inside the house shouldn't be such a problem."

We sped back through the wall of ice, to the place where we ate whatever was put in front of us. The trees were coming back into first bud, the grass still brown.

4: "This Is a Kind of Opportunity"

West of Tomioka and formerly at the edge of the forbidden zone lay Kawauchi Village. You may remember* that here was where three years ago a police checkpoint had marked the inner ring, and I'd wandered among those homes abandoned just days before, their curtains drawn and their houseplants just beginning to die. Last summer's newspaper had reported that *some 480 of the 1,061 houses in the emergency evacuation advisory zone still had atmospheric radiation doses over 1 millisievert [per year] following decontamination.* In other words, decontamination had failed to achieve its target maximum. This was discovered only when suspicious villagers had performed their own gamma surveys. *Since the [environment] ministry hasn't explicitly promised to do such work again, municipalities are increasingly concerned about the future of the decontamination process.*

All the same, the village was already partially reinhabited. The Prime Minister had called it *a front-runner in rebuilding,* and the headmaster of the elementary school predicted that he would soon have more than 20 pupils, although the student body appeared *unlikely [to] return to three digits.* That was in March 2013. A year later, the place now pondered permitting its residents to sleep there overnight—another heartwarming indication that there was "no immediate danger."

On a recent printout, courtesy of the Japan Broadcasting Corporation, Kawauchi's radiation level was listed as a mild 0.059 micros per hour—half of

---

* From pp. 313ff.

Iwaki's reading, and for that matter less than I received at my kitchen counter in Sacramento.* Were this situation to endure, a steady resident would take in only half a milli per year. Unfortunately, if that Tokyo anti-nuclear activist Mr. Yamasaki spoke rightly, Kawauchi actually presented the second highest radiation level of any inhabited place in Japan, the worst being the infamously poisoned settlement of Iitate, which during my visit in 2011 had always been an ugly spot on morning television's news-maps. For all I could tell, Mr. Yamasaki and the JBC might both have been correct. "Kawauchi varies according to the place," said the former. "If it's far from the plant, the residents have already returned."

On a crisp cloudy Saturday morning, we rolled cheerfully in a new taxi out of Iwaki toward Tomioka, since the direct route to Kawauchi remained snowed in. The sea was a paler blue than before, and my third driver, older, greyhaired and elegant, wore a white pinstripe shirt, a grey vest and a grey shirt.—He had believed the same as the others; he never fled. About the radiation he said: "There's nothing to do but trust them." But unlike the first driver, he blamed Tepco as much as bad luck.

Just as we passed those black-clad bundles of radioactive waste, the dosimeter coincidentally registered another digit.

Here came the familiar sign for Tomioka Town, which must have merely indicated the border of Tomioka Township, although just now as we began to descend through the forest, the only structure yet visible was the stack of Plant No. 2 looming over the trees to our right. Then came the broken Grand Hotel.

The driver remarked that only recently had this sector been opened. He believed that the radiation was lower than before. Well, why shouldn't life be getting better?

So we came into Tomioka, past that same decontamination crew still working on the same house as two days before, then curved around west on Highway 36, the Joban-do, heading toward Kawauchi. No one else was on the road. Here and there in the center of the straw-colored fields stood those neat black bags of contaminated matter. The driver remarked that these places had not yet been "scraped" as he put it.—"The bags will be taken to that incineration place," he said.[†]

The Joban-do had just today reopened "due to lowered radiation levels," he said. A sign announced 20 kilometers to Kawauchi. So many bags in field after

---

* On October 22 of that year the Nuclear Regulation Authority website reported 34 monitoring stations for Kawauchi. The side of the prefectural highway read 0.357 micros, while a certain bus stop was 0.584.

[†] Perhaps the one whose photograph appears on p. 456.

field! There was snow on the side of the highway. We passed these silent yet not entirely unkempt houses on Tomioka's outskirts, ascended toward the snow-crusted mountains and crossed a bridge into a mountain tunnel whose flashing sign proclaimed 6° Celsius. It looked very lovely and natural now, that snowy forest, those leafless trees and evergreens. There came another long tunnel, with still no other vehicle in sight; we passed a dam, took tunnel after tunnel and presently began to go down along the side of a steep snowy river gorge, with the river frozen and snowed over far below. Telling the driver to stop for a moment, I stepped out, breathing a lovely fresh-smelling wind, which happily was not blowing from the poisonous direction. Orange and golden leaves twitched on a sapling down at the edge of the steep snowhill, where a narrow channel of brown river was open. I gazed at the lovely snowiness of the Takikawa* Mountains, finding nothing about which to spread my harmful rumors, except for snow and clouds.

A sign informed us that we were now entering Kawauchi Village, which, by the way, lay 25 kilometers from Nuclear Plant No. 1. I did not yet recognize any part of the place, for we were coming from the opposite direction I had taken three years ago, when this side had been prohibited due to radioactivity.

I saw silent houses piled with snow. The driver said: "Younger people won't return yet. Only recently some factories started up again."

Here hunched a house covered with snow, then trees covered with snow, a gate soft with snow—almost romantic. The snow appeared clean, and house-icicles shone crystal-bright. Now on the snow-lined highway we finally came into car traffic, not much—all from construction workers, said the driver; they were preparing to widen the road.

A snowed-over tatami store, a side road of virgin snow a foot high, snow smooth and high across the little bridge over the Kawauchi River, snow heaped around and on top of a cemetery's dark steles, such was the village on that day. Presently I recognized the park where I had seen a policeman walk toward the toilet three years earlier; it was snowed over. Stopping the taxi, I turned around to face the former checkpoint, which had left no memorial.

Excepting some workers in the streets, even downtown looked deserted; but outside the Kawauchi Village Residence to Promote Young People's Settlement I saw a path shoveled through the snow, almost clean, with only a few icy stretches, and the foyer was unlocked. There came a dripping of snow from the roofs. The interpreter and I rang doorbells. Failing to interview first a teenaged

---

* "Waterfall River."

girl whose mother had a cold, then a man who had no time, we set out in search of better prey, the spoor being wet newspapers in a mailbox, a few fresh paths carved out.

Snow-heaped entrance of abandoned home, Kawauchi

Here was the barbershop by which I had stood watching the line of policemen at the checkpoint. On that spring day in 2011 it had been closed, and likewise every other shop on the street, whose clean emptiness had been far more disturbing to me than its present snowy quietude, for I tend to think of winter as the season when businesses shut down. Gazing at the barbershop, I tried to determine how changed it appeared today. For one thing, there was now a ridge of icy slushy snow between the street and the front door, whose curtained window was mostly occluded by two posters, one of which bore no characters; the other was an advertisement for the Perm hair pack, above which the tiny stripe of exposed window was very dark. Above that was another dark window, which appeared to be frosted halfway up the inside. Between the snow and the door, the threshold had been swept or shoveled clear, not today but recently, and a few footprints darkened the thin white dusting that had accrued there like an extra microsievert on the dosimeter—which I pulled out of my shirt pocket, pleased to see that its reading remained unchanged since Iwaki. The barber pole was clean. I could not perceive anything at all sinister. As an Iwakian might say: "It's invisible, so I don't feel anything."

Kawauchi barbershop, 2011 and 2014

Indeed, the dosimeter held steady for the entire hour and a half I spent in Kawauchi, which surprised me after the newspaper's claims about the failure of decontamination, but Mr. Kanari had told me that whenever Iwaki city workers took scintillation counts over snow, that reading likewise would be lower. And perhaps Mr. Yamasaki had exaggerated; maybe the Japan Broadcasting Company was spot on.

Walking farther eastward on that street which used to be the dividing line between restricted and voluntary evacuation zones, the interpreter and I reached a house which I remembered as empty. Inside, a very little old woman, bent over, peered at us through a window. When I rang the bell, she came to the door and pretended to be deaf, but while the interpreter was sweet-talking her, there came a moderately elderly man (he was 55, a year older than I) in a dark down jacket, with kind eyes and creases in his tight thin face. He answered my questions there on the snowy porch.

"That's my mother," he said. "My parents live here. They came back in November 2012."

(It is worth reminding ourselves here that at this time officially *nobody* was supposed to be staying here overnight. Thus the leniency of Japanese enforcement, which we shall have occasion to see several more times in these pages.)

"How much of Kawauchi is inhabited now?" I asked.

"About half, but that actually means four days a week; then they return to their temporary housing in Koriyama and elsewhere. Only about 20% are settled for good."

That explained why the place was so quiet. On the weekend most people would stay away.

"So your mother feels safe?"

"Because she is elderly, it's no use worrying about radiation. Of course she feels lonely; that's why I am visiting her. The children will never come anymore. When there was a funeral" (I think he meant his father's), "ordinarily we would have held it here, but we had to arrange it for a different place. The grandchildren wouldn't come; that is the result of some *fear information*."

"And how do you feel here?"

"Me? To carry out ordinary activities it's fine here. In Koriyama the radiation is higher than here! That's on account of the wind. So there are those who return here, but those who have settled in Koriyama and other places, they get used to the life and the conveniences there; now it's already been three years . . ."

"Would you yourself ever come back?"

"No," he said. "Actually, I would like to take my mother to where my wife is. When the road is widened, we will have to take down this house anyway."

"What are your conclusions about the nuclear accident?"

Until now we had been standing outside, but at this he invited us in, and the old mother began to make green tea. "Because my mother doesn't have many friends and the grandchildren never come, she's happy to serve tea."

"I used to work for Tepco," he began. "But I was not directly involved in the nuclear power plant. My work had to do with buying and leasing of land for construction. Basically, my personal opinion is that given that original 1960 design, it was risky to choose that location at all. The major problem with the emergency power generator was that it was installed at sea level, not in the correct place. All the contract workers were saying so, including me."

"When the accident occurred, were you surprised, or did you think this was to be expected?"

"The upper level management and I all entered the company at the same time, so we consider ourselves classmates. All of us can talk honestly in spite of rank differences. None of us could expect this bad situation."

"Do you consider nuclear power safe?"

"A nuclear power plant, yes, you can make it safe. I share Prime Minister Abe's opinion: It's good to gradually reduce the use of nuclear energy, but now it's like our blood, our life. All the nuclear plants are currently stopped. You cannot maintain this situation forever," he said as our emerald green tea was served by the grandmother in her green kerchief and baggy quilted clothes.

"What will Kawauchi look like in the future? Will it come back?"

He laughed. "Probably it will eventually die out. The families with young children won't live here."

"What should people in America and other outside places learn from what has happened here?"

"The major problem, well, for Tepco it was only a matter of cost that prevented them from building a dike or siting the emergency power generator in a higher place."

"When will it be safe to walk up to the reactor and touch it?"

"In my opinion, from Highway 6, which they opened just today, all the way to the reactor site, probably that entire area will be closed, and maybe in a hundred years there will be a sign stating **THIS IS WHERE THE DISASTER OCCURRED A CENTURY AGO**. That area will be used for nuclear waste. That's just my idea."*

"Do you feel any grief or regret about Kawauchi Village? What do you miss most?"

The interpreter said: "He doesn't answer directly. But he says: I have land in the Difficult To Return Zone. Due to the road construction, we knew that we must demolish this house, and we had hoped to relocate onto this other land. When my mother heard—you see, she's 88 years old . . . and when I become as old as she, in 30 years, well, I just feel . . . Well, you can't help it. There is nothing I can do, nothing in my power. Therefore, one can only change one's attitude. I have to be positive, considering that this is a kind of opportunity.† What is lucky is that since the road must come through here, we have to move. The other houses will stay, but no children will ever come."

Presently he added: "The Kawauchi village head wishes to increase the population, of course, so he proposes to settle decontamination workers here . . ."

"How long have your family lived in Kawauchi?"

"Maybe a hundred years."

---

* As might be expected, the public relations men whom I interviewed later that year at Tepco's headquarters in Tokyo were more optimistic about Plant No. 1. Mr. Togawa Satoshi of the Corporate Communications Department told me: "Regarding Reactor No. 4, it will be over within this year. For 1 to 3 it's more difficult. Inside the reactor, the melted fuel has to be removed. This is difficult work under very high radiation. Unlike in regular work, we need to develop new technology. For 30 to 40 years we'll have to keep doing this work, and then finally we will demolish the building."

† How often in Japan have I heard such nobly cheerful aphorisms! Let me remind you of Mrs. Utsumi Yoshie, the lady in Ishinomaki who had lost so much in the tsunami and who told me (p. 290): "The passage of daily life will create another sense of value. Unless you think that way, you cannot advance."

The old lady, who no longer pretended to be deaf, had become ever more animated as the subject turned to the family and its history. Her son said: "Even from the time of her grandparents we have been here. We once moved to Tokyo, but we moved back."

The old woman interjected: "My husband was born four kilometers from here. He died last May."

"What have you liked most about this place?"

She smiled. Her teeth were lovely white. She said: "The air is clean!" (I checked my dosimeter.) "It's a good place. I went to Koriyama after the disaster, but I came back here."

"Is it safe to grow vegetation in the soil now?"

Her son replied: "I think so. I'm not sure. The health bulletin warned us not to eat mushrooms or honey. We were all so shocked! People no longer go deep into the woods to gather mushrooms. After the disaster, the number of wild boars increased. But if you kill one, you will find that the meat is shockingly radioactive."

In that small tatami room, with the old lady half tucked in under the blanket, I watched the clouds and dripping snow through the window and thought what a pleasant home it was. I told them so; the mother seemed happy.

Like so many others, they did not wish to be photographed. He wished his name to appear here as Mr. M. Y. An engineer for 15 years, he had worked for Tepco for 33—until 2011, I presume. I asked him how he was getting along financially, and he said that he was managing.

MY THIRD TOMIOKA IDYLL

Now the sun had come out; it was warming a little; and there were wavering shadows of a railing upon the sunny snow from which it barely emerged, drops of water sparkling on a young pine before a snowed-in deserted house, blinding heaps of snow.

As I said, the dosimeter remained unchanged during that entire hour and a half in Kawauchi. A red-cheeked municipal worker was standing by the snowplow, directing traffic with his signal flag right there where the checkpoint used to be.

I asked the taxi driver what he made of it all. "This is the first time I've been here since the disaster, so I don't really know. Certainly some companies start to come in gradually and along with that some people will come back, but young people haven't come back."

Presently we were returning down the snowy hill into Tomioka Township, with neatly numbered bags of radioactive waste at one snowy roadside point between tunnels; then I got a glimpse of the blue and doubtless radioactive sea to the left.

As we reentered the town itself there were so many bags of waste. In the driver's opinion we were maybe 15 kilometers from No. 1. The dosimeter was still at 0.278.

The first few houses looked better than the others, perhaps since the ground beneath them had been firmer, the driver said.

Here came more of those rectangularly organized bunches of waste on the fields beneath the clouds, then a broken greenhouse.—Pointing at a rectangle of bare dirt, the driver explained: "This was all vegetable field. Where you see grass, they haven't decontaminated yet."

We had now reached what must have been the city center, where streetlights were glowing even then in the middle of the day, probably to deter theft, which the driver said was frequent. Later that year I attempted a night visit, and found the whole district barricaded off; that was how I learned for certain that this was the yellow zone. A frozen convulsion of blinds hung in a shattered window whose shards lay dustless in the street, so that the damage must have been recent and therefore caused by robbers or wild boar.

Central business district, Tomioka

I requested him to stop. Wandering here and there along that long still highway with its dead shops and homes on either side, and all of these shuttered and darkened, some of them smashed, with weeds standing dead before them on the snowy sidewalks, I tried to learn something that I could tell you here, but my thoughts and speculations resembled the broken glass breaking again beneath my heels.

I asked the driver what he thought of Tomioka. He said: "When I see this, I have no words."

Interior of garment shop

We entered a garment store, where some of the forms lay cast down, the rest still standing like sadly eerie silhouettes of headless legless armless women; and a vacuum cleaner had fallen over on its side. A portion of the rack nearest to the road was stripped, as if the robbers had begun there, then gotten scared and rushed away. In the back office with its old telephone, most of the books and ledgers were still in place, but papers lay all over the floor. Had someone been hunting for money, or was it the earthquake that had done this? Returning to the dark cold show floor, gazing out at the snowy street, I stood for awhile, trying to understand. I cannot certainly say why it was that of all the places in Tomioka I visited, this felt the eeriest. A tinsel wreath lay at my feet, and on the pillaged rack was a sign for a 50% off sale. I almost found myself listening for something or someone, but that was nearly as futile as expecting there to be life within the dark forms of these dressmaker's dummies. It was too dark to read the dosimeter, so I stepped out into the snow, wind and broken glass of the main street, where weeds marched in twin rows, one on each edge of the sidewalk. Across the road, more of those disarticulated blinds were twisting down with all their metal fingers gaping open and apart like frozen broomstraws or great pine needles. A broken shutter was trembling, creaking and clattering ever more loudly, and some other metal thing was hissing and whispering almost like leaves.

In those cold dark doorways and burgled storefronts, as in the deceitful snowy softness of Kawauchi, I should have been able to find something that would help me know *how bad it really was*. As I paced here and there, longing to understand this, my feet crushed more of those great and small shards of glass on the sidewalk which lay reflecting clouds. Who could tell me what I wished to learn?—Not the driver, perhaps not Mr. Kanari and Mr. Kida, maybe not Mr. Yamasaki and certainly not one of those dressmaker's forms cast down in broken glass.

The dosimeter remained at its Kawauchi reading. I had the driver continue on to **TSUBA** and wait.* As I got out of the taxi, on the loudspeaker there suddenly came the voice of the *robot girl,* less distorted and echoey than before, probably just because I was used to it. She was saying: "If you incinerate something or use incense, be careful not to start a fire."

---

* For posterity I would like to report that near us was a sign for the History and "Forkways" Museum.

Across the street from the garment shop

As always, I thought to go inside **TSUBA** and explore, but there were many weeds and shrubs between the place's gashed facade and me; I wondered how much fallout they held.

Down an empty road the interpreter and I began walking east, toward the ocean. There came a scuttering of darkness in a deserted greenhouse. It was just the reflection of my moving knees and lower legs, but my heart jumped.

The *robot girl* was now reminding Tomioka's former inhabitants: "For temporary housing people, make sure to kill the breaker before you leave, and lock the door to prevent thieves." This was what Mrs. Kuwahara, Mrs. Yoshida and all the others would hear on those rare occasions when they came to check on their desolate homes.

Pointing, the interpreter said: "This must have been a very nice house. The owner must have been very proud."

There were dark clouds over the working class houses with their yellow danger tape. Perhaps it would rain. I tasted metal in my mouth; I don't know why.

Wandering over one of the barriers, I reentered the forbidden zone, but the weeds were high and it seemed foolish to touch them. Half a kilometer away, a traditional Japanese house stood on the horizon. The roof was damaged. For some reason the place drew me, but I could think of no good reason for going there. The landscape looked more or less the same; all I would have done would be to subject myself, and, if she came, the brave and faithful interpreter, to a greater risk of cancer. Never once had she ever said: "I won't do this; it's too dangerous." Who was I, to abuse her loyalty like some Tepco manager? So I stepped back across the barrier where she stood patiently waiting. Then we resumed our eastward promenade, passing a lone sewer camera worker. It began to seem more like a country walk; there was even a single bird. The forest was more unkempt than wild. Presently we reached two more bags of radioactive waste. One was black and one was grey. They made me worry, but the dosimeter did not increase—oh, yes, exactly now it went up another point (coincidence again).

Returning to **TSUBA** where the driver waited, we began to roll home to Iwaki, back through the wall of mental ice. To please us the driver made a brief side trip to the second checkpoint of Nuclear Plant No. 2, where we had to turn around before the gate. Then we continued south. Not far over the crest of the hill was a large half-wrecked Japanese-style house, almost a mansion, and on the grounds toiled a squad of decontamination workers. There were patrols and platoons of others all about, slowly scraping dirt and leaning on their shovels beside the black and grey bags there in Naraha.

On a lovely bluff, we stood looking down at the Kido River, whose scenery was now decorated with black bags. The tsunami had come here, but I could see no sign of it. The driver said: "This river is very famous for salmon. Now they still come upstream but we're not allowed to catch them."

On the bluff behind us there were so many black bags and workers swarming around them—evidently a joint venture of Maeda, Konsike and Dai-Nippon Civil Engineering. It was disgusting how the black bags went on and on.

5: The Harvest Festival

"No, I'm not worried about any thieves," answered Mr. Endo Kazuhiro,* "since my home is not gorgeous. I wish the thieves would clean it up! I was born there. I'm the fourth generation—more than a hundred years, almost 200. The house is not gorgeous, just old. The roof is traditional farmer's thatch."

"In what district is it?"

"To the west, along the Kawauchi road. It looks all right from the outside," he laughed. "I was there last year in summer. I only went to visit my ancestors' tombs."

"When was that?"

"On the first of August. I think it was about four microsieverts"—per hour, it

---

* There are countries, not to mention organizations such as the U.S. Army, where it is best to ask questions at the bottom of the hierarchy. In Japan it tends to be the opposite. At Takaku I finally listened to people's advice and went to the head of the residents' association, who was Mr. Endo.

must have been, although if he had mostly stayed in a vehicle that could also have been his total accrued dose. In company with everyone else, he was unsure of the measurement units. Thus one of the reasons that Iwaki people threw up their hands and shut out harmful rumors. I might have done the same.

"What did you most enjoy about your childhood in Tomioka?"

"I lived close to the mountain, and in my childhood I seldom came to the sea. The landscape was prettier then than now. In the past there were almost no roads like this"—as we sped by a row of black bags on an Iwaki beach. "The kids played hide and seek and kick the can. We waded in the river. There were wild boars and foxes—also raccoons, although we didn't see them so often. Now most of the humans are gone, so the animals might be more rampant."

"You don't think they die from radiation?"

"I don't think so."

"So they're stronger in that way than humans?"

He laughed. "I don't know. There are some abandoned pets..."

He always liked to drive alongside the ocean as far north as he could. He did not recommend the swimming nowadays. In between black bags the shorescape appeared lovely and even harmless. I asked him to tell me more about his life. He said: "In my 20s I went to Tokyo for 10 years. I learned a lot of bad things, like drinking!" He laughed again. "In Tokyo I was a truck driver. After I came back here, I was employed by a waste disposal company from my 30s, collecting garbage, sewage and so on. The nuclear power plant had been built even before I was 20."

"Based on your experience in the field, is decontamination effective?"

"Not at all! It's the most wasteful, useless activity."

"Who's making the money?"

"General contractors from Tokyo City and this prefecture. If we all work together and voice our opposition strongly, we might be effective, but as it is we're too weak."

On this first springlike day when suddenly the birds were singing, I would have rolled down the window to breathe fresh air, but Mr. Endo liked the heat on, and anyhow the breeze seemed to be blowing directly south from No. 1. Halfway to Tomioka the dosimeter had already accrued an extra microsievert. This made me pull it out of my shirt pocket more frequently, but after that it remained at 0.287 accrued millis.

I inquired regarding relations with the Iwaki people, and he said: "The area around our temporary housing has good households, and the local elders take good care of us. In that area they're nice to us."

Then I asked, just to see what he would say, whether he considered Iwaki people selfish,* and he readily changed his tone. "It's just rumor, but I heard that one former Tomioka resident purchased land in Iwaki and built a house. Then he went to greet the neighbors as is our custom, but they said: *No, we cannot be friends with you because you are from Tomioka.*—I hear that the city of Iwaki has accepted from the central government forty or fifty thousand yen per refugee. So it's not right that Iwaki people be so mean."

According to his information, the temporary housing was supposed to close "in about five years, although this may be postponed."

"After five more years of radioactive decay, will Tomioka be safe?"[†]

"Regardless of natural decay, I believe that it will take 20 years." For an instant he fell silent, as if he might be contemplating those 20 years. (As you and I now know, 20 years would fall short of even one cesium half-life. But decontamination and rainfall might work wonders.) Then he said: "Elderly people will die naturally before radiation cancer affects them, so they wouldn't mind returning. I myself won't return. I don't know if my children will come back."

His eldest child was already 33. "Regrettably, none of them are married yet, so I have no grandchild," he laughed.

"How is your wife managing nowadays?"

"So-so. At least she's not sick."

"Do you know anyone who has come down with radiation sickness?"

"I believe that some of the workers at the No. 1 site are sick, but this information won't be leaked; the families will be paid off to keep it secret."

"What do you miss the most about Tomioka?"

"There was a shrine at the mountain, and we always held a torch festival there. This event I cannot experience again," he said in his gravelly voice, steering one-handed now to gesture at the air. "And I used to fish in the river. We played in the mountains in our generation . . ."

"How did Tomioka get its name?"

"Ah, I don't know, but the character containing *oka* means 'hill.' *Tomioka* may mean 'rich hill,' but since the township is based on the consolidation of several small villages I cannot say exactly."

Then he remarked: "When we were told to evacuate, everyone thought we could come back in a few days, so we didn't take so many things."

"When your friends and neighbors return to Tomioka, how long do they feel safe staying here?"

---

* Last month some kind soul had visited several temporary housing sites in order to break car windows, and in December a certain city office had been grafitti'd as follows: EVACUEES GO HOME. I'm sure they wanted to.
[†] See II:625–26.

"Six hours, and only during the daytime. You're not allowed to stay at night. There's electricity, but no water, so you can't stay. If you pay for electricity, it will come. But no water, not yet."

"What's the longest anyone has stayed?"

"Six to eight hours."

"How do you feel about Tepco?"

"When we were young, when the Tepco plant was built here, we were very thankful to finally be able to work locally. In this area, when you say, I got a good job, that means, I'm a Tepco employee or a civil servant."

MY FOURTH TOMIOKA IDYLL

As we reached the border of Tomioka Township, that familiar tingling metallic taste came into my mouth—surely psychosomatic. The snow had almost all melted as we came down the hill, slowed by a crush of construction vehicles. Now we had already passed No. 2.

"What do you think about nuclear power nowadays?"

"At this stage, I'm against it, because the response to the accident has been so poor. If the government cares to restart any of these nuclear plants, they'd better do a much better job."

When we rolled into Tomioka the dosimeter was still at 0.287 millisieverts, and I expected it to accrue digits rapidly, but in fact it registered only one additional microsievert during our two hours there, the most non-radioactive visit I ever had to that place, at least so far as gamma rays were concerned. If I claimed otherwise I would be spreading harmful rumors. The fact that the breeze now blew from the northwest must have borne on our pleasant situation, and it might have advantaged us that we soon turned off Highway 6, which Mr. Kanari had measured for me just the other day with his scintillation meter: 3.72 microsieverts per hour, as you may recall—doubtless the result of all those decontamination trucks grinding in and out of that main checkpoint. While we were still south of the pachinko parlor **TSUBA**, Mr. Endo swung east, and much more quickly than I had expected we came in sight of the sea, which, to be sure, still lay some distance away, across the tsunami-leveled plain of grassy foundations. We parked at Tomioka Station, and I asked to stroll about for a moment. A television crew happened to be there; its members wore paper facemasks which might at least have kept the alpha and beta particles out. As for us, we amused ourselves with the old joke that there was "no immediate danger."

So we looked across the ruined railroad tracks at the forgotten foundations and high grass; we gazed at the green-blue sea.

Remains of Japan Rail station. For a view of this place in 2017, see II:623.

"I'll bet there are deformed fish over there," he said with a grim laugh.

Now he wished to go. Begging his indulgence for another moment, I walked up the earthquake-damaged stairs, following the helpful arrows in the direction of the former Iwaki line. Fortunately the stairs did not give way, so I crossed the high enclosed bridge over the tracks, stepping on broken glass and whole panes, all of which lay fairly dustless. I stood looking at the sea. It would have been pleasant to wander through that tall grass and come to the shore, but Mr. Kanari's vegetation measurements the other day had made me more cautious, perhaps because the freshening breeze now changed direction and blew dust in my face, so instead I returned across the bridge to the near side of the station, where cables dangled down like fingers over the platform, and I breathed in that cool sea breeze, no matter how poisonous it might have been. Which plutonium isotope had escaped from No. 3 in 2011? Was its half-life 253 minutes, 88 years or 24 millennia? The latter, warned a Japanese muckraker—for No. 3 was a plu-thermal reactor designed to produce plutonium-239. When I returned home, Ed Lyman consoled me: "According to measurement, the plutonium seems to be a relatively small fraction of the radiological hazard—one ten-thousandth or less."[*] But he added that all measurements of this disaster's nuclear contaminants

---

[*] The highest level of this element in 2012 was 11 becquerels of plutonium-238 per square meter, measured at Namie—only 1.4 times greater than what fluttered down in average everyday fallout,

remained incomplete. Well, well; in any case we must admit that this famous "pluthermal" reactor proved itself to be a true credit to the industry.

Now more than at any other time during this return to Fukushima I was reminded of how the coast of Tohoku had looked three years ago, just after the earthquake and tsunami, for here by the seashore remained wreckage and the broken-open ugliness of things, cars slammed into crazy positions . . .—but even in this place I felt a slight sense of mellowness, since the tidal wave's mucky filth was gone, and grass had begun to soften the dead streetscapes into something forlorn and shabby rather than outright raw. In some of this weed-grown decrepitude there had even begun to shine, at least in my eyes, those classical Japanese "rusty beauties" called *wabi* and *sabi*,* but of course they must have been poisoned.

On the west side of the street stood a Chinese restaurant with a square hole in it.—"I think there was a second floor," said Mr. Endo.

"How was the food there?"

"I don't like Chinese cuisine anyway. I was not impressed, but then I was not even impressed by Yokohama's Chinatown."

I expressed my pleasure at the excellent radiation reading, and Mr. Endo remarked: "They say it's pretty low over there."

"See that one?" he said, smiling as if it were a joke. "That building had just been built!"

We stood on that still street with the brown weeds growing up between wrecked cars, and daylight shining through a hole in the back of an apartment building. Then we reentered his car and began rolling up west again, with the high grass thickening as we went. I glimpsed yellow tape over an especially ruined house, a patch of snow beneath a shaded house, down through this sad weediness and up again, the visible damage rapidly lessening now that we were out of the tsunami's reach. When I remarked on this, Mr. Endo said: "They may look good on the outside, but once the roof tiles are gone, the rain comes in."

A single policeman stood by a checkpoint at a narrow street. The light had already begun lowering, and as we ascended westward there were more patches of snow in the brown grass.

"Here in the beginning along the river there were a lot of loose cattle," Mr. Endo said. "They had to be killed because they entered houses."

---

courtesy of the nuclear testing club of nations. How fortunate that it wasn't Pu-239! But the time for the latter isotope's radiation to reduce to 1/1000 was a mere 877 years.

* *Wabi* is the beauty of loneliness and infinity, as in stars in the night sky or some wild island promontory. I saw this in the distant radioactive ocean. *Sabi* is pleasant decrepitude, exemplified by an old peasant hut. For more discussion see my *Kissing the Mask*.

Rolling past what appeared to be a vast vacant lot in front of some decrepit and vaguely official edifice, he said: "There's the elementary school from which I graduated. At that time the building was still wooden."

There was never any self-pity in anything he said. He was such a strong and cheerful man, so Japanese in his bright brave patience.

Here came a ruined neighborhood, two black bags by the roadside, and then we turned down a boulevard of leafless cherry trees whose tops reached toward each other across the road and nearly met above our heads. Mr. Endo said: "When they're in bloom, it's so beautiful, like a cherry blossom tunnel! But now they're not maintained."

This district was called Yonomori (which perhaps means "night forest").

He pointed out the branch railroad station. I could hardly see it through the weeds. Then came more black bags, snow and brown grass. Two workers were loading black bags into a truck; we passed a lovely bamboo forest, and then a man in white coveralls was feeling the ground with some sort of wand, perhaps a scintillation counter. Now we had driven into Chuka District—which was an old name, he said, not a formal one.

"There's my house," he said, pointing through the snow. "But it's overgrown now, not really presentable."

I said that I would very much like to see it, so he turned in past a lovely dirty mound of snow, carefully driving over slushy ice on a narrow uncleared road, the car swaying and laboring; I wondered whether we would get stuck. Then over the snow-covered road a bird flew up as the brassy female voice roared about protective gear.

"That's our garden," he said. "But the bamboo grew! It's a forest!* Three years ago there wasn't any."

Walking through soft snow, pushing our way through waist-high weeds, I wondered how contaminated these surfaces might be. One block away the red zone began; later I would stroll up to that barricade, but without a scintillation counter, what could I perceive there but harmful rumors? There loomed a dead freeway underpass, out of which an unpleasant breeze blew into my face. I preferred the windless thicket around Mr. Endo's house. Meanwhile the dosimeter remained at 0.287, but that meant nothing; even at 72 times Iwaki's radiation level it might not turn over another microsievert for a quarter-hour.

---

* Such thickets did not even offer the saving grace of sequestering CO_2 in proportion to their thriving. From the 2015 Japanese greenhouse inventory: "Carbon pools in bamboo in forest land remaining forest land are all reported as 'NA' because annual growth and death of bamboo trunk in established bamboo are equivalent."

Mr. Endo Kazuhiro in front of his former home

So we entered the bamboo grove, which had grown up around three parked vehicles, and was pretty enough in its way; and presently we reached the house.

Mr. Endo led the way around two fallen lockers. Once upon a time they had secured his tools. Removing a spacer from the track of the glass door, he slid it open; there'd been no reason to lock it.

Rat droppings speckled shoes, boots, a fan, a vacuum cleaner, shelves and chairs, which were all tipped and tilted by the earthquake into a waist-high jumble within the dankness of his former home. He did not invite me in and anyhow I saw no point in further trawling through that melancholy disorder, so I took a photograph or two; then he closed the door.

Brown wisteria had grown up over the house. The *robot girl* was roaring and echoing.

"What did you take when you evacuated?"

"I got my bank book and family seal, and my parents' portrait."

"When was your wife's last visit here?"

"She came just once to visit the tombs of our ancestors. Otherwise she prefers not to see this."*

---

* For all I know, it may have proved easier for her than for him to start over. She was not a Tomioka native; he met her in Tokyo. "She was working in a grocery store. My friend was the go-between. She had trouble at first in Tomioka. The old people asked her so many questions."

There he stood, with his cap tilted upward on his forehead, staring at the wisteria whose creepers aimed at him, bullwhips frozen in mid-crackle, reflecting themselves in the front window, hanging off the roof. I wondered whether he felt any impulse to fight them.

He showed us the family persimmon tree, remarking: "It was here when I was born." The fruit was of the bitter type, most suitable for drying.* Beside it grew their loquat, whose fruit, he confessed, was also not so delicious.

There in the fallen leaves beneath the bamboo lay his daughter's abandoned car along with two other old vehicles which he had used for storage back in the day. The doors were unlocked, of course. Mr. Endo thought for a moment, then took out the young woman's fishing rod from the back seat, remarking, as if this feature had once pleased him: "This rod extends for 10 meters."

We wandered around his archaeological site. I could tell that he wished to be gone, and I felt guilty to have dragged him here, but I also thought that coming here with him was the right thing, for my sake and yours.

He was now 62. When he was 60 the accident had occurred. He said: "I wanted to be a subsistence rice farmer in my retirement. Right here by the road, this was my field, 300 *tsuba*.† The other one was my neighbor's. Elsewhere I had two other ricefields; one of them was near a pond. Well, that was my dream."

"Will you ever live here again?"

"The house is contaminated with radioactivity. Once the waste site for the location has been established, we will demolish the house. Owners are compensated in proportion to tsunami and earthquake damage. My house is less than half collapsed, so I must take care of the demolition myself. It would have been better if it had collapsed."‡

I wish you could have been with me in that tall, brown grass, with those head-high weeds all around and his house half vanished under the bamboo. (Call it another manifestation of *sabi*.) There we stood in the cool wind and the snow, and that home of his so shabby and forgotten with the wisteria crawling up the roof and the bamboo already three times as high as the house.

"Soon the wisteria will crush it," he said, almost smiling. "It's more dangerous than the bamboo."

---

* Japanese persimmons are of two kinds, sweet and bitter. Only the sweet tastes pleasant straight from the tree.

† About 991.8 square meters.

‡ It is peculiar to me that meanwhile the central Japanese government had allowed my first Iwaki taxi driver, the one who called himself dull, to collect some reimbursement from fire insurance for his damaged house even though a tsunami has little to do with flames. Perhaps Mr. Endo had no fire insurance.

Across the snow-drifted street from his house was a modern brick build-ing where the neighbors once dwelled: a husband and wife with three chil-dren. "They live in Iwaki. Sometimes we see each other and drink together."—I remarked that their house looked pretty good. He said, "Inside, it's as messy as mine."

Then we began to drive away, with the car sliding and almost sticking in the snow. I felt a quiet horror . . . but hordes of crows were cawing over the fields. With their rapid metabolisms, shouldn't they have quickly died? Well, perhaps they did, and new ones flew in. I remembered how Mrs. Kuwahara at the To-mioka Life Recovery Support Center had insisted that she never saw dead ani-mals except those that had starved to death. The rats in Mr. Endo's house certainly survived long enough to breed. And, as Mr. Kanari had said, "Inside the house shouldn't be such a problem" with radioactivity.

So what if my 10 half-lives calculation were faulty? Weren't Kawauchi and Hirono partially open to return, with Naraha regaining some daytime use? Ac-cording to that intrepid nuclear experimenter, the U.S. Army, after the first 48 hours *the extent of radioactivity will be reduced a hundredfold, enabling you to leave the shelter . . . At the end of two weeks, the radiation level will decrease a thousand times.* If this were true (as I wished it to be), then the Iwaki people would be correct in their quotidian calmness.

"The issue," remarked Ed Lyman, "is how uniform are the hot spots?"

He believed that the levels were dropping more rapidly in urban areas, whose non-absorbent surfaces caused the cesium to wash away (into the water table, perhaps); while forests and open meadows retained their higher counts. It seemed only right to him, and I agree, that every returning resident be given a good do-simeter, and that anyone who did not wish to accept a one-thousandth concen-tration of poisons over prior background levels be compensated and permitted to leave.

Mr. Endo meanwhile asserted: "The government says they are trying to force the people to return."

If I were him, I would stay away.

We drove along. Certain deserted homes sometimes looked less unkempt, so I inquired: "Are the decontamination people actually pulling up weeds?"

"It simply appears that way because the weeds die in winter and then new ones grow up."

Mr. Endo now pulled in at one of those Tepco establishments in which the *robot girl* kept advising decontaminators to shed their protective gear before leaving the exclusion zone. This station, the only one in its area, was near the

former Futaba Rose Garden. The Tepco men here called themselves the "voluntary checking forces." Mr. Endo wished to have his daughter's fishing rod tested.
As I stepped up onto the sidewalk to photograph this action, which involved
a device similar to the scintillation counter I had seen, a Tepco man immediately cautioned me to move; they had not yet decontaminated that spot,
and while the street was a mild 0.95, the "background" read 500. When I asked
what the units were, he replied that they were cpms, counts per minute. I wished
to know how cpms related to microsieverts per minute or per hour, but the
man didn't answer, perhaps because he didn't know. Five hundred micros
per minute would not have been too healthful, so I certainly hoped that cpms
were a different animal. When I asked the Tepco man what kind of radiation his counter measured, he answered: "Beta waves." This of course was what
mine could not measure. The limit was 13,000 cpms, they said; "above that
we clean."

Frisking the fishing rod

I asked one of them to test my shoes after all these times in Tomioka. He did.
I asked him what the count was, and he wouldn't say; he merely told me that I
had no problem. He scolded me for not having worn shoe covers and a mask.
Anyhow, he and his colleagues were all wearing naked tennis shoes.

Wondering how to interpret the Tepco units, I asked Ed Lyman, who said:
"Assuming that it's beta decay from cesium-137, which is most likely, and that it's
not contributing to a whole body dose, which is just an assumption, you could
convert cpms to millirems by dividing by 1,000."

In other words,

SIDEWALK:

500 cpms = 0.5 millirems = 5 microsieverts / hour = 43.8 millisieverts / year*

and

MINIMUM LEVEL REQUIRED FOR "CLEANING":

13,000 cpms = 13 millirems = 130 microsieverts / hour = 1,138.8 millisieverts (or 1.14 sieverts) / year

I could certainly see why Tepco might find it advisable to "clean" pollution of more than a sievert a year! Perhaps the interpreter had misheard, or the man had misspoken, and the correct figure was 13 micros an hour, how would I know?[†] Forty-four annual millisieverts was not so good, either. Upon me, of course, the sidewalk had bestowed perhaps [5 micros per hour × 1 hour / 3,600 seconds × at most 15 seconds, or] 0.0208 micros. Even an hour's worth of radiation, 5 microsieverts, would have constituted "no immediate danger."

But as for the sidewalk's level of 5 micros per hour, to me personally that was unpleasantly informative. Tomioka might be more unhealthy than I had supposed.—Despite that possibility, I let myself remain an honorary Iwakian, "because it's invisible." Since the alpha and beta in the windborne dust failed to register in whatever enumeration of my cumulative cellular damage appeared upon the narrow grey screen at the dosimeter's end, I managed to live the good life, only occasionally wondering what might be on my shoes. Mr. Kanari had said that he never worried much about radiation; when he got home he shampooed his hair and called it a day. Of course, Tomioka had been different even for him. Never mind! "I eat what is in front of me." "I don't think twice about it." "I'm kind of dull. I just think this area is safe," and I remembered a photo of the horrid hourglass of a beta burn on the neck one month after exposure—but hopefully that derived from an acute dose from a nuclear weapon, not from chronic exposure after a reactor accident.

By this time, poor Mr. Endo would have really liked to get away, but I asked

---

* A few months later, when I returned to Fukushima with that pancake frisker, which measured in both cpms and microsieverts per hour, my data did not verify this conversion. Instead of dividing by 1,000, I needed to divide by 300. But I was measuring alpha, beta and gamma all at once. For details, see the source-notes. This being said, I should add that plenty of sidewalks I frisked in 2014 read 5 micros an hour or higher.

† Based on what I was told in Okuma later that year (p. 484), cpms = microSv × 558, and the same cutoff of 13,000 cpms was given—which would have been a far more plausible 23.29 micros per hour.

if we could please stop by the shrine he had mentioned, so he kindly drove there; it was not far, the road was empty, of course, and the dosimeter remained steady. The characters on the gate read "Hayama Shrine."*

Leaving the car, we walked toward a wide hill of trees. Given what Ed Lyman said about radiation declines in porous *versus* nonporous surfaces, not to mention Mr. Kanari's measurement of that bamboo grove, I would imagine that the lovely shady place could have dosed us nicely with beta rays. And how much of a dose had we accrued in our earlier weed-meanderings? Fortunately, I prefer to think of radiation as a sort of vitamin. That way I won't spread harmful rumors.

"Whom is this shrine to?"

"I don't know. We just go there on the festival to pray for the good harvest. Young men climb to the top of the mountain; and then if you see the flame at the top of the mountain from your house, the harvest will be good."

The festival took place every August fifteenth.

"We used to circle this shrine building and then use the path to circle the mountain. At the beginning we'd run," he laughed, "but then we'd get tired. It takes 45 minutes or an hour. When we get close to the top we start running again."

Only men carried the torch. Mr. Endo had done it every year from age 16 to 40.

One of the two stone guardian dogs had fallen; and not far from his cracked corpse rose the wide stone steps, across which a heavy branch had lately fallen. A great block of joined stones had also toppled, partially smashing the concrete wall upon which it lay, half sunken in the earth behind.

At the top of the stairs I saw weeds on the edge of the shrine porch, darkness under the eaves.

"What happened to the priest?"

"There is none. On festival days we borrow a priest," he chuckled.

We stood looking down from the torus at a plain of black bags. Then we looked at the forested hill above the shrine, the place of dead leaves and lovely tall pines.

"It's hard to get up there!" he laughed.

In his harsh nasal voice he told me a little about bygone festivals, and like the old grandmother in Kawauchi he appeared to grow very happy, smiling as if with real pleasure.

"When I was a child there was some closed place up there, and I was mischievous, so I opened it, and lots of bees came out and stung me!"

I smiled back at him.

"In the summer there's lots of rehearsing," he said. "I can hear the drumming

---

* *Yama* [or *yoma*] means "mountain."

from my house, the drumming and the flute. Then I feel that I'm coming into festival season."

I asked whether anyone celebrated here nowadays.

"Last year I tried to get the young guys to practice with the drum and flute, but it didn't happen in time, so we played a CD."

"Up here?"

"At the temporary housing. Served some drinks . . ."

Again we were looking down from the chilly shadowed shrine. There was much snow, and a single bird. He said there were thrushes, sparrows, doves and pheasants hereabouts. He used to hunt pheasants with a certain friend, but the heart for that went out of him after his friend died.

Hayama Shrine

We descended to the car and began to drive away, across that plain of black bags and brown grass, with crows flying up everywhere. "Both my parents were from here," he presently said.

We turned onto the empty toll highway: 15 kilometers to Hirono. As a victim of the accident, he could drive it for free—lucky fellow! The sky was already performing its spurious brightening into evening light. Soon we would be on the safe side of that still unbuilt wall of ice, where people lived their lives as best they could, while past and future tragedies resembled those tsunami-murdered homes whose foundations were only now slightly more distinct than the boundaries of ricefields. Mr. Endo drove silently. When I thanked him for his trouble, he assured me that it was all right, since no one else had been willing to go, and, besides, he didn't really feel anything. I sat in the back seat, so I could not directly see his face; his reflection looked really sad in the rearview mirror.

6: "It's Not That They're Not Worried"

One Sunday afternoon the orator, a psychiatrist in dark clothes, stood frowning, raising his voice, fanning and chopping the air, leaning on the podium, waving, circling, nodding. About 300 people listened there in the auditorium of the sixth floor of the department store tower (which also contained the Iwaki city library). There were far more men than women. Few wore suits, for this event had been organized by the National Railway Mito Motive Power Union, some of whose members might or might not have been Communists. There came loud applause, a few listeners now raising cameras over their heads to videotape other speeches, and one man clambering up a small painter's ladder which he would later carry to the parade and set up at this or that intersection so that he could photograph from the perfect angle.

A Mr. Saito Eiichi from near Koriyama was saying: "There is no place for Fukushima people to throw their anger against. We feel the administration has abandoned us. In my dairy farm I cannot feed my cows with my own grass. It is forbidden. The grass, the hay and the compost are contaminated with cesium. They are just sitting there. Finally the city of Iwaki claimed to clean them, but just moved them from one place to another. The cows are having more difficult births."

A lady in black named Ms. Shiina Chieko said: "This is our fight for survival against all the contradictions," possibly referring to "capitalist contradictions." "In the temporary housing where I used to live in Namie, in two years, out of 20 residents, three* died."

Then came a lady who said: "Those who make the decisions are not affected by radiation because they live in Tokyo." She mentioned radiation-deformed azalea leaves. I requested her name for *Carbon Ideologies,* but she wished to be anonymous.

---

* I might have misheard this translated number.

(Left) Watching the anti-nuclear march

(Below) Anti-nuclear woman

Presently they went out to the street. There were whistles, drums, chanting, banners pink and yellow and blue and crimson. Some appeared almost ecstatic as they banged their drums or raised their clenched fists, others shyly or listlessly touching their hands together.

A banner said: **STOP ALL NUCLEAR POWER PLANTS**. There was a red labor union banner: **WE WILL NOT ALLOW ANY-ONE TO BE FIRED**. A banner said: **3-11 ANTI-NUCLEAR FUKUSHIMA ACTION**.

I saw hardly any spectators. In a travel agency, a young couple who were being waited on turned and frowned over their shoulders through the glass, while the clerk looked at us as if out of politeness because the couple were looking. A restaurant proprietor stood unsmiling in his doorway. Three pretty women peered out of a second-storey window, waving down at the parade. Two little girls were walking somewhere, and the younger wished to join but the elder was against it; the marchers tried to coax them with dancing and smiling, calling out: "Come with us, just to the station, only there!" but the children would not.

A man was shouting: "You guys did it, you large corporations and Tepco!

Marchers with police escort

Tepco men watching the protesters present their demands

Why should we have to suffer? Solidarity, everyone! Work together!"—a sentiment underscored by furious drumming.

There was a musical chant about poison rain and our children being guinea pigs: *"Stand up, Fukushima!"*

They went to the unmarked Tepco building and delivered their demands to a man who stood waiting on the sidewalk. He bowed slightly when he received their document. I felt a little sorry for him. Behind the glass doors, two other men in suits stood watching. I took their photograph, since someone on their side had probably taken mine.

Then what? The procession made a circle, and so we returned to that long six-lane boulevard* running south and skyscraper-shaded from the train station where on those February mornings I sat in the chain-brand coffee shop writing up my notes, watching the buses come and go, delighted by the gaggles of giggling schoolgirls in their short plaid skirts who were ascending the escalator to take their morning train; accompanied by the young men with parkas and backpacks, the salarymen with their briefcases and dark suits, the longhaired woman in a dark coat and white paper mask over her mouth—just people living their lives. If I have ever sounded harsh about the complacency of Iwakians, I apologize, for I never meant to be. In my home town I had a friend who skipped all news of global warming, "because what can I do?" In West Virginia many good people assured me that climate change was merely a harmful rumor. I preferred not to dwell on my own death. There was no immediate danger. I sat safe and happy in my city library and read old books by reactor men: *Criticality accidents are extremely unlikely in all Western nuclear designs,* promised one. *The likelihood of an accident leading to a breach of the containment is extremely remote,* assured another.

There was a gathering afterward in a smoky restaurant; perhaps 50 of the marchers attended. One was a longhaired redfaced decontamination worker; in the U.S. I would have thought him a sunburned Native American. He said that in Okuma he had measured 30 microsieverts in six hours.[†] He said: "I used to work at No. 1, and in six months I got 6.62 millisieverts, working 20 days a month.[‡] I was working there for 14 months, up until July last year. My job was to measure the vehicles' radiation. In one day, there were approximately 13,000 ppp"—but he could not explain what that was. Since this number exactly equaled

---

* Japanese streets rarely have names. Intersections are named instead.

† Or 5 micros an hour. I find this figure quite believable, having measured far hotter spots than this. (Thanks to being a vehicle-protected journalist, not an active decontaminator, my own dose in the Okuma red zone measured merely 3.033 micros per hour.)

‡ This would be 6.13 micros an hour (see source-notes). Had he been exposed to this level without relief for a year, he would have received 53.70 millis. But given his actual working hours, he would have accrued only 13.24 micros—within the ICRP maximum for professional workers.

the "13,000 cpms" above which the Tepco people at Tomioka also had to "clean," I will guess that ppps means cpms.—"If it shows that value," he continued, "it has to be cleaned, but even after high-pressure cleaning it only comes down by half.* In the mountainside we dug a groundwater contamination measuring well, and when we could not get the radiation value down on some cars we would just leave them there on the mountain. Finally Tepco changed the policy and stopped letting so many cars in. Because of them, the radioactivity of the road itself will not go down. So now J Village uses a bus, which will not go inside the restricted area."

"What does it look like at No. 1?"

"If the seashore is the edge of this table, then it's all tank, tank, tank like this, right at the edge" (or, if you prefer harmful rumors, "*leaking* tank, tank, tank"), "then it slopes upward into a pine forest. The trunks of the pines are more reddish than natural pines.† A lot of wild boar. They do like it near the decontamination areas, eating bamboo shoots! And there are eagles. But you know the phrase *eagle eyes*? Well, they don't have such good eagle eyes. I don't care to tell you further; you'd have to see for yourself.

"In my experience they are so sloppy at Tepco. The pipes should be color coded so you know which end should be tightened, but often both ends are the same color, so you can't tell right from left. That's a child's mistake."

He said that decontamination workers were allowed *800 microsieverts per day.* I thought that impossible: 100 micros an hour in an eight-hour day! But in a few more pages† I will introduce you to a former worker at Plant No. 2 who assured me: "When I used to work at the site before the accident, it was not unusual at all to get 100 micros within a nuclear facility building." You will also meet a refugee from Iitate who shouted about getting dosed, together with his family, at 95.1 microsieverts an hour.§—Back to my informant in Iwaki. If those decontaminators drudged away for a realistic period of twelve 20-day months,

---

* Confirming what Ed Lyman said about decontamination on p. 353.

† My college physics book, published 37 years before the accident, notes that "pine trees are particularly sensitive to radiation damage" and showed "deleterious effects" at 1–2 roentgens per day; 20–30 roentgens per day allegedly killed them. Since a rem is about a roentgen, and 1 rem = 10 millisieverts, then 10 to 20 millis per day (3.65 to 7.30 sieverts per year) would do the job. This would be a horribly dangerous amount of radiation, and whether Plant No. 1 was still generating such a field I cannot say.—As it happens, I have heard of red pines at Chernobyl, and also at Hiroshima. One of the two decontamination workers I interviewed in Iwaki had worked at Plant No. 1. He failed to notice any red pines since his visor restricted his vision and he had more important things to see. When I went to Okuma I got a pretty view of No. 1 from about 2.5 kilometers away. In conspicuous evidence was *tank, tank, tank,* but from that distance I could perceive no irregularities among the pine trees.—A Greenpeace study released in 2016 did not mention pines, but found "apparent increases in growth mutations of fir trees."

† Pages 448ff.

§ On p. 438.

that would be 192 millisieverts—one milli more than the annual level of that Tomioka drainpipe which Mr. Kanari had measured for me. (Could these figures be off? The legal limit was 20 millis a year.)*

"You're very brave," I said.

"Someone has to do it," he replied.

He claimed to have measured 1.6 to 3.8 microsieverts in any one of those black bags—"6 microsieverts if you stand beside it." These numbers likewise sounded high, but then I remembered that Mr. Kanari's readings for the black bags in Iwaki—0.35 to 0.38 microsieverts—were at least 30 kilometers farther from No. 1 than the place where the redfaced man worked. I cannot verify anything he said. Besides, who am I to know what harm the black bags caused? Did they or would they leach into the ground?

About the waste bags in Tomioka Mr. Kida had remarked: "Our test results prove that if you are five meters away there will be no effect."

"What if you stand against a bag?"

He smiled. "How can I say? Even if it were only two microsieverts an hour, maybe one month would hurt you."

The decontamination worker remarked, as if to himself: "There are hot spots, of course, from one block to another . . ."

He said: "Thirty to 40 die at the nuclear plant every year. This is secret.† They do not die at the site, but upon coming home. So many in Iwaki, too. High blood pressure or heart attack is the diagnosis they get. High blood pressure is a clear indicator of working there. In Minamata the disease is clearly shown by deformed hands, but here the sign is stroke. In the 14 months I was working there, two people died."

Who knows if that is truth or just another harmful rumor?—Not I.

"What should I call you?" I asked him.

"Anonymous, please."

When I asked him to smuggle me into the forbidden zone, he insisted that we would be arrested.

"Why do you think the Iwaki people are not worried?"

"It's not that they're not worried. It has to do with their income. Those who get some benefit commensurate with radiation damage are more interested. The ones not compensated are not interested. But deep in their hearts they are

---

* See below, pp. 450–51: By 2016, 176 Tepco workers had accumulated more than 100 millis.

† According to *The Japan Times,* "the first official fatality of the plant's decommissioning process" was a subcontractor who was buried alive on March 28, 2014.

worried. The Fishermen's Union is not controlling anyone's actions anymore, so they are free to fish. But they don't. So it's not that they don't worry."

They worried, or not.* The harmful rumors thinned as they spread, beginning to fade away. Two months later, for the first time since the disaster, Iwaki seafood was being offered for sale at the Tsukiji market in Tokyo. As an "intermediate wholesaler" explained: "Since they have been thoroughly checked . . . it's not good to remain swayed by bad rumors."

That was reported on May 18. In the very same issue of the paper, I learned that between the water intakes for Reactors No. 2 and 3, tritium levels had increased to 1,900 becquerels per liter. Between the intakes for Reactors No. 1 and 2, strontium-90 levels were up to 840 becquerels. Both of these set new records; hence the headline: **Fukushima radiation at all-time high**.

Would it be reasonable to wonder whether those radiocontaminants might also leak into the ocean? It might be more reasonable to ask what could possibly stop them from leaking.—Why, the ice wall, of course! That would save the fishes and seal away our radioactive future.† Perhaps that explained the demeanor of people in Iwaki, where by six a.m. the sky and high-rised streets had all appeared below my hotel window in shades of pearl, and seagulls called unseen. Rows of light-globes still shone on either side of each thoroughfare, and the dosimeter had registered a single digit of increase—the same as it would have done in Tokyo.

---

* A letter from the interpreter, March 2014: "After I spoke with you over the phone the other day about our possible internal exposure, I felt that my throat was somehow soar [*sic*]. I'm sure it was something psychological. I think I understand now a little better the anxiety of those people who are in the contaminated area."

† Two years later *The Japan Times* ran the following item: "Tepco admits ice wall efficacy still not known." "Not being able to provide an assessment" of that project, its head official said: "It's really unfortunate and I am very sorry." But why not hope? We were experts at that, back when we were alive. "If the wall succeeds," the daily radioactive outflow into the ocean "should reduce to about 70 tons . . . , from hundreds of tons."

Shinjuku, 2014

October 2014,
with a Hanford excursion in August 2015

THE RED ZONES

1: Over Ten Times the Level Measured the Previous Week

First those faraway Americans had forgotten about the accident*—after all, its fallout barely showed up in Oregon milk—and presently most residents of Tokyo (excepting steadfast Friday night protesters and a few tired self-educated Cassandras such as Mr. Yamasaki, whose NGO had tested my Iwaki tomatoes) took heart again, because Reactor Plant No. 1 had matured from a national calamity into the merest affliction of distant country cousins—three cheers for the wall of ice!—while in Iwaki people diligently defended themselves against harmful rumors by eating whatever was set before them, even as the decontaminators improved everything into neat islands of black bags—and by the time of my third visit to the disaster zone, when I managed at last, thanks to months of effort on the interpreter's part, to penetrate legally or quasi-legally well into the exclusion zones—yes, I even got to see and photograph Plant No. 1 from a convenient radioactive overlook!—it turned out that the people who guided me, which is to say those with knowledge and local experience—a district head, municipal employees, a former reactor worker who had been on duty at Plant No. 2 on and after March 11, 2011; and I should also mention the various taxi drivers and rental car personnel daily or almost daily associated with travel through the red zones—had become as blasé as the rest of us. At the Iwaki gas station where an old man and a young boy were cleaning our hired vehicle, which had just returned from Tomioka, Okuma and Namie, I asked the former reactor worker to warn them that they were scrubbing away fallout, which must be particularly

---

* A qualification, courtesy of *The Japan Times:* "But restarting reactors remains crucial because Washington, as reported in *The Wall Street Journal* in May, is pressuring Tokyo to do so."

spicy around the wheel wells—to which the old man courteously replied: "Never mind."—He and the boy went on scrubbing.—All hail! At last (and so it must be for you who live in my future) *radiation contamination had been normalized.**

It was invisible, you see. There was *no immediate danger.* They'd done it for the nation. One had to fight harmful rumors.

As for me, I felt the opposite, because at last I possessed an adequate toolkit, and measuring the real-time scintillations of the objects around me in those red zones proved unnerving.

THE PANCAKE FRISKER

Because the dosimeter (as I have so often complained) could only show accrued gamma radiation, not alpha or beta, and because it could not reveal hot spots with sufficient timeliness to prevent them from quite possibly hurting me, I decided to call Clyde up in Richland, Washington. He told me that what I required was a *pancake frisker.* It looked like a little yellow golf club, he said. He took his everywhere; it seemed to be a pretty swell machine. Scrap metal dealers appreciated it because it enabled them to weed out the odd radioactive nightmare in a pile of junk—an increasingly frequent situation, Clyde confided, at which point he and I shared a millisecond of telephone silence, to mark the eternal wickedness of this world. Extolling the device's sensitivity, Clyde now entertained me with the truckload of anecdotal hay that had alarmed some ingenuous soul who imagined himself into an emergency; one needed to remember, Clyde said, that alfalfa fertilizer, like bananas, contained a naturally occurring radioisotope of potassium.—By now I liked him considerably. He was as cheerful as a cricket. He even offered to loan me his very own frisker, but only if the results stayed off the record, since he had been too busy to recalibrate it. I replied that I might as well just buy one. The price was $750. Clyde gave me the direct number of the main office in Sweetwater, Texas. I fretted aloud that some other salesman would deprive him of his commission, which he had certainly earned with his amusing stories, but he said, "Don't worry; they take real good care of me."

First things first: The Sweetwater people warned that their product was subject to export license regulations, so I wrote a note to whomever it might concern that if I happened to carry my new toy outside the United States, I would exert every muscle to bring it home. That was good enough, they said, and which two units did I wish the display to read in? Out of loyalty to the dosimeter, I chose

---

* John O'Hara, 1961: "Those little bastards blame our generation for the state of the world. I think they're taught that in school and college. So they hate us. Really hate us, Ned. I don't think there's a God damn one of them that ever stops to think that we weren't responsible for 1929."

microsieverts; the other unit might as well be cpms, which the Tepco friskers had employed at Tomioka. So I placed my order, hoping that the pancake frisker could fulfill at least some of Clyde's claims. The company was busy; orders got backlogged (another omen that radioactivity would scintillate ever more brightly in our collective future); but a week sooner than they had promised, I received my Model 26-1, which was truly yellow, and did look sort of like a golf club. Eli in the technical support department advised me to slowly sweep it back and forth when I was in the field. I asked how well it could measure food.—"It should be very helpful, I would think," he said. He was an *au courant* and tolerant fellow, whom I ended up calling several times. He said that gamma rays should definitely register in any meal; as for beta, at least one of Fukushima's two common cesium isotopes would announce itself in a friskable sort of way. The iodine-131 might or might not be revealed; I consoled myself that hardly any of that *should* remain in the red zones, for it had been an immediate product of the accident, and its 10 half-lives occupied only 80 days. As for Nuclear Plant No. 1's continuous flow of ocean-bound tritium, those emitted beta particles were minuscule and might not be friskable. To measure their concentrations I would need, among other apparatus, a cart to carry my own private tank of radioactive potassium-10 gas. Beset by a peculiar feeling that my nation's guardians of airport safety might not look favorably upon such luggage, whether I checked it in the hold of the plane or brought it on board to stow in the overhead bin, I gave up on monitoring tritium. Who could say whether I would even reach the ocean? (As a matter of fact I did, in Okuma. According to the pancake frisker, the air dose there was astonishingly low. But I abstained from chasing waves.) What then about a river or a glass of water? Remembering what that lying salesman named Ray had promised me when I bought the dosimeter, I now inquired, with more curiosity than hope, whether I could simply sweep the frisker over either of those. Sadly, Eli explained that the hydrogen atoms with which water is afflicted tend to mask neutrons. But I hereby testify most heartily that the thing did everything that Eli and Clyde said it would. In Japan its readings correlated extremely well with those of municipal and prefectural officials; moreover, it was, as I'd hoped the dosimeter would be, *fun*. I immediately fascinated myself by measuring the radioactivity of my daughter's cat (0.12 microsieverts per hour, which was twice as high as my darkroom coating table and exactly the same as my kitchen counter). My neighbors in Sacramento took pleasure in the frisker at first, although the low and stable levels around there rapidly bored them; you in the future doubtless enjoy more interesting readings.

The pancake frisker was about 13 inches long from the top of its rubberized handle (which concealed two AA batteries) up to the forward point of its hexagonal head. Three buttons decorated it. When I pressed the leftmost one, the machine uttered a three-tone chirp not unlike the sound one of my sweetest girl-friends used to make when she climaxed. Then its screen lit up, and in NORMAL mode the frisker began to express the moment-by-moment scintillations of, say, the dining room table of an apartment in San Francisco: 46, 30, 35, 40, 26, 19, 27, 34, 22 counts per minute—and with each new figure the thing stridulated ador-ably. Anytime I wished, I could push the middle button, to toggle the units into microsieverts per hour, and from that dining room table I now present several of those: 0.022, 0.118, 0.50, 0.32, 0.20, 0.102, 0.122 and 0.118. These altered with such rapidity that in the red zones, where I was always in a hurry, I sometimes scribbled down only the first digit or two of each item in this ever altering series.

My friend Jay loved the pancake frisker, and said: "It's like having an-other sense!" This comment was on the money, for I can and will tell you what such places as Tomioka, Okuma or Iitate looked like to me, and observations of abandoned decrepitude do convey something useful, but what about an innocent-looking thicket or meadow? Please recall Ed Lyman's aphorism: "The issue is how uniform are the hot spots?" Without a scintillation counter I could not have ad-dressed that issue.*

In sum, the best way for me to complete my word-pictures of the red zones is to overlay my descriptions with numbers: Here is how that meadow appeared . . . and in that meadow a plume of pampas grass was emitting so many nasty microsieverts.

I ask your pardon if I now devote a further couple of paragraphs to the frisk-er's modes of measurement, and to how I interpreted them. Since these numbers are of extreme importance in my accounts of the red zones, I owe it to you to say where they came from. For any reader who is repulsed by arithmetic (and also for me, because the results bemused me), there is that basic comparative chart back on page 244, so that rather than wearying yourself with converting from counts per second to microsieverts per year, you can simply look up an interesting shrine torus or sidewalk-stretch and see how many tens or hundreds of times more ra-dioactive it was than my kitchen counter in Sacramento.

So. NORMAL produced a string of numbers. Well, in practical terms, how "hot" was that dining room table in San Francisco? Whenever I cared, I could push the frisker's righthand button. On my first click, the device would go into MAX

---

* Do you remember those hateful black bags behind the silvery wall in Iwaki? In March my only way of measuring them was to approach them at a distance, watching the dosimeter to see whether its displayed accrual would immediately increase. Since its righthand digit represented microsieverts, my local real-time perception of radiation was awfully blunt. Had my total dose display upticked by one within, say, 15 seconds, that would have implied a field of 4 micros a minute . . . 240 micros an hour . . . more than half a sievert per day—far more biological damage than I wished to absorb. And had the numbers in that narrow grey window remained the same—as happened—the radioactivity of the black bags would remain out of my ken.

mode. The display then showed only the highest reading. If the emitter were stable, the local maximum generally ceased to alter within 30 seconds. Here in this dining room the radiation usually got up to around 0.2 micros per hour, or, if you prefer to toggle into other units, 65 cpms. This proved convenient and useful in its way: Whenever officials of the Transportation Security Administration, a well-meaning but arrogant institution that sometimes bullied American air travellers back in the days when I was alive (their agents specialized in slicing the linings of my various suitcases), refused to tell me how powerful their X-ray was, all I had to do was forget to shut off my pancake frisker, which accidentally happened to be in MAX mode as it rode down their conveyor belt.* That told me just what I wished to know, and if you are curious you can look it up in that same comparative chart.

A second push of the rightmost button, and then a touch of the lefthand button, and the frisker entered what came to be my favorite mode, SCALER, which averaged all scintillations over the course of a minute: The dining table was 30 counts per minute, or 0.002 micros per minute. That worked out to 0.12 micros an hour—the same as my daughter's cat, and indeed my average reading for San Francisco. In actual fact, this was two and a half times higher than Tokyo's Shinjuku district had been on the day before the nuclear accident. In 2014, San Francisco and Tokyo were (by the frisker's measurement) nearly equally radioactive, which is to say well below normal background for many parts of the world.

Multiplying this innocuous number by 24 gives 2.88 micros per day. You may recall that in San Francisco and Tokyo the dosimeter usually turned over a single microsievert[†] in 24 hours. To have one device reading nearly three times more radiation than the other concerned me, and I spent some time considering it, because once I reentered the hot zone, my health would depend on the accuracy of my measurements. Calling the ever patient Eli, I told him about the granite countertop in the bathroom of my hotel suite in Charleston, West Virginia. I had measured it 10 times for one minute each in the SCALER mode. Usually the display read 0.005 or 0.006 micros per minute, but once it showed 0.004 and once it came back 0.007—a considerable variation, I thought.

"Well," said Eli, "there's bounce and sway on the needle, especially in the low values. Make sure you let that thing run on NORMAL in that energy field and let it settle."

I had done this in Charleston. So I asked: "Is it less accurate in a low-radiation environment?"

---

* Legal notice: This sentence was a work of fiction. And, by the way, West Virginian TSA officials were invariably friendly and permissive to me.

† Or 0.1 millirem in its original units.

"The hotter the field is, the easier it is for that thing to do its job."

He looked up the calibration test for my serial number. That particular frisker had measured 3,000 counts in a 10-microsievert field. The correct figure was 3,200 counts. "So it's just a little under, but within the typical range," he said.

I inquired what their typical background radiation might be out there in Sweetwater. He said that at the factory they took in about 10 microrems per hour, which would be 0.1 microsieverts—more or less what I received in San Francisco or Tokyo, and half of what Ed Lyman got in his apartment in Washington, D.C. That sounded reassuringly plausible.

Perhaps the frisker read higher than the dosimeter for low values because it was picking up alpha and beta while the dosimeter could not. More likely, one or both suffered from a higher margin of error as the field approached zero. At any rate, the dosimeter's measurements had corresponded, more or less, with officially reported values in Tokyo, Koriyama, Tomioka and Iwaki. The pancake frisker readings I took in the United States were comparable to the numbers that Ed Lyman and Eli had given me. Hoping for parity once I had entered a hotter field, I decided to be warned by whichever value of the two instruments was higher.

I further decided to use the SCALER mode in preference to the other two. There were many occasions in the red zones when I had to fall back to NORMAL, because it was discourteous to keep others waiting for a full minute time after time, all of us meanwhile absorbing radiation; and when I ascended the unpleasant stairs of the White Bird Shrine in Iitate I frisked my surroundings continuously, and thereby both got to record and to avoid the unhealthiest spots. But as often as I could do so considerately, I would make a one-minute timed walk in SCALER mode, or take a scaled measurement of such interesting objects as the drainpipe in Tomioka that Mr. Kanari had measured earlier that year with his municipal scintillation counter.

So I whiled away the months, hugely entertained by my pancake frisker. When I reached Japan, the first readings were no eerier than at home.

Tokyo. Japanese risotto restaurant, Kabukicho red light district.
33 cpms / 0.12 micros [once again, the same as my daughter's cat].

As you might imagine, the red zones were hotter.

I used to get impressed by the difference between 0.06 and 0.18 micros an hour. At 0.36 micros, that granite countertop in Charleston certainly beguiled me. In Moscow, Molotov's grave was three times hotter than Gogol's—which in turn proved equivalent to the National Orchid Garden in Singapore. After going to the red zones I forgot to care about such piddling variations.

GRANITE AND PLUTONIUM

In terms of average radiation levels, Portland, Oregon, was certainly the healthiest city I visited (come to think of it, the little California town of Dunsmuir was just as good). As for the unhealthiest, well, the red zones more than overshadowed every other place I went in those days—including Hanford, Washington, where in September 1944 our government most secretly commenced to make plutonium. At first there had been three 800-foot-long separation facilities. Columbia River water kept the irradiated slugs cool at a depth of 16 and a half feet, flowing through the aluminum cylinders of *the largest plant Du Pont had ever constructed.* Downstream from there, eddies, bends, sediments and creatures grew "hot," but war aims exempted our G-men and their pet scientists from responsibility from all such secondary issues. Had Japan been on a war footing 67 years later, Tepco might have been equally carefree about the contaminated outflows from Plant No. 1; I guess some entities are luckier than others.

(Here I wish to reiterate that no one with a heart can read unmoved the stories of those Tepco people who tried to deal with the catastrophe—for instance, Mr. Yoshida Masao, the chief of Plant No. 1, waiting for the containment vessel of Reactor No. 2 to possibly rupture, meditating on the floor, going through the names of his best-known colleagues: "There were about 10 or so. I thought those guys might be willing to die with me.")

The atomic weapon that struck Nagasaki was named "Fat Man." It contained 9 kilograms of plutonium. *It is a thing of beauty to behold, this "gadget,"* wrote the journalist who flew in the plane that dropped the bomb. *We removed our glasses after the first flash, but the light still lingered on, a bluish-green light that illuminated the entire sky all around.* So satisfied with that result was our government that production of the lovely silver metal continued until 1987; eventually there were nine reactors at Hanford.

An Atomic Energy Commission pamphlet from 1969 boasted that the Hanford reactors, then still creating or for all I know even satisfying demand, enjoyed *environmental conditions more favorable than most power-plant . . . sites,* being *relatively remote from towns and . . . adjacent to the Columbia River, with its high volume of flow.* The *activated impurities* got held for a good *1 to 3 hours,* which *reduces the activity by 50 to 70%.* Then they got *diluted by the streamflow.* In the case of *unusual radioactivity because of leaks of fission products,* the water was *discharged into trenches and seep[ed] into the ground. This percolation is very effective since the soil retains the radionuclides.* And a diagram of the "underground crib," decorated by "monitoring wells," showed the plutonium on top of the rare earths, then the strontium, followed by the cesium, ruthenium and NO_3.

Decades and trenchloads of "hot" river water flowed by under the slogan "the peaceful atom"; who am I to argue against such a delightful sentiment? For one thing, plutonium is *less poisonous than arsenic,* and perhaps even less hazardous than cesium-137 or strontium-90—although that judgment is more relative than it sounds, due to plutonium's scarcity and chemical immobility.*

In practical terms, *Hanford became the primary source for plutonium,* thanks especially to the zealous industry of the N Reactor (built 1964), whose design approximated Chernobyl's,[†] *and except for small experimental quantities, it produced all the American-made plutonium that eventually reached the Northern Rio Grande system in New Mexico*—through contamination, of course.—Let me now anticipate this book's section on fracking, in which a consultant for resource extraction companies explains in Greeley, Colorado: "When you're doing oil field, it's not the fracking, it's the piping. Things leak, things go wrong, when they have thousands of feet of piping in the surface. That's human nature. I can't go through a day without making a mistake; neither can they."[‡] Thus Hanford; thus Fukushima.—A guide for river paddlers called the site *the most serious long-term threat to the Columbia River and the livable communities along its shores.* Meanwhile *The Japan Times* praised Hanford thus: *It's among the most toxic nuclear waste sites...* Apparently some 212 million liters of radioactive poisons had been *cached in aging underground tanks... Gravel fields cover the tanks themselves,* six of which had already been *found to be leaking 1000 gallons... a year of highly radioactive stew, possibly reaching the groundwater*—but for all I knew, they might have been more secure than King Tut's tomb, thanks to the "big five"[§] miracle called "concrete cocooning." *The Japan Times* article was even headlined: **Hanford offers Tepco lesson in cleaning up Fukushima.**

Suitably inspired, I decided to take a little frisk. Clyde in Richland had insisted that I would detect nothing of dramatic interest, but back when I was alive I used to be a *curious* sort.

According to a certain William L. Graf, whose *Plutonium and the Rio Grande* charmed me in the library stacks, *sediments always contain the largest quantities of heavy metals,* so that *soils and sediments are the major repository for plutonium.* Continuing his train of thought, Graf remarked that inhalation of plutonium-spiced sediments seemed *probably the most important from the standpoint of human contamination.* Well, then I would put on gloves and a painter's dust mask, after which

---

* Its most stable compound is plutonium dioxide, which has been called "an attractive reactor fuel."

† That Soviet reactor was a graphite-moderated 950-MWe RBMK, fueled with uranium oxide enriched to a rather low 2%.

‡ See II:389.

§ See p. 134.

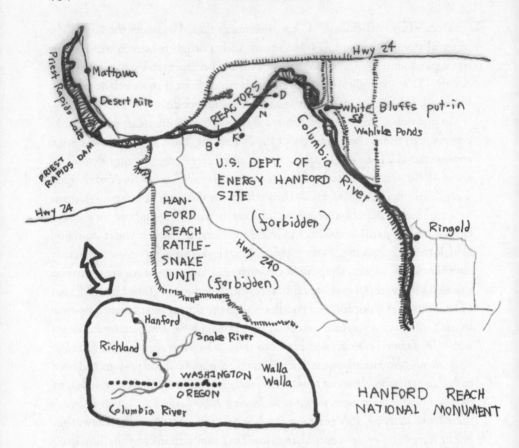

HANFORD REACH
NATIONAL MONUMENT

I could measure a couple of streambeds or gullies, with the frisker's pancake head close to the ground. A Japanese nuclear engineer* advised me to measure both with and without the plastic cover, which was thick enough to prevent alpha and beta particles from reaching the device's scintillating crystals. The difference between the two readings would reflect alpha and beta levels. Trusting myself not to detect that extremely scarce beta emitter Pu-241, which has no relevance to *Carbon Ideologies*, I settled for hunting the alpha in Pu-239, which had been employed in the MOX reactor at Fukushima, and should be Hanford's dominant isotope—and in Pu-238, which in 2012 was detected in the Japanese red zones.

My detector was rated, said its spec sheet, at an efficiency of 11% for plutonium-239, so I figured I would need to divide whichever alpha number I got by 0.11 to obtain a rough plutonium level in counts per minute.[†] Later on, some nuclear

---

* Aki, whom you will meet on p. 448.

† According to the World Nuclear Association, "Pu-238 (half-life 88 years)" is "a strong alpha emitter and a source of spontaneous neutrons"—in other words, not the right toy for children. It "is increased

Samaritan might provide me with the proper factor to convert this number to microsieverts.—Ed Lyman informed me that it would be more complicated than that. "You need to know the isotopics," as he put it. I should send a soil sample to a lab for spectroscopy. That was the only way to learn the proportions of different plutonium isotopes, which would determine subsequent calculations.

I telephoned a dozen soil testing companies, and none of them would or could sample plutonium. I wish I had written down the stern and horrified things they said; it might be entertaining to learn how many of them then launched good-citizen e-mails to the Department of Homeland Security. Well, so why not just go to Hanford and see what I could see?

The drive from Portland took five hours. Arriving at a gate with a darkened booth, and a sign which ran: WHEN FLASHING TUNE TO 530 FOR HANFORD EMERGENCY INFORMATION, we turned south on Highway 240 toward smoke-hazed, dust-hazed Richland, where for all I knew Clyde was right then on his hands and knees in the back yard, happily frisking the grass; the temperature was 95° Fahrenheit. Rolling across the sagebrush flats, with the Hanford Reserve on our left for many a mile (there were only *some 1,500 sq. km. of scrubland* to search in—1,518 square kilometers, to be precise), we glimpsed well in from the fence a red-and-white radio tower—frequency 530, I assume—then two other slender white towers and a distant row of boxy white buildings—almost certainly not the reactors, which necessarily followed the south bank of the Columbia as it curved northeast before twisting sharply southeast along the White Bluffs. We ourselves were now angling more southeast than south on Highway 240, passing the brilliant whitish-yellow grass that was silver-pocked with sagebrush, and that long white fence so cheerfully decorated with yellow warning signs.

Below the reserve, not far past the southern extremity of Richland, we crossed the Columbia, came into Pasco and swung back northwest and north, paralleling the reserve's eastern edge. Within another half-hour we had arrived at the put-out by the Ringold fish hatchery—one of the few places where our government would allow us to camp. This spot had received the highest fallout dose from Hanford—although most of it had been the extremely short-half-lifed iodine-131. My companion proposed stopping here, which was fine by me.—The Mormon crickets must have died off; for he said that two weeks earlier the river had been thick with them. The chirring of cicadas vibrated almost continuously through the buggy underbrush, and there was a fishy smell of muck, while there in the smooth lavender-grey river the sun lay reflected as a long orange

---

in high-burnup fuel. Pu-239, Pu-240 and Pu-242 are long-lived and hence little changed with prolonged storage."

bulbous-tipped wand not unlike the pancake frisker itself; and I felt (as I certainly never did in the red zones) so joyous and free in this novel part of my native land; no doubt in your time the heat must be worse at Hanford, and perhaps civil order has so far failed to let all 212 million plutonium-contaminated liters of progress seep out; I hope that those who cannot read remember from their grandparents not to go there. In my time it was a sweet enough place; since it was open to the public, I credulously believed myself safe, and my credulity got rewarded.

Directly across the river lay the widest sector of the reserve. I saw nobody on it, which failed to amaze me, because according to a pamphlet from the U.S. Fish & Wildlife Service, *the Department of Energy's side of the river, south and west shores, are closed above the high water line.*

There on the shore the humidity had already begun turning cool, but when I strolled the few steps back up to our grassy campsite where Tom was looking over the motor of his beloved *Water Witch* it still felt almost sweltering.

Now it was nearly dark. Walking toward the trees, I commenced another measurement and then strolled back to the aforesaid Mr. Tom Colligan, who was a classicist and could explain exactly how difficult it might sometimes be to parse a sentence of Thucydides in Greek. While the frisker shone like a glowworm in the sweet-smelling grass, counting down to zero, I requested his opinions on

Catullus, the coyotes meanwhile howling like young girls at a carnival, with a ring around that bright moon which tonight was so distinctly bat-winged with continents. I went to collect the frisker, which read 0.24 micros an hour, 72 counts per minute—identical to the library interior at Estes Park, Colorado.— And by the way, the presence or absence of the plastic cover never made a difference at Hanford.

Where trucks and boats were parked, and Tom had parked tonight, the level held steady at around 0.30 micros and 88 or 90 counts per minute (the same as a brickyard in Bangladesh).* Down toward the trees the dose fell off by 20%. It could have been that decades of contaminated Columbia River water dripping off of boats had increased the radioactivity, but that was the merest night hypothesis, and in the morning it proved incorrect.

I thanked Tom again for bringing me, and he passed me a beer and said: "Always happy to be out here where there's nobody else."

Mosquitoes kept biting my wrists, so I pulled my gloves on. Now it was chilling down pleasurably. We talked about Homer and ate meat with our nut-buttered bread until Tom felt ready to snooze in his truck. I fluffed out my sleeping bag in the mildly radioactive grass; Clyde might have reminded me that a truckload of bananas was worse. Tom had brought an inflatable mattress for me, but the grass was so soft that I used it only for a groundcover. The coyotes fell mostly silent. The moon was a fruit hanging from the black tree-spider just beyond that outspread its black arms in limp curves.

Mosquitoes annoyed us for most of the night, but before dawn the air chilled down farther, and before I knew it, morning woke me. Tom made us coffee and I brought jerky, berries and more bread with nut butter to the occasion. Then my friend strolled back down to the put-out. He said the river had strengthened enough that his motor might not be able to take us upstream; evidently Priest Rapids Dam had let out some water.

So we drove farther north and west, toward the White Bluffs put-out where (so Tom had heard from an old lady) there used to be a bridge, and before that an Indian crossing; once upon a time there had even been a town (the old lady claimed to have lived there), but the Manhattan Project expelled the inhabitants and razed the buildings, most likely for reasons of perpetual political security, although radiation levels might have had something to do with the action. The day was warming quickly. Could we have gone straight our route would have been 10 miles northeast, but it took us more than 15, and I was just as happy—for

---

* See p. 410.

we were in a lovely land, following R-24 almost west through the yellowish-greenish sagebrush, crossing a concrete irrigation ditch, heading toward the lavender haze (some of which must have been from forest fires); and to the north lay a low glowing ridge the color of grass.

Entering Grant County, we soon neared the entrance of the reserve, where a helpful sign advised us: FIRE DANGER EXTREME. A dirt road brought us suddenly to good pavement—installed, I should guess, for the benefit of the U.S. military—and then we reached our turnoff above the White Bluffs boat ramp, with the Columbia a wide grey line to the west. According to the map, we were four miles as the crow flies from D Reactor, which lay across the 90-degree bend above which the Columbia flowed northeast.

So we backed down to the ramp. The *Water Witch* easily entered the water; Tom knew his business. When I waded into the lovely cool water and took hold of the painter, a dead salmon goggled at me. Tom drove the truck back up the road to park and lock it, and I gazed at a squat pale building across the river, wondering if it might be the lowermost reactor, while a white heron or egret bowed its neck far away. Up where Tom had gone, a doe slowly crossed the road, stopped, looked back over her shoulder, then carefully pranced into the underbrush.

I felt happy, not only because it would be *fun* to learn the frisker's sensitivity to plutonium isotopes, but also because according to Tom's river guidebook, *the leg from Priest Rapids Dam to Ringold Springs, if done from the Vernita Bridge, runs along the longest wild and free-flowing, non-tidal section of the Columbia River in the United States . . . Keep in mind the U.S. Department of Energy prohibits public access to many areas of the Hanford Reach National Monument . . .*

Tom returned, and seated himself in the stern. I waded out a little, pushed us off and crawled into the bow. He started the motor, and we were off.

I had a dosimeter, too, of course; I kept track of every stray emanation I could.* In our 20 hours at Hanford, we accrued 1.6 micros of gamma radiation, or one micro per 12.5 hours; for that I could have stayed in bed! In my 39 and a half measured hours in Portland I took in 2.3 micros—a micro every 17.2 hours. So at Hanford I soaked up gamma rays one and a half times faster than in Portland—a minuscule difference, especially since at Hanford I was almost always outside, in Portland mostly inside, which certainly reduced my dose. So how did one factor that in? The proper thing to say was that from the standpoint of gamma rays, at least, Hanford was insignificantly "hot."

Looking down through the dark water at the many-fingered hands of weeds, I listened to the motor, which was just loud enough (and we were far enough

---

* The old one continued to accrue between 1 and 2 micros a day at the interpreter's house in Tokyo.

apart) for Tom and me to need to shout at each other if we wished to be heard; so mostly we let each other be. He looked pleased to be back out on the smooth glassy river, whose calm-appearing substance flowed so powerfully around the *Water Witch* that in places she could barely move upstream at all. A great silent fish splashed downstream near the restricted bank; and almost level with us sat a grey car, possibly containing a security guard. If I could, I would keep you here with me in this glorious if merely semi-wild place, instead of luring you into the red zones. But in my day I had to earn my keep, which meant informing my fellow planet-spoilers why it was hopeless; so at 8:57 a.m., with the building that Tom thought might be the lowermost reactor now enlarging off the bow, I set the frisker on MAX and held it out into the wind until 9:13. The highest reading was 1.184 microsieverts an hour, a laughably safe level which would not have been out of place as a one-minute timed SCALER average in a jetliner still ascending to cruising altitude. Multiplying this figure by 24 yields 28.416, which estimates the amount of radiobiological damage one would have accrued at this exposure over a full day. Receiving a month's dose in a day might be considered vaguely unfortunate—but it was not a month's dose, merely the extreme upper bracket of what the frisker sampled. I remembered from earlier that summer a certain walk of approximately 10 minutes' duration in Saint Petersburg, whose levels were slightly higher than Moscow's but still within the healthful or at least "normal" range; and going west along the Nevskii Prospekt on that breezy June afternoon, with the frisker set on MAX, my readings never exceeded 150 counts per minute, or about four times the average air dose of Sacramento—until as I crossed the Anichkov Most the count suddenly jigged straight to 480 counts per minute, or 1.579 micros per hour. Perhaps a cesium particle from Chernobyl had flown up against the frisker's recessed circle-grid, whose crystals waited to scintillate—but the previous evening, when I had held the frisker straight into a wind on a Neva cruise boat as we neared the Peter and Paul Fortress, a full one-minute timed average measured only 26 counts per minute, 0.001 micros in that period (the lowest possible reading), which is to say 0.06 micros per hour, so I pronounced that Russian breeze salubrious. The likely cause of that scintillation was the Anichkov Most itself. It was a fine granite bridge, and some of the highest radiation levels I found in "safe" cities all around the world derived from granite:*

---

* Although 90% of the uranium available for mining lies in certain sedimentary rocks, igneous rocks abundant in silicates—which category includes granite—tend to contain more uranium *in general* than most other kinds of stone. This fact might help explain the frisker readings.

LOWEST AND HIGHEST MEASURED ONE-MINUTE AVERAGE RADIOACTIVITIES IN SELECTED SAFE CITIES, 2014–15

[in microSv/hr]

All readings were one-minute timed SCALER counts. By their nature, MAX doses read higher. [The highest MAX in Saint Petersburg was 1.579.] These and NORMAL readings are omitted since they were more ephemerally accurate.

Where several values were tied for highest or lowest, I picked whichever might interest you.

The *range* above each city name (for instance, "4" for Barcelona) is simply the highest reading divided by the lowest.

Range: 1.3

Singapore

Marble-tiled bathroom of suite, Grand Park City Hall Hotel	0.18
Same suite near window	0.24

   Reason for high reading: Unknown.

2

Dunsmuir, California

Sacramento River, from bridge	0.06
Nearly all locations	0.12

   Reason for high reading: It was delightfully low, actually.

2

Poza Rica, Mexico

Entrance to PEMEX refinery during small oil burnoff	0.12
Table at Enrique's Restaurant, near tile wall	0.24

   Reason for high reading: The ceramic tile, I suspect.

2.5

Dhaka, Bangladesh

Interior of room, Pan Pacific hotel	0.12
Bricks in brickyard	0.30

   Reason for high reading: Bricks always measured high. See San Francisco.

2.5

Lausanne

Vineyard above Lake Geneva   0.12

Granite wall, Jardin de Veant   0.30

  Reason for high reading: Three guesses. First one begins with "g."

4

Barcelona

Subway L3 in motion toward Plaçade Catalyuna   0.06

Air dose inside massive stone entrance tunnel, Montjuïc citadel 0.24

  Reason for high reading: Almost surely the stone. More usually, enclosed
  and especially subterranean spaces (such as the subway L3 noted above)
  measured very low. This is the rationale for a fallout shelter.

4

San Francisco

Intersection of Haight and Ashbury streets   0.06

Brick tiles in backyard garden   0.24

  Other objects at this garden were up to 4 times "cooler."

4

Pineville, West Virginia

Cow Shed restaurant, plastic bench   0.12

Granite "Ten Commandments" tablet at courthouse   0.48

  Reason for high reading: Probably not any one commandment.

4.5

Denver, Colorado

Interior of Irish Snug pub on East Colfax   0.12

Granite boulder in Denver Botanical Garden   0.54

  Reason for high reading: Not much of a puzzler.

4.7

Estes Park, Colorado [7,522 ft]

Trunk of blue spruce tree, bank of Fall River   0.18

Lichened granite boulder off Highway 34   0.84
  Do you see a pattern yet?

7

Greeley, Colorado
Interior of Mad Cow restaurant   0.06
Concrete loading dock near railroad tracks, old downtown   0.42
  Reason for high reading: Some mineral in the concrete aggregate, I would
  guess. The air dose in the same place was nearly 40% less.

9

Moscow
Marble floor of cafeteria in Tretyakov Gallery   0.06
Granite curb of Molotov's grave   0.54
  Reason for high reading: Well, he was a glowing old Bolshevik.

16

Saint Petersburg
Air dose in wind on cruise boat, Neva River near Peter and Paul
Fortress   0.06
Granite windowsill outside fancy cafe Kutsov Eliseevi on the Nevskii
Prospekt   0.96
  I rest my case.

Typical air dose off the Hanford Reserve, downstream of the B Reactor

Slapping mosquitoes on my face, I toggled the frisker to a one-minute timed SCALER count, with a building that Tom thought might be D Reactor now off the bow at the 10:00 position, and got an air dose of 26 counts per minute, 0.06 micros an hour—the latter being once again the lowest reading possible.* My other frisks of river wind at Hanford yielded comparable results.

Looking upstream toward the chalky ridge on the northeastern bank—one of the White Bluffs, in fact—and the low olive grasses on the other shore where the reserve was (I saw a pelican on the forbidden side, and then mincing long-necked egrets, none of them diving yet, just watching the dark water), as Tom's boat hummed over the water-weeds, which sometimes resembled corals or rock concretions, I now began to be pelted by caddis flies, which clung in hordes to my arms and shoulders, crawling inside my spectacles, hiding in my hair, sun-bathing on the pancake frisker and thronging in the oarlocks. Tom was brushing them off, too, grinning a little; and slowly the *Water Witch* overcame the current of that creamy river. My friend had remarked on how deserted it always was out here; he'd called it a *silent river,* although come afternoon the fishermen might be out. Everything was hazy, muted and gentle, the river crabbed and crowded with ripples, caddis flies crawling on us, the pallid reactor (if that was what it was) growing and growing off the bow almost like a farmhouse. Dark vehicles crawled along what appeared to be the baseline of that structure (there must have been a road flat against the horizon). I could faintly taste wildfire-smoke on my tongue. Drawing level with the building, I took a MAX reading, then another timed count, followed by a series of NORMAL readings: 0.103, 0.051, 0.124, 0.079, 0.099 micros per hour; another near-identical timed count, and then we neared two yellow signs side by side, one in English and one in Spanish. Through the binoculars I made out the English-language one:

WARNING
HAZARDOUS AREA
DO NOT ENTER

Past the two signs was a dull-looking pale shed behind some power poles, with a white truck parked in front, and that long low graveled bluff before it, gently descending upstream; to tell the truth, the scenery was much prettier off to starboard with those cliffs of soft greyish-pinkish white, underlined by a long

---

* Because 0.06 micros an hour is 0.001 micros per minute. The lowest number of counts per minute I have ever measured was a leaking natural gas rig in the Kanawha State Forest by Charleston, West Virginia (see II:1). Expecting titillation, I'd held the frisker directly in the gas stream—and read 15 cpms.

island of olive-green which resembled parts of the off-limits side. I saw a huge white bird, probably a heron. (As I revise this paragraph on an electrically-lit January night, with my words hoarded in a battery-powered device whose consumption of energy I have quantified on page 68, I remember how as we motored upstream that island seemed to be sliding by itself, leaving the sand-cliff behind it untouched, because the latter was so much farther away.) Two MAX counts both read 81 cpms and 0.266 micros—nothing to write a book about. Squishing cold caddis flies on my forehead, with the boat droning upstream toward the bend where the White Bluffs and the reserve's olive-green seemed to meet, I immediately took another one-minute timed SCALER sampling, with the usual results, my feet pleasurably cool in their dripping shoes; while on the Hanford side eight deer grazed close together just above the shore, watching us without fear, right by a tall sign labeled 14; then we passed more deer almost silhouetted on the bluff's flat top; there looked to be a road up there. In a black tree sat a silhouetted osprey. There came a yellow sign too small to read, after which the bluff top had been fenced off for a good hundred yards; maybe waste tanks had been buried there, so that might have been a good place to frisk, but the bluff's gravel remained smooth and even, with no sign of any creek or gulley to sample from. Here the bluff began to rise, its gravel still evenly sloped and now overlined with a long dark stripe that must definitely have been a road. A sign announced that we were entering Area 13. The bluff momentarily dwindled into a marsh, rising again up at the bend; then far back on the plain were more of those odd whitish buildings; and Tom now thought that *these* might be the reactors.

There was a pumping station, with a curving road behind it, on which two white vehicles were parked side by side, until one followed the other down the bank, and they both vanished behind the station. A vast flock of tiny black birds—cliff swallows, I should guess* (all silhouetted, too tiny for my binoculars to make out)—rushed over the reserve. It was 10:59, and the frisker read 0.06 micros an hour, 26 counts per minute. Just past the pumphouse, by another English-Spanish pair of yellow warning signs, with my three-minute MAX reading almost the same as before, Tom announced that we had better turn back; the current was getting too strong. It became evident that we had never even reached D Reactor, having come about four curving miles, with another three to go. But since the entire Columbia was supposed to be contaminated, anyplace downstream of the reactors should do.

Two hundred yards below the pumphouse we might have made a hypothetical landing, but that would have been illegal, so let us call the rest of this paragraph a work of fiction as Tom paddled us close, and I stepped out, frisking

---

* Said to live in "countless" numbers on the White Bluffs.

pebbles on the beach while the *Water Witch* slowly began to descend the stream. The one-minute timed count was about as I had expected: 45 cpms and 0.12 micros. After all, the pebbles were river-licked and not particularly porous. In the dirt up on the bluff, where I never went because that would have been even more illegal, was deer scat and yellow grass, with a reading of 65 counts per minute and 0.24 micros—pretty close to the edge of the trees at our campsite. Chasing after Tom's driftboat, I reentered nonfiction.

At 11:50 we could have made another hypothetical landing, although of course I love authority and must respect all laws. Up on the flat plain where mulleins grew in towering isolation and that mysterious boxlike structure looked out at me, with a tall skinny entity behind (Tom had thought it an emergency siren) and the predictable sign proclaimed Area 13, I could fictionally have measured sandy soil at 52 cpms and 0.24 micros, after which, covering my face, I might or might not have troweled a two-inch-deep hole and frisked again, in which case the frisker would have read 66 cpms and another 0.24 micros.* There were many more hypothetical caddis flies here on land, and I was glad to get back onto the *Water Witch* and push us off. Then for awhile Tom happily fished on the river with a red lure that dove and jittered, as he told me about the big steelhead that got away.

About three miles downstream lay a good muddy place I would have liked to sample, since it lay on the shore of the reserve, where radiocontaminants might tend to wash down from the bluff and maybe even stop and sink—because plutonium dioxide, you see, is heavy, its density being 11.46 grams per cubic centimeter, while common quartz is only 2.5; unfortunately, just then a busload of approved individuals appeared, so it seemed best to keep this landing even more hypothetical than the others.

At 1:00 p.m. exactly (the do-   Tom on the river

---

* Perhaps the difference between 52 and 66, divided by the frisker's 11% plutonium alpha efficiency, would have revealed a meaningful number; that quotient was still a modest 127 counts per minute.

simeter at 578.6, up from 578.3, when we had embarked at 8:41 a.m.) we emerged from the river at the White Bluffs boat ramp. I had not heard of any law against "fishing around," so I did just that. Remembering the following rule to live by, courtesy of William L. Graf: *Heavy metal concentrations are usually highest in the finest stream sediments,* I took my trowel and commenced to scoop up black mud and water at the river's edge, just below the surface.—Why the water? Because, at least at Los Alamos, *plutonium content of suspended sediment is greater than that of bedload sediment.*

Two days later, when that watery mud and muddy water had dried somewhat, in the kitchenette of my Portland hotel room I opened the bag and frisked it: 35 counts per minute, 0.024 micros an hour. Clyde was right. I might as well have frisked my own back yard!

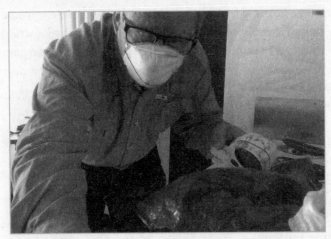

Frisking a bag of
mostly dried
Hanford mud

I consoled myself with this finding of William L. Graf's: *The grand mean concentration for plutonium-238 in river water from the six major regional sites (on Rio Grande) is nearly zero . . . and the mean concentration for plutonium-239 and -240 is 0.0041 pCi/l, a value close to the minimum level of detection.** In which case, why not fight global warming with plutonium fuels?

---

* In other words, no immediate danger!—Two years later, "hundreds of workers . . ." had to "take cover" at Hanford "after the collapse of a 20-foot-long portion of a tunnel used to store contaminated radioactive materials." Back in 1997, a Department of Energy report had advised that the tunnels could not be inspected "because of radiation in excess of five roentgens *an hour,*" which was "*the annual limit* [both of these are my italics] for a U.S. nuclear facility." (In the language of *Carbon Ideologies,* this equated with an "excess of 50 millis an hour.") Did those cowardly inspectors expect to live forever?— "Officials requested that the Federal Aviation Administration put a temporary flight restriction in place" over this happy graveyard.

CONCERNING BIOLOGICAL CONCENTRATIONS

The Japanese anti-nuclear engineer Hirose had noted (and in his book he spoke specifically of Hanford) that even after the radioactivity of waterborne particles decreases, the danger (or, as energy technologists might call it, the benefit) will continue to concentrate in plankton, and accumulate still more in the plankton-eating fish.

Hanford's plutonium, then, might well have built up in the tissues of various organisms. I did not know how to measure those.

The following illustrates this phenomenon in the case of other Hanford pollutants:

CONCENTRATIONS OF RADIOACTIVE PHOSPHORUS AT HANFORD NUCLEAR SITE,

1954–58

[in milligrams per gram]

In Columbia River	0.00003*
In egg yolks of ducks and geese	6

CONCENTRATION FACTOR = 200,000†

OTHER CONCENTRATION FACTORS AT HANFORD,

1954–58

[also in milligrams per gram; original water concentrations not given]

Cesium-137
From waste pond water to muscle tissue of waterbirds	250

Iodine-131‡
From "desert vegetation" to thyroids of jackrabbits	500

Strontium-90
From waste pond water to bones of waterbirds	500

---

* That is, 0.003 parts per million.

† The original source has 2 million, which must be a decimal point error.

‡ The isotope of the "radioactive iodine" is not stated, so I am assuming, as is most likely, I-131.

Sources: Eugene P. Odum, 1971, citing Foster and Rostenbach, 1954; Hanson and Kornberg, 1956; Davis and Foster, 1958; with calculations by WTV.

The pancake frisker failed to detect *any* of these radiocontaminants at Hanford—doubtless because they had been dissolved or suspended in such dilute concentrations.—The case of potassium was sobering.—The concentration factors for iodine, cesium and strontium bore grim relevance to Fukushima, where all three of those had been detected after the accident.

As Eli had told me, "The hotter the field is, the easier it is for that thing to do its job." "That thing" certainly found its job easier in the red zones. When you consider the radiation levels I discovered there, please consider concentration factors. Imagine being encouraged to return to your home in Tomioka. Remember that the pancake frisker was optimized for cesium-137. Consider that much or most of the radiation I measured derived from this isotope. Multiply my readings by a factor of 250, and imagine bearing that amount of radiation in your muscle tissues.

And my failure at Hanford raised another unpleasant question, which will be germane to all subsequent chapters of *Carbon Ideologies:* What other dangerous substances right in front of me was I unable to detect?

All I could do was my best. On that subject, I had already learned from Mr. Kanari* that in the red zones, bricks and granite dwindled in importance; drainpipes were a better object of the frisker's attentions:

Frisking a grating in Tomioka

---
* See pp. 358ff.

MEASURED RADIOACTIVITIES OF SELECTED DRAINPIPES AND SEWER GRATINGS, 2014–15

[in microSv/hr]

In "normal background" areas

Sacramento
- Sheet metal pipe down the side of my studio **0.12**
  Sidewalk 5 ft from same 0.18

Charleston, West Virginia
- Grating in asphalt parking lot, Budget Host Hotel **0.12**
  Asphalt curb 2 ft from same 0.18

Williamson, West Virginia
- Drainpipe on side of Ryan & Ryan, attorneys, at ground **0.18**
  Grass 1 ft from same 0.24

Estes Park, Colorado
- Metal drainpipe, measured at sidewalk **0.15**
  Concrete sidewalk 1 ft from same 0.36

Barcelona
- Grating at Passeig de Sant Joan (frisker turned on its side) **0.18**
  Maximum air dose (sampled over 5 minutes) in vicinity of same 0.08

Saint Petersburg, Russia
- Thick, painted, cast iron (?) pipe down courtyard wall of Hotel Nevsky Forum **0.18**
  Sidewalk directly below same 0.18
- Painted drainpipe down side of cafe Kutsov Eliseevi on the Nevskii Prospekt, frisked where it crossed exterior granite window-ledge **0.54**
  Window-ledge adjoining same 0.96

In Japanese green zone

Hisanohama [16 km north of central Iwaki]
- Drainpipe of traditional Japanese house being decontaminated **0.49***
  Rain channel in concrete driveway of same residence 0.16*
- Next door neighbor's drainpipe **0.17***

In Japanese red and yellow radiation zones

Iitate
- Decontaminated drainpipe   **4.00**
  Roadside vegetation near same   1.048

Tomioka
- Grating on border of green and yellow zones, frisked at 1 ft   **3.01**
- Drainpipe near **TSUBA**, frisked at base, February 2014   **22.1***
- The same, frisked 1 ft from ground, October 2014   12
- The same, at ground level, October 2014   32
  Air dose 2 ft from same, October 2014   5

Okuma
- Grating near Ono Station, frisked from 3 inches above   **4.20**
  Air dose nearby, continuously sampled   6–10
- Another grating in vicinity, frisked at 1.5 ft   **11.52**
- The same, frisked at 10 inches   **21.9**
- The same at 8 inches   **23.3**
- Rusty grating at fish hatchery (deep drainage well beneath),
  frisked at waist level   **20.0**
- The same at 3 inches   **30.0**
  Air dose in vicinity   9.46

* Frisked by Mr. Kanari Takahiro of the Iwaki Department of Radiation Countermeasures, using an Aloka TCS 172B gamma survey meter. All other measurements made by WTV with a Ludlum Model 26-alpha-beta-gamma "pancake" frisker.

And now, reader, you are as numerically prepared as was I to visit the red zones.

THE RULES

Regarding these interesting places I should express my admiration as an American citizen—for when I lived we believed in "states' rights": In certain zones two women could marry each other,* or a cancer patient could legally drink marijuana tinctures for his pain, or a gun owner could stroll down Main Street loaded at the ready, while in other parts of the same nation any or all of those

* Halfway through the writing of this book this became a national right. Bitter letters to the editor persisted in coal-loving West Virginian newspapers.

things were forbidden—that their rules of entry so colorfully varied. The red areas of Tomioka could be permissibly entered only through a national call center whose policy seemed less than favorable to non-residents. As for Okuma, that seaside paradise now rendered even more salubrious by the presence of Nuclear Plant No. 1, one could apply directly to the municipality, a written process that took months and required an exact route to be pre-approved, a vehicle (in this case another $600 taxi), a guarantee that each member of the party would possess protective gear and a dosimeter, and of course the construction of the most satisfactory answers to certain questions. In my case, for instance, it seemed better for me to be an author than a journalist, and best of all for me to be a photographer, as indeed I was. An official car would escort me, to ensure that I kept all my promises. So I anticipated a blinkered, commanded experience. But when push came to shove, nobody cared if my driver or interpreter had dosimeters; the driver even declined my offer of shoe covers, a mask and a paper painter's suit. He was supposed to remain in the taxi at all times, so he got out whenever he felt like it, taking happy breaths of sea air and tramping through the radioactive weeds. Smiling, he said: "Just write that I stayed in the car." (Now I have.) Once we had exited the red zone through the final security gate, the interpreter and I were sent for decontamination, while the driver and the two municipal officials (who had led rather than followed, and obligingly allowed me to wander around wherever I liked) stood passing the time, unmasked in that presumably particulated breeze. My painter's suit, I sadly report, immediately tore across the crotch, while my shoe covers, which had soon become pincushions of significantly radioactive stickleburrs, stayed on until we were in view of No. 1, at which point the left one finally tore off. So there I was, ponderous and ludicrous, frisking the world with one hand while I photographed and scrawled notes with the other. I suspect that those two officials, so elegantly unencumbered, quickly lost whatever respect they might have felt for me, not that I much cared. Soon enough I had unzipped my paper suit halfway down the chest; except when we were admiring No. 1, from which an ill wind blew; I left my mask dangling around my throat (the prudent interpreter wore hers, although when we dropped by Tomioka later that day she put it on backwards, presumably inhaling whatever Okuma fallout had accrued to it, at which we both had a good laugh); so I was an unkempt shambles of a foreigner, but at least I could not later accuse myself of protecting my health more than did the people who were helping me. That was Okuma. In my university Russian language class our instructor, a Russian émigré, once joked that Russians were Italians trying to be Germans; and comparably peculiar were Okuma's rules: strict in advance, mild in practice.

Iitate and Namie occupied still another category, for each of these places opened itself to me by means of a private tour. My guide for Iitate headed the Nagadoro subdistrict, which owned the unfortunate distinction of being the only part of the village actually in the red zone. When we approached the security gate, I asked whether he had obtained permission from the central government to bring me in.—"You're not really allowed," he explained, "but I will take responsibility." How I loved that answer!—As for Namie, there too a kind of tour materialized—this one with a former Tepco worker who believed in nuclear power.

BUSINESS AS USUAL

In the summer of 2014 I entertained myself at home, cooking soup on my fossil-fueled stove, flying to the West Virginian coal country and generally doing my mite to burn more carbon while I awaited the approval of my Okuma application. Wishing to someday become a canny journalist, I even read the newspapers. That was how I kept up with Japan.

In April, Tepco's attempt to freeze radioactive water in a tunnel between Reactor No. 1's turbine building and a contaminated trench achieved disappointing results. I wasn't awfully surprised.

A June headline ran: **Tepco's ice wall runs into glitch.**

No matter how hard they tried, they could not seem to lower the temperature far enough. Since July they had *injected,* they said, *more than 400 tons of ice and dry ice.*

In August, an equal 400 tons of groundwater were still flowing into the reactor's basement every day. There it got spiced up with cesium. That month Tepco petitioned to construct *a facility to dump radiation-tainted groundwater . . . into the sea*—but don't worry, it would be *after filtration,* of course.

Then came another headline: **Ice wall at No. 1 plant fails.***

Two weeks before my return, Typhoon Phanfone added its mite (168 millimeters of rain) to the groundwater around the four reactors of Plant No. 1, so that a monitoring well between Reactor No. 2 and the ocean now presented 150,000 becquerels per liter of tritium, *a record for the well and over 10 times the level measured the previous week.* That was very special.

The day before I arrived, what seems to have been the same well set a new record: 251,000 becquerels per liter of cesium—a concentration 3.7 times greater

---

* More than two years later, with the east side finally 100% frozen, and the west side 95% frozen, the headline ran: **Tepco admits ice wall efficacy still not known.**

than four days earlier. It was *the highest recorded in water samples from any of the wells*. And, oh yes, strontium-90 and other such beta emitters had also increased their presence, to 7.8 million becquerels per liter.

I interviewed some Tepco P.R. men that month, and one of them, a Mr. Hitosugi Yoshimi, set my mind at rest on this issue: "We do sampling of the ocean water, and all of the results are publicized on our website. From immediately after the accident, if you compare to now, it's much lower. It's almost unmeasurable; that's the fact."—When he was saying these things I felt as sorry for him as I had for my hometown utility spokesman who could not bring himself to mention global warming.*

And poor Tepco kept working away. What else could they do? They could hardly sit pretty, as the U.S. Department of Energy had at Hanford. The plan was to clear away *explosion debris* from Reactor No. 1 in the winter of 2015. In 2013, they had attempted this at Reactor No. 3, incidentally contaminating various rice paddies with fallout. Maybe it would go better this time. If it did, then sometime in late fiscal 2017, they could attempt to withdraw No. 1's fuel rods.

The tsunami that wiped out the backup cooling defenses of the four reactors at Plant No. 1 had been 15.1 meters high. Certain number-crunchers now projected that a 26.3-meter tsunami might strike the site—perhaps tomorrow, maybe 10 or 100,000 years from now. If that happened, said Tepco, 100 trillion becquerels of cesium could reach the ocean. It would certainly be nice to decommission the reactors before that took place.

They hoped to clear away Reactor No. 4's fuel rods by November. And on another wonderful note, **No. 1 fuel rod removal to finish ahead of time, Tepco chief says**.

In fact the news grew so good that the Nuclear Regulation Authority approved the restart of a two-reactor plant in Kagoshima Prefecture.

As for me, I bought spare batteries for the pancake frisker, and stocked up on those cheap painter's suits I have mentioned; frisker tests would establish that they could indeed keep the alpha and beta particles off my clothes. Perusing *The Japan Times,* I learned enough to hunt for swallows with white spots and peculiarly sized butterflies. (I saw the second but not the first.)

It was October, and the cities and ricefields grew ever foggier as the bullet train shot northward. In Koriyama the white fog resembled smog, and the buildings looked as ugly as ever. The air dose from the train platform was no worse

---

* See above, p. 152.

than Tokyo's, although a few days later my walk around the block measured 0.30, .36, .42 and .54 micros an hour with the frisker out before me at waist level. Even the granite flagstones of the station plaza were less radioactive than some boulders I later frisked in the Colorado Rockies.—Proceeding to Fukushima City, where the average value of my 40 frisks would be 0.24 micros (about two-and-one-third times higher than my home town), I lay in my hotel bed, listening to the rising and falling of a negligibly radioactive wind.

The following table is expressed in multiples of the average radiation level in my home town, Sacramento. It may be interesting for you to compare this table with the one on page 244.

The figures here seem less alarming than the ones in the other table. They are also less accurate. By making claims about the average radioactivity of a given municipality, as opposed to merely reporting what is emitted from this or that specific object, this table pretends to do what only an army of friskers could achieve in fact. To record a few measurements here and there, many of them from within a moving vehicle, is merely to suggest. So be it.

Red zone border marker in Tomioka (2.34 micros per hour)

COMPARATIVE AVERAGE RADIATION LEVELS, 2014

in multiples of the Sacramento average

(from pancake frisker data)

All levels expressed in [microSv/hour]. Headers over 10 rounded to nearest whole digit.

1

Sacramento [0.08 microSv/hr]. *This equals 0.7008 millis/yr.*

1.43

1 milliSv per year. Maximum dose for ordinary citizens, per the International Commission on Radiological Protection. [0.11416].

1.63

Tokyo and San Francisco [0.13*].

1.75

Poza Rica, Mexico [0.14].

2.13

Namie, Japan [0.17]. *Based on only 3 measurements along decontaminated highway. Had I been given more time, I probably would have found significantly higher levels.*

2.25

Aizu-Wakamatsu, Japan [0.18].

2.38

Hirono and Singapore [0.19]. *Many of the 11 Hirono readings were taken in vehicles en route to Tomioka. Since Hirono lies north of Iwaki, I suspect that a more thorough and accurate sampling would have found higher radioactivity than in Iwaki.*

---

* The average in Shinjuku, central Tokyo, on March 6, 2014, was 0.0339 microSv/hr, "about the same as the day before" the nuclear accident. The high reading in Tokyo on the morning of March 18, 2011, was 0.049 microSv/hr. It is possible that the pancake frisker's measurements read a trifle high in the low fields of Tokyo.

2.50
Iwaki, and various places in West Virginia, U.S.A. [0.20].

2.85
2 milliSv per year. Japanese national target air dose ("1 additional milli"). [0.22832].

3.50
Fukushima City [0.28].

4.00
Koriyama and Naraha [0.32].

5.14
3.6 milliSv per year. Alleged average worldwide dose. [0.41095].

7.14
5 milliSv per year. "For an individual steadily receiving 500 millirads per year, the chance of dying from cancer or leukemia is increased by 30 percent." Disputed. [0.570776].

16
Iitate outside red zone [1.24].

18
Commercial airline flights [1.43]. *These readings include takeoffs and landings but exclude any runway measurements. At cruising altitude, where airplanes spend most of their time, but where I frisked less often, my measurements generally fell between 2 and 3 micros an hour, so a truer (less eclectic) sampling might have averaged more like 30 times the Sacramento figure.*

29
20 milliSv per year. Lower limit of yellow ["residence restriction"] zone designation. [2.283].

29
Okuma outside red zone [2.35].

35
Tomioka [2.77].

39
Iitate red zone [3.12].

71
50 milliSv per year. Upper limit of yellow zone designation; lower limit of red [no-go] zone. [5.708].

89
Okuma red zone [7.14].

143
100 milliSv per year. 0.5 percent increase in probability of fatal cancer, if this dose is received for a year. [11.416].

7,441
100 milliSv per week. Emergency exposure limit for nuclear workers in Japan, 2014 (susceptible to upward revision, which the government was considering). [595.238 micros/hr].

12,673
U.S. Transportation Security Administration X-rays [1,013.82].

2: No Crime in Nagadoro

Although Iitate Village lay a good 40 kilometers from Plant No. 1, and might have been expected to be as safe as anywhere on the outer ring, the place proved unfortunate in its winds. Perhaps this was the case which persuaded the Japanese government to replace that convenient system of the two rings with a more realistic fallout map.

It was another "combined village," like Tomioka. "We took one sound from each included area to make this name. First four villages became two, Odate and Iiso; so we took *Ii* and *tate*," explained a man from Iiso; the two towns had joined "30 years ago, or maybe 50."

You may recall that the pre-accident population of Tomioka had been 16,000 at most. As for Iitate, in 2016 someone mordantly remarked: *There every day 7500 workers are "decontaminating" the village where 6500 lived.*

Iitate apparently received its first cesium, iodine and tellurium* in a snowfall after the explosion of Reactor No. 3. It was the most distant of the 10 localities to be blessed from plutonium from the accident (Okuma, Minamisoma and Namie also got improved in this way). I remember in 2011 watching the weather reports on Japanese television; this village always figured in them. Some of its subdistricts grew more contaminated than others, Nagadoro being the most dangerous—which is why it owned the lonely distinction of being a red zone.

Mr. Shigihara Yoshitomo (the man from Iiso) was the head of Nagadoro. He had once been a welder. His household had contained six persons, including two grandchildren. On his dairy farm there had been four breeding cows and three calves—all sold on June 22, 2011.

He wrote that on March 15, 2011, when they were making riceballs, the radiation level was 44.7 micros an hour—slightly higher than my own worst reading ever in Fukushima. A day later it was 49, and the day after that it had reached 95.1. On April 6, a Mr. Takamura *found 28,000 becquerels from soil in Nagadoro.* Come the end of that month, the government announced that radiation accumulation in Nagadoro now exceeded 20 millisieverts.[†]

At that time, people were still allowed to stay in their homes, but not to eat what they had planted. As for the cattle, eventually they slaughtered nearly 3,000 head. All the same, everyone in Iitate had remained in Mr. Shigihara's words *free to come and go* for 14 months after the accident. At the end of May 2012,

---

* See p. 540.

[†] I presume this meant 20 millis in the month and a half since the accident. Had the 95.1 micros per hour kept up for a year, the accrued dose would have been more than 833 millis. [Converting from becquerels to millisieverts is difficult; see pp. 545–47.]

compulsory evacuation measures fell upon them. Each of the Nagadoro people was to receive 6 million yen* a year for five years. When Mr. Shigihara told me this, I remarked on the difference between his compensation and Mr. Endo's— for as you may remember, the latter got nothing, his house being insufficiently damaged by the tsunami. Mr. Shigihara replied: "The government policy is to isolate all the evacuees. If they are too powerful, they are against the government. At the beginning we were all united against Tepco, but not now."

His family left home on June 22. On July 17, Nagadoro became a no-go zone. Mr. Shigihara believed that the subdivision of contaminated areas into red, yellow and green further separated the people from each other.

On July 4, a whole body counter found his exposure of cesium-134 to be 1,200 becquerels; he also bore 1,400 becquerels of cesium-137.—How badly was he poisoned? At Chernobyl a certain Colonel Vodolazhsky, who died from his injuries, is said to have received 600 becquerels, evidently on a single overflight of the reactor; and from context this appears to have been a very high level of irradiation. But I should note that the colonel made many other flights; hence this dose might not have been lethally carcinogenic in and of itself.—At any rate, Mr. Shigihara's 1,400 becquerels could scarcely have been good news.

By 2013 he had begun to wonder whether decontamination were even possible. In a typescript which he prepared at that time, he noted the immense amount of time it required to decontaminate a place. *In addition,* he wrote, *the* resulting *radiation level decrease is only to half.*

Here are some other thoughts from that sad document:

> What hurt me most is that I left my grandchildren exposed to radiation . . . I often say that the truth will be revealed when my grandchildren have babies. This is unbearable . . .

> I feel we are now being tamed. There is no dream or hope. Nagadoro residents and Fukushima people are all supposed to be angry. I don't really like "social movement[s]," but after experiencing this, I now think that we need to voice against what we think is wrong . . .

> My house was built when I was a child, carrying logs from the mountains, and [construction] took 2 years. After all these years, you can't easily leave the village, which was constructed through our ancestors' efforts generation after generation.

---

* Around U.S. $55,000.

His "young son" announced that he would never return to Nagadoro. The same went for his grandchildren. Saying that she could not imagine life without them, his wife informed him that in that case she would not go back, either. Mr. Shigi-hara: *So I thought, there is no point for me to return there alone.*

> Think about it. Why [do] people gather and hold festivals at the shrine? Why do they dance? What is your home? I can't explain well myself, but perhaps gratitude to the ancestors . . . I realized for the first time after I left my home. It is like you realize how you appreciate your parents only after they die.

That September he measured the levels again, reading 5 to 7 micros at the "Nagadoro intersection," 4 to 6 micros both in front of and inside his house, and 8 to 15 micros in his ricefield and his vegetable garden; the latter two, he noted, had certainly fallen off.

I was to meet him on October 23, 2014. On the twenty-second, at 4:30 in the afternoon real time, according to the Nuclear Regulation Authority website, which a lovely Japanese angel accessed for me, it had transpired that of Iitate's 40 monitoring posts the highest official value was the senior high school at 2.338 micros per hour; the junior high school was the merest 1.441; there even appeared to be a few zero points in the yellow zone. Then the angel's cursor froze up; certain station values did not appear; should I have labeled that human error or was it the machine kind? I calculated that 2.338 micros an hour meant 20.48 millis a year, but I would hardly be in Iitate for a year, so who cared?—As it happened, 2.338 micros was an understatement. (Shall I say that the NRA lied, or simply that certain unmentionable things happen?)—On that cool afternoon of light grey rainy sky and dark grey concrete buildings it came time to consider the practicalities of one's painter's suit and half-face dust mask which was recom-mended for use against harmless dusts only; one also carried the shoe covers and blue gloves at the ready.

The twenty-third was a grey morning. I wondered if it were drizzling; then a cyclist rode by with an uncovered head, overflown by two crows. The next thing I wondered was whether sunscreen would adhere alpha and beta particles to my skin or in fact keep them from blowing down inside my clothes. Unable to guess, I skipped the sunscreen.

The dosimeter read 0.571 millisieverts accrued over more than a year in Tokyo—the usual a micro a day, or very rarely two. Before tonight, several more digits might turn over.

In the lobby of my business hotel, men bowed over their breakfasts at small round tables, one fellow, perhaps a student, for he wore cheap black clothes, rapidly chopsticking sticky rice from his plate into his upturned bowl. The other men were all dressed at least in stiff white shirts and black business slacks; some were in full suits, one of those latter, a balding chubby specimen, having just finished eating, wore a four-faceted paper mask over his nose and mouth as he sat with folded arms, rapidly tapping his feet; while behind him at the reception desk the plump young lady in the yellow pseudo-military uniform complete with black and gold epaulettes as required by that corporate chain stood waiting to give or take a key; and two meager-faced elder women in white blouses, green aprons and green kerchiefs carried dirty dishes to their den on the other side of the room, where they now began washing them. Faced down by a scene so stoutly quotidian, the red zones faded into the merest harmful rumors.

The railroad clock displayed 11.9° Celsius at 8:00 a.m. The bulky, jolly, head-shaved taxi driver sped us over the half-dry river. He said: "I don't mind about the radiation. It's the younger ones who have to worry." And so we arrived at the apartment block.

Mr. Shigihara was a roundheaded grandfather with very short spikes of grey hair. He said: "It *hurts* me when they say that radiation causes mutation. It *hurts* me that the government evacuated Okuma and Futaba but not Iitate. They first said that even 10 microsieverts per hour was no problem at all. At our community center it was 15 or 17 . . ."

To his young-looking wife I offered to frisk anything in their abandoned place that she might wish to have with her, but she calmly replied that the neighbors would be too frightened if it came out that she had brought anything back from *there*. She served tea. They seemed to be a loving couple. Their story had a typical village beginning: Her family's home had been only 1.5 kilometers away from his. I thought it kindest not to inquire what had happened to her parents.

Mr. Shigihara remarked that the Nagadoro people hardly kept in touch anymore. "People from Okuma and Futaba, 70% of them are in the same place. From Iitate they are spread all around, because the other places were evacuated right away, and by the time we were told to leave, no good facilities were left."

We got in his car, and I turned on the pancake frisker: the merest 55 counts per minute, 0.18 micros. That was higher than Tokyo and San Francisco but just barely lower than Singapore.

I asked what he used to do when he was a boy.

"We went after birds. There were no toys, you know. No swings! We made swings using big ropes. We played marbles. We just went into the mountains to make charcoal. At that time we didn't grow much rice," evidently for lack of flatlands, "so mountain jobs were more important. Forestry keeps you busy for the whole year . . ."—and we were winding up into the mountains, paralleling the sparkling grey-green river.

"We would sell the charcoal for cash," he said. "You put the charcoal on the back of a horse and you walked 30 kilometers. That was what we did until the Tokyo Olympics in 1964. Then we started to go to other cities to work. In Iitate we earned 400 yen per day, and in Tokyo we could get 2,500 yen per day. From January to March there was nothing to do here; that was when people went away."

We passed steep rock outcroppings, higher than in West Virginia. There were lush banana trees, rainy sky, high round hills of yellow-green forest. I wondered how contaminated they were.

"After the Olympic Games there was a company set up in Iitate to make electrical parts," he was saying. "Women worked there and also in a garment factory . . ."

"When did you first hear that nuclear power was coming near you?"

"Forty years ago. I was about 20. That was when I heard about it. At that time Iitate Village was against it, but we had no say since the plant was more than 30 kilometers away. Futaba and Okuma* received many subsidies, but we got nothing."

He might or might not have been a jealous soul, but he paid most definite attention to what he and his were given in relation to others.

He said: "In my opinion, the people who benefitted would like to go back by any means, but I think they should compromise a little bit; my grandchildren can't go back."

I saw bamboo and something like honeysuckle and a few trees turning yellow.

"The government says they are going to make a facility for temporary storage. I think that temporary and final storage should be the same," he said.

"Whom do you blame?"

"The government, and nationalists, including me. I personally don't feel responsible, but everyone blames Tepco and the government alone, and that's wrong. Our parents, the earlier generation, they accepted it. Everybody including me is responsible. The government shows the least intention to be responsible. It doesn't even want to decide threshold radiation levels for whenever we should go back to Iitate. They say that we should decide it, because they're just trying not to take responsibility."

"When do you think your home will be safe?"

"The head of the NRA†, Mr. Tanaka, he came to my house to carry out an

---

* By 1978, nearly 90% of Okuma's tax revenues derived from Nuclear Plant No. 1.

† The Nuclear Regulation Authority, a Japanese government body formed after 3-11 to replace the Nuclear and Industrial Safety Agency, which functioned up until the Fukushima incident. NISA was

experimental decontamination. At that time, he said that it would take at least 20 years. It was as if he were talking to himself. It won't be good in 20 years . . ."

He added: "Around then the front of my house read 8 micros an hour. Mr. Tanaka said they would make it one-tenth, but they only reduced it by 50 percent. That made him so disappointed . . ."—and we were passing through the high town of Kawabata, whose cemetery steles glittered on the hills; just here (in the car, at least) the radiation was as low as in my kitchen back home.

"Up to here we used to sell charcoal," he said. The dosimeter remained at 0.571.

Every little flat space had long since become a ricefield. We turned onto our mountain shortcut, lurching up into the red and yellow leaves.

"So Japan is going to deteriorate," he said. "If you don't take care of these fields and forests at the prefectural level, the water and air will not be remediated."

We passed a field of scarlet maples, rolling uphill past the terraces.

He said: "I believe that if the nature is beautiful, then even if the living level is economically lower, it's still a better life.

"There is no name for the road," he said. "We are almost at the border of Iitate."

Suddenly we were in a forest of cryptomeria and pine which to me appeared quite wild, although Mr. Shigihara said that it had been planted only some 30 years ago. He said: "I don't wear a mask, nothing. The government, I don't see many of them wearing masks . . ."

At 9:49 we passed the sign that said IITATE VILLAGE. "This is the best week of the year for color," he said. "You're lucky," and doubtless I was, although the pancake frisker was distracting me from my good luck, for almost immediately as we descended into the town limits the one-minute timed reading I obtained there in the back seat grew significant, displaying 226 counts per minute and 0.78 micros an hour. Were I to set up housekeeping around here, this level would present me with nearly seven times the international (not Japanese) recommended annual civilian dose, but since our tour was intended to last no longer than three or four hours, I felt no unease just yet; my venerable "incident guide" for nuclear first responders had set the worry threshold at 0.1 millirems, or a full 1 micro; you may remember that that lichened boulder in Estes Park, Colorado, was "hotter" than this—but I did feel it incumbent on me to pay attention, insofar as our situation was not invisible to me—what truly did I have but the frisker, the dosimeter (which had not yet turned over a single digit that day), and

---

subordinate to the Ministry of Economy, Trade and Industry "whose goal had been to promote and expand the use of atomic energy."

Mr. Shigihara's confident experience?—"These fields are damaged by wild boar," he was saying, and just then I was too occupied in writing down radiation measurements to look, but presently he pointed again (he was a lonely old man whose thick Iitate dialect, which sometimes baffled the interpreter and others, further isolated him, which might have been why he often repeated himself), and I saw how parts of the green abandoned fields had been pawed down to black dirt. "They eat worms inside the soil and also roots. That's why they dig. You see them early in the morning..." In NORMAL mode the frisker was chittering between 1.0 and 1.1 micros an hour—the average 2015 air dose for Iitate (8 or 16 times higher than my home town's)—after which it plunged to 0.5, immediately ascending to 0.6, where it stayed for a moment, then chirped to 0.7 and 0.8. In other fields, decontamination workers were servicing those horrid black bags. Some wore half-masks and others let them hang around their throats. Asking Mr. Shigihara to stop the car, I walked up to the security rope around a neat army of bags and thrust out the frisker in NORMAL: 1 to 2 micros only.

A quarter of an hour later we had reached the gate in the empty road where stood the sentry, who must be getting his innards tanned in recompense for controlling access to the red zone. He made a fine impression on me, that white-helmeted guard before the red zone of Iitate, with the paper half-mask tucked under his eyes and his gloved hands crossed and characters marching down his white suspenders.

Guard at entrance to Iitate red zone

"See, this guy has been here for three years," Mr. Shigihara said. "I wonder why the government doesn't care about him."

The inevitable day when we Americans experienced our own true nuclear disaster would bring out our own national comicalities. The perimeters of our forbidden zones would be patrolled by self-important federal police officers or their bully hirelings. Some national agency would have all the authority; the locals would have none. In Japan, of course, it did not work like that. Mr. Shigihara was, as you know, the head of Nagadoro Subdistrict, a position which must have been less powerful than that of the village mayor; while the man at the gate, so I would have thought, had the whole central government at his back. Mr. Shigihara remarked that as a matter of fact he was only permitted to bring in journalists on the first and third Sundays, and this was a Thursday, but the guard asked no questions, checked no papers; withdrawing the central bolt from that two-section metal gate, he swung the pieces apart, and in we drove. I liked that.

It had come time to use the NORMAL mode more frequently, in order to save myself and potentially my companions from long minutes of being frozen in place while getting irradiated, so that is what I did. The frisker chirred quite busily, displaying 4 or 5 micros an hour, and once in awhile even 6. Six microsieverts—now, that was a hundred times higher than what I got on the Amtrak train in California. Still, it wasn't all that bad, compared to the 95.1 that Mr. Shigihara had experienced on March 17, 2011.—Just now the frisker read 5.71 micros. That would be 50 millis if it went on for a year.

Mr. Shigihara saw me checking my dosimeter. In disgust he said: "If you have 10 of them, the results are all different."

We arrived at White Bird Shrine, and I asked to get out and measure. Where we parked, a one-minute timed count measured 2.94 microsieverts per hour.

He bitterly disdained to wear a mask, and I saw him watching me, so I did not put mine on, either, and that was how I fell into bad habits in the red zones. Since he was willing to expose himself for my sake, I would rather take chances with my health than offend him. (The interpreter wore her mask in Okuma and sometimes in Tomioka—not today.) So nobody put on anything, not even shoe covers. It was a dryish day, with particles blowing in the faint breeze. Trying to play it cool, I asked what the shrine was dedicated to.

"This is to be grateful for the harvest," he replied. "I would like to say thank you."

He pointed out two votive figures. The one on the right had been erected because no villagers got killed by the Americans during World War II.

I checked the dosimeter again.—"People see these values and they get pale," he said. "But we come here every three days."

The two shrine figures (3.08 micros)

I read 3 micros beneath the torus. We ascended the front steps, and the level climbed to 4. I wished I had had the pancake frisker when I was standing at the ruined shrine in Tomioka with Mr. Endo.

"In the past we used to sing together and dance together," said Mr. Shigihara. "This stone lantern came down. We fixed everything last year."

There had been moderate damage from the big quake. He showed me the new plaque with his name and others' on it to commemorate the repair; they had paid for the stone, then laid it in themselves, hired laborers being expensive and afraid.

I decided to explore the place's radioactivities. Mr. Shigihara stood by the interpreter while I began to climb; and what I later most remembered from Iitate, after the bright red berries and lovely geraniums and all the other bright flowers of so many colors, each of them radioactive in its own way, was ascending that steep and narrow flight of shrine steps up through the darkness of the pine and cryptomeria trees,* my shoes sinking into the softness of reddish-brown cryptomeria needles on those long unswept steps, so that I wondered how contaminated they would be; holding the frisker before me like a torch (toggled to NORMAL so that its numerical flickerings would keep me updated to poisonous realities), I watched the readings go steeply up, from 3 microsieverts per hour to 4 to 5 as my head drew closer to those hideous branches—hideous, of course, only because the frisker made them so—and because I, an occasional sufferer from tree allergens in

---

* See the photo on p. 428.

my home town, have sometimes found myself sneezing or even choking when I pass directly under a pollen-raining branch, I could not help but wonder whether in this enclosed space alpha and beta particles might be likewise sifting down to tickle my lungs with carcinogenic scintillations; but near the top of the steps, the ceiling of that tunnel of branches finally drew off from me a trifle, and the frisker began to chirp more slowly. The decrepit shrine stood right ahead of me now, so I climbed up to it: 3 micros, then 2, then 3 again. In spite of my creeping anxiety about the radiation, I paused on the topmost step to take a one-minute timed count: 721 counts per minute and 2.4 micros per hour. Well, so what? On one of my transcontinental business flights out of Washington, D.C. (cruising altitude, 36,000 feet), the frisker had read 733 cpms and 2.46 micros, so aside from the issue of inhalable particles, standing here was no more unsalubrious than airplane travel, which hardly anyone worried about.

Presently Mr. Shigihara and the interpreter came up after me. He bowed and thanked the god "for the land, the ancestors and everything." I do not remember whether he clapped. He left no coins. On the way down he asked me if I were a Christian, and I said that I liked to express gratitude for my life. He approved of that; he said that he did much the same.

At 10:25 the dosimeter turned over a digit: another microsievert in my body. Down at the first torus the air dose had reached 5 micros an hour. Inside the car it was only 1.196 micros.

As we departed, Mr. Shigihara said: "Sometimes foreigners ask me why we don't riot. It's very difficult to make an action. I wouldn't do anything myself. I know that's not a good thing, but that's reality. I just can't."

We stopped at the carp pond, where he came every three days. There were no fish in sight. When I checked my exposure, he said: "Your dosimeter is just for your peace of mind. No one will tell me what is safe, so it's good for nothing."

All the same, he kept pulling out his own scintillation counter.

Driving past another field torn up by wild boar, he said: "The monkeys eat chestnuts and persimmons. Over there is my land."

At an abandoned intersection, probably the one mentioned in his writings, a red cross had been painted on the road to mark the point of highest radioactivity at some prior period. "They used to come every day to measure, and now they come every week," he said bitterly. The pancake frisker was reading 4.81 and 4.91 micros an hour.—"Look at this!" he shouted, and from the logbox pulled out the radiation records. "March 17—95.1 microsieverts an hour! We were making *onigiri** here—"

I suppose that if I had lived through such terrifying levels, I might have

---

* Flavored riceballs. This must be the incident mentioned in his statement on p. 429.

repeated myself, too—wondering all the while how long I would walk this earth before the radiation caught me out.

Calming down, he said: "You're not supposed to be walking here anyway. Prohibited to ordinary people! But it's nothing."

Here on concrete supports stood a grand air dose monitor behind a heavy security grating, its top an angled slab of solar panels. He said: "I still think the government underreports the radiation. The sensor should be on the ground."

I remember a certain dusty-windowed building of unknown purpose right there in the heart of Iitate's red zone; the light industrial district of my Californian home town held many such former auto body shops and defunct printing establishments on which the weeds kept closing in; in Japan, of course, such sights were more out of place. I could see the reflections of trees and clouds in those dark windows whose grimy sashes quite stimulated the chirpings of my pancake frisker; and grasses licked up against them as the bright green leaves (now just beginning to go yellow or rusty with autumn) of what impelled itself to become forest groped toward the sun; their stalks were becoming trunks and one window was already smothered.

Going up the narrow road to the decontaminated community center I measured 0.8 to 1 micros. He said: "There used to be black bags here."

The rusty leaves on the tree behind the community center, the do-not-enter sign on the empty blacktop at the intersection, the cloudy sky and the moss on graven stones made for a fine post-human picture. I walked around and measured 0.48 micros an hour—the same as that Ten Commandments granite tablet in front of the courthouse in Pineville, West Virginia. That had been the highest reading I got in the eastern United States. (Doubtless I could have fished around for radon in Connecticut basements, but why bother when Iitate put them to shame?) A few steps farther, the levels dropped to 0.42, which replicated my reading in a small commercial plane at 17,000 feet over West Virginia, and the air dose at Koriyama Station's street plaza, two feet from the granite flagstones.— Although it was not so bad, I found myself trying not to breathe too deeply of the fresh chilly air. It was quite still. There were blue shadows on the rolling green hills. Yellow leaves twitched.

On the hood of the car hunched a very strange butterfly, greyish-orange and asymmetrical. I asked if he had ever seen one like it. "No," he said angrily, "but don't write that it's a mutant because of the radiation!"

This was one of his *bêtes noires.* At his temporary home back in Fukushima City I had asked him, meaning no harm, to point out any plant or animal abnormalities for me to photograph, since the only other two ways for me to represent the radiation visually were to record abandoned decrepitude, or photograph the numerical display of the pancake frisker. He grew almost enraged. It seemed that he had conceived the idea that documenting mutations was the trivializing or even sadistic entertainment of journalists. Not wishing to provoke him, I dropped the subject. And I made a point of not photographing the deformed butterfly.*

Driving away from the community center, we encountered mild weather: about one microsievert per hour. When we reached his land, the frisker was reporting 1.8 micros, because exactly here the forces of good had established their usual multitudes of black bags; and on that green field now cut into berms and terraces, white-clad figures strode mysteriously yet purposefully, with a bulldozer and a crane behind them and clouds in the lovely sky. A big man in knee-high boots was standing at the lip of the well they had made, with his feet spread wide and his gloved hands hanging loose. A slender young woman whose white suit was too big for her stood facing him, bowing her half-masked face as if waiting to receive some order; from beneath the white hard hat her rich black hair fell loose across her shoulders; I wondered how radioactive it was. And Mr. Shigihara stood in the middle of the empty road, calling out to them.

I went up to measure a black bag, and the reading went down to half a micro! Mr. Shigihara laughed at me. He explained that the outer rows consisted of sandbags. I could have crawled onto the real black bags, but somehow I didn't want to; come to think of it, back when I was alive it always put me out whenever I had to irradiate my testicles.

Ten meters back from the bags, the dose was 1.74.

Mr. Shigihara stood chatting with the smiling and laughing decontamination workers, none of whom were now wearing masks. After all, they said, alpha and beta could not get out of the bags! Aside from Mr. Shigihara himself, and the decontamination traffic cop who stood there on the road, directing nonexistent traffic, they were the only human beings I saw in Nagadoro. I strolled across the deserted highway, frisking weeds, flowers and pampas grass. They read one, two and three micros. The fallen leaves made a sweetly pre-radioactive sight: as they decomposed they would emit ammonia, which *is known to draw cesium from the soil and make it more absorbable by plants.* And now Mr. Shigihara had completed his pleasantries with the decontaminators. I could not judge how sad it made him to see this former field of his whose only present crop was black bags. They had compensated him 190,000 yen per 10 *ar.*†

---

* Two years later a Greenpeace study of the disaster area found "heritable mutations in pale blue grass butterfly populations."

† A unit of area measurement (conversion unknown to me).

A 21st-century crop on Mr. Shigihara's land

We drove up the side-road to his house, which he predicted would read 1.2 micros inside. They had decontaminated his rain gutters to about 4, he said. They had taken five centimeters of topsoil from around the house, and pruned one of his favorite trees nearly to death; he still felt cross about that.

It must have been a pleasant home once. Even now it appeared far more live-able than Mr. Endo's place. There was a pond, a dairy barn, a rich-looking pasture and many trees. Mr. Shigihara had been prosperous. The house itself remained in excellent repair.—Pulling out his own counter, he stood in the driveway, then said: "It's the lowest reading yet." I made my own measurement: 1.14 micros.* If the radiation remained at that level, staying here might be no worse than work-ing as an airline stewardess.† —Where was all the cesium going?—I am guessing that a great deal of those 95.1 micros which had afflicted this place on March 16, 2011, derived from iodine-131, whose 10 half-lives of dwindling to 1/1000 strength would already have elapsed by early June of that year; by now the stuff's concentration must be nearly indetectable. As for the cesium, whose ability to spread itself at significant distances from the locus of any atomic explosion was infamous (being the decay product of certain radioactive gases, it could travel high and quickly on any nuclear wind), it probably made up the bulk of Iitate's present radiation burden. The two and a half years since the reactor failures

---

* Our counters agreed.

† Dr. Crownover interjects: "The US has no formal limits on radiation exposure for flight crews. My recollection is that the European Union has limits for them that [are] higher than what they (and we) allow for nuclear plant workers."

comprised a mere 12% of cesium-134's first half-life, and 8% of cesium-137's. It must still be causing wickedness somewhere.

Inside the house it was chilly and there was a faint mildew smell. We sat down at the table. "We used to dry mushrooms here," he said. The frisker settled down at 0.8. Why, that was hardly more than 7 millis a year!*—If he wore a mask, and never went out, perhaps he could have dwelled there awhile before the cesium built up in his muscles. "They used to say 1,000 becquerels is fine, but now they say less than 100 becquerels . . . ," he wearily murmured.

CESIUM CONCENTRATIONS IN IITATE MUSHROOMS, 2014

(in becquerels per kilogram)

Japanese government's safe limit "for regular food items"	100
Found in matsutake mushrooms	"up to" 3,032
Found in shiitake mushrooms	"up to" 8,839

Source: The Japan Times, January 2014.

His grandchildren's portrait hung on the wall. I frisked the air in front of it; it was only 1.084 micros. Sitting there on the floor in his house, he seemed, like Mr. Endo at that shrine in Tomioka, to have become younger; this was his place; for a few moments his lost world reorganized itself around him; when we went out he puttered across one of his fields almost happily.

He had a camera set up to record night visitors. He showed me photos of the monkeys and wild boar that had visited. When I asked how often he found sick or dead animals, he answered, as people from the red zones always did, that no creatures showed any signs of harm.[†]

---

* A Tepco official once told me: "The level is 0.8 microsieverts per hour in Gifu Prefecture naturally. There are no health problems in that prefecture." I contacted the prefecture. It turned out that the actual level was 0.08 micros.

† Nearly half a century before the accident, scientists at the Brookhaven National Laboratory irradiated a forest of oaks and pines with cesium-137. They must have had a blank check; they blasted that place for two years. Do you remember Mr. Shigihara's 95.1 microsieverts an hour? The experimenters would have been unimpressed. A tenfold increase over that dose—which is to say a full millisievert per hour—was where they first began to notice environmental degradation. Where the measured exposure remained near zero, there was, as might be expected, no perceptible biological effect on insects and vegetation; then the curve of radiation *versus* damage began to rise, shallowly at first. One to 2.5 millis

Mr. Shigihara's grandchildren (1.084 micros)

His back window measured almost 4 micros—higher than the highest timed reading I ever got in an airplane.* But the rest of the house never registered much over a single micro—call it 20% of what he had measured the year before. I wish I had thought to frisk the floor. As always in those red zones, all parties preferred to get out sooner rather than later, and the sense of pressure I felt to hurry, less for myself than for him and the interpreter, who were deriving very little benefit from their radiation exposure, rendered thoroughness impossible. Had the floor been "hot," the mystery of the disappearing cesium would have been solved.

Pursuing this question, I asked Mr. Shigihara whether he would let me dig a hole in his field, maybe over there beneath the big tree. He was hospitable; he even lent me a shovel. The air dose was nowhere near the 8 to 12 micros he had

---

stunted plant growth and diminished animal diversity. Pines suffered before oaks, which "persisted at rather high dose levels" (10 to 20 millis per hour). In the ring where the oaks had absorbed 2.5 millis, aphids infested them in populations more than 200 times greater than where oaks were left to themselves. Closer to the cesium source, at a dose of more than 50 millis, only fungivorous, xylophagous and saprophagous insects persisted in significant numbers; above 150 or 200 millis an hour, the insects were described as "occasional transients." At 50 millis the oaks had already given way to huckleberries and blueberries; at 100 millis only sedges hung on; but not until somewhere beyond a horrific 250 millisieverts an hour did the diagram read: HIGHER PLANTS DEAD.

* My highest MAX reading in 2014–15 was at an unknown cruising altitude (from the look of the landscape below me, I should estimate we were no more than 35,000 feet) over the Great Salt Lake in Utah: 4.17 micros. My two highest one-minute SCALER measurements, taken over that same general area on two separate flights (one at unknown altitude, the other at 37,000 feet), tied at 3.3 micros. The next highest SCALER reading was 3.17, measured at 37,000 feet over Minsk, Belarus.

Mr. Shigihara in the parlor of his former home. His dosimeter hangs from around his neck.

measured 13 months ago; in his pasture and over his pond I could only find 3 or 4 micros. He and the interpreter stood back as I began to dig, perhaps because they entertained considerations of radioactive dust, or maybe just to give me room; well, they continued to refrain from masking their faces, so I would follow suit. As I dug, I frisked. The topsoil was about 3 micros. From 6 inches to a foot deep, the levels went up from 3 to 4, then 4.5 and 5, after which they began to fall off again. At two feet deep, they read 3 micros. That was as far as I went.—As good old William L. Graf once put it: *In the Savannah River area . . . 84 percent of the soil plutonium is within 5 cm of the surface, and 90 percent is within 15 cm . . . In Japanese soils, more than 80 percent of plutonium-239 and plutonium-240 is within 10 cm of the surface.* Perhaps he was onto something applicable to all the heavy radionuclides. Happy to re-inter what my old chemistry textbook used to call *an extremely active member of the Sodium Family*, I shoveled the dirt back in and thanked Mr. Shigihara.

He said: "Many people came to Iitate to evacuate. I thought it was safe. About a thousand of the Self-Defense Forces were here, and suddenly they had to go! The American government policy was 80 kilometers from Daiichi. I've heard the American soldiers are suing . . ."

I wandered into the reddish-brown dimness of his dairy barn, where a small bright window of tree-light bestowed on it some of the tones of an Andrew Wyeth painting; the walls were a little mossy and maybe moldy; nothing looked amiss; the cows could have come home at any minute, and the levels varied between 3 and 4.3. Then I elongated my arm (I don't know why), and the frisker chittered up to 8.70 micros—the highest reading I got in Iitate. After that, the place felt less inviting, and I hastened out of there. Mr. Shigihara and the interpreter were ready to be gone. I played the frisker over the edge of the pond, and then across the leaves and red berries that were there, finding anywhere from 2.88 to 4.29 micros.

By now the dosimeter had turned over two digits, so we had absorbed 2 micros.

I had asked to visit the village's most beautiful place, which of course meant the cherry trees of Iitate, which were green-mossed on their dark trunks and branches, almost but not entirely leafless—seven months out of blossom— growing out of the wet grass and ferns on a hill where, as Mr. Shigihara pointed out, we could see all the way to the ocean—which meant that cesium-laden ocean breezes had a straight path to that hill, and indeed it was somewhat radioactive—3 and 4 micros by the cherry trees, the car up winding around past hydrangeas (3.94). At our first stop the air dose was only 1.3, but as I stood looking down through the cherry trees, a fresh breeze arose from the ocean, and the frisker instantly chittered up to 5 micros. Higher up I took another measurement, on the face of the hill which must receive that particle-laden wind; certain glossy leaves read 6.79 micros; the air dose by a road sign with a curvy arrow on it measured 3.81, and then I drew back the frisker a little, and it read 5.80. There must have been gamma rays about (a fine indication of cesium), because the level inside Mr. Shigihara's car had tripled, to 3 and 4 micros an hour, although it occurred to me that we might have let in too much fallout as a result of all the times we had opened the door; but in the end my cesium hypothesis seemed the more plausible, for as soon as we departed the red zone, the radiation inside the car dropped by itself, all the way back to 1.3, so there must not have been many conspicuously irradiating particles riding with us.

Mr. Shigihara and some others maintained these cherry trees. On July 6, 7 and 8, some 40 people had come here, representing about 70% of the households; they also spruced up the cemetery with a "weeding machine."

I asked how radioactive it had been, and he grandly replied: "We didn't measure it. We don't care about the level."

I did. It was 6.37 and 6.80 and 5.80—not too bad, compared to the next two red zones . . .

Amidst the dark, green-splotched curvature of Iitate's cherry branches he stood looking out at the slit of ocean (which must have been almost in sight of Plant No. 1) and said: "For Fukushima City people it was lucky. We were sacrificed, but other people were saved. So we ask for more help. Futaba and Namie suffered

Frisking the hill with the
ocean view (6.79 micros)

a lot, but they evacuated before the explosion, while we kept getting told, *no problem, no problem . . .*"

Black branches curved gently against the sky, while below them the moist grass and the many flowers were as vibrant as if they had never been poisoned. Their slight decrepitude epitomized the Japanese aesthetic of *sabi.**

Now it was time to go. We drove out through another gate; the radiation field dwindled, and then we were back in greater Iitate, which was merely a yellow zone, with islands and continents of black bags going on and on; none of the ones I could reach with my frisker read more than a micro and a half.—Where would those bags finally go?—Two months ago, after what I was told had been strenuous negotiations, the prefectural governor of Fukushima finally granted consent for the central government to store contaminated "debris" for 30 years in exchange for 301 billion yen. The politicians of Tokyo decreed that each radiopoisoned prefecture must find somewhere to store black bags within its own borders. And so a local headline ran: **Tochigi town protests nomination**

* See above, p. 379 fn.

as atomic dump site. Fortunately, Shioya was also "a candidate"—how lucky for Shioya!

At the mostly deserted town hall, on a walkway of hand-laid stones, just before a pair of cutesy stone figures, there stood on a metal pedestal the two stout metal legs and the two glowing red lights and the square metal face of the current air dose display, with its two readouts of red numbers composed of glowing squares; one said 0 and one said 44; no one could tell me what they meant; perhaps the 44 was the daily accrual in microsieverts, although according to my pancake frisker, a day's worth of radiation at this spot would only be about 13 micros; it might have referred to the red zone, for it worked out to about 4 micros an hour, which correlated somewhat with my readings. Mr. Shigihara liked the two stone figures. They must have reminded him of the good old days.

At last we came out onto Highway 399, leaving Iitate, the dreary armies of black bags bivouacked in mud, with green fields all around them, and the dosimeter accrued another digit.*

Mr. Shigihara said: "How I like to pass the time is have a draft beer and talk about Iitate . . ."—so that is more or less what we did. Remembering him now, I hope that by then he felt friendly toward me, since I had refrained from asking about mutants, never embarrassed him with pity, and breathed in the sweet radioactive breezes as masklessly as he. I must admit that at White Bird Shrine, going up that long dark flight of steep steps beneath the cryptomeria trees with fallen needles very radioactive and soft underfoot, I had wished for shoe covers. But Mr. Shigihara was testing me, and I meant to pass the test. Besides, he *was* still alive. The field would have had to be something like 117,000 times stronger to bring on radiation sickness, and if I got lung cancer in five or 10 years I could blame three cigarettes that I had half-smoked in my life.

He said: "Evacuees, they are suffering, but at the same time they are the objects of envy from those who do not get any subsidy. That hurts us."

He said that 30 out of Nagadoro's 70 households had already purchased real estate in Fukushima City. I think that made him feel pretty hopeless.

Lean, active Mr. Shigihara, who despite his grey-fringed baldness had the bright eyes and mobile mouth of a younger person, liked to keep busy. He must have been a conscientious farmer, back when he had a farm.—He patrolled the

---

* In five and a quarter hours it accrued 4 digits, which I calculated out at 1.285 micros per hour, most of that time having been spent in the relative protection of Mr. Shigihara's house or car or else the decontaminated community center. Between 11:27 and 12:37, when we might have been outdoors for nearly half that time, the dosimeter turned over 2 micros; hence my approximation of the real air dose in Nagadoro would be 4 micros an hour.

red zone every three days. I asked him if he encountered any crime, and he assured me that there was no crime in Nagadoro.

3: "Your Risk Is Close to Zero"

"No problem at all," said Mr. Kojima. "Even Okuma. None at all." He was referring to the rental car, which would carry us 180 kilometers round trip. The highway had been decontaminated to pretty low levels, but in Okuma Town one ought to wear a mask and protective gear—although since he and Aki didn't, I didn't, either.

"For a short time you don't need a mask," said Aki. "This is how small the actual effect is. However, if you are active outside, then you need a mask."

I nodded cheerfully, and he elaborated: "If in a box is radiation, then you don't need a mask. But if there is no box, then it is possible that radiation comes into your mouth. Then you need a mask. But even without a mask, your risk is close to zero."

He estimated that in Okuma, one square centimeter might register up to 40,000 becquerels. I being ignorant, that failed to scare me.

"If there is dust, you need to wear a mask. Think about dust. The dust itself is radioactive. So the risk is higher then."

It was 9:10 in the morning, and my brand new dosimeter stood at 1.5 micros, which made me feel almost young again.

Aki and Mr. Kojima living the dream in Tomioka (0.5 micros)

"If the meter catches a scintillation, that's one count per second—if it catches it," he said. As you can see, he knew his physics. He was the guide; he was a former Tepco engineer. Mr. Kojima was the go-between.

Aki's family name was Yoshikawa, but since I told him to call me Bill, he invited me to call him Aki. The next night we got tipsy together, and I felt that I had done a good thing for international relations to watch him stagger off to his monorail. Mr. Kojima had already gone home; hopefully I had pleased him with the contents of the gratitude envelope.

As the interpreter explained the setup in a letter to me:

> Mr. Kojima organized these tours as a member of a private organization, Isshin-Juku, that fosters human resources that may promote democracy . . . The graduates . . . include politicians, venture capitalists, . . . etc. Mr. Kojima works at Fujitsu for a living, but spends his private time . . . promoting a project of his own within the framework of Isshin-Juku—organizing tours to Fukushima.

> Mr. Yoshikawa was graduated from a senior high school managed by Tepco . . . He worked at Tepco from 1999 to 2012. He quit the job as he wanted to work from outside to do something about the poor working conditions of the Fukushima Nuclear Plant workers . . . [and their] excessive bashing from others. He tried alone until November 2013, but to no avail. So he set up an organization named "Appreciate FUKU-SHIMA Workers" hoping to make a larger impact . . . Since 2014 he has been working toward the following:

> • To create an environment where people who had to rely on [the] nuclear power plant can live maintaining their local characteristics and . . . "traditional" industry . . .

> • To create an environment where people who are involved in decommissioning work that will continue over 40 years can work safely.

This document reveals the unique situation of Japanese nuclear energy in this period. Of the other three carbon ideologies—coal, oil and natural gas—only coal found itself significantly embattled; and in the two places where I studied it, Appalachia and Bangladesh, coal miners, if not always the coal industry, were members in good standing of their own communities. As for oil and natural gas, proponents of those energy categories might feel somewhat prickly about environmentalists and criers of climate change, but they feared no existential threat.

The Friends of Coal and their counterpart organizations did vigorously defend their fuel against what they claimed to be disingenuous, politicized attacks. The enemies they described were "outsiders," leftists, secular humanists, President Obama, the U.S. Environmental Protection Agency. The few homegrown anti-coal activists were, as you shall see, a hated, intimidated minority.

But in Japan, nuclear power was on the skids, and Tepco workers the cowed minority. To be sure, some Japanese were beginning to turn against the evacuees, and everyone longed for the Fukushima catastrophe to be "solved." Business as usual! If only that would come back!

I asked Aki: "What is your opinion about nuclear power? Is it desirable?"

"Currently in Japan under this situation, personally I think it is necessary. Whether it is good or bad is another story."

He remarked that this area was deficient in flat land for rice. "Because the land is very narrow here. It's hard to live. For the people of the locality, the most important thing is to survive. They need some industry, like nuclear power."

Here we were again in Yotsukura, with the ocean coming back into sight as we rolled north on Highway 6. Aki said that the tritium here was so sparse as not to be measurable. Just then we saw a raccoon dog. At 9:34 the dosimeter remained at 1.5 accrued micros, as well it should have, and I asked whether the melted fuel at Plant No. 1 could be reused; Aki replied that the uranium had fused with the metal cell that once contained it, making it good for nothing now. We ascended the green forested hill to the tunnel, approaching Hirono, the northernmost habitable town.

"What's the best way to decontaminate food?"

"No way."

At 9:39 we arrived at Hirono, with the dosimeter remaining loyally at 1.5. Aki said that only some 30% of the original population had returned. There had been 1,700 people here. "During the day about 3,000 Daiichi workers pass here," he said.

How could I not wish those decontaminators well? By 2016, according to Tepco, a recorded 32,760 toilers at Plant No. 1 had accrued more than 5 millisieverts a year. (*A read of 5 millisieverts is one of the thresholds for whether nuclear plant workers suffering from leukemia can be eligible for compensation . . .*) Of this lucky multitude, 176 individuals showed cumulative doses of more than

100 millis, *a level considered to raise the risk of dying after developing cancer by 0.5 percent.* The highest reading on any of their dosimeters was 678.8 millis.*

Their counterparts, the ones employed by entities other than Tepco, were often spared the tiresome duty of wearing dosimeters. In 2016 *The Japan Times* would report:

> They typically work on three- to six-month contracts with little or no benefits, living in makeshift company barracks. And the government is not even making sure that their radiation levels are individually tested . . . Nearly 70 [of "more than 300 companies doing Fukushima decontamination work"] committed violations in the first half of last year, including . . . failure to do compulsory radiation checks . . . The ashes of half a dozen unidentified decontamination workers ended up at a Buddhist temple in this town [Minamisoma] . . . They were simply labeled "decontamination troops" . . .

Fortunately, I was still living in 2014, when these bad things were unthinkable. So I activated the pancake frisker quite cheerfully. As might have been expected, here the air dose within the car remained at most 43 counts per minute, 0.12 micros per hour. By that token we could have been in Barcelona or Veracruz.

I asked Aki: "What does the cesium actually look like?"

"Even a large particle of cesium you need an electron microscope to see."†

Today was Sunday, a lovely warm day; and we were now passing the Hirono town office, with the pastel blue of the sea off to the right. "The middle of November is the best time for colored leaves," he said.

We stopped at a farmer friend's home. With that admirably typical Japanese communitarian spirit, Aki had begun helping him and others to grow olive seedlings, since the produce of those trees absorbs hardly any radiation. On that subject, the air dose (NORMAL reading) was already up to 0.30 micros, while inside the car it had dropped to 0.10—both of these being trivial, even safe.

We drove on. On the parking lot beneath the thermal plant a few children were riding their bicycles round and round. Aki remarked that it was now

---

* Had the winning worker been irradiated here over the entire five years since the accident, his yearly dose would have been 135.76 millis—15.48 micros an hour. That was one of the more elevated levels my frisker would record in Okuma, but by no means the highest. Had he accrued his 678.8 millis in a single year, however, his hourly dose would have averaged 77.38 micros—not atypical for outer space.

† A jolly element collector supplies a more vivid description: "The cesium in this ampoule melts if you hold it in your hand for a minute, yielding the prettiest gold liquid."

unusual to see them. (The local weather was only 0.14 micros.) "On such a sunny day before 3-11 you would have seen several hundred kids here," he said.

We stood taking the air at the edge of the municipal park, part of which had been employed to house decontamination workers. No children played here; their mothers had instructed them to avoid the place, because (as he said with scorn) they foolishly believed the workers to be radioactive.

At 10:40 we reached J Village, which demarcated the beginning of Naraha Town. The air measured 62 cpms and 0.18 micros. Here we spied the first weed-outliers—mostly goldenrod and pampas grass as usual, growing unhindered from crevices between the asphalt and the concrete divider, with their shadows pleasingly distinct upon the empty highway. In an inhabited Japanese town they would never have been tolerated; and one of the most consistently striking signs of the red and yellow zones was this vegetative overcoming of stone-hard rigid constructions. I rather liked them even though they made me sad.

Aki was saying that he considered Tepco's ice wall a good idea. It would be needed for only 10 or 20 years, until the fuel rods could be removed. In an enthusiastic military metaphor, he described it as "a thousand-ton tank." Who was I to say that it couldn't win the final victory? (By the way, every year it would require 45.5 million kilowatt-hours*—as much as 13,000 Japanese households used.)—I

---

* To generate this much electricity, a conventional power plant would need to burn about 3.6 million gallons of heavy fuel oil.

asked why Tepco could not separate the tritium from the water by means of simple distillation. Smiling, he reminded me to look at the atomic diameters; H_3O was much more compact than, say, cesium-137; those radioactive hydrogen atoms would be too small to catch. Of course the Americans knew the secret, he continued; they had learned it when preparing their gifts for Hiroshima and Nagasaki. Since the technique continued to bear military applications, they had so far refused to release it.—And just then the dosimeter accrued that day's first tenth of a micro, turning over to 1.6. While he joked around with Mr. Kojima, I took a one-minute timed walk with the pancake frisker, which read the air dose at 82 counts per minute, 0.30 micros per hour: I could have been measuring a brick sidewalk in West Virginia.—J Village continued silent, and a breeze rolled in at 0.35 micros.

Now we were passing the sign for Naraha, the frisker showing 0.2 inside the car, and we got a lovely view of the ocean, made still more splendid by tall waves of black bags, one of which appeared to already be leaking. The roadside was characterized by pigeons and goldenrod. Here and even in most places I saw in the red zones the highway continued to be in perfect repair, so that travellers could experience loneliness and get irradiated without any but financial inconvenience.—Aki said: "This used to be ricefield. Now, grassy marsh."

The berms of the former ricefields were a slightly darker green than the rectangular pits within them. Since rice is an annual, the three years since the accident would certainly have sufficed to erase most of their vestiges.* Had I been more intrepid I would have clambered down to inspect those grasses, but the thought of radioactive stickleburrs deterred me, so that I postponed my weed-hunting for the next day in Okuma, when I was not only resolved but also would be required to wear protective clothes. Driving away from the ocean and into the tall pampas grass with its white plumes—the goldenrod again everywhere—we discussed those strange red pine trees that the decontamination worker had mentioned to me last March in Iwaki; Aki said that all around the nuclear power plant in Niigata he had seen many of them even before the Fukushima accident—which, by the way, had apparently deposited some fallout over there. He had not visited Niigata lately.—We now reentered the main highway, with central Naraha appearing more quiet than decrepit; and at 10:50 we stopped at the only functioning convenience store, in the parking lot of which the levels read slightly lower than before. I was surprised to see customers; Aki called them "those who used to live here"; they had come home for a few hours, since it was Sunday. Children aged 16 and under were not legally permitted to be in town, not even

---

* At least these paddies were no longer warming the atmosphere with methane (p. 116)—for which we should praise nuclear power. About eight times the normal yearly figure for cropland leaving circulation was recorded for 2011. Of this cropland, almost 89% consisted of ricefields.

in this convenience store—and as a reminder why, just northward lay an organized myriad of black bags roofed with a green tarp.

We drove up to a vantage point and parked there, looking down on Highway 6 and the thermal plant. Here the air count was 116 cpms and 0.36 micros.
Aki repeated: "Once we get into Okuma, none of us are allowed to get out of the
car."—Hoping that I could jolly him into letting me pass the frisker out the
window when we passed the infamous Plant No. 1, I kept quiet and agreeable.
So we turned our eyes northward past the empty playground to the sea-cliff at
Naraha's edge, enjoying the fresh breeze as slow blue-green waves crawled toward
us from beneath two distant clouds, and then we gazed down on the empty playground. Had it not been a Sunday the sight would have affected me less; for then
I could have supposed that the older children were in school, and the younger
ones perhaps being carried here and there on town-errands by their busy mothers. It was a pleasant-looking field, well groomed, with the sea foaming white and
clean against the rocks below. The swings hung still. A conical web of climbing-
net rose around a central pole, with no child crawling up it. My companions were
silent. Bending down to frisk a gutter from three inches, I read about 200 counts
per minute—0.67 micros per hour. That would be about six times the average air
dose of San Francisco, but after all it was a gutter, not the air. Aki said that immediately after the explosions this place had reached 6,000 cpms and more,
which would have been a good 20 micros an hour—175 millis if that had kept
up for a year, but it had not. Next spring the ban on living in Naraha would be
lifted, "but maybe there will be some red places here for another 20 years." I asked
about Namie and Okuma. He thought that in 20 years their values might be as
safe as Iwaki's. Considering what Ed Lyman had told me about fallout disposition, I agreed on the "might be"; it might be unhealthy for a child to dig a hole in
the back yard, or for a farmer to turn over the soil. As I would see on my final
night visit to Tomioka, it "might be" worse than that.

Even past the cereal factory, the frisker (on NORMAL) flickered mostly between 0.1 to 0.276 inside the car, only very occasionally touching 0.3; but as soon
as we entered greater Tomioka, our radiation level rose to between 0.270 and
0.30. At 11:38 the dosimeter read 1.7 accrued micros. Now as we approached
Plant No. 2, where Aki used to be employed (he nostalgically informed us the
jellyfish used to be the enemy, because they clogged up the cooling intakes), the
frisker scintillated up to 0.38, then 0.4; once we had passed that place our average
readings fell back to 0.20. Everything looked very summery; it was an unseasonably warm October; the interpreter once remarked that nowadays summer came
a month earlier and stayed a month longer than when she had been a girl.

Deserted playground in Naraha

Immediately north of No. 2, we turned in toward the seashore, where the government had begun building great incinerators which would reduce the volume of waste in the black bags to one-fiftieth. I had never been here. From the scaffolding, the structures looked to be seven-storey buildings, and the cranes dwarfed even them, rising ugly and crazy like broken weeds, while black bags marched shapelessly across the tall grass, and waves smashed against broken concrete. Looking back along the shoreline, I got a last glimpse of those two misty thermal stacks in Hirono, where those children had been riding their bicycles. This was a different world, but my one-minute timed walk still measured but 116 cpms and 0.36 micros, with the frisker extended from my chest. Investigating several bags, I read around half a micro each. (It was now 11:54, and the dosimeter went up to 1.8.) Aki mentioned that his job at Plant No. 2 had been to burn waste, especially paper suits. Perhaps this was why the new incinerators were of special interest to him. I wished the burners luck, but could not help but wonder how many fallout particles might spread out with the smoke.*

The tsunami had torn away the ground floor from what used to be a *ryokan*, a traditional-style Japanese inn. It was passing ugly now. Not going out of my way

---

* There would be another cost, of course—combusted carbon, the focus of this book. According to the Ministry of the Environment, one of three methods for "volume reduction of contaminated soil" was "heating processing," by "using rotating heater and reaction accelerator together to separate cesium from soil[.] [A]lmost all cesium is collected by a bag filter. Decontamination rate is 99.8–99.9%," which sounded splendid, although maybe less so from the standpoint of "volume reduction" since "the volume of purified soil is doubled because reaction accelerator is needed to add the same amount of the soil." Now for the price: "It costs about 200,000 yen/t[on] (400 t/day, *for 10 years running*)" [my italics]. That would be an impressive amount of CO_2—to say nothing of the energy required to run the reaction accelerator for that decade.

to frisk it, I wandered along the seashore, among gravel and concrete where the beach used to be, with black bags endless off the west side of this narrow road which, said Aki, "used to be a sightseeing place." Goldenrod grew tall behind the bags. Lacking the expansive brilliance of the acres upon acres of that I would find in Okuma, it staked its claims in individual stalks, but whether those were fated to sooner or later destruction from the machines of humankind I could not tell. The beaches as far south as Hisanohama remained as bleak with their tsunami-broken concrete hunks as if D-Day had just happened there.—"My exposure was one of the smallest," said Aki, as if to himself, and just then, but only for a moment (I had it on NORMAL), the frisker scintillated up to an unpleasant 10 micros an hour, the highest I had ever seen it go so far. But the dosimeter accrued no new digit, and as I strode forward to photograph a blasted tree which towered behind black bags, the glowing numbers in the rectangular window fell back to 0.452 and 0.3 micros.

Half-built incinerators, Tomioka

Continuing up the highway, we soon arrived at the Japan Rail station where Mr. Endo had taken me that spring. A pigeon waddled very slowly up the road (no worry about traffic) and many more goldenrod grew up colorfully lovely against the faded red and green houses. A troop of black bags now stood on the seaward side of the station—more than I remembered. The station premises read only 94 counts per minute and 0.30 micros. I walked half a block down the street, where it was 111 and 0.36, and then found a well of standing water which perhaps had once been a storm sewer. Squatting down on the asphalt, I extended the bright yellow protuberance of the frisker over that white-lipped rectangular hole, whose cover must have been removed either because the earthquake-tsunami had cracked it or because some official had thought it wisest to let radioactive liquids drain freely. Down in the blackness was a bright reflection, whose boatlike shape and dark lines (perhaps the representations of power wires) reminded me of some many-paned arch-shaped stained glass window. But although it did in fact reflect the sky, the most churchlike association it engendered in me was with a crypt. Sticks and leaves floated unmoving on its stagnant brightness and its stagnant darkness. One foot over the hole, my frisker chittered back and forth between 7 and 8 micros, which was about as high as anything I had measured in Iitate. I refrained from measuring any deeper. The air dose in that spot was 7.52 micros.

At 11:38, when we first entered Tomioka, the dosimeter had turned over from 1.6 to 1.7; at 11:54 it read 1.8; at 12:20, 1.9; and at 12:28, 2.0 micros, so we were now accruing something like half a micro an hour—still less than we would at cruising altitude on a typical long-distance flight. On the other hand, much of our time had been spent inside a vehicle, so the situation for any unprotected person would have been worse. Once again I congratulated myself on not living in Tomioka.

We drove inland for five minutes, over the river into a yellow zone, the radiation rising to .7, .8, .9 in the car, and then I asked to measure the invading goldenrod, which was almost harmless: 148 cpms, 0.48 micros. I frisked vine-overgrown crevices in the asphalt; they were three times worse.

As we stood looking down over the tall white plumes of pampas grass, into a ditch of gently running water, I frisked the silent street; then, since Aki liked that gadget of mine, I passed it to him to play with, while the *robot girl*'s voice echoed and echoed, conveying new instructions. It was as if the fields had come closer to the houses; they retained their rectilinear shapes, but were not ricefields anymore.

Aki said that in the first year the air dose in this place had more than 2,000 cpms, which calculated to be about 6.6 micros an hour. Right after the accident, he said, the air dose at Plant No. 2 had been 8,000 cpms—26.667 micros—and his car read 300,000—in other words, a cool thousand micros an hour. (You may recall that the experimenters at the Brookhaven National Laboratory first began to detect adverse effects on plant growth and animal diversity at a thousand micros.)*—Aki was grinning and laughing as he told me about his radioactive car.

---

* See above, p. 442 fn.

Tramping through mildly radioactive weeds, I got down on one knee to measure the ground dose, because he had assured me: "You don't have to worry at
all." Once I finished frisking, I inquired: "You mean that there's no internal exposure from breathing here? And it doesn't stay in the lungs?"—"Probably yes,"
he replied, "but take me as an example. I have an internal exposure beyond comparison with yours, just by working at the nuclear plant, and I have checked
myself thoroughly. I was below the detectable level."—I thought my thoughts
about genetic damage, but my role was to learn from him, not to hurt or challenge him. His devil-may-care position might for all I knew be justified.

We returned to the car, speeding down the empty roads. Our interior dose,
the frisker restlessly chittering in NORMAL mode, was 0.7, 0.8, 0.4 and 1.04. I
had wanted to return to Yonomori, the cherry tree avenue of which Mr. Endo
had been so proud, so we paused there among the silent houses. While our earlier
halts had appeared strangely summery thanks to the tall and handsome weeds,
Yonomori looked like autumn. Fallen leaves lay here and there on the clean
blacktop, with many more of them along the curbs. The shadows of branches
crossed the highway like wrought ironwork. To the north and south the highway
dwindled beneath bare trees, with nobody coming or going, and the branches
appeared to touch each other across the road, hiding the abandoned homes behind them. Whatever leaves still remained on the branches were silhouetted
against the clouds. I should not have liked to walk alone down that long avenue
at night. Strolling down the center line toward the ever receding vortex of that
tunnel of trees, with the frisker extended at waist level, I read an average air dose
of 1,226 counts per minute, or 4.08 micros an hour—nearly a micro more than
I'd measured in a jetliner over the North Pole. (Or would you rather know that
before the accident Yonomori's air dose had been 0.042 micros—97 times less?)
By now I was getting used to such numbers. The *robot girl* was saying: *"The garbage has not been appropriately segregated."*

Aki said: "The center of this town was pretty busy. Even though my place was
20 kilometers away, I often used to come here. There was a large scale supermarket, and I liked the electrical shop . . ."

It had taken him three years to complete the cleanup at his home; in those
days it was not so simple even to throw away the broken dishes. For three years
no one would accept such garbage. (Hearing this, I imagined that I felt a burning
in my nose.)

Driving off that cherry tree boulevard and past an army of goldenrod, we
soon reached one of the meandering local borders between red and yellow zones,
where accordion-grated barriers closed off ever so many side streets; the company
that manufactured those must now be rich. In the red zones the weeds were often

taller than elsewhere, and more ivy grew on the trees. I remember the way that in Tomioka the glossy-leaved tree branches and the pale green ivy could cradle a pole so lovingly that no explorer who lacked a pair of pruning shears could tell whether this had been a streetlamp or the support of power lines; it leaned, and the pavement around it was cracked and broken, evidently from the tsunami; it seemed well on the way to metamorphosing into a tree-trunk in a peculiarly indeterminate forest whose floor was grooved concrete—although even in the yellow zone's commercial districts I often found pampas grass bursting out of planters.—About that latter plant an American guide to invasive weeds said: *Growth is astoundingly fast—8 feet in one season is not uncommon—and it easily shoves aside other vegetation . . . Eradication is daunting.* A gardener's encyclopaedia described its genus as *hardy to frost hardy . . . perennials forming clumps of narrow leaves . . . over which soar plume-like flower panicles in white, silver, gold or rose-pink in late summer or autumn. Cortaderia should be positioned in full sun and cut to ground level each spring to maintain healthy, vigorous plants.* In the red zones it was getting plenty of sun, all right. Whether or not anyone was cutting it down (last February Mr. Endo had said that the weeds of Tomioka simply died back in winter and then came up again), it was healthy; it was vigorous—the pride of any radioactive garden. And so Tomioka's yards had turned into vacant lots, never mind the shuttered houses within them.

Often the red zone announced itself simply through one of those vertical signs held against a guardrail by a red-and-white horizontal signal bar each of whose looped ends was affixed to a road cone. No fence or other physical barrier prevented anyone from stepping over the guardrail into the pampas grass and the darkness between ivy-grown trees; in fact it would have been more difficult to step through the pampas grass that occluded the patio of a certain decrepit house from whose eaves scraps of torn cloth inexplicably dangled; perhaps the householder had tried to curtain off a window with what he had, after which he fled and then the wind tore it off and hung it from the rain gutter. Some houses and driveways were in perfect repair, while their yards were hills of pampas grass and of saplings grown roof-high. I frisked the edges of a few of those yards, which then became for me as distinctly inimical as the borders of any red zone.

There was a pedestrian overpass flowing with vines as we drove through the yellow zone, toward the Futaba police station;* it was 101 cpms and 0.36 micros within the car when we came to the Tomioka Bridge; here came another overpass vine-grown and silent. The barricade on Highway 6 that had vexed me last

---

* In case you are wondering, the Futaba Town Recovery Promotion Committee "compiled a long-term plan," but the plan failed to state when people could come home.

A house in Tomioka

Commencement of red zone. As the frisker shows, the air dose here in the yellow zone was 3.75 micros. To the immediate right of the warning sign, it was not significantly higher. In several other Tomioka locations the air dose rose within two or three steps after crossing into a red zone.

February came down in September; so we rolled unhindered into the red zone, with the understanding that we not only could not stop until Minamisoma but also must keep our speed at 60 kilometers an hour. I didn't mind; tomorrow I had clearance to walk around in Okuma. So we rushed past the thick grasses and goldenrod, finding no one else on the road, and the frisker's numbers danced around 2 micros an hour: 2.0, 2.71, 2.67, 2.07, 1.8 (what must the level be outside?), and again and again, despite my horror and sorrow, I could not but find myself beguiled by the freshness, not to mention the vegetative silence that had overcome that corporate monoculture of convenience stores and shining vending machines, beautiful though it could sometimes be, as evinced by nighttime Iwaki, which offered the red and white glow of a long horizontal restaurant sign whose blood-color echoed behind the shutters and reflected brilliantly off the back window of a dark car; in my way I loved Iwaki, and of course I hated the red zones; I worried about cesium in my muscles, strontium in my bones and gamma rays stabbing through me; more than that I grieved for the people who had lost so much, and I worried about the worse atomic accidents that might at any time happen . . .—but these red zones were new places, not entirely ugly; when those oceans of goldenrod rose up in their own gentle tsunamis to drown the abandoned homes of nuclear refugees, I experienced something refreshingly non-human, reminding me that however much we might alter our planet, the planet itself would survive awhile, bearing its own loveliness, just as Jupiter or Venus must be lovely, even if lethal to us. Here came another ricefield now transformed into goldenrod, and the frisker chirped 2.18 and 1.95. Five kilometers into the red zone, we saw some red maple leaves and yellow goldenrod; the fall colors made Aki and Mr. Kojima happy. Now at last we were approaching the infamous Plant No. 1, whose power wires ran to our right, and as we got closer the frisker chirped faster, until I muted it; our interior air dose was 2.46, 2.65 (here came the hollow scaffolds of No. 1's power towers, not unlike ideograms), 3.48, 4.14, 4.69, 5.46 micros; then we got a view of a stack at No. 1; Aki said that we were "two kilometers from the boundary of the premises," and more than five from the No. 1 reactor building, rolling along at our legal 60 kilometers an hour. What wondrous readings could I have gotten now? Tomorrow I meant to find out. As soon as we passed that well-goldenrodded place, our dose fell to 5 micros—no harm done: from 1:02 to 1:19 we'd gained only another third of a micro.

By the time we reached Futaba our interior level had fallen to 0.8 micros an hour. Sometimes it even dipped to 0.6 (181 counts per minute).—At Plant No. 1, by the way, Reactors 1 through 4 lay in Okuma; Reactors 5 and 6 in Futaba.

Aki remarked: "My wife's parents used to live around here, but now they can't come. I don't have a permit myself."

A few cars now enlivened the highway; a policeman was running across the road, wearing a light mask; our air dose within the car went up to 1.3, then down to 0.45; and presently a simple blue marker informed us that we had reentered the yellow zone. We could slow if we cared to, or even roll down our windows.

"Here," Aki said. "I used to live here. This is Namie Town. The apartment I used to live in, there. You need an identity card to visit."

The external air was 112 counts per minute, 0.36 micros per hour; and the dosimeter read 2.9 micros at 1:31 p.m. Thus Namie, which according to that Nuclear Regulation Authority website offered 94 monitoring stations (I remembered from a day or two ago the senior high school, showing off its 5.969 hourly microsieverts).

I saw a few houses half crumpled into a soft russet meadow—tsunami relics— and then the radiation declined to 0.2 and 0.1 as we came into Minamisoma, which had once been a center of silkworm weaving, then became a textile wholesaler; now it was a moribund junky sprawl.—Aki pointed out the Odagakawa Bridge.

"And in this area, you can survive by raising rice," he said proudly. "This is the same distance from Daiichi as Hirono, but the situation is so much better! Enough flat land."

He said that in the "central area," the population used to be 50,000; 43,000 people declined to evacuate. The area appeared quiet and grungy but by no means abandoned. A woman was carefully cleaning the windows of her half-shuttered retail establishment; most neighboring stores were utterly closed but we did see a few people walking, and here came a girl on a bicycle . . . These signs of life made me happy after Tomioka and Okuma, but Aki was sad; he remembered how much more vibrant it used to be. We stopped for lunch at a kind of automat where souvenirs were sold. Then we began our return.

In the car our dose was 0.060, 0.041 and 0.114 micros per hour. It was 2:46, and we had accrued only 3.00 micros since morning.

At 3:05 the dosimeter accrued another 0.1 micro, and the former ricefields were brilliant with goldenrod. (Praising its *racemes or spikes of golden-yellow flowers,* the previously cited gardener's manual calls it *a genus of vigorous perennials,* and I do consider *vigorous* the best possible adjective). Some ricefields were indicated by dark grass and white pampas separated along an invisible line of peculiar straightness. In central Tomioka one could easily perceive the berms as raised frames around the bygone crop. Here the reversion had gone farther.—At 3:35 we reentered the red zone. This time a uniformed man stood at the checkpoint, although he did not require us to halt. Aki said that some of guards were former nuclear workers, especially maintenance men. I toggled the pancake frisker into NORMAL, and it showed 0.3, 0.4, 0.5, then almost immediately 1.3, 1.4, 1.8, 2.10 micros an hour; now the goldenrods were thick along the roadside. We were on the outskirts of Okuma, with another 0.1 micro in our bones and tissues to show for it.—As we once again approached the accident's nasty locus, my pancake frisker indicated 4.8, 1.3, 1.8, 1.9, 2.10, 3.10, 3.52, 4.71, 4.99, 4.88, 5.05 and 4.94 micros per hour, with our windows cranked up tight, and goldenrod carpeting more ricefields, and the white plumes of pampas grass almost translucent in the late afternoon sun. I would not care to imply in any way that this time Aki proved willing to bend the law for me, and allowed me out of the car once or twice to indulge in a frisk, but if he had, and I did, both of which are preposterously hypothetical, I would have registered NORMAL roadside measurements of a stiff 10 to 12 micros: 100 times higher than my kitchen counter in Sacramento. Very soon we had returned to Tomioka's yellow zone, alone on the clean and empty highway behind whose guardrails rose tall pale plains of pampas grass, some of whose heads glowed with great distinctness against the shaded wall of forest.

At 3:44 we were back in Yonomori District of Tomioka Town. We were all pleased: I myself especially enjoyed being able to stroll around in eerie freedom on the highway pavement, metering the warning signs and gratings of the red zone's much-indented, many-gated frontiers to my heart's content, with no one to hinder or befriend me but my own shadow. Sometimes the roadside was as richly disorderly, thanks to ivies and grasses, as some overgrown pumpkin patch, the pampas head-high and more. In the town itself, cracked, taped houses, up whose walls some ivy-tendrils had already nosed right up to the eaves, lay dark-windowed behind overgrown trees; from their unkempt yards, spears of pampas grass bristled outward at the highway. But I never saw more than a day's worth of fallen leaves upon the asphalt. Sometimes the house windows were fogged up inside, and the notices taped within them looked waterstained or even moldy. Through the dark panes whichever objects I could see—they were usually jumbled, either by the earthquake or by panic—resembled underwater detritus perceived at the extreme end of a flashlight's range. Weeds busily widened and elongated the cracks in driveways. Where the road had been wounded by the earthquake, flimsy articulated barricades whose tops and bottoms were defined by two long parallel pipes had been wrapped across the pavement in angular approximations of semicircles. From any distance, Tomioka's houses, hidden behind their weeds, withdrew into the darkening forest that would quickly conquer them if humans kept away. Goldenrod resembled water-streaks on certain houses' shadowed sides. Roofs sometimes missed tiles, and ivy kept growing up toward power wires.

Strolling down the expressway on a luxurious one-minute timed walk, I measured 261 cpms and 0.9 micros per hour—a mere 15 times higher than the radioactivity in Sacramento. My NORMAL readings fluttered around a micro.

In a neighborhood where I had never been, I wandered around with the frisker, foliage reddening for no one's pleasure beside the weedy parking lot of a shuttered office in Tomioka, persimmons shining in the evening light, hanging in clusters over the wall of their jungle which was once someone's garden in Tomioka; my companions stood around the car, waiting for me to finish.

Aki now proposed to take us to the substation whose failure, he said, had made the explosions at Plant No. 1 inevitable. Just then the frisker showed 429 cpms and 1.446 micros.

Persimmons shone against their almost silhouetted trees. The sky was gentle with stratus clouds. On the lawns some leaves had gone orange, and their reflections were vaguely pale in the dark windows of those houses around which the vegetation ever more tightly pressed. We came to a white sign almost overgrown with ivy; I frisked it, but the dose was merely 1.054 micros.

There on the forest ridge a humanoid, insectoid figure with three pairs of downturned latticework arms stood letting out wires forward and backward from its six wrists. It was one of those high voltage electrical transmission towers

Tomioka street sign (1.054 micros)

that so many of us used to take for granted, back in the days when we were alive. Beside it stood a truncated cone of girders, with three round platforms and two Mickey Mouse ears on top; that must have been a microwave tower. Then there was a double-bridge conduit, built of the same latticework.

Just before dusk we began walking down the power plant road, whose air dose was 0.83. Then I spotted three of those wild boar I had heard so much about—young ones, dark brown, trotting toward me without fear, then veering into the weeds. I called my companions; they rushed to take pictures with their cell phones, then said we had better get out of there in case Mother Hog lurked in the bushes.

At 8:23 we reentered greater Iwaki with the dosimeter at exactly 4.0; so in hours it had turned over merely 2.7 micros—about 0.2 micros an hour—about three times more than in Sacramento. We returned the car, the greyhaired gas station attendant with his longhaired boy assistant bowing low, then diligently wiping radioactive dust from our windshield. I had told him that it must be contaminated, but he smilingly told me not to worry.

4: As Innocuous as Plant No. 1

The taxi driver said: "Nuclear power plants, I wish they would all be abandoned, because there was a safety miss. They promised nothing dangerous would happen, but then this accident clearly showed there was a miss . . ."

The new dosimeter read 4.7 microsieverts as we rolled out of central Iwaki, and the van's interior radioactivity was a homelike 0.12 micros an hour—appropriate for Tokyo or San Francisco.

"The fact is, the government is trying to restart other nuclear power plants," he said. "That move is unbelievable. With all these reactors turned off here, still electricity is not short at all. Renewable is better."

It was a cloudy morning, promising rain to the west. The ricefields were stubbled green and brown, most of the crop having been harvested in mid-October. Occasionally a very few yellow-green shaggy fields awaited the gathering. Water was sparkling on young trees, and in several yards ripe persimmons glowed on the trees.

The driver's notion was Okuma would be safe to live in after 60 or 70 years.*

At 8:07 we entered the expressway, with the frisker reading nearly unchanged. With the driver now silent, I gazed down on ricefields and an occasional scarlet maple in the light and dark green of forest. At 8:16 we could see the twin thermal stacks of Hirono, the frisker showing 0.24 micros an hour. Naraha read comparably

---

* The Union of Concerned Scientists asserted an official return date of 2022, although, as will be seen, different zones would actually be decontaminated at different times.

at first; the sun shone there on a small hillside cemetery in a clearing, so that the stones looked almost cheerful. The mountain forest remained mostly uncut except for the so-called "laydown areas" where green tarps overlined the black bags.

In the four minutes that it required for us to pass through Naraha, the frisker readings (on NORMAL) climbed from 0.23 to 0.6 micros per hour—with ups and downs, to be sure. Then we entered Tomioka. Just here a digital sign advised us that the radiation was 2.34 micros outside. Departing the expressway so that the interpreter could use the washroom, the driver parked a few steps from the toll taker, who wore white protective gear in his booth; and I took a one-minute timed walk to frisk birdsong and pine-smell: only 374 cpms, 1.26 micros an hour. Why the radiation level was so much lower here than it had been at the digital sign was one of life's mysteries; one could blame the frisker, the digital sign or local variation.

We now took a certain forest road whose air dose by brown pools and tall sedges was 1.644. High over a reservoir I read a happier 0.356. It was 8:40 when we entered the city limits of greater Okuma, whose name means "Big Bear"; the dosimeter read 4.9. Continuing onward, we came to the sign which warned: NO GO ZONE AHEAD, and presently arrived at the Okuma Town office, whose anteroom a NORMAL frisk found to be a remarkably salubrious 0.22 micros—not out of place for Moscow if on the high side for Poza Rica, Mexico. We took off our shoes and went upstairs. Especially for me the two officials had prepared their daily weather report:

Situation map at the Okuma Town office

They said that in one spot in central Okuma where the radiation used to be 100 micros it was now only 40. When I asked about Tomioka, they replied that the downtown there was typically 0.3 micros, which was far lower than the pancake frisker indicated. Perhaps they meant the air dose. Anyhow, they were full of good news.

According to the regulations, they were to follow our taxi and ensure that we did not deviate from the route plan, which I had been required to propose and clear several months earlier, at which time there had been an additional implication that I might not be permitted to get out of the car. In fact these two men were wonderful guides and hosts. They drove ahead, leading the taxi through the central downtown and to a temple, then to the ocean, a river, a shrine, a highly radioactive place and finally to a vantage point from which Plant No. 1 would be visible. Whenever I wished, I told the taxi driver to stop, then walked about and frisked to my heart's content. We were all supposed to have dosimeters, including the taxi driver, so I had an extra one for him, but nobody cared about that. By the way, he was forbidden to get out of the car. Following the rules, I now supplied myself, the interpreter and the driver (who declined) with shoe covers, painter's union suits and masks. Those items were all manufactured with pride in the United States of America, and they began to tear almost immediately. Duct taping my imperfections at crotch and thigh, shuffling about ludicrously in the flimsy, wrinkled shoe covers, I watched myself fall in the estimation of those dignified officials. Fortunately, years of disappointing and even disgusting the Japanese with my American gaucheries had made me an expert in looking ridiculous with extreme tranquility, so on that understanding we went downstairs and set out down the road. It was 9:09, and the dosimeter had accrued 5.0 micros exactly. I felt quite happy.

Approaching the red zone almost at once, we reached a narrow vertical warning sign of red characters on white; ahead stood a sign whose red and blue characters were facing in three directions. There were three sentries. Each one clutched a bright orange baton in his white-gloved hand. Unlike Iitate's, the obstacle course of barriers they oversaw was of a merely suggestive character; anybody could have stepped over or driven through. Behind a tall tripod from which a dark lamp-bulb depended, a pale accordion gate, evidently for night hours, stood collapsed into irrelevance on the righthand side of the road. Stepping out of the taxi, I frisked the air, and was pleased to find the very moderate level of 1.54 micros an hour. One of the officers checked our permits, and my passport. Then they bowed us through. We entered the red zone at 9:21.

We were now in Ogawa Ward, where a one-minute frisk down the street captured 750 counts per minute, or 2.58 micros an hour—a bit "hot," to be sure, but hardly exceptional for a red zone; even in the yellow parts of Tomioka it would have been in place. I remember another gate, and a lovely lane overgrown with the usual pampas grass and tall goldenrod, houses pleasantly secluded behind the trees. I saw two men in protective gear at an abandoned Esso station. Inside the taxi van the radiation was already more than 2 micros. We drove on, and then I asked to stop again. The air dose by some pampas grass was 3.27. The officials waited patiently in their car.

There were places where weeds were just beginning to break through the asphalt of what had evidently once been magnificently maintained streets, while at the roadside a clamor of ivy, goldenrod and other weeds almost obscured the houses behind them, with only a few roofs still showing, like the forecastles of sinking ships. A meter above one bit of weedy pavement I measured a cool 5.0 micros.

The officials wanted to show me their town hall. They proudly considered it to be "a model decontamination," and I do admit without reservations that it read only 0.826 and 0.816 micros—800 times "hotter" than Portland, Oregon. They said that cesium was now found "normally," as they put it, at three to five centimeters down, but no further, "because the nature of it has a particular affinity for clay." My little excavation on Mr. Shigihara's land in Iitate had detected what must have been cesium at a greater depth than that—but then the radioactivity had obediently fallen off. Perhaps the clay ran shallower in Okuma.—The officials remarked that removing five centimeters of farmland was easy, but expensive here, due to the asphalt, but (I detected understated pride) they had persevered in this spot all the same, to a depth of two to three centimeters. To decontaminate a garden, which they had also accomplished, one must excavate it all by hand. They were still "just learning," they modestly said. I cannot now remember whether I complimented them on their hard work; I hope that I did.

They showed me the former health center. This edifice I did assure them now looked very nice.

The more talkative of the two was named Mr. Suzuki Hisatomo; the interpreter remarked that he was "very cultured." I asked him which accident had been worse, Fukushima or Chernobyl. He said that in terms of the amount of radiation released it was Chernobyl, by far. Neither one of them showed any worry about today's excursion. They wore dark galoshes, perhaps to keep from tracking home radiocontaminants, and they put gloves on and off at will, but declined to trouble with masks; as the morning warmed up they rolled down their hazard suits to the waist, revealing the crisp municipal tunics beneath.

CESIUM-137 RELEASED IN THE WORLD'S TWO WORST NUCLEAR DISASTERS

(in terabecquerels)

1

Fukushima (2011):

- Estimate by Tepco 13,600

1.29–1.51

- Estimate by the Fukushima University Institute of Environmental Radioactivity 17,500–20,500

6.25

Chernobyl (1986): 85,000

Source: *The Japan Times,* May 2014, with calculations by WTV.

The Okuma air dose monitor was not grated off as Iitate's had been, but likewise crowned with a slanting plane of solar panels. It appeared to be turned off, for its display showed four zeroes in microsieverts per hour; behind it, dead leaves huddled against the curb, weeds grew up out of the sidewalk, and the hedge hung shaggily over them. And why not? All 11,000 residents remained evacuated.

We drove on. In a certain long commercial street, which strange to say had one parked car every block or so (perhaps the owners had fled by other means), the rectilinear geometries of sunlight and shadow emphasized its forsakenness more than did the relatively few weeds and vines; the place was being cared for after a fashion; and the shards and flotsam of the earthquake had been raked to one side. The radiation was 2,060 counts per minute, which is to say 6.9 micros an hour, so that was getting up there; a year of it would make for 60.44 millis— well above the maximum for nuclear reactor workers. Like an eager puppy I frisked about the central district's shuttered shops. The almost immaculate pavement was cut by multiple jagged shadow-diagonals, and sometimes pierced by tall weeds. Broken pots lay on certain sidewalks. Fewer windows were broken than in Tomioka, perhaps because the higher radiation discouraged thieves. I sometimes saw tattered scraps of cloth hanging from abandoned

facades, broken boards and bricks heaped on sidewalks here and there—but the streets were clean save for those weeds. (In one street, it is true, I discovered a sort of beach of broken tiles, all swept up against the curb but on the asphalt nonetheless.) From behind an air conditioner or space heater on blocks on the sidewalk grew one of those ivy-vines I was always seeing in Tomioka; it crept up the side of a shop, gripped a drainpipe to whose radioactive effluent it must have been partial, then insinuated itself through a door's crack and behind some establishment's dark window. What it did in there I cannot tell you. Ivy flourished over and through a barbershop's barred gates. Other weeds bowed, spreading their many fingers over the asphalt. The air dose there was usually around 4 micros— 35 millis a year. We drove to another part of the same district, and here several windowpanes had gone; behind wrinkled curtains lay books, crates, rectangles of corrugated siding. In one place that reminded me of the retail block across from the garment shop in Tomioka, the lower part of the blinds had been twisted down at a 90-degree angle and then mostly torn away, so that in the dark cell just behind it I could see a table on which sat a teacup beside several closed laptop computers. What had happened there? Had the proprietor drunk a last cup of tea before he evacuated, and had he feared that his computers might be contaminated? How had the window gotten broken? Most lives are unfinished stories (suicide perhaps comprising our best chance to "complete" a life), so the red zone's plethora of such scenes of interruption as this was in a way extremely ordinary . . . but to have all these lives so interrupted at once . . . !—A crate lay beneath shuttered lattice windows. A great weed in full summer flourish rounded off that block, and then more closed doors and weeds decorated the next. It looked slightly tidier here than the neatest street of Tomioka.

Now proceeding to a more suburban-looking neighborhood, we came upon more hidden and thus more apparently similar interruptions, for here no windows had been shattered to reveal whatever wreckage, panic or quiet sadness lay within. As I paced the empty street, with a fresh-trimmed lawn giving way to pampas grass on the side of it, and a castle-like apartment tower rising behind everything, the air dose measured 2,060 counts per minute and 6.9 micros— almost as much as the interior of Mr. Shigihara's dairy barn in Iitate. A power pole leaned over a parking lot. Over a weedy drainage grating by some decrepit apartments the level was 8.79.

Just before ten we were at Ono Station, the only Japan Rail stop for Okuma; for some reason there had not been an Okuma Station and might not be for awhile now. Mr. Suzuki said the target reoccupation date for Okuma's less contaminated areas was three years; here it would take 10.

In the year previous to the reactor failures, Ono's air dose had varied between 0.041 and 0.042 micros an hour. Above a grating in the street before the station I now measured 4.20 micros. Two steps away, another grating, frisked from about eight inches, read 23.2 micros, which in one hour would have given me what it took nearly a month to absorb back home. I was a little shocked; a year's worth of that would be 203.3 millis.*—But when I raised the frisker slightly, to 10 inches, the count dropped to 21.9. At a foot and a half it fell to 11.52, which I considered spicy enough. Meanwhile the stairs within the earthquake-damaged station were only 1.604. I think that if I had to dwell in a red zone I would find a thick-walled place and keep within it as much as possible. In this connection I might note that during those three hours in the Okuma red zone my dosimeter accrued 9.1 micros of gamma radiation.† Call it 3 micros an hour. I should say we were in those two vehicles for at least half the time, so of course the "real" average dose one could expect to accrue in Okuma would have been significantly higher. But if one mostly lurked inside and managed on a budget of 3 hourly micros, which might not have been much worse than the working hours of an international airline stewardess, then 26.28 annual micros would be one's portion.

The weeds around the overgrown tracks were 9 and 7 micros. Here they had truly been left to themselves, which encouraged my appreciation of the work which was being carried out against nature in the radioactive city; for in the long narrow zone of fenced-off tracks the pampas grass rose high above masses of whiskery weeds. As soon as I approached the fence to frisk it (about 10 micros), contaminated stickleburrs festooned the legs of my paper suit.

---

* But there were sunnier ways to calculate those 23.2 micros. *Magill's Survey of Science,* published in 1993, advised me that "those working with radiation are required to keep their dosage below 5 rems [= 50 millis] per year, which is . . . 25 microsieverts . . . per hour for a 40-hour work week. No ill effects have been observed at several times this dosage."

† Between 9:21, when we passed through the gate, and 12:25, when we departed the red zone, the display altered from 5.0 to 14.1 micros.

Mr. Suzuki remarked that he saw wild boar around here every day; yesterday they had trapped three; I presume they killed them.*

At one corner where weeds were growing mightily from broken pots and from right out of the asphalt, a white hard hat lay in the street. I pointed it out to Mr. Suzuki. With a smile, he said that the crows had carried it there.

The officials had explained that "former resident families" could return 15 times a year for up to four hours each, but I never expected to see them. Not far past the station, however, the officials stopped their car to introduce us to an evacuee named Mr. Tazawa Norio, who had once been a colleague of theirs. (Later I learned that his wife was inside the house cleaning, but I never saw her.) He was all dressed up in a mask, gloves and a white hazard suit so that he could trim the weeds in front of his home. Something like a shower cap crowned his head with a mushroom slit. From a narrow stripe of exposed but shaded face his eyes squinted sadly at me, although when he attacked the weeds with his clippers (they were taller than he), his eyes widened as he gazed upward. His dosimeter, Japanese-made, hung from a lanyard around his neck. It was an odd sight to see

Mr. Tazawa Norio   Mr. Suzuki Hisatomo

---

* Under the category "treatment of contaminated wastes," the Ministry of the Environment included "captured wild harmful animals," for which the "executing agency," Kyowa Kako Co., Ltd., stood ready to supply a "demonstration of the safe composting system for treatment of dead bodies of captured animals."

him beside Mr. Suzuki, who was barefaced but for sunglasses and whose coveralls had been rolled neatly down to his hips. Since that latter person entered the red zone nearly every day, one would think that he would be less raffish about his exposure. But perhaps radiation damage is merely another harmful rumor.

"Three years ago my house was almost new," said Mr. Tazawa. "What I now have left is nothing but a mortgage." He was trying to pay it off as quickly as he could, to avoid burdening his descendants. "Birds come here, and their feces contain seeds."

While I talked with him, those obliging officials began pruning and weeding for him with their ungloved hands. That seemed very sweetly Japanese. I could not imagine some American ex-colleague of mine ever troubling to do the same for me.

He said: "I was a worker in public relations promotion of nuclear power."

"And now?"

He looked up at the sky. "If you have this kind of accident, then . . . I wish there were any kind of renewable substitute for nuclear."

In his baggy white suit, with his paper mask covering him from around his chin almost to his eyes and his headgear resembling a shower cap, Mr. Tazawa stood on the street by the white line, determinedly working his pruning shears while the weeds rubbed against his legs. When he took a step forward, he was in those weeds all the way to his armpits. The stone gateposts of his home were nearly sunken in vegetation.

Glancing down at my own so-called "protective gear," I saw that my torn paper pants, like the interpreter's, now bristled with radioactive stickleburrs (1 to 3 micros). I was glad to keep them away from my inner clothes.

Weeds and their perfect shadows were conquering the asphalt, guarding what must have been the entrance to an apartment building (no weeds yet grew upon its stairs). We got back into the taxi van, where the frisker read 1.7 and 1.9 micros per hour, then turned onto a narrow weed-lined road, the empty fields looking the same as before. At 10:16 our interior count began to increase: 3.8, 3.95, 4.16, 4.37, 4.76, 4.41, 5.60 NORMAL micros per hour—"since we are approaching Daiichi," said the taxi driver when I told him. He smiled; he too was enjoying the adventure.

Our next stop was a temple called Hen Jo, meaning unknown. The flight of entrance steps ascended a sort of inlet in the vegetation-crowned stone wall, some of whose bricks were disarrayed. Shrubs had begun to take over the steps, although someone had trimmed them partially back. At their summit were two character-engraven pillars, and then, set back within its flat yard, the red-roofed temple itself, whose white facade stood unevenly decomposed down to the inner wood. The place felt peculiar: abandoned and yet not exactly neglected; consider for instance the temple grounds, stripped down to sand, or perhaps stripped down and then sanded, by well-meaning decontaminators, with armies of gold-enrod standing at attention in tall close-packed array just behind the wall. No weeds grew *here,* at least not for the moment. The decontaminators had aimed to make the tombs sufficiently safe for former residents to come and briefly pay their respects to the ancestors. Mr. Suzuki informed me that the air dose here had been 19 micros but after decontamination it became "officially 5.05 as of September." A sign from September recorded a reading of 5.06, and today the frisker found even less to chirrup about: 3.9 micros an hour. Behind two metal-lipped incense wells, a stone statuette stood clasping together its palms and dreaming, with a tall tomb-slab at its back. Everything was as still as the folds of its stone robe. Close-eyed, serene and baby-bald, inhumanly patient, it waited for nothing that I could ever imagine. Bending down and extending the frisker to-ward that figure, I encountered the unpleasant value of 7.0 micros. Had I been

Hen Jo Temple (7.00 micros)

condemned to stay here until I reached the nuclear worker's five-year maximum of 50 millis, I would have served my sentence in less than 10 months.

Now we drove past house-islands in the great rich sea of goldenrod. Some of the homes had been swept away to their foundations by the tsunami. Disobeying a do-not-enter sign, we descended a narrow asphalt road as goldenrods towered on either side, more and more of them. So we arrived at the ocean, about three and a half kilometers from Daiichi. It was 10:45. The air dose had become an almost benignly mild 1.374.

How radioactive was the water? In May an unnamed site "outside the port" and three kilometers away (there was a fifty-fifty chance that it was right here) had measured 4.3 becquerels of tritium.* Remembering the difficulties that Eli had laid out before me when I asked to sample that very same contaminant, not least of them the fact that water is a neutron shield, I forebore to frisk the waves. Perhaps I could have scooped up some mud, waited for the water to evaporate, then measured what was left—but if H_2O evaporated, why wouldn't H_3O? Wishing not to harm myself and others through ignorance, I abandoned that project.

The seashore at Okuma. Note the sheared-off crown of the pine tree.

The breakwater was wrecked, of course. A pine leaned toward the bright blue sea, its top pollarded by the tsunami. Birds flew up in flocks from the river's mouth. Seeing a small dark beetle on the sleeve of my paper suit, I asked my usual question about whether the radiation was killing any creatures.

---

* "Inside the port," of course, the levels read at monitoring stations were breaking records again: 1,900 and 1,400 Bq (their prior respective prize-winners had been 1,400 and 1,200).

"I don't know what your impression was," said Mr. Suzuki, annoyed, "but no animals died."[*]

The driver, wearing no protection at all, was happily wandering the seashore, which he had not seen since before the accident.[†]

I still remember the smooth grey rocks and pebbles of that beach, with here and there a paler stone, and a line of wet sticks and even a little kelp, and then the foam where those low slow waves of greenish-grey came in. Sometimes a jade wave was a little higher than its cousins, and spray leaped up from its shining white shoulder just before it struck. Even then the impression I got was one of gentleness. This looked to be a place to wade with small children. Perhaps it used to be. It was clean. The pancake frisker showed 4.16 micros an hour as I stood facing the ocean breeze, then rapidly went down to half a micro. To the landward, russet marsh grass struck an appropriately autumnal note on this magnificently clear day with the forest ridges very blue and distinct to the west. A raptor glided slowly above a broken tree. On a little rise a few steps from the shore, several other trees (pines, I believe, although I did not have time to go see them) seemed to flourish, never mind that their crowns had all been evenly sheared off; Mr. Suzuki explained that the tsunami had reached just so high, and as I gazed up at them, trying to imagine being right here and watching the approach of a wave of that height, some of the horror of March eleventh came back to me. The tidal wave had killed 11 people in Okuma; the 12th had not yet been found, and so there was an ugly mound of broken board, sheet metal, rags and other detritus on the beach where the bulldozers had gone corpse-hunting. Another of Fukushima's incomplete stories was told by two dark sodden sneakers, and a single white shoe. What had happened to the other one, and was its owner alive or dead? Dark birds went swarming in a low flock over the blue lagoon. Among those mismatched shoes lay a woman's purse, miserably sodden, and a framed photograph glittering with moisture; I had neither the heart nor the right to invade any of these in search of information. Clambering up what remained of the wrecked breakwater, I measured 5 micros and more in a puddle of

---

[*] In 2016, in Namie, whose radiation was supposedly "15 times the safe standard," the internal organs of irradiated cows "so far have shown no significant abnormality particularly linked to radiation exposure, . . . but it's too early to draw conclusions about thyroid cancer and leukemia." Meanwhile, a Greenpeace study mentioned "DNA-damaged worms in highly contaminated areas."

[†] If you would like to envision the expression on his face, I refer you to the booklet "Atom Fukushima No. 86" (November 1990), in which a wide-eyed cartoon couple admires the ocean view at one of Tepco's Fukushima plants, perhaps even Daiichi, the pigtailed girl clasping her hands to say: "Wow, beautiful!," to which the boy brilliantly replies: "What a wonderful environment it is to have to nuclear plant here, isn't it?," after which a helpful old man remarks: "That's right. All nuclear plants in Japan are located at the coast."

rainwater or seawater on the steps. Angling the frisker up into the sea air as I neared the top, I encountered the dislikeable value of 10.14, and turned away. Mostly the levels there at the shore were less than a micro. I had asked to see a river, and so we all strolled to the mouth of the Kuma-kam. As we neared its wide bend in the marsh grass, myriad white birds arose, almost silently. Caught between my obligations to the frisker and to my companions, whose every remark must be tediously interpreted to and fro, I had not the time to make out what species they were. They ascended to no great height, then quickly settled back; evidently our presence did not much disturb them. Mr. Suzuki said that this place was famous for salmon, and indeed in that shallow, grassy, gently curving river, which in my country we would have called a creek, oblong palenesses wriggled in the crinkling water: spawning time.

I was astonished to learn that people could sell some fish from here—but of course the bottom feeders remained off limits.

I strolled up beside the taxi driver, who was taking deep breaths of the sea air, looking out across the white sand at the lovely lagoon and the low blue mountains beyond it.

Returning to our vehicles, we proceeded inland, more or less following the river. Within the taxi van the frisker within three seconds went from 1 to 1.8 to 2 micros. On a bridge I asked to stop. We could see spawning salmon wearily swimming upstream, and often simply weaving in place against the current, like long dark windblown leaves attached to some invisible stalk; a few were dead and drifted down; one kept turning over and showing its bright belly. In one minute the frisker scintillated 1,286 times: 4.26 micros per hour.

Now for a brief distance we retraced the route we had taken to the ocean. Along that immaculate empty road, on which a puddle vaguely reflected the clouds, and weeds were just beginning to rise up along the concrete blocks where cars had once parked, a glorious plain of goldenrod underlined the mountains, and in that yellow lake stood a few lonely white islets: abandoned houses.—How does one know that no one is at home?—When there is no way to it.—Silver-white plumes of pampas grass reached higher than their roofs. One three-storey white house with a fine balcony rose more distinctly from the goldenrod, in part because it was especially close to the road, and also because the tsunami must have hissed through here, for between the house's wide-splayed legs was a dark cave where most of the ground floor had been carried away. When I walked up toward it, I began to see sky and pampas grass within the jaggedly peeling lips of that vacancy. Upstairs, one window-half was curtained, and the other dark; perhaps that darkness was the inside of the house. The frisker read between 5 and 6 micros. Less than one of my three allotted hours in this red zone remained. I had

stopped too often. Approaching the roadside, I aimed the frisker at some goldenrod, and read 6.73 micros—only 112 times higher than my studio back home.
I took two steps into the goldenrod: 7.40 micros.

Tsunami-wrecked house
(7.40 micros)

Hastening back to the taxi van, I asked that we drive a little more quickly,
in order to see the reactors if we could, and we soon reached another checkpoint, with men in white protective gear bowing us through on either side of
the road.

Everywhere the lovely weeds were more beautiful than anything humans
could do.

Inside the taxi van as we rode up a low hill our radiation level climbed: 3, 4,
5 and all the way to 9.23 micros as we crested that hill. Now we were rolling
through the lovely goldenrod wilderness of a former industrial park. To the right
lay decrepit and sometimes broken building-cubes, and that pale blue ocean.

At 11:18 they showed me a tsunami-destroyed shrine: 2,551 counts per minute, 8.52 micros. Here was another of those places where the grass and flowering
weeds massing along the edges had begun to creep onto the pavement itself.
When I frisked the road, extending that pancake head at chest level, I found a
patch that measured more than 20 micros, and I felt as I had when I first saw that
gnawed-away house with the blue mountains showing through its missing first

storey. Mr. Suzuki, unimpressed, reminded me that the level used to exceed 100 micros at the time of the accident.

He pointed, and I got my first glimpse of those infamous blue and white tanks: Daiichi. There were grey tanks also. Grey meant bolted while blue meant welded, which leaked less; Tepco was trying to replace grey with blue.

At 11:25 we reached the former fish hatchery, or, to give it its due, the Fukushima Prefecture Aquaculture Association, where the air was 3,260 counts per minute and 10.86 micros. Once upon a time the two officials had been quite proud of this establishment. Mr. Suzuki explained that the water used to be warmed for the hatchlings with waste heat from Plant No. 1. I agreed that that had been clever. Ruined houses grinned at me from the weeds.

The fish hatchery

Trolling the emptiness of the cracked pavement by the weedy buildings there where the land slanted down toward the seashore, I performed my usual involutions. The wounded half-cylinder of the hatchery gaped open high above the pampas grass. Sometimes the air dose was 6 micros and sometimes it was nearly 10. The pampas grass at the roadside read only 2 or 3 micros. I knelt down and frisked the air above the pavement: 29.5 micros. The white heads of pampas grass were shining beyond the bridge's guardrail, which had been half pulled away like the top of a tin can. Wild thickets of pampas grass towered as high as the new trees, suffocating lost walls and foundations.

A grating by the
hatchery (18.03 micros)

Over a well within a rusty grating I lowered the frisker from waist level to
about three inches, and its count rose from 20 to 30 micros.

Not far from here I took my highest measurement in Japan: 41.5 micros. This
very nearly reached the lower boundary of the radiation in outer space.*

On the bright side, by 1973 Okuma had achieved the highest per capita

---

* You may remember from p. 308 that the maximum recorded radioactivity at any inhabited place in
mid-April 2011 was 16,020 microsieverts accrued over 21 days, or an average of 31.79 micros an
hour.—But here I must quote from the [August 22 of this or last year's] blog of Mr. Yoshikawa—that
is, my friend Aki from yesterday's tour—so that you will see how the other half lives: Tepco had
brought "Appreciate FUKUSHIMA Workers" out to Plant No. 1. At Reactors 4 and 5, "we did not
need a mask or gloves. Several hundred workers were taking a rest . . . as if they were at an ordinary
construction site. This is outdoors . . . In front of No. 4 reactor, the radiation level was about 50 micros
[per hour] *in the bus* [my italics]. Ordinary people may find it tremendously high, but . . . I regard it [as]
surprisingly low for a reactor that exploded. For your information, when I used to work at the site
before the accident, it was not unusual at all to get 100 micros within a nuclear facility building."

My highest measure-
ment in Japan
(41.5 micros)

income in Fukushima Prefecture, all thanks to nuclear power! Wasn't that worth a few gamma rays?

Inside the taxi van at 11:36, heading straight toward Plant No. 1, I found our air dose to be fluctuating between 10 and 12 micros. The two officials had planned this tour superbly, for in four minutes we had arrived at my final requested point of interest: an overlook on Plant No. 1. This proved to be the grounds of an old age home, and here I finally lost one of my torn and wrinkled shoe covers. Consoling myself that I could hardly make a less dignified impression on my hosts than before, I resolved to keep that foot out of any vegetation for the duration.

In the three-quarters of an hour from the river bridge to the old age home, the dosimeter had accrued 4.6 micros. That calculated out to an unpleasant irradiation rate of 5.75 micros an hour. But thanks to dosimeter, frisker and this moving vehicle, I felt more or less in control of our exposure. Although the plant was merely 2.2 kilometers away, the air dose here rarely exceeded 2 or 3 micros an hour.

In the courtyard, goldenrod grew higher than the windows, sometimes bending and leaning against the walls.* The grass was not wildly overgrown, so the place seemed almost cheerful. One window was open, and the white curtain pulled back to show off its darkness. Given 10 more minutes' time I would have gone inside, but it was already the stated departure time. In the other windows, trees, weeds and sky reflected themselves. Some of the grass was golden. A tennis shoe lay in it, sideways. The seedheads and flowers of those tall weeds blocked the doorway, invading the parking lot and reaching up toward the dark window beneath a roof overhang that was vertically streaked with blackish grime and fallout.

Proceeding to the road on the northern edge of the hill where the two officials waited, I read 2,200 counts per minute, or 7.38 hourly micros. Down below through the waving pampas grass I could see a horizon of ocean, cranes, tanks and low, wide buildings. They were guarded by a belt of dark green trees, which perhaps were those famous pines of the reddish trunks.† Between the trees and our hill lay a few houses and some fields whose verdant yellow-green I suspected must be goldenrod.—Mr. Suzuki now very precisely gave me the lie of the land: "On the right is an exhaust tower; next is Reactor No. 4, and left of that, two pillars in, then below that is Reactor No. 3. The white building to the left with the blue pattern is No. 2, and to the left of that is No. 1."

Like so many culprits, they bore an unimpressive, even innocuous appearance. If I could only have gotten closer I would have seen the pipes, opened walls, rubble and crumpled latticework; and then, still unseen but conjectured, the liquefied and resolidified reactor cores, lumped and twisted around the reactors' skeletons. And what a thrill it would have been to frisk Tepco's underground trenches! Three days ago the poisoned cloaca beneath Reactor No. 1 measured at 161,000 becquerels of cesium, which once again made the highest reading ever. Tepco blamed the recent hurricane.

Since I had detained the two officials for an unscheduled half an hour, we now sped out of the red zone, passing a place where the taxi driver had seen wild boar four or five days ago, then a row of beautiful trees in red and yellow leaf, a

---

* I sometimes wondered whether a longer established radioactive community such as Chernobyl would show greater variety in its plants and animals. "As with all kinds of stress," says my college ecology textbook, "reduction in species diversity is associated with radiation stress." All the same, provided that the dose rate was less than obscene, it seemed plausible that plants, animals and some insects would adapt. Surely the trees would come back; if it were too "hot" for pines it might not be for oaks.—Or do you from the future for whom I write dwell mostly upon rolling plains of goldenrod? Were I a biologist I could tell you more; and you would surely rather read scientific observations which might somehow ease your predicament than my merely descriptive emotings. But when I was alive, those were what they paid me for.

† See above, p. 392.

Plant No. 1, Reactors No. 1, 2 and 3

Reactors No. 3 and 4: Pacific Ocean on right*

plain of goldenrod with grey berms in the former ricefields, and to the right a cemetery surrounded by goldenrod; then we departed the last gate.

Mr. Suzuki and his colleague drove straight back to the office, but the interpreter and I must now undergo decontamination screening at a roadside checkpoint. With great pleasure I tore off my coverall, mask and remaining shoe cover. Then they frisked me with their magic wands. They remarked that today's surface contamination was 240 counts per minute, or 0.43 microsieverts per hour.[†] They measured me at a mere 230 cpms, which exempted me from abandoning

---

* Left of Reactor No. 1 lay Reactors No. 5 and 6, which were screened from us by pines and pampas grass.

† You may recall that in Fukushima I found a fairly close correlation between counts per minute and 300 times microsieverts per hour. These officials' factor was 558, not 300. In other words, my conversion of their 240 cpms would have been 0.8 micros instead of 0.43. On this subject let me note that the arithmetical average of the 92 readings I made in the Okuma red zone was 7.14 micros—16.6 times higher than their 0.43 micros. Of course the frisker was measuring gamma waves in addition to alpha and beta particles; all these people presumably cared about was the particles, which could continue to do harm after being removed from the zone. Here I want to say that whenever I had a chance to compare my frisker's reading with that of a Japanese government scintillation meter (as at Hen Jo, and in the 41.5-microsievert patch by the fish hatchery), the measurements closely agreed.

any of my possessions or taking an immediate shower. The central government standard was 13,000 cpms for any object's surface, not for the human body, which rated a flat 20 millis per year. (To me this sounded like apples-and-oranges obfuscation.) If a car was above 13,000 cpms, it must be washed or abandoned. Of course they did not inspect the taxi at all, nor even the driver.

"Yes, I never got out of the car," the driver laughed.

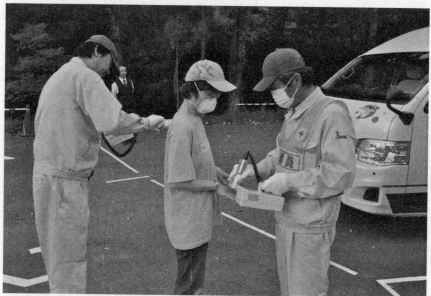

Decontaminating the interpreter

MY FIFTH TOMIOKA IDYLL

Since the taxi driver was willing to earn more money (he cost me something like $700) and Tomioka, my metonym for Fukushima, lay so conveniently near, I proposed to take more measurements and photographs there—only in the yellow zone, of course; we lacked permission for the other. The ever agreeable interpreter acquiesced, and as soon we stopped she slipped her mask back on. That was when we got our laugh, to see that this time she'd worn it inside out! Some of those radioactive particles from Okuma which the mask had previously filtered out must now be in her lungs. Such are the amusements one finds in nuclear zones. As for me, wishing to emulate Aki, Mr. Kojima, Mr. Suzuki and Mr. Shigihara, I went maskless here as in Okuma; so it was an even bet whether she or I would get cancer first. Well, despite those harmful rumors there was *no immediate danger.*

Afternoon rush at the
Night Pub Sepia

There on that side road I had come to know so well was the pachinko parlor **TSUBA** (today its facade and the weeds around it ran about 4.8 micros) and the Night Pub Sepia, which put out a mere 4. (By the way, these black-and-white reproductions will not reveal that the place actually had a peach-colored paint job.) Unimpressed by the scenic pulchritude of Tomioka, the driver remained in the taxi van, considerately driving after us as we strolled here and there. Within a few paces I came to a gap in a hedge of what might have been laurels, and then some fingers of a cryptomeria below and in front of me, the richly gloomy silhouettes of the same trees above and behind. The middle of that ovoid gap allured me with a glowing meadow of tall grass pale green and russet, silvered with pampas grass and illuminated with goldenrod, with hints of sky in those cryptomeria trees like tiny stained glass windows, and I longed to saunter there, all the more since there had been no time to make such excursions in Okuma; but as I introduced the pancake frisker through the hole and ducked my head as I prepared to follow behind it, the displayed numbers chittered upward, from 5.52 to 7 micros per hour, and the frisker not even half a meter in yet; then I felt a loathsome spiderweb on my forehead and the dusty leaves seemed on the verge of brushing against my face, so that in revulsion I drew back, cleaning off my head with my relatively uncontaminated sleeve, and that was when it struck me anew how dangerous last spring's Tomioka idylls must have been. That morning the two Okuma officials had asserted that the level in downtown Tomioka was "typically" 0.3 micros. So this was not typical—although the average of my 151 frisks of Tomioka worked out to 2.77 micros. Such discrepancies do make a fellow wonder. All the same, as I think back on this meadow hole experience, it seems to say more about my state of mind than about the place, for what should a micro and a half's difference truly have been to me? Given my ignorance and inexperience, those excursions into the red zone *were* a strain—and rightly so; it was the people who had been compelled into familiarity (precursors of you in my future) whose perceptions were distorted into dullness. (We were all fools.) Those spiderwebs and leaves, with their tactile content, affected me much as had my lonely walk up those dark shrine steps in Iitate, with my shoes sinking into leaves, and vegetative particles tickling my nostrils; probably they were no worse for me than the gamma rays shooting from the ground by the fish hatchery at Okuma, but as so many taxi drivers had explained, the invisible and unfeelable could be set aside.

HOW RADIOACTIVE WAS IT?

or, Extracts from an Official Website

Tsunami zone

Here the first reading listed is the first available.

ISHINOMAKI
Point No. 70, distance from Daiichi not stated
    Aug. 1, 2012, at noon: 0.06 microSV/hr.
    Oct. 28, 2014, at 8:00 am: 0.05
    July 20, 2017, at noon: 0.05

KESENNUMA
Point No. 79, distance from Daiichi not stated
    Aug. 1, 2012, at noon: 0.04
    Oct. 28, 2014, at 8:00 am: 0.04
    July 20, 2017, at noon: 0.04

SENDAI

Point No. 74, distance from Daiichi not stated
    Aug. 1, 2012, at noon: 0.04
    Oct. 28, 2014, at 8:00 am: 0.03
    July 20, 2017, at noon: 0.03

———

Red and yellow zones

IITATE

Point No. 109, 38 km NW of Daiichi
    April 1, 2011, at noon: 7.54
    Aug. 1, 2012, at 8:00 am: 0.28
    Oct. 28, 2014, at 8:00 am: 0.28
    July 20, 2017, at noon: 0.28
Point No. 736, 37 km NW of Daiichi
    April 1, 2011, at noon: 2.85
    Aug. 1, 2012, at 8:00 am: 3.06
    Oct. 28, 2014, at 8:00 am: 0.73
    July 20, 2017, at noon: 0.35

KAWAUCHI

Point No. 650, 19 km WSW of Daiichi
    April 1, 2012, at 8:00 am: 0.18
    Oct. 27, 2014, at 8:00 am: 0.09
    July 20, 2017, at noon: 0.07

TOMIOKA

Point No. 3955, 10 km SSW of Daiichi
    Dec. 27, 2013, at 6:00 pm: 1.15
    Oct. 28, 2014, at 8:00 am: 0.44
    July 20, 2017, at noon: 0.23
Point No. 3966, 9 km SSW of Daiichi
    Dec. 27, 2013, at 6:00 pm: 1.63
    Oct. 28, 2014, at 8:00 am: 0.61
    July 20, 2017, at noon: 0.35
Point No. 645, N of police box and W of post office, 6 km SSW of Daiichi
    Sept. 8, 2011, at noon: 4.96
    Aug. 1, 2014, at 8:00 am: 4.05
    July 20, 2017, at 12:10 pm: 1.08

The two monitoring posts nearest to Yonomori were Point 645, 6 km SSW of Daiichi, whose reading on Oct. 28, 2014, at 8:00 a.m. was 1.92 micros; and (for "real time dose measurement") Point 3954, 7 km SW of Daiichi, which at the same time read 3.82 micros. No readings for this or any other place in this table were posted for Oct. 27, the date of the visit just described. But on Oct. 28 I did make another visit to Tomioka, as detailed below.

Nuclear accident zone

VICINITY OF NUCLEAR PLANT NO. 1

Point No. 1373, 2 km west of Daiichi

    April 1, 2012, at 8:00 am: 26.04

    Oct. 27, 2014, at 8:00 am: 10.80

    July 20, 2017, at 11:00 am: 6.20

Source: http://fukushima-radioactivity.jp/pc.

Engaging in my hobby of measuring drainpipes and culverts, I recorded 5, 3.5, 12, 32 and 5 micros. Over in the red zone, a lovely russet-golden sea of grass and young goldenrod and tall pampas grass was already halfway to drowning the closed-up little houses.

Departing from the hedge and from the repellent graspings of the ivy upon the Night Pub Sepia, I directed the driver to that uncanny garment shop in the business district, where the *robot girl* was echoingly advising us: *"When you enter the town, please be very careful about wasps. Contact the Environmental Section for information."* We were already inside the garment shop when we heard this. Since February, heaps of rodent feces had been swept up into piles here and there; the rest appeared unchanged.

The radiation in there was 0.36 micros. Perhaps the "typical" dose for this area had been computed from interior samplings such as this. The sidewalk immediately outside varied from 0.7 to 1.3 micros measured at waist level. I walked a few steps down the street in both directions: 1.0, 2.58, 1.86 and 2.0. A fallen goldenrod on the sidewalk was 2.51.

On October 22, the Nuclear Regulation Authority website listed Tomioka's 67 monitoring spots. The highest radiation reported was at the "multipurpose meeting place, east side": 2.218 micros.—According to my frisker, the base of a pole for an asphalt parking lot's barrier chain read 12 micros and more; the asphalt was 11.

In the garment shop I stood looking out through the darkness into the

The hole in the foliage
(from 2 to 7 micros)

yellowish-white light of the glass doors, with the mannequin's elegant silhouette gazing down through the open doorway at the two bright lines of grass that sprouted up from the sidewalk and the rubble before her and the ivy-grown ruined shops across the street.

5: And You Can't Live

Not far south of Iwaki stands Ganjo Temple, the site of Fukushima's only National Treasure. I am told that it dates all the way back to the Heian period, although the Japanese have fooled me that way before; once in Nara I saw the oldest wooden building in the world, and felt quite impressed until I learned that they replaced the wood as often as it wore out—sometimes every 30 years. Well, who was I to say how others ought to count time and genuineness? I could not even tell whether Tomioka would be safe to dwell in by 2017 as the central

Central business district, Tomioka

government predicted;* that was only one-fifth of a cesium half-life; 10 half-lives would have left the town depopulated until 2311. So let me be magnanimous to Ganjo-ji, however ancient it actually might have been; I liked the 12 lightning Buddhas of Amida-do Hall. There was also a standing statue of Kannon Bosatsu, the all-gendered deity of mercy.

As I learned from my pancake frisker, the temple grounds were a trifle more radioactive than downtown, although not unhealthily so: a quarter of a micro an hour, two millis a year.† I remembered Ed Lyman's assertion that the fallout would remain longer in evidence in absorbent places. On the pond's bank, dead lotuses stood crowded together like the chambers of crazy multicylindered rusty

---

* On this very question Mr. Togawa Satoshi of Tepco said to me: "Even after six years, it is feared that the annual radiation won't go below 20 millisieverts annually. That's in this red zone"—indicating a map of Tomioka. "In the yellow zone, the annual dose is between 20 to 50 millisieverts. In the green, they are preparing to lift the evacuation order for this area. In this green part the infrastructure for carrying out normal life has almost recovered, and also the good decontamination work has been begun for where kids are. When all these conditions are cleared, people can return."

I asked: "If people were in the green area, could fallout from the reactor come back to them?"

His colleague Mr. Hitosugi Yoshimi replied: "We have evaluated the level in our premises, and basically, there's no new source of radiation. We are working so the people can come back as soon as possible. For that, Fukushima has to be in a stable condition. So we have to go forward with the demolition of the reactor."—I took his meaning to be: Under certain conditions the green zones could get recontaminated.

† The Japanese national target air dose.

revolvers. This mud was no more radioactive than anything else in sight. Some of the Japanese maple leaves had already turned crimson and purple. Just now the foliage must be lovely in Iitate. Shadows lay long on the grass, and just below the bridge great carp kept opening their mouths, as greedily expectant as the grumbling and chuckling ducks. The bright sun was in my eyes, and birds were singing. It was nearly four o'clock.

We commenced driving north for the last time, with undulating tree-ridges sharply illuminated in the low sun so that one could see each pallid trunk and each irregular oblong vertical of shadow between the trees, sunshine gleaming and dazzling on certain angles of white house-wall or shoulder of silver car; and back in Iwaki many ribbons of power wire dangled down. I glimpsed yellow foliage against the dark glass of some office building. The gas station and strip mall signs and the traffic lights and the glowing cans on the shelves within the vending machines were all pleasantly lurid. Harvested ricefields all caught the light, their texture like that of flattened corncobs. Within the taxi it continued to remain less radioactive than at the temple; from that standpoint I could have been on the street in front of the hotel.

Now we were already in Yotsukura, the pavement on which we traveled slipping into shadow; the roofs of the cars ahead of us had lost their sunlight, although the forest ridges to the north were still as brightly picked out as ever. The sky had grown pale yellow over the ocean, and up on the horizon, seeming closer than they were, those twin white thermal stacks of Hirono arose as if out of the water. According to the frisker, our interior levels were still between 0.132 and 0.308 micros. Pampas grass began to flourish along the roadside, but all the little cemeteries were still neat and clean. Then we passed through the four hill-tunnels, none of which had collapsed during the great earthquake of 2011, and at 4:31 we entered Hirono, having not yet accrued even one-tenth of a micro-sievert since leaving Ganjo-ji, the levels within the cab more or less the same, the two thermal stacks even yet small ahead.

Flashing quickly through that town, the highway now passing darkly through a deep cut, after which the white plumes of the pampas grass began to go grey, we made fine progress, since hardly anyone was going north from here on. A few last ricefields remained under cultivation (I would have wished to frisk that rice), and we reached the sign for J Village, the ocean now out of sight for awhile, cars with their taillights shining red as we came into Naraha right at 4:40, where the levels remained the same both outside and also in the dreary convenience store, whose only customers at that hour were decontamination workers with longtailed *hachimaki* tied around their foreheads, samurai style (I have been told that sometimes even high school students about to take some difficult exam will fasten these on, in order to nerve themselves up for some especially demanding work).

Then we were rolling down, down, everything vaguer except for one glimpse of flat calm sea, a crimson maple tree going dirty-brown against the pale sky. Even now we had received only a tenth of a microsievert since the drive began; after all, it was just 4:54 now as we entered Tomioka. To our right, the bridge across the ravine and up to Plant No. 2 glowed with the headlights of stalled traffic, including a busload of decontaminators. Our interior levels remained trivial. We continued down that hill, passing a long line of headlights going south: decontaminators off shift, most likely; as before, no one at all was going our way. Almost immediately the pancake frisker began chirring more busily, and the illuminated head, which made a pretty good flashlight to take notes by, displayed 0.664, 0.904, 1.748 and 0.896. A one-minute timed count measured about three-quarters of a micro per hour. Well, that still wasn't much.

Tomioka was awfully dark. Right away I found myself more disoriented than ever before. I had been under the impression that the municipality kept streetlights on in order to deter theft, but this did not seem to be the case. The driver, who had not been here since before the accident, could not be prevailed on to slow down (perhaps he was nervous), and so before I knew it we had sped past the left turn to that main street where the garment shop was, and came all the way to the former checkpoint, which now of course allowed through traffic to Namie but was still manned by police who stood guarding the barricaded side streets. Then at last the driver made a U-turn, and it became clear that the street I had wished for had been likewise closed off; this must be the nightly practice. So we asked a policeman how to get on the Kawauchi road, and he directed us a good kilometer back to the south. He was masked, helmeted, in greatcoat and boots; his orange light-stick flashed coldly. I wondered whether he would die from cancer. Thanking him, we went as we had come on Highway 6, the driver going too fast as usual, not that it mattered, for there was only one possible westward turn through the dark empty town, a mound of black radioactive bags darker than the dusk in a strip mall's parking lot.

This road skirted the urban core, and after two or three dark blocks of abandoned offices there was not much to see but weeds in the headlights. For the moment, our interior levels rose no further. That meant that we might as well have been anywhere or nowhere—for all I really had to see by in this dark empty place was what the frisker saw. The driver kept speeding in terror.

In the blackness in the direction of the town office, our headlights pricked very dark thick masses of weeds on either side, and then a ridge of silhouetted blue against a pale yellow sky—the very last light. This place seemed very rural and lovely. We must be near Yonomori, and Mr. Endo's house, and that shrine with the broken stone lion. If we continued any farther on this road, we would leave Tomioka, and be rushing uselessly down the Joban-do toward Kawauchi. It took half a dozen requests before I could get the driver to slow down. The next problem was to halt him entirely. His anxiety was catching, of course, which made me cross, but at last he did stop, and if there was no more light left in the sky, at least there was some kind of purplish pallor.

Stepping out of the vehicle with the interpreter, with the frisker on NOR-MAL, I encountered pleasant night weather conditions of 2.6 micros per hour—less than 23 millis a year! . . . the Emporium Kindergarten in Koriyama had been significantly more radioactive than that in March 2011. Scribbling into my note-book words I could not see, I noted a crescent moon, cool air, no crickets; perhaps it was already too cold for them. Unable to perceive very much, and unwilling to dwell on nasty realities, my mind soothed me with a plausible projection of a happier place, so that this spot began to seem to me almost like high ranch coun-try in its loneliness, with just a single distant blue light here and there in the darkness, as if I were back in the California desert where I had once been safe and young. Now the stars began to come out.

Walking forward on what must have once been a two-rut dirt road along the north side of the ricefield, I found the radiation levels rising almost at once, to 3.9 and 4.7. There were more digits than this, of course, but they kept changing, and I had to write blindly, and the trend was plain enough: 4.07, 5.34, 4.63, 4.80, 4.90, 5.74, 5.75, 6.96. Now that my eyes had begun to accustom themselves to the night, I could see black bags on my left. That sight did not please me. I was walk-ing down a lane of pallid grass. Between the lanes I could feel the grass rubbing against the sides of my paper shoe covers, which I trusted not to tear, although just the other day in Okuma one of them had come off. Sooner or later some taller weed would brush against my trousers and deposit radioactive stickleburrs. Well, in Iwaki I would put all my clothes in a plastic bag and then take a shower. My shoes I could always frisk. There were tall dark trees on the right, and on the left, that crowd of black bags. Stepping to the edge of the road, I extended the frisker down toward them, as far as my arm could reach, and although it must have still been a good three or four meters away, it instantly read 12 and 13 mi-cros. If the statisticians were correct, it might take a good year at that exposure to raise my risk of fatal cancer by half a percent; all the same, those measurements increased my uneasiness, being a couple of hundred times higher than back home. I can still remember the horrid crowd of black bags down there as I went along that road that was shaded by tree-darkness, an equally horrid sighing of wind, and then columns of pallid light between the otherwise solid-dark trees, hands of plants, weed-hands with many long thin twitching pallid fingers, hands of pale grass splayed out, slits of pallor in the dark tree-wall—but here the average timed value was only 3.12 micros.

"Forward or back?"—The brave interpreter replied: "I will follow you." I decided that we might as well go back, and seek out another place. So we returned to the taxi, neither rushing nor dillydallying by the black bags, and there sat the poor driver with his lights and engine on.

The next site was not far away: a dark intersection, the road empty but for an orange-and-silver-striped cone with a flashing red light on it—the centerpiece of a knee-high barrier of cones and cross-poles, marking the border of the no-go zone, into which I slowly, cautiously, illegally stepped, with the frisker ahead of me (a small marker proclaimed this to be Hometown Appreciation Road), and on my left hunched another fat black bag. Standing in the center of the pavement for a good long timed minute, with the frisker in my left hand and the rest of my body as far away from that bag as possible, I presently learned that the level was 1,098 scintillations, 0.061 microsieverts—a mere 3.66 micros per hour. I almost laughed. There were ever more stars in the sky ahead, and on either side were trees, and cold autumn wind licked my sleeves.

What in fact I could have learned, and what I am telling you here, consists of little more than blindness, uneasiness, helplessness and ignorance. The frisker and the dosimeter were all I had to go by, and what they told me was disputable. *For the general public, it is recommended that the total radiation dosage (exclusive of medical dosages) for an individual by age 30 not exceed a specified dosage which is about 50 times the annual natural background dosage. For radiation workers . . . the guideline is . . . about 100 times natural background dosage per year.* So my old college physics book had it, back in 1974. But a hundred times natural background had not worried Aki, not one bit. As for me, I began to feel uneasy in the windy darkness, watching the frisker's numbers go up. That crescent moon and those stars lost their charm. Advancing a trifle more down that silent dark street, I received notice from the frisker that I was now taking in 4.08 micros. Then, from the trees on my right I heard a deep loud grunting—surely another of those radioactive wild boar, who were said to be aggressive (I felt more afraid of them than of the radiation), so I returned to the taxi, within which the level was only about a micro per hour.

When I asked the driver whether he could find the ruined Japan Railways station to which Mr. Endo and then Aki and Mr. Kojima had guided me, he, enthusiastic to finally be set a more straightforward goal, quickly keyed this destination into the electronic guidance system, so that once again we began speeding, much too fast as usual, through these dark and empty streets which were all the darker for being walled in with radioactive vegetation; to tell the truth, he was more confident or at least hoped more wishfully than I, who began to suspect that his software might not have been updated since the accident; indeed, we soon reached another roadblock. Stepping out to sample the darkness, I found an ugly pit ahead, followed by scaffolding; for once the difficulty was not radiation but simply earthquake damage; in any event we could go no farther. Taking a timed measurement, I found the radiation quite low—only 0.78 micros.

Almost desperately the driver sped down this or that other dead road, so that it occurred to me to ask him if he could find the way back to Highway 6 if his software did not pan out; as for me, sitting in the back seat, scribbling notes and staring at the pancake frisker, I would have been well lost had he asked me how to go, which he did not. It began to strike me what a nightmare it would be to get trapped in Tomioka all night in this taxi, rushing purposelessly from street to street, with nothing to show for it but maybe a month's worth of genetic damage. As it happened, the numbers around here continued to be nearly pleasant— mostly around a half-micro as we whipped through the dark forest; then came a hateful line of black bags on our left; even in the taxi the frisker's numbers went up. But that was a mere anomaly, like the accident itself, so we continued stubbornly down that empty road, a single yellow traffic light blinking far ahead, our levels now 0.8 and 0.7, after which we reached the next National Treasure, which is to say a dark field of black bags, the driver stubbornly faithful to his navigation system, still speeding no matter what I said, often coming up against another roadblock; once he admitted that we had been on this particular dark road before; but finally we were back on Highway 6.

My next intention was to have him stop on the shoulder near the barricaded side-road that led to the garment shop; I was pretty sure that I recognized it now even in the dark, but as usual he overshot it. (He was the worst taxi driver I ever had in Japan; in the end I paid him $600.) So I gave that up, contenting myself with a final excursion to that ruined pachinko parlor called **TSUBA**.

On the east side of the highway as we rolled back down into town, I glimpsed a small house in which greenish-yellow light glowed behind two curtains. Who could possibly be staying here? Remembering the tale of the stubborn farmer who cared for abandoned pets and cattle (but didn't he live in a red zone?), and the woman who looked after her paralyzed relative, I wondered if this light marked some kindred story of small defiance; I never found out. Here came that nasty carnival of whirling lights upon a police pole at the former checkpoint, the headlights of a police car muted, somehow not carrying very far in that darkness;

and we neared the double line of glowing poles, policemen standing lonely in the dark. Just before that was our familiar sharp right, which the driver miraculously failed to miss. Promising not to leave him alone in the car too long, I stepped out, and the interpreter with me. Now for another one-minute timed stroll in view of **TSUBA**: 3.9 micros an hour beneath the pale dangling plume-fingers of the pampas grass, that crescent moon momentarily bisected by a power wire; we passed a streetlamp, then approached the pallor of the shuttered nightclub on the right—sickly greenish weeds across the front of that white, white building, yes, the Night Pub Sepia, and on NORMAL the frisker read a mere 3.7 or 3.8, the pampas grass pale and reaching, while on the left loomed a broken sign nearly devoured by bushes, then the tall vertical **NO ENTRY** sign on account of radiation; the interpreter had translated it before. Past the Night Pub Sepia rose a dead building with black windows and shadows black upon it; the space between it and the Night Pub Sepia was *black, black,* solid black, and the frisker chittered up to 4.59 and then higher.

As the U.S. Atomic Energy promised us back in 1966: *Atomic energy is revolutionizing life today . . . The men and women who enter the world of the atom invariably find that they are exploring a world more exciting than . . . the dreams of Marco Polo . . .*

There were places where a building seemed to hang above the darkness rather than rise out of it, because as I later (but not then) remembered from daytime excursions there, swells of radioactive vegetation rose up around the bottom of it, almost obscuring the main window, and vines grew thickly across the entrance steps and up the front door, while pampas grass stood pallidly behind.

Here came that vegetation which had scared me the other day; well, well, it was just a wall of grass and spidery weeds, and pine trees up ahead. This time I refrained from poking the frisker into that poisonous stuff. Ahead stood a familiar house with broken dark windows and a porch; that was where the decontamination workers had been so busy last February. How much good had they accomplished? Toggling the pancake frisker to SCALER, I walked forward for a timed minute, measuring an unexceptional 4.02 micros—twice the national target air dose. That pallid dry cold grass on my left was something to keep away from. Farther ahead, at the edge of the road rose the next looming pale wall of house and then a dark tree above and behind it, there where the dead street curved leftward into darkness—a place of death, one might have supposed, and yet there came the piping of a single frog. Staring into the darkness behind the pampas grass, I saw nothing, of course, then turned back to the decontaminated house. As I swung the frisker on NORMAL at my hip, it began by showing only 3.75, 3.4, 3.44 micros, but then I thought to lower it toward the pavement itself, and right away it shot up to 11, 12, 13.5, 15, 18, 21, and at 21 it stayed. When I reported this number to the driver, he cried out. For all I knew the decontaminators had done their job, but trailed fallout from their truck, or very possibly some dry breeze this very day had sifted cesium-137 from the vicinity of Reactor No. 1 right to this spot. Once more I marveled at the unpredictability of fallout distribution; here for instance was water running under a dark culvert, 4 micros at waist level and still only 4.8 at culvert level, perhaps since it flowed so rapidly.

The interpreter and I stood watching that roadside pampas grass which was so much higher than our heads; there we stood and there it stood, pale and still in the dark. What was there to understand? Thus Tomioka: dead grass with stickleburrs, and the curving road, that tall pale pampas grass on the left and the grim nightclub on the right with its darkness, this night of dead houses overgrown with weeds—a creepy, vile experience.

The instant we reentered the taxi, the driver began to roll homeward without further instructions.—"What do you think of Tomioka?" I asked him.

He replied: "I feel that three years have passed and you can't live."

"Do you ever worry about going into a place like this?"

"I don't worry too much, because I'm over 60 . . ."—but he sped and sped until we were up the hill and past No. 2 and coming into Naraha.

NORMALIZATION ON THE ROCKS

Some people were scared, but mostly they resembled the bartender in Fukushima City who performed his mixings quietly and magically as if the show would never end, the edge of the bar itself illuminated so that the glasses glowed alone and mysteriously in that one segment where he stood quietly constructing them; the gin glowed bluish and the ice cubes were like great diamonds, while the bottle of gin shone red.

He was from Fukushima and owned the bar, rented the space. He looked very young to be the owner. Bowing over his mixologies, wearing large spectacles in which the drinks were reflected blue, he took great care.

He bought a Geiger counter in 2012 but grew bored with it after a year. In his opinion, nuclear power was here to stay, and about the accident he didn't really care, although he realized that those from Okuma or Tomioka might feel differently.

He said that this area was considered a bit "hot," the levels varying (just as my frisker indicated) between about 0.12 and 0.5 micros an hour—and you from the future for whom I write may well be jaded at such trivial values—even when I lived, half a micro was not unheard of; those stone Ten Commandments tablets in the courthouse lawn of Pineville, West Virginia, measured in at 0.48, and I have told you how in the Rocky Mountains I frisked a boulder at 0.84—furthermore, you may not believe that for me it was novel, unnerving and downright disturbing to *need to* guide myself with a scintillation counter; for you, I suppose, there will be hot spots here and there, and perhaps even your coolest spots will read, like Iitate's average air dose, a full micro per hour (8.76 millis a year)—eight times what I got at my kitchen counter at home—so that you will mask yourself as needed, rinsing off alpha and beta particles at your destination, perhaps with recycled grey water,* and slicing off your tumors every few years until they get you; back in the years when I was alive, such a world was not looked for.†

---

* On the other hand, the Intergovernmental Panel on Climate Change had predicted that "precipitation extremes are *very likely* to increase over the eastern Asian continent in all seasons and over Japan in summer." So who can say that you future Japanese will not live happy lives, trusting in extreme typhoons to wash your cesium away?

† Do you remember that unstable and poisonous "uranium legacy site" in Kyrgyzstan? [See p. 185 and II:557.] Here was another place in that country:

Yet even this bartender in Fukushima City now said, laughing a little, that in consideration of the high levels the authorities had considered making this area a no-go zone, but refrained on account of the expensive bullet train that passed through here—although perhaps that was merely another harmful rumor. He laughed, and I smiled back, for it is always better not to let little things such as nuclear disasters crack the containment vessels of our happiness—because, as the Celtic sprites used to sing in Fairyland, *things that have grown sad are wicked.*

In Fukushima City I once came across a sign that commemorated an aerial count of 0.30 hourly micros as of a certain day of the previous month; but when I measured the spot, my pancake frisker registered 160 counts per minute, or 0.54 micros an hour. That would have earned me 4.73 millis, but only if the dose had kept at exactly that level and I'd camped on that corner for a year. Which exposure assumptions did one care to make? In Ernest Hemingway's final home in Ketchum, Idaho, there was a nice fireplace in the master bedroom, and the hearth-tiles, which seemed to be made of clay brick, read the same 0.54 micros.* But I had measured those tiles with the frisker lying directly on them; the air dose in the house was only 0.18. My belief was that an air dose signified more to human health than did the levels of some object on the ground—in which case that corner in Fukushima City was more dangerous than the Hemingways' bedroom. And that bedroom was more radioactive than mine—but Sacramento lies not far above sea level, whereas Ketchum is more than a mile high. In short, my calculus went:

Provided they are both comparable, the air dose deserves more attention than the level of any local emitter.

The higher the altitude (and the more granitic rock), the higher the normal background radiation.

• Most high contaminated horizons characterized by gamma dose rates in range 50–200 [microsieverts]/h[r].
• The total amount[s] of the tailing materials having different level[s] of radioactive contamination are still uncertain.

"Uncertain" was the word, all right!

I imagine a future (your present) in which megastorms tear open inadequate "containments" for nuclear waste, which then washes into rivers. I imagine a time before you and after me when people terrified of climate change rush to build reactors in order to cool their sweltering cities and pump away rising waters, and in their haste they build poorly, so that new disasters occur. No doubt in your time hurricane winds have carried away the warning signs and eroded the walls of plutonium tombs. *Radio-active contamination . . . still uncertain.*

* Or 165 cpms.

Hence I had two reasons to believe that the 0.54 micros in the air on the street corner in Fukushima City was more dangerous than the 0.54 micros coming up from the Hemingways' hearth.

So that was how I looked at things, and it might have been plausible but it was also complicated. Other assumptions led to other complications. For instance, the way the Ministry of the Environment calculated their persuasions and reassurances, the "additional" annual milli that Japanese should accept with equanimity in these "remediation zones" was predicated on being outdoors for 8 hours and indoors for 16. (They estimated the indoor dose at 60% of the outdoor one.) Their one milli was actually 1.2 millis. A forester or construction worker, not to mention a decontaminator, would find himself drinking down extra helpings of that hearty soup of becquerels, curies, rems and microsieverts that is our life; someone who was outside all the time would absorb 2.02 millis. But how many Japanese now lived like that? Perhaps the Ministry of the Environment was more right than wrong—or maybe not. Everyone conspired with me to make me think the red zones less inimical than they actually were; the poison of their desolation could be perceived only through numbers on dosimeters and scintillation counters; the sad people who guided me there longed to return home, and therefore might have been disposed to accept them even in their present terrible condition, but for the grandchildren. And it is tiring to be anxious all the time. Moreover, who was I to increase anyone's bitterness, never mind the divide between nuclear refugees and me, by harping on the perils of their landscape?

The Ministry of the Environment went on to state: *Individual additional exposure is approximately 1 mSv/yr for the residents living in the area where the air dose rate is about 0.3–0.6 microSv/hr . . .*

A continuous exposure to 0.3 micros an hour would win a lucky citizen 2.628 annual millis. And 0.6 micros an hour added up to 5.256 millis.—But never mind! For the Ministry of the Environment also said: *Japanese institutions are encouraged to increase efforts to communicate that . . . in remediation situations, any level of individual radiation dose in the range of 1 to 20 mSv per year is acceptable and in line with the international standards.*

Any level, from 1 to 20 millis! Well, now, I couldn't quite agree with that.

In January 2017, with Tomioka's radiocontamination significantly remediated,* I telephoned Ed Lyman of the Union of Concerned Scientists to ask him: "Would you move back there? Suppose you had a young wife and young children . . . ?"

---

* For my measurements and observations from that time, see II:619–26.

"Well, that would just be me," he said.

"But you know more about that than I do; I'm guessing you have an opinion."

"Well, children are the most sensitive biological receptors. I wouldn't want more than a doubling of my children's previous background exposure. The international recommendation is one millisievert per year *from all sources*. The problem is, after a nuclear accident the authorities in their wisdom will have a weaker standard. Up to 20 millisieverts a year? For children that's unacceptable."

LOWEST AND HIGHEST RADIATION MEASUREMENTS IN SELECTED RED ZONE CITIES, 2014

[in microSv/hr / **milliSv/yr**]

The two readings not asterisked were one-minute timed SCALER counts. Asterisked readings were made in NORMAL mode to reduce radiation exposure for myself and my companions.

Readings within a vehicle, or within a city's official limit but outside the red zone, were excluded. So were MAX measurements.

Millisieverts per year were calculated by multiplying [microsieverts per hour] × [24 × 365 / 1,000 = 8.76].

The range above each city name (for instance, "20.7" for Iitate) is simply the highest reading divided by the lowest.

20.7

Iitate
Air dose a few steps from decontaminated
community center   0.42 / **3.68**
Interior of Mr. Shigihara Yoshitomo's cowshed   8.70* / **76.21**

88.9

Tomioka
Timed walk by seaside incinerator for
contaminated debris   0.36 / **3.15**
About 100 yards south of abandoned Night Pub Sepia, drainpipe
The next highest measurement was a few feet away: an air dose frisked
at 3 inches above pavement.   32* / **280.32**

99.8

Okuma

Air dose on beach by ocean (upwind of Reactor No. 1)	0.416* / **3.64**
Grass at pavement's edge by abandoned fish hatchery	41.5* / **363.54**

Back to 2014, and the Ministry of the Environment's convenient suppositions. It might, I confessed, be true that 1 milli a year *is not a limit of exposure or a boundary between safety and danger.* How should I (or the Ministry) know? Meanwhile, whether it was or wasn't, *an additional individual dose of 1 mSv per year is a long-term goal . . . that . . . cannot be achieved . . . solely by decontamination work.*

To sum it up, that additional 1 milli was more than a milli, was not relevant and might not be achievable. Anyhow, whatever cannot be avoided must be lived with, unless it kills us. In the meantime, never surrender to harmful rumors! I remember the grey-green sea rolling cleanly over the white-green pebbles of Okuma, and the taxi van driver joyfully inhaling the breeze—why not? The air dose had been only 0.416 micros.

Proceeding another block toward the Fukushima City train station, I frisked the air and found its level lower by more than half: only 73 cpms, 0.24 micros—about the same as the tall grass at Hanford). At the next block, passing the Bar Plein Rêve, it had risen again, to 143 cpms—not quite half a micro an hour. One more block of slightly wet street and yellowish lightbulbs in convenience stores, a lovely longhaired woman on her bicycle at a corner, waiting for a walk signal while a sake bar shone a rainbow on her face and the air dose was back to 77 cpms, 0.24 micros. Around two sides of the hotel building it was 100 cpms and 0.36 micros. Perhaps more relevant than any of these findings was that in those entire four hours in Fukushima City the dosimeter had not turned over another digit. (Never mind the uncounted alpha and beta.) So I went to bed in that delightfully mild environment of a room, radioactively speaking: 21 counts and 0.001 micros per minute. I wasn't worried; just like the locals, I'd long resumed eating seafood—for instance, fried Tohoku oysters at 0.12 micros and 37 cpms.

"THE STANDARD IS DIFFICULT"

Still hopeful that someone in authority could tell me what was safe, I waylaid a junior public relations official in the Fukushima Environmental Recovery Office, requesting his personal experience. Even back in the days when I was alive I

quickly forgot whether he was the one in the white shirt or the other one in the
suit and the pinstripe shirt. But I do remember a stuffy, windowless conference
room where he bestowed his quarter-hour on me, in one of those office towers in
Fukushima City.

"I do live in Fukushima now," he said. "I only moved half a year ago. I am in
public relations, so we respond to journalist inquiries just like this. We have a
decontamination information plaza and we advise what to display there. Regarding radiation, there are some explanation materials . . ."

"Are you married?" I asked, wishing to find out whether he worried about the
health of his hypothetical children, but he replied: "I am separated from my
family."

"Where did you live before?"

"I was in Shizuoka Prefecture."

Remembering what the farmer Mr. Hamamatsu had told me in Iwaki,* I said
to him: "I heard their tea was affected by the accident."

"Mount Fuji is the east side of the prefecture, and the tea is grown on the west
side, so I don't think I have heard that the radiated particles came to that area."

"Do you yourself ever have any worries about radiation exposure?"

"Not at all. Because next door is the co-op store, and they sell Fukushima
produce, and I purchase from them. Rice is inspected a hundred percent before
shipment!"

"Do you have a dosimeter?"

"Normally I don't carry one, although when I go to the dangerous zone it is
mandated."

"When you put on your dosimeter and go to the red zone, what do you do
there?"

"When you enter the nuclear power plant premises, you put on some protective gear, and there is no problem. I have been there once. There was some cabinet
reshuffle on the third of September, and some ministers came, and because our
job is public relations we have to take care of them."

"What was your impression when you went there?"

"The area close to the reactor, the situation is quite unchanged since the accident, so it's vivid. Although that closed area is vivid, the efforts toward the
abandonment are being done, and people are working hard . . ."

"What did it look like?"

He put his hands on his hips to think, then said: "How should I answer?
Because we cannot see inside. It's covered with a blue sheet for the works, and the

---

* See above, p. 343.

workers, they are removing a screw with nuts, just steadily working toward the abandonment."

"Did the ministers seem at all afraid?"

"Not at all."

"Do you remember what your dose was?"

"Ah, no."

"Was it more or less than a millisievert?"

"Of course it was more than a millisievert."

(This seemed very unlikely to me. Receiving a milli within a few hours would surely be cause for anxiety. More likely the young man simply did not know his units.)

"What dose would make you comfortable? Suppose it were 20 millisieverts…"

Again he put his hands on his hips to think. "In my case, personally, when I entered the plant, I remembered seeing air dose 16."

"Millis?" I asked, just to see if he knew the difference.

He had to think; he wasn't sure. "I suppose it must have been microsieverts." How sad that he couldn't sort it out! One unit was 1,000 times greater than another, and he was an official! He finally said: "When we travel in the bus, we are told, the value is so and so; and I remember 16 . . . Basically in the air, there are no radiation particles."

"If you walk around in this city, will you find the same radiation in each block?"

"No, different"—which of course had been validated by the pancake frisker.

"What's the highest and what's the lowest?"

"Ah, I don't have that information. There are monitoring posts. There are rules for decontamination. Annually, 50 milli, or 100 milli for five years. If you cannot clear even one of those, then you cannot continue."

"What is the annual maximum dose for a resident of this city?"

"This is a law about decontamination, so the regular residents are also under the same number."

"So if you lived in the city and you received 49 milli, then in your opinion that's no problem."

"Well, in a way, but an ordinary resident, in order to show that, he or she needs to wear the dosimeter all the time."

I thought this answer evasive.

"Actually the standard is difficult," he said. "The municipalities have different policies. We are in the position of the national government, so in the long run, our target is to make the air dose of 1 millisievert."

"And that's one additional milli?"

"*Absolute* one," which, as you know, was not at all what the Ministry of the Environment had said.

"And for now, if someone told you, I got 49 millisieverts, that's okay?"

"You keep saying 49, but it's so unrealistic that we cannot discuss about that. If you are going to write something in your book, then . . ."

At last he allowed that there was no formal standard. He was very young looking; I had irritated him; his name was Mr. Abe Masahiko.

THEIR STANDARD IS AS ARBITRARY AS OURS:

Some allowable cesium concentrations in food, Ukraine and Japan

(from published government sources)

The Ukrainian standards contain many more subcategories and other specific oddities than their Japanese counterparts. Of the former I include only a sampling.

All levels expressed in [becquerels per kilogram].

	Ukraine*	Japan[†]
Water	—	10
• "Drinking water from underground sources"	2	—
• "Mineral water from underground sources"	10	—
Bread	20	100
Root & leaf vegetables	40	100
Milk	—	50
• "Raw milk for industrial commodity"	100	—
• Milk and "concentrated cream"	300	—
• Infant milk formula and babyfood	40	50
• "Dry dairy products," including infant formula	500	?
Potatoes	—	100
• Fresh and processed potatoes	60	—
• Dried potatoes, dried mushrooms, etc.	240	—

*For Cs-137 only. Ukraine's permissible levels for Sr-90 were in a separate category and were usually much lower.

[†] For Cs-137 and Cs-134 combined.

Meat	—	100
• "Fresh slaughtered meat" & poultry	200	—
• Dried poultry	400	—
• Wild game	400	—
• Bacon, etc.	100	—
Rice and most other items	—	100*
Honey & bee products	200	—
Wild mushrooms and berries	500	—
• The same, dried	2,500	—

Up through 2011, the Japanese government simply "set the annual maximum permissible dose from radioactive cesium in food as 5 mSv and assigned[ed 1 mSv] to each food category": water, milk and dairy, vegetables, grains, meat, eggs and fish. "Food safety is basically assured. However, to achieve further food safety and consumer confidence, Japan is planning to reduce [the] maximum permissible dose from 5 mSv/year to 1 mSv/ year."

---

* Ministry of Environment, 2014: "All the rice crops of the year 2013 were monitored with Becquerel Monitors and test results read 99.93% ... contain less than 25 Becquerels of Radioactive Cesium per Kilogram in the shape of Brown Rice. By polishing brown rice into polished rice, ... cesium content will be lowered ... to 40% ... By cooking polished rice with water, ... Cesium density [by weight] will further be lowered [to] ... less than 2.5 Becquerels per Kilogram."

PART OF QUOTIDIAN LIFE

And who would *not* rather be as unworried as a cabinet minister at a poisonous nuclear reactor? The savoir faire of this Mr. Abe was enviable. Others wished to emulate it. I remember stopping off in Koriyama one morning of that year, and it was fine weather of mostly 0.3 microsieverts per hour by the train station (two and a half times what it would have been at home), but all the same trivial; indeed, I was quoted a similar figure by the clerk at the camera emporium, where up on the second floor, among other interesting household appliances, friskers and dosimeters offered themselves for sale, helped along by an advertisement explaining that once upon a time dosimeters had been unusual but now they were a part of quotidian life . . .—and so I bought a new one, which happily read to tenths of a microsievert. That was how I learned that in my first full hour in Koriyama, spent mostly indoors, I had taken in a tenth of a micro, just as I might have at home; while in the next hour, during some of which I strolled outside in the unseasonably hot sun, I took in another 0.2 micros—less than the pancake frisker kept showing, but after all the new dosimeter, like the old, could read only gamma, not alpha and beta. At Koriyama Station the granite flagstones read 0.42 at two feet away, and 0.54 at two inches. Well, that was granite for you—nothing to fear. The air dose on a one-minute timed walk around the block measured in at 0.36—a little "hot," but surely within normal range. (In 2016 a woman in Koriyama wrote: *I very much hate my psychological state, endless anxiety for 5 years. But now so many people just live normally. They say loudly: "It [this] is rumor damage! We need reconstruction!" . . . [T]hey look at us coldly. They look [at] those anxious people with prejudiced eyes.*)—In my two hours in Koriyama my body absorbed a very mild 0.17 micros an hour. That wasn't much worse than my home town. Perhaps without the pancake frisker, my walk around the block might have lasted longer, and I might have taken in a few more gamma rays. But that level was still insignificant. For the sake of this book I carried my dosimeter in my pocket, and I frisked here and there; it mostly felt like keeping up with the stock market.

When I first arrived in Japan with the pancake frisker, I remained half-uneducated as to which levels should even be expected; and I remember that at 7:30 on my first morning in Fukushima City the lovely flocks of chubby school-girls in their navy blue or plaid miniskirted uniforms were in the fullness of their swarm from the caverns of the railway station, confiding, giggling or bowing their heads alone beneath their tight-packed umbrellas as I, not yet knowing how radioactive this drizzle might be, felt polluted by it and so hooded myself, re-membering that rain in Kesennuma three years ago when the interpreter had grown so uneasy that we had to duck into some cable-tentacled ruins while she found her umbrella; the radio had been warning about fallout even as it went on promising that there was no immediate danger. Now the count must surely be lower; and for a fact it was now quite easy for me to check for myself, although I must confess to hesitating out of what must have been shame at the prospect of deploying that bright yellow pancake frisker in front of everyone, for it might recall certain realities sometimes called harmful rumors. Eventually I nerved myself up to do it, and it almost seemed plausible that no one in that crowd was looking at me. It was a long one-minute timed count. Rain fell on the frisker. Finally I turned it over, and was relieved see a mild 54 counts per minute, 0.18 micros per hour—slightly less than Ed Lyman got in his Washington, D.C., apartment. Several days' experiments proved that (unlike in Barcelona) rain did not increase the dose. So I got used to 0.18 micros, and even 0.54 micros—after all, in Saint Petersburg the curb of Dostoyevsky's grave was 0.48 . . .—but the red zones still felt different for me—although if I were to patrol my former home every three days, the way that Mr. Shigihara did, I'd surely get used to that, also! Do you remember what he said on that cherry tree hill in Iitate? "We didn't measure it. We don't care about the level."—As Aki had said on that drive to Namie, everyone had grown so cavalier about it!

(From that same Ministry of the Environment pamphlet, section 7-4, "Future [A]ctions for the Areas Where it is Expected that Residents will Face the Difficulties in Returning for a Long Time . . .": *Improve the measures to start a new life* . . . In other words, forget about going home.)

Most of us neither measured it nor cared about the level, because experts did that for us so that we could dream the urban dream of Shinjuku, where the air dose was commonly 0.12 hourly micros (on the Marunouchi subway line, deep underground, it went as low as I could measure: 0.06). Sometimes I dreamed my wishes straight to that bright grass shimmering like true gold through the dark-framed hole in the laurels and cryptomerias of Tomioka, although I also had unpleasant memories of that dressmaker's shop . . .—but in Shinjuku there were female pop singers crooning amplified from skyscrapers, and ever so many young women attached to extreme high heels, because thanks to electricity they would not need to walk very far:—See! a pair of green-and-silver trains was crossing overhead, and above them a young giantess on a sign pointed with a forefinger that must have been at least half man-sized; then towers both grey and white, rectilinear and curved, rose above her. A grubby man in an unbuttoned shirt was whispering slack-mouthed into his cell phone; a girl in a miniskirt and a pale yellow sweater was smoothing her long red hair as she came up the sidewalk past middle-aged ladies chattering and pointing forefingers ahead; they were all looking into shopwindows that promised electric-produced goods. A thirtyish couple did the same, happily holding hands, while traffic noises and competing amplified songs döpplered past, and up high, a silver train crossed an orange one at wondrous speed. Behind two girls in white face masks, and before an old cyclist who was pedaling wobbly with the cigarette crooked in his mouth, came two *rorikon** girls, one of them in white, white shorts and big black leggy boots, the other in pink, both of them infantile and cute; thanks to electricity they could sell souvenir cards and stickers with their likeness. I wandered into a department store where Chopin played through echoing crowd-noise, his piano-notes re-performed and simultaneously amplified by electricity.

We normalize our lives in order to diminish our pain. Most human beings

---

* "Lolita," *rori,* plus *kon* for "complex." These were the ladies, many of high school age, who "sold spring." Some of them vended only walks around the block with their older male admirers. At this time the Electric City district was famous for them.

Rorikon girls

endure what they cannot help, and try to adapt to it. Some happen to be policy-makers, dutybound to protect, support and advance the interests of the rest. The people whose mercantile capital builds a nuclear power plant might well intend (I'm all for good faith!) to protect us from heat, cold, darkness and other some-times undesirable circumstances, powering our magical appliances at a reason-able cost, sweetening our existences with tax incentives and jobs, and all the while staving off global warming. To be sure, we might also wish of them to guard against radioactive accidents. But they too wish to diminish their pain, which so often seems to be financial. Raising their seawall in case the next tsu-nami should be higher than the last would cost real money, and possibly scare us into wishing them and their projects away. Relocating their backup generators onto higher ground would be expensive, and unappreciated by us (especially if that raised our utility rates). They normalize their business-lives by increasing their benefits and decreasing their costs. They decrease their costs by spending no more on risk management than they deem necessary. Normalizing the unexpected, denying the unthinkable, benefits short-term profits.

From the comprehensive accident management plan for Fukushima Daiichi: *The possibility of a severe accident occurring is so small that from an engineering standpoint, it is practically unthinkable.*

Ms. Allison Macfarlane, geologist and professor of environmental policy: "What I would say in terms of lessons learned from Fukushima... is that setting reactor design ... just above recorded human experience is turning out to be really shortsighted."

And whenever experience did surprise us, why not look on the sunny side? Consider this: Before the reactors exploded, the average air dose throughout Fukushima was 0.035 to 0.046 micros an hour—well below the minimum my pancake frisker could measure. Six years later, when I finished *Carbon Ideologies,* the whole prefecture remained 10 times "hotter."—But that was merely an average, upwardly distorted by those inconvenient little hot spots at Okuma, Tomioka, Iitate... And even though the Union of Concerned Scientists would have called these levels unacceptable, to any positive-thinking nuclear ideologue they were trifling! A level of 0.35 micros corresponds to one Ph.D.'s definition of *an average radiation background dose for a human being.* As for 0.46 micros, that was exceeded by the stone Ten Commandments tablet I frisked by the courthouse in Pineville, West Virginia.*

Thus the theme of this book. As we contaminated our homes, warmed our atmosphere and acidified our seas, whatever would happen next stayed comfortably unthinkable, or at least potentially acceptable, back in the days when I was alive.

---

* Both comparison readings from the table of Comparative Measured Radiation Levels on p. 247. headers **5.7** and **8**.

Is this really not enough? He knew that it wasn't enough. He knew that billions and billions didn't know a thing and didn't want to know and, even if they did find out, would act horrified for ten minutes and immediately forget all about it.

Arkady and Boris Strugatsky, 1972

Postscript: Japan Sees the Light

On April 11, 2014, the Japanese cabinet announced that nuclear power and coal would now be equally significant constituents of the nation's energy policy. How grand, to see one of my homeland's carbon ideologies finally receiving its due! *Japan's 10 power companies consumed 5.66 million metric tons of coal in January, a record for the month and 12 percent more than a year ago.* (Most of it came from Australia; 1 or 2% was American coal.) And the smoke went up!*

Fools and children might have inquired: What about renewables? Well, once upon a time, in a burst of environmental sportsmanship, Chibu Electric installed 11 wind turbines alongside the three-reactor Hamaoka atomic station, whose chief counterattacked as follows: "They ran at 26 percent capacity" in 2012. "We'd need 6,000 of them here to replace Hamaoka." And coal came into its own. As Japan's greenhouse gas report told the story, looking back from 2014: *Carbon dioxide emissions from fuel combustion in the energy industries increased by 60.6%*

GLOBAL DISTRIBUTION OF NUCLEAR REACTORS IN 2014,

with added statistics

Number of "operable"* reactors operating in Japan: 0 out of 43.
Until the meltdowns at Fukushima, reactors had produced 30% of Japan's power.
Number of countries with operating reactors: 31.

---

* This phrase was presumably inserted to exclude Fukushima Nuclear plants No. 1 and 2.

---

* In 2015, it came out that *about a third of 45 new power-generation units fueled by coal won't face the government's environmental assessment because their small size puts them under the review level.*

Number of reactors worldwide: 391, *totaling 337 GW net capacity [= 19.2 billion BTUs per minute,] or 10.8% of all power generation.* Percent of nuclear electricity produced by "big five" countries: 69%. *These were, in decreasing order of nuclear power production, the U.S., France (which two generated half of all nuclear electricity), then Russia, South Korea and China.*

Sources: Heinrich Böll Stiftung, 2015; The Japan Times, 2015.

since FY1990 . . . The main driving factor . . . is the increased solid fuel consumption for electricity power generation. After 2011, how could that trend not continue?*

Although one "conservation group" responded in a tactfully worded statement of disappointment that *the energy plan failed to present the spirit of innovation,* not everyone was such a stick-in-the-mud. For instance, Yasui Akira, "an official in charge of coal policy at the Ministry of the Economy," said more or less what a West Virginia Republican might: *Our basic stance is to use coal while caring for the environment as much as possible.*

---

* The European Union's greenhouse report made the impetus clearer: "Substantial [CO_2] emission increases between 2011 and 2012"—in other words, immediately after the *faux pas* at Fukushima—"were reported for the source categories listed below . . ." Europeans looked over at Japan, and shuddered. "In Germany, power production from coal increased mainly due to lower nuclear power production . . ."

This ends the text of Volume I.
Volume II will examine the three major fossil fuel ideologies.

DEFINITIONS, UNITS AND CONVERSIONS

Readers should feel free to skip this section. In the main text the difference between various types of coal may be sensed from context, or ignored at pleasure. Conversions from counts per minute to millisieverts need not occupy inmates of any low-radiation paradise. However, while researching *Carbon Ideologies* it became my duty to teach myself about these things, and what I found useful might be the same to you.

Furthermore, in this book I make various assertions about the inherent energy efficiencies, power consumptions, etcetera, of this and that. Some of my claims could seem provocative or peculiar. Some may (I hope not) be wrong. The following pages will allow anyone with arithmetic and common sense to verify a carbon ideologue's calculations. If one source expressed a figure in horsepower and another expressed a second figure in kilowatt-hours, I probably converted both numbers to British Thermal Units (BTUs) in order to compare them both to a third figure in my grandfather's *Mechanical Engineers' Handbook* from 1958. If you lack a better way to pass the time, this section will help you do the same.

I for one hope always to approach units and conversions with humility. It would be all too easy to compare apples and oranges—or, worse yet, to imagine that I knew something.

For instance, in 1991, the world consumed 1.04819×10^{13} net kilowatt-hours of energy. This equals 3.578×10^{16} BTUs (or 35.78 quadrillion BTUs)—of which North America's share of was scarcely more than 5%.

However, to *produce* those kilowatt-hours the world consumed 3.469×10^{17} BTUs of energy, of which North America's portion was nearly 28%.

Reflection suggests why this might be so. As we have repeatedly seen in *Carbon Ideologies,* to generate one unit of energy output takes more than one unit of energy input. How much more in this case? Well, what about the world's nuclear energy? How were its 2.124×10^{16} BTUs calculated? Was waste heat taken account of? And so I decided to abandon this tantalizing calculation.

As for you, why should you care?—All too often when I was alive, generalists who could look at overarching meanings and patterns (and therefore most thoughtfully consider where we were going and why) lacked proficiency in math and science. Meanwhile, some of the scientists and mathematicians I met were naive, or worse yet, indifferent, concerning our where and why. *Carbon Ideologies* strives, however unsuccessfully, to bridge the gap.*

A few items especially relevant to the text, such as ton of oil equivalents, kilowatts *versus* kilowatt-hours, coal slurry and global warming potentials, have been greyed off and boxed for easier finding.

---

* Most of the material in this section was verified by Prof. Anna Mummert of Marshall University.

To save the reader trouble I have duplicated a very few items. For example, the definition of watts appears in both sections 1 and 3E.

Cross-referenced terms are **boldfaced**. In some cases I have added page numbers for the reader's convenience.

1. Energy, Electricity, Efficiency, Power, Work, Manufacturing and Engines

Ampere (amp or A)—"The number of electrons passing a given point in the circuit per unit time is called the **current**. Current is measured in amperes . . . One amp corresponds to the flow of 6.24×10^{18} electrons per second."

Amperes = volts / ohms

For ohms, see **resistance**.

1 ampere = 1 coulomb/sec

A device's rating in amperes may be multiplied by the available voltage (as I write, residential U.S. voltage is often 115 V) to determine its power consumption in watts. See the equation under **power.**

Battery—See **fuel cell**.

"Big five" materials—Cement, paper, steel, plastics and aluminum, whose joint manufacture (*ca.* 2015) emitted 56% of the world's industrial carbon dioxide. See p. 134.

British Thermal Unit—See p. 531.

Coulomb—A unit of electrostatic charge. "Comprises approximately 6 million million million electrons."

Current—"A movement or flow of electricity." "The movement of charges past a particular reference point."

Efficiency—"The ratio of output to input." "The efficiency of an ideal machine is 1.0 (100%). However, all real machines have efficiencies of less than 1.0." "When the rate of **work** is constant, either work or power can be used to calculate the efficiency. Otherwise, power should be used." For the generation of electric power with traditional steam turbines, be it from nuclear or fossil fuel, about ⅔ of the heat input accomplishes no direct work. Hence one could say that steam turbines are only about ⅓ efficient. Efficiency in manufacturing has to do with how much thermodynamic work must necessarily be accomplished to create a specific material, and how much more work than that minimum is actually being ("wastefully") done; further, efficiency reflects how much of the input material and the manufactured material are wasted during processing, as in the "trim scraps" cut off of paper in a printing plant. **High heating values** of fuels can be considered the light of efficiencies. From the table of Calorific Efficiencies beginning on p. 208 we see that a pound of gasoline will give up 5.1 times more combustion energy than a pound of green cottonwood. To reduce climate change one should obviously strive for maximum efficiency and minimum greenhouse emissions.

Electricity—"A form of energy that consists of mutually attracted protons and electrons . . ." "Any manifestation of energy conversion of charge that results in forces in the direction of motion those charges carry." "The electron theory states that all matter is made of electricity." "Electricity cannot be generated . . . It can, however, be forced to move." Of course we talked about electricity generation all the time. By this we meant the generation of **power**.

Energy—Since in and of itself it cannot be measured, energy is a relative term. A high-temperature system has more energy than a low-temperature system. One common definition is the *ability to do work*. "Energy is a measurement of power over a period of time. It shows how much power is used, or generated, by a device, typically over the period of an hour . . . It is measured in **watt-hours** (Wh) and **kilowatt-hours** (kWh)." As it happens, energy is also measured in **calories, joules, British Thermal Units**, **curies**, etcetera.

Energy flow—An attempt to total up all various energy expenditures in the manufacture, packaging, shipping, preparation, consumption and disposal of a certain item. Thus the energy flow for a food item would include the energy units (such as BTUs) in its planting, fertilization, harvesting, cooking, waste collection, etcetera. "An important concept underlying the measurement of energy flows throughout an economy is the 'conservation of embodied energy' . . . , which states that energy burned or dissipated by a process is passed on, embodied in the products of that process."

> **Embodied energy**—(**i**) "The energy required to produce a material from its raw form, per unit mass of material produced. The energy is usually measured as the **low heating value** (see p. 535) of the primary fuels used plus any other primary energy contributions. These energy requirements are dominated by two main steps: (**i**) harvesting and (**ii**) refining." (**ii**) Another definition is: Manufacturing energy + use energy.
>
> > **Manufacturing energy**—The energy needed to manufacture a product. More or less equivalent to embodied energy, (i). For instance, the power used to fabricate a refrigerator's parts and then assemble them.

Use energy—The energy expended by using the product. For instance, the power consumed by running a refrigerator over its working lifetime, or "use phase."

Erg—A unit of **work**, comparable to **joules**, but dimensionally more confusing, and now thankfully not much met with. Often used in older literature to express radioactivity, solar energy, etc. 1 erg = 1 gram per square centimeter per second per second. Or, to be more definitionally regressive:

1 **erg** = 1 dyne-cm = 1 g-cm^2/sec^2

1 **dyne** = 1 g-cm/sec^2

Here are some values visually estimated from their placement on a bar graph "Range of Energies": "Earth's annual energy from Sun," more than 10^{32} ergs; H-bomb, more than 10^{24} ergs; first atomic bomb, more than 10^{20} ergs; fission of a uranium nucleus, about 10^{-4} ergs.

1 **erg** = 9.4805 × 10^{-11} BTUs = 7.3756 × 10^{-8} foot-pounds = 1 × 10^{-7} joules = about 0.012 roentgens [See section 3, p. 550.]

1 **erg-sec** [nowadays more correctly written **erg/sec** or "erg per second"] = 1.3410 × 10^{-10} horsepower = 1 × 10^{-7} watts

1 erg per square centimeter per second [**erg/cm^2-sec**]= 1 × 10^{-3} watts per square meter

Exajoule—See **joule**.

Exergy—A word often employed by Prof. Timothy Gutowski of MIT in his writings on manufacturing and use efficiencies. I avoided it in *Carbon Ideologies*. In this section it may be worth introducing: "Exergy of a material flow represents the maximum amount of work that could be extracted from the [energy] flow considered as a separate system as it is reversibly brought to equilibrium with a well-defined environmental reference state . . . [I]n exergy analysis, work and heat are not equivalent, as they are in First Law analysis." A detailed understanding of specific exergies would obviously assist the fight against climate change.

Foot-pound [ft-lb]—The amount of energy required to lift a 1-pound object 1 foot. The non-metric (and nearly obsolete) counterpart of a **joule**. If a weight of 7 lbs is lifted 2 ft, then the **work** done is 14 ft-lbs.

1 ft-lb = 1.356 J = 0.001285 BTU

Ft-lbs ought to be distinguished from lbm [pounds-mass] and lbf [pounds-force]. "The units of pounds-mass and pounds-force are as different as the units of gallons and feet, and they cannot be canceled" in standard dimensional analysis.

Fuel cell—"An *electrochemical device* in which the chemical energy of a *conventional fuel* is converted *directly and usefully* into *low-voltage direct-current* electrical energy . . . In a battery, the chemical energy is in the cell. Fuel cells keep converting [by combustion] with the addition of oxygen and fuel," which is usually a carbon-dioxide-emitting **hydrocarbon** (see p. 571), but can be clean hydrogen. "For short periods, conventional batteries are generally superior to fuel cells, which come into their own when electrical energy is required over long periods."

Gigawatt [GW]—A billion **watts**.

1 GW = 10^9 watts = 10^6 [= 1 million] **kilowatts** = 56,884,000 BTUs per minute

Horsepower [hp]—Originally based on the amount of work an "average" horse could do, this archaic unit continued in my day to be applied to motors. ("The real thermal efficiency of an average horse, even at heavy, continuous work, is probably not more than 6 to 10 per cent," based on BTUs in of horse feed and BTUs out of work, and counting BTUs out of digestion and other metabolic processes as waste. "Under laboratory conditions" it might be 20 percent.) "You may assume that 1 HP equals 3 ft lbs of torque at 1750 rpm (a common motor speed)."

1 hp = 33,000 ft-lb in 1 minute (or per minute, or / min) = 550 ft-lb/sec = 0.7068 BTU/sec = 42.408 BTUs/min = 745.70 watts [absolute]* = 0.7457 kilowatts = 76.0404 kg-m per sec

1 horsepower [U.S.] = 1.0139 metric horsepower

Internal combustion engine—[Sometimes abbreviated ICE.] "A machine for converting chemical energy to mechanical energy by burning a fuel with air in a confined space and expanding the products of combustion, extracting energy as work." There are two kinds. Gasoline engines are of the spark ignition type, oil engines of the compression ignition type.

Joule [J]—One of several units of **work**. "The practical unit of electrical energy." Also, as it happens, "a unit of mechanical energy," since it can be defined as 1 newton [1 kilogram being accelerated at 1 second per second] for a distance of one meter. As Prof. Gutowski wrote me: "Pick an orange up off the table and raise it over your head. The work you do is about one joule."

---

* "746 watts are theoretically equivalent to 1 h.p., but because of losses in the conversion of the electrical energy into useful work about 800 watts are normally required for 1 h.p. with electric motors . . ."

1 J = 1 watt-second = [1/3.6 million or 2.778×10^{-7}] kilowatt-hours = 0.0002388 [= 2.388×10^{-4}] kilocalories = 0.7376 ft-lbs = 3.725×10^{-7} hp-hrs = 0.000948067 [= 9.48×10^{-4}] BTUs = 6.242×10^{18} electron-volts.

One milli[joule]= 0.001 joule/kg. Obsolete.

One **megajoule** [MJ] = 1 million J = 948.067 BTUs

Energy content of a fuel [see **high heating value,** p. 534]: 1 MJ/ liter [a common metric way of expressing it] = 2.78×10^8 BTUs/gal

1 kJ/kg = 0.4299226 BTUs/lb

1 GJ/kg = 429,922.6 BTUs/lb

One **gigajoule** [GJ] = 10^9 [= 1 billion] J = 1 million kJ = 948,067 BTUs

1 ton of TNT = 4.184 GJ = 3,966,712 BTUs

One **kilajoule** [kJ or KJ] = 1,000 [10^3] joules = 0.9481 BTUs

One **terajoule** [TJ] [= 1 trillion] = 10^{12} joules = 948.1 MBTU [million BTUs]

Therefore, *1 metric ton of greenhouse emissions per TJ of combusted fuel* [a typical formulation in non-American greenhouse gas inventories] converts to *1 pound of emissions per 430,955 BTUs.*

One **petajoule** [PJ] = 10^{15} [= 1 quadrillion] joules = 9.481×10^{11} BTUs

"One petajoule is enough to run the Montréal subway system for a year" (*ca.* 2009).

Total Canadian energy consumption, 2009: 7,650 PJ.

One **exajoule** [EJ] = 10^{18} [= 1 quintillion] joules = 9.481×10^{14} [or 0.9481 quadrillion = 948,000,000,000,000] BTUs = 0.948 **quads**.

365.93 EJ was the amount of energy used by the entire planet in 1991.

1 EJ consumed in a year is equivalent to 1.902 billion BTUs/ min (or 33.45 billion watts), used without letup for that year.

One very rough conversion from EJ per year to million barrels of oil per day: Divide by 2.23.

Kilowatt [kW]—A thousand **watts**. "The term kilowatt . . . indicates the measure of **power** which is all available for **work**."—*American Electricians' Handbook*, 2002.

1 kW = 1,000 watts = 56.884 BTUs/minute = 3,413.04 BTUs/hour = 44,254 foot-pounds/minute = 1.341 hp = 3,600 million J/hr (or 1,000 J/sec)

Since ⅔ of the energy combusted in a traditional fossil fuel power plant is lost, 1 kW takes 170.652 carbon BTUs to generate. *See next page.*

Kilowatt *versus* **kilowatt-hour**—The first is a measure of power, which means rate of energy consumption, or energy used per unit of time (an hour). The second, in spite of its name, lacks a time dimension, because it expresses the absolute amount of energy that has been, will be or would be used in that unit of time. A physics textbook from 1971 explains it more mathematically: "The amount of energy used during a certain operation or process is frequently expressed as the product (power) × (time). For example, the amount of energy generated by a 1-kW power plant operating for 1 hr is 1 kW-hr." In other words, 1 kWh, for instance, = 1 kW × 1 hour = [56.884 BTUs/minute] × 60 minutes = 3,413.04 BTUs. The time units cancel out. This anti-intuitive distinction figures in conversion calculations relating to the tables in *Carbon Ideologies*.

Kilowatt-hour [Kwh or KWH or kWh]—Power consumption, *not* measured in kilowatts per hour. Prof. Anna Mummert notes: "Comparable to BTU (not BTU/time), ft-lb (not ft-lb/hr)." "The common engineering unit of electrical energy." "If you play a radio (80 watts) an average of 1 hour a day, then during a month's time (30 days) it would consume 30 times 80 watt-hours or about 2.4 KWH." "To store 1 kwhr of energy in a reservoir at 1 m (3.28 ft) requires 367,000 kg of water or 96,900 gal . . ." 2,650 sacks of sand, each weighing 100 lb, falling for 10 ft, will produce 1 kWh.

1 kWh [no time unit given or needed] = 3,413.0 BTUs [or, according to one conversion which I did not follow, 3,412.14 BTUs] = 3,600,000 [or 3.600×10^6]

joules [abs] = 3.6 MJ = 2.6552 × 10⁶ foot-pounds = 1.341 hp = 3.6 mega-joules [no time unit given or needed].

1 kWh/day = 2.37 BTUs/min.

According to the U.S. Energy Information Administration, to generate 1 kWh [the standard power plant wastage of ⅔ is implied (see text, p. 152)], require the following *[figures in italics are my conversions*]*:

- 1.04 lbs coal, or
- 0.01011 Mcf [= thousand cubic feet] natural gas *[= 0.421 lbs]*, or
- 0.07 gal. petroleum *[= 0.492 lbs]*

Kilowatt-day [Kwd or KWD]—Power consumption, measured in kilowatts used per day.

1 gram U-235 yields 8 kilowatt-days of energy.

1 langley = 3.69 BTUs/ft². [See section 7, p. 578]

Megawatt—1 million watts. "The power consumed by roughly a thousand households. It's the equivalent of 10,000 100-watt lightbulbs." "An average human body exudes about the same heat as" a 100-watt bulb, ". . . so a megawatt . . . can be compared . . . to the rate of bodily heat emitted by 10,000 people . . ."

1 megawatt-hour [MWh] = 1,000 kWh = 3,413,000 BTUs

Ohm—"The **resistance** through which the fall of potential is 1 volt when the current [electromotive force; see **volt**] is 1 amp."

Power—"Power measures the rate of energy conversion. It is measured in **watts** . . . It equals **volts** times **current**." [**Watt-hours**, **kilowatt-hours**, etc., measure the total amount of energy that was used *in* a given time.] Power is "the amount of **work** done [per] unit time. It is a scalar quantity." "Output power per unit volume is directly proportional to speed. Low-speed motors are unattractive for most applications because they are large and therefore expensive. It is usually better to use a high-speed motor with mechanical speed reduction . . . The efficiency of a motor improves with speed."

For a given device:

Watts of power = volts × current (amps)

---

* Calculated using the densities beginning on p. 581.

"If you have a mobile phone charger that uses 1.2 amps at 5 volts, you can multiply . . . to work out the number of watts," in this case 6.

> **Primary power**—Power used to supply primary energy, which is "the energy required to produce a material from its raw form, per unit mass of material produced. The energy is usually measured as the **low heating value** (see p. 531) of the primary fuels used plus any other primary energy contributions. These energy requirements are dominated by two main steps: (i) harvesting and (ii) refining."

Quad—A quadrillion BTUs.

Resistance—"The opposition to **current** flow." Measured in **ohms**.

Ohms = volts / amperes

Terawatt—1 trillion **watts**, or 56,884,000,000 BTUs per minute.

Terawatt-hour [Twh, TWh or TWH]—Power consumption, measured in terawatts [trillions of watts] consumed in an hour.

1 TWh = 1 billion kWh = 3,413,000,000,000 [3.413 trillion] BTUs.

1 TWh = 0.086 mtoe [million **tons of oil equivalent**]

"However, the primary energy equivalent of nuclear electricity is calculated from the gross generation by assuming a 33% efficiency, i.e. 1 TWh = (0.086 ÷ 0.33) mtoe." [I suspect this last is an error for 0.086 *times* 0.33.]

Use phase—See **embodied energy: use energy**.

Volt—"Voltage is a measure of the 'pressure' that is trying to force electrons down the wire. Increasing the voltage across a given load will increase the current through the load." Or, if you like, "voltage . . . is the electromotive force (EMF) which causes electrons to flow." More technically, voltage expresses the potential difference [in **joules** per **coulomb**] between the sourcepoint and the endpoint of electrical flow. "An electric potential or voltage is the *work* done on a unit charge to bring it from some specified reference point to another point." "1 V shall be taken as that emf [electromagnetic force] which will establish a current of 1 A through a resistance of 1 [ohm]."

Volts = power / current

For example, a 48-watt motor operating in a 4-amp current must be running at 12 volts.

Volts = amperes × ohms

Watt [w]—A **joule** per second. "One watt is produced when one ampere flows at an emf [electromagnetic force] of one volt." "A watt is not a unit of energy per se; . . . it's a unit describing how rapidly energy is used." "Power measures the rate of energy conversion. It is measured in watts . . . It equals volts times current."

A 100-watt lightbulb consumes 100 watts per second (360,000 joules per hour).

> 1 watt [absolute] = 0.998 watts [IT] = 1 J [abs] per sec and = 0.000948 BTUs per second = 0.056884 BTUs per minute = 3.413 BTUs per hour = 0.00134 horsepower = 1×10^7 ergs per second

Prof. Anna Mummert inserts here: "The unit watts per second is comparable to joules per second per second, or joules per second per hour."

Since ⅔ of the energy combusted in a traditional fossil fuel power plant does no useful work, 1 w takes 0.170652 carbon BTUs to generate.

> 1 kilowatt [kW] = 1,000 watts = 0.2388 kilocalories per second = 1.341 horsepower = 0.9478 BTUs per second = 56.884 BTUs per minute. See also **terawatt-hour**.

1 terawatt [TW] = 1 trillion watts = 56,884,000,000 BTUs per minute. See also **terawatt-hour**.

1 watt per square meter = 0.317 BTUs per hour per square foot = 0.005288 BTUs/min/ft^2 = 2,777.63 BTUs/yr/ft^2

Watts = amperes × volts

[See also **kilowatt** and **megawatt**. See section 3E, p. 551, for watts expressed in units relevant to nuclear power.]

Electric goods shop in Dubai

Watt-hour [Wh]—Electricity consumption, measured in watts used in each hour. "A 100-watt bulb burning for 5 hours uses 500 watt-hours or 0.5 KWH of energy." For discussion, see **kilowatt-hour**.

1 watt-hour [Wh] = 3,600 J = 3.413 BTU = 0.001341 hp-hr

1 Wh/kg = 1.548 BTUs/lb

Work—"Work is the overcoming of mechanical resistance through a certain distance." "If a system undergoes a displacement under the action of a force, *work* is said to be done, the amount of work being equal to the product of the force and the component of the displacement parallel to the force." More simply put, *work is force × distance*. But since matter and energy are always conserved, and all energies are mutually convertible, heating a liquid (for instance) is also doing work. "Work is a signed, scalar quantity. Typical units are inch-pounds, **foot-pounds**, and **joules**," not to mention **ergs** and **calories**. "Mechanical work is seldom expressed in **British thermal units** or **kilocalories**," but in this book, which focuses on fuel combustion, I have done just that.

For a given explosive, one may calculate the work ("the mechanical equivalent of heat") it can accomplish, in kilogram-meters (a precursor of joules), by multiplying the heat of explosion, in calories, by 425. According to this procedure, the work of detonating nitroglycerin is 671,500 kg-m, or 3.85 times the work of detonating mercury fulminate (174,250 kg-m).

2. Heat and Refrigeration

Blackbody [or "black body"]—"For theoretical purposes it is useful to conceive of an ideal substance capable of absorbing all the thermal radiation falling on it. Such a substance is called a *blackbody*." This concept is useful in solar engineering. "The amount of electromagnetic radiation emitted by a black body depends on its surface temperature. Its intensity distribution is dictated by the laws of quantum mechanics and is a manifestation of a state that, in physics, we call thermodynamic equilibrium[: a] state of a physical system where all the energy inputs equal the outputs."

1 British Thermal Unit [BTU]—The amount of energy needed to warm a pound of water by 1° Fahrenheit. This is contained in one match tip. "On average, coal contains 25 million BTUs per ton." Prof. Gutowski wrote me: "The world uses joules and watts for energy and power. The US (alone, as far as I know) uses BTUs and horse power. Even the British don't use BTUs any more. We tried to make this transition in the 1980s but Reagan killed it. I suppose the machine tool companies complained loudly . . . European readers will find it quaint if you use BTUs . . . In my work I have completely abandoned the BTU but many industries still use it and the Department of Energy does too." In 2016, so did the U.S. Geological Survey; see **high heating value of a fuel**. (From a build-your-own-solar-powered-home manual, 1975: "The selection of the BTU as a standard energy unit was made reluctantly . . . The rest of the world has adapted to . . . metric . . . and the United States will undoubtedly follow suit in the next decade." *Plus ça change* . . .) I decided to use BTUs because even in 2017 they were comprehended better than metric units by my fellow Americans, who like me were energy wastrels who avoided the themes of *Carbon Ideologies*.

Note:

1 International Table BTU = 1.055×10^3 joules [J]

1 mean BTU = 1.056×10^3 joules [J]

1 thermochemical BTU = 1.054×10^3 joules [J]

Gutowski notes here: "I use 1 BTU [approx. equals] 1 kJ."

1 BTU = 778.98 foot-pounds = 0.252 kilocalories = 1,054.8 joules [abs] = 1.0548 kilojoules = 107.56 kg-force meters = 251.98 [mean gram] calories [abs] = 251.996 IT calories. Another conversion, not used here, accepts 1 BTU at 252.16 cal.

1 IT calorie = 1/860 IT watt-hours = 4.187 abs J = 4.187×10^7 ergs

IT = "international"; used before January 1, 1948.* New system is abs = "absolute."

1 kilocalorie [kcal or food Calorie] = 4,187 [sometimes calculated 4,186] J

1 BTU per minute = 17.579 watts

1 BTU per minute per ft^2 = 189.226 watts/m^2

Energy content of a fuel: 1 BTU per gallon = 3.597×10^{-8} megajoule [MJ]/liter [the common metric way of expressing it].

1 BTU [IT] per lb = 2.326×10^3 J/kg

1 BTU [thermal] per lb = 2.324×10^3 J/kg

---

1 MBTU = 1 million BTUs = 1.0551×10^{-3} terajoules (TJ) = 2.931×10^{-4} gigawatt-hours (GWh)

1 Q-BTU = 1 quadrillion BTUs = 1 quad

1 BTU/sec = 1.415 hp

1 BTU/hr = 0.01731 watt/cm^2 = 2.930×10^{-4} [or 0.000293] kilowatts = 3.9292×10^{-4} horsepower

1 BTU/hr-ft^2 = 3.155 watts/m^2

1 BTU per pound (BTU/lb) of a given substance = 2.33 kilojoules per kilogram (kJ/kg).

Caloric or calorific—Refers to the inherent energy of a fuel, ready to be released through combustion. Calorific efficiency can also be expressed as **high** or **low heating value**.

Calorie:

1 [thermochemical gram-mean] **calorie** [cal or gcal]—The amount of heat needed to increase the temperature of 1 gram of water by 1° Celsius. Written with a small "c."

1 calorie = 4.184 joules = 3.9685×10^{-3} BTU

1 **Calorie** [**food calorie**]—The same as a kilocalorie [kcal].

---

* As in my grandfather's *Machinery's Handbook.*

1 Calorie = 1 kcal = 1,000 cal = 1 "food calorie"

> And why not use food calorie units to indicate the high heating value of coal or oil? "The availability of ammonia and straight-chain paraffins may permit future production of food from fossil fuels."

1 kcal = 3.968 BTUs

1 kcal/g = 1,798.7961584 BTUs/lb

1 kcal/kg = 1.7987961584 BTUs/lb

Efficiency of a heat engine [e.g., a steam turbine power plant or an internal combustion motor]—Work output divided by heat input, both of them being expressed in the same units of energy [e.g., BTUs or joules]. Or, more explicitly:

Thermal efficiency = 1 − [heat rejected / heat absorbed].

As *Carbon Ideologies* often points out, about ⅔ of the energy in a nuclear or fossil fuel steam turbine plant accomplishes no useful work.

Fuel—"Fuels, whether for the furnace, automobile, or rocket, are energy-rich substances and the products of their combustion are energy-poor substances."

Heat—The temperature-related energy which can be transferred between two objects or systems of different temperatures.

Heat of combustion—A more techno-chemical approach to **HHV**. Although I avoided using this term in *Carbon Ideologies,* it may be helpful to introduce it, in case you should come across it while comparing fuel energies. When I was alive, it usually signified **kilocalories** *released or taken in per substance oxidized to water and/or carbon dioxide* at constant pressure and 25° Celsius. A negative value implied that heat was given off, as would always be the case for a combusted fuel. [A positive value, not relevant to our purposes, showed that the reaction was endothermic; the substance actually cooled as it oxidized.] Heat of combustion was expressed not in pounds or standard volumes, but in **moles.** Simply and inadequately put, one mole (now sometimes written "mol") of a substance (6.023×10^{23} atoms per molecule, or the basic unit of a chemical reaction; a mole of carbon weighs 12 grams while a mole of hydrogen weighs only 1, but those two moles are considered as equivalents) is calculated as the sum of the atomic numbers of its component atoms:

Molar weight of H_2, or hydrogen:

> (hydrogen's atomic number = 1.0) × 2 (since as the formula shows a hydrogen molecule exists as 2 atoms) = 2 grams per mole

Of CH_4, or methane:

(carbon's atomic number = 12.0) × 1 (since there is only 1 carbon atom in the formula), + (hydrogen's atomic number = 1.0) × 4 (since the formula shows that 4 of them are present) = 16 g/mole

C_4H_{10}, or butane:

(12.0 × 4) + (1.0 × 10) = 48 + 10 = 58 g/mole

C_8H_{18}, or octane:

1,307.53 (12.0 × 8) + (1.0 × 18) = 96 + 18 = 114 g/mole

The heats of combustion of these substances are:

Hydrogen:	−68.32 kilocal per mole [= kcal/2 grams]
Methane:	−212.80
Butane:	−687.98
Octane:	−1,307.53

Thus, to convert methane's combustion heat back to a more familiar form (see next entry), one would multiply [212.80 kcal/mole] × [1 mole/16 grams] × [453.5 grams/lb] × [0.04163 lbs/cu ft] × [3.968 BTUs/kcal] = 996.34 BTUs per cubic foot.

For a discussion of molar ratios in climate change computations, see **mole** in section 12.

High heat[ing] value of a fuel = HHV (in BTUs/lb) = 14,544C + 62,028(H − O/8) + 4,050S

    C = carbon content [in %]
    H = hydrogen content
    O = oxygen content
    S = sulfur content

("I never use the term hhv in my studies. My colleagues and I at the USGS used Btu as the measure of the heat content of coal. From a quick check on the Internet the terms seem to be comparable."—Robert Finkelman, United States Geological Survey, 2016.)

The high heating value is also called the **gross calorific value**. This is the energy emitted in combustion. [See **heat of combustion**.] In many countries, among them Bangladesh, the GCV is expressed in kJ/kg.

1 BTU/lb = 2.326 kJ/kg
1 J/kg = 0.0004299226 BTU/lb
1 kJ/kg = 0.4299226 BTU/lb
1 MJ/kg = 429.9226 BTUs/lb
1 GJ = 429.9226 million BTUs/lb
1 kcal/kg = 1.7987961584 BTUs/lb

For HHVs of many fuels, see the table of Calorific Efficiencies beginning on p. 208.

Low heating value [LHV], or **net calorific value** (sometimes known as **recovery heat**), is the fraction of emitted energy which actually warms the target material. There are several reasons why HHV and LHV are not the same; one is that a frequent product of combustion is water, which then absorbs some heat. [LHV is measured "in the absence of water condensation."] According to the International Energy Agency, "for coal and oil, net calorific value is usually around 5% less than gross and for most forms of natural and manufactured gas the difference is 9–10%." In the Argonne National Laboratory's 2010 list of LHVs and HHVs of various fuels I note a spread closer to 10%. Since combustion energies are sometimes expressed only in HHVs, I have used only those in this book even though the LHVs are more realistic.

Quad—A quadrillion BTUs = 10^{15} BTUs = 1.055×10^{18} **joules**. [See section 1 beginning on p. 521.] "Roughly the energy contained in 200 million barrels of oil."

1 quad = 1.054 **exajoules***

Rankine Cycle—A mathematically idealized description of the compression, heating, vaporization, superheating, expansion, condensation and then recompression of water in the "closed loop" of a steam turbine power plant. Heat losses in the condenser stage explain much of this system's typical inefficiency (30–40%). [A counterpart model for an internal combustion engine is called the Otto Cycle.]

---

* Conversion calculation from Oak Ridge National Laboratory, 1982. Prof. Mummert writes here: "For readers who are less comfortable with numbers, it would make more sense if the numerical values for joules and exajoules were identical." And of course 1.055 and 1.054 are really the same number, computed differently.

1 Refrigeration ton [U.S.]—"Represents the amount of heat that must be removed from a short ton (909 kg) of water to form ice in 24 h[ours]."

= 3.51 kilowatts = 12,000 BTUs per hour = 12.7 megajoules per hour

A British refrigeration ton (mentioned here only for completeness) = 14,256 BTUs/hr.

Specific heat—Generally speaking, the thermal energy needed to alter the temperature of a given mass of a substance by a given amount. In the units BTU/lbm-°F, the specific heat of water conveniently equals 1.0. Often abbreviated "c," and expressed in BTU/lbm-°F or J/kg-°C. [For a definition of lbm, see **footpound** in section 1, p. 523.]

Some specific heats:

water:	1.0 BTU/lbm-°F
gasoline:	0.53
light oil:	0.50
benzene:	0.41
pure aluminum:	0.23
pure gold:	0.031

Therm—My [natural] gas bill is expressed in this unit. One therm = 10^5 (100,000) BTUs or 1.055×10^5 joules (= 25,200 kcal).

Thermal efficiency—The LHV divided by the HHV.

The thermal efficiency of natural gas is 84%, of propane 85%.

Thermie—1,000 kilocalories.

3. Nuclear Energy

For some 2014 Japanese air dose figures, see p. 548.

A. Radiation types; half-life, fallout

Alpha particle—"The nucleus of a helium atom consisting of two protons and two neutrons. It is ejected at high speed from the atomic nucleus." Can be halted by a sheet of paper, "but on being stopped" it "produce[s] a very large amount of ionization locally." In other words, "the alpha particle is particularly damaging, because it is doubly charged, with two positively charged protons in its structure," which is why one source estimates that "alpha radiation is 20 times worse than X-rays or gamma rays." Like beta, alpha particles can be permanently harmful if ingested. Plutonium-239's alpha particles will radiate for thousands of years (see Table 2, p. 540).

Beta particle—"An electron ejected at high speed from the nucleus of an atom during certain transmutation events." Or, if you like, "the conversion of a neutron into a proton and an electron"—which, yes, gets emitted at high speed. Beta is considered both a particle and a wave. Unlike gamma, it can be stopped by flesh and clothes ("may travel . . . up to a couple of centimeters in tissue"). If ingested, however, it can remain dangerous, whereas gamma simply passes through the body, does its damage and continues on its way.

> *"When an element transmutes itself through radioactive decay it shifts its position on the periodic table two places to the left if it emits an alpha particle . . . one place to the right if it emits a beta ray . . ."*

Fallout—"The radioactive dust that falls to earth after atomic explosions."

Gamma ray—Similar to an X-ray. Only a thick lead shield will stop this wave. Many accrued dosimeters, such as the one I used for this essay, measure only gamma. [A slightly different animal, the neutron ray, relates more to atomic bombs and will not be considered in this book.]

Half-life—The time required for half the atoms of a radioisotope to decay. Note that it takes 10 half-lives for an initial radioactivity to reduce to 1/1000. Note further that the daughter element (what the original element decays into) may be itself radioactive, requiring another 10 of its own half-lives to decay into something else.

Table 1
COMMONLY MENTIONED RADIOCONTAMINANTS IN FUKUSHIMA

Isotope	Half-life	Years to 1/1000 strength	Comment
Cesium-134	20.65 years[1]	206.5 years	Gamma ray emitter, with some beta. When I visited Iwaki 3 years after 2011, it seemed to have decayed much more quickly than the Cs-137. *The Japanese Ministry of Health, Labour and Welfare lists the half-life at 2.1 years.*[2]

Table 1
COMMONLY MENTIONED RADIOCONTAMINANTS IN FUKUSHIMA

Isotope	Half-life	Years to 1/1000 strength	Comment
Cesium-137	30.2 years[3]	302 years	Gamma ray emitter with "higher penetration power" than medical X-rays.[4] Also emits beta. The human body interacts with it much as if it were potassium. Often absorbed through plant foods. "Preferential deposition in muscle . . . Much less of a biological hazard than . . . strontium-90."[5] But in fact cesium may be correlated with stomach and heart disease.[6] "Has gaseous precursors and is unfortunately a significant component of the more soluble 'long distance' fallout."[7] Of all the isotopes released at Fukushima, this is, says one scientist, "the most dangerous because of gamma plus quantity."[8] Quantities of Cs-137 (and -134?) soon took up residence in sub-surface clay deposits.
Iodine-131	8.05 days[9]	81 days	Implicated in thyroid cancer. Also used as a medical tracer for thyroid problems.[10] Beta emitter, with some gamma. Especially problematic for children. An early contaminant at Fukushima.
Strontium-90	28.8 years[11]	290 years	"One of the best long-lived high-energy beta emitters known." Frequently absorbed through plant foods. Because Sr-90 is created from the decay of one radioactive gas (Kr-90) into a second gas (Rb-90) before becoming strontium, it can be carried far away from the initial explosion site.[12] Its "properties similar to calcium"[13] make it excel at causing bone cancer. "The major portion . . . is excreted with a biological half-life of 40 days during the first year . . . less than 10 percent . . . is tightly bound to bone and is excreted very slowly with a long biological half-life of about 50 years."[14] At Fukushima, less *immediately* dangerous (and apparently less prevalent) than Cs-137 since its emissions were contained within the marine environment.[15] But don't eat the seafood.
Tritium	12.5 years[16]	125 years	A form of hydrogen. Can enter the body in several ways, concentrating in no particular organ. "Because of the low energy of the beta particles it emits and its relatively short half-life, tritium is much less of a long-range radiation hazard than the radio-isotopes already considered."[17] But some disagree.

1 *CRC Handbook*, p. 11-118 ("Table of the Isotopes").
2 "New Standard Limits for Radionuclides in Food," Department of Food Safety, Pharmaceutical & Food Safety Bureau, Ministry of Health, Labour and Welfare (Japan), www.mhlw.go.jp/english/topics/2011eq/dl/new_standard.pdf, accessed by Jordan Rothacker on December 10, 2015.
3 Loc. cit.
4 Hirose, p. 58.
5 Glasstone and Dolan, pp. 604–5, 606.
6 Interview with Mr. Hisataka Yamasaki of No Nukes Plaza, Tokyo, Feb. 18, 2014.
7 Odum, p. 463.
8 Edwin Lyman, Union of Concerned Scientists, phone interview by WTV, 2014.
9 *CRC Handbook*, p. 4-18 (entry on iodine); pp. 11-112-14 ("Table of the Isotopes"); Tilley and Thumm, p. 23.
10 Tilley and Thumm, p. 25.
11 *CRC Handbook*, p. 4-34 (entry on strontium).
12 Odum, p. 463.
13 Hirose, p. 72.
14 Glasstone and Dolan, pp. 605, 606.
15 Lyman interview.
16 Not listed in *CRC Handbook*.
17 Glasstone and Dolan, loc. cit.

Table 2
OTHER ISOTOPES OF INTEREST

Isotope	Half-life	Years to 1/1000 strength	Comment
Cesium-135	2.3 million years[18]	23 million years	Presented here as a reminder that undifferentiated references to "radioactive cesium" could be dangerously misleading. (As it happens, most other cesium isotopes have half-lives of days, minutes or seconds.)
Iodine-129	17 million years	170 million years	Another such reminder. "Radioactive iodine" may not always be I-131! (Again, other iodine isotopes have half-lives of days, hours, minutes, seconds or milliseconds.)
Plutonium-235	25.3 minutes[19]	253 minutes	Most Pu isotopes are alpha emitters (Pu-241 emits beta[20]). They are "produced in extensive quantities in nuclear reactors from natural uranium."[21] In 1994, "the total amount of fallout over the entire earth surface" was around 24,000 curies [24 kCi] "of plutonium 238 (16 kCi which is from SNAP-9 ['a navigational satellite lost in 1964']), 154 kCi of plutonium 239, and 209 kCi of plutonium 240."[22] *"After the explosion at #3 reactor . . . TEPCO . . . announced that, on 21 and 22 March [2011], in soil collected on the plant site, they detected plutonium."*[23] I wish it had been Pu-235. In 2012 the highest concentration of this metal appeared to be Pu-238.[24] Unfortunately, the primary isotope in **MOX**[25] reactor fuel is Pu-239.[26] Note that although Pu-238 decays much more rapidly than Pu-239, "its radioactive energy is about 270 times greater. The relative danger . . . of a radioactive substance cannot be determined by its half-life."[27] As for Pu-239, it has "a long biological half-life in the skeleton (about 100 years) and the liver (about 40 years).[28] Let us just say that any isotope of that metal is dangerous—if less poisonous than arsenic.[29] The U.S. Department of Energy standard for plutonium is any amount above "background levels."[30] (By the way, it is "a toxic silver-colored element.")[31] Pu-239 and -240 "are the most common isotopes found in the natural environment, being 21 times more often found in sediment than is plutonium-238."[32] *The Japanese Ministry of Health, Labour and Welfare lists the half-life of "plutonium" at "14 years and more."*[33]

18 *CRC Handbook*. These longer half-lives are given in scientific notation, which I have recast in more familiar forms for the benefit of non-mathematical readers.
19 *CRC Handbook*, p. 11-191 ("Table of the Isotopes").
20 Graf, p. 27.
21 Loc. cit.
22 Graf, p. 110. "A navigational satellite lost in 1964" appears on the same page. How expansive of us! According to this source [loc. cit.], the maximum fallout is between 40 and 50 degrees N: 0.079 mCi/sq. km for Pu-238 and 2.2 mCi/sq. km for Pu-239 and -240.
23 Hirose, p. 51.
24 *The Japan Times*, Thursday, August 23, 2012, p. 1 (unattrib., "Plutonium traces detected at 10 locations in Fukushima").
25 Mixed-oxide fuel, a mixture of uranium and plutonium that can be used in nuclear reactors" (Tabak, p. 57).
26 Edwin Lyman to WTV, phone interview, March 13, 2014.
27 Hirose, p. 59.
28 Glasstone and Dolan, p. 601.
29 Graf, p. 28.
30 Graf, p. 31.
31 Tabak, p. 41.
32 Graf, p. 27.
33 "New Standard Limits for Radionuclides in Food," Department of Food Safety, Pharmaceutical & Food Safety Bureau, Ministry of Health, Labour and Welfare, (Japan), www.mhlw.go.jp/english/topics/2011eq/dl/new_standard.pdf, accessed by Jordan Rothacker on December 10, 2015.

Table 2
OTHER ISOTOPES OF INTEREST

Isotope	Half-life	Years to 1/1000 strength	Comment
Plutonium-238	87.7 years[34]	877 years	
Plutonium-239	24,100 years[35]	241,000 years	
Plutonium-240	? years	? years	
Plutonium-241	? years	? years	
Polonium-210	138.39 days[36]	1,384 days	"Most readily available" isotope of this alpha-emitting element. Polonium is 250 billion times more deadly than cyanide.[37] One of its unspecified isotopes appears in anti-static brushes used to clean photographic negatives. The label advises "DO NOT TOUCH STRIP UNDER GRID."
Polonium-209	102 years	1,020 years	
Polonium-208	2.9 years	29 years	
Tellurium-130 and -28	2×10^{21} and 7.7×10^{24} years*	2×10^{22} and 7.7×10^{25} years	Mentioned only once in the Fukushima literature I have read. The "tellurium" [no isotopes stated; here I supply the most abundant] contaminated Iitate Village.[†]
Thorium-232	140 billion years[38]	1,400 billion years	Alpha emitter.
Thorium-234	4.5 billion years[39]	45 billion years	
Uranium-238	4.4 million years[40]	44 million years	Alpha emitter, reducing first to thorium-234.[41] "Decays very slowly, which is another way of saying that the amount of radiation it emits per unit time is very small. As a result, it has a very low level of toxicity . . ."[42] "Enriched" uranium contains this fissionable stuff, which given the relatively short half-life is rare in nature.
Uranium-235 (a)	70.4 million years[43]	704 million years	These two isotopes of U-235 are much more common in nature than U-238.
Uranium-235 (b)	10 quintillion years[44]	100 quintillion years	

34 Hirose, p. 57.
35 *CRC Handbook*, p. 4-27 (entry on plutonium).
36 *CRC Handbook*, p. 4-27 (entry on polonium). Tabak, however, says 5 days (p. 108).
37 *CRC Handbook*, p. 4-27: "Weight for weight it is about 2.5×10^{11} times as toxic as hydrocyanic acid."
38 *CRC Handbook*, p. 4-37 (entry on thorium).
39 Tabak, p. 108.

40 *CRC Handbook*, p. 4-39 (entry on uranium). But Tabak says that "uranium-238's half-life is roughly equal to the age of Earth" (p. 32).
41 Tilley and Thumm, p. 13.
42 Tabak, p. 108.
43 *CRC Handbook*, p. 11-189 ("Table of the Isotopes").
44 Ibid.
* Ensley pp. 428–29.
† Lochbaum et al., p. 118.

B. Radiation meters used for this book

A dosimeter is necessary for anybody wishing to avoid exposure above a certain amount. A scintillation counter is essential for exploring and perceiving the varying levels of a place. I should have had both all along.

Dosimeter—An instrument to measure accrued gamma radiation. When this personal protection device is transferred between individuals, or otherwise left away from the designated user, it cannot report someone's dose correctly. When the foregoing happens by some superior's order [see II:589], exploitation should be suspected.

In 2011 my dosimeter arrived calibrated in millirems. In 2014, for reasons explained in the text, it was recalibrated in millisieverts.

1 millirem [mrem] = 10 microsieverts [microSv]

Rather than go back and alter the units into a spurious consistency, I have left them as they are, except in comparative tables, where all original units are noted. As you will see in part E of this section beginning on p. 545, the conversions are simple. Moreover, I have drawn equivalents where needed in the text. Hopefully you will agree that this is better than rewriting history.

Scintillation counter (or scintillation meter)—A machine to measure scintillations (cpms; see "Units and Conversions," p. 545). Similar in function but not in makeup to the familiar Geiger counter, which it has mostly superseded. Radiation excites flashes in its crystal or liquid chamber. The best of these toys (among which I count my beloved pancake frisker) will measure alpha, beta and gamma. The display reveals how "hot" an area is, in scintillations or equivalent converted units per interval of time. A scintillation counter will show in real time what specific spot or thing is "hot," whereas a dosimeter can only show the dose already received.

From the pancake frisker, let me offer up the following measurements, presented as a special service to my fellow bookworms:

RADIOACTIVITY OF SELECTED LIBRARY INTERIORS, 2014–15

[in microSv/hr]

In "normal background" areas

San Francisco, 3rd floor	0.12
Charleston, West Virginia, windowsill	0.12
Madison, West Virginia	0.12
Welch, West Virginia	0.12
Iwaki, Japan	0.12
Denver, Colorado, 3rd floor [elev. 5,280 ft]	0.12
Salt Lake City, public main	0.12
Salt Lake City, University of Utah	0.18
Sun Valley, Idaho [elev. 5,840]	0.18
Estes Park, Colorado [elev. 7,522]	0.24

In Japanese yellow zone

Greater Iitate, community center	0.54

In Japanese red zone

Iitate, decontaminated community center	0.90*

---

* Average value of NORMAL readings. All other measurements in 1-minute SCALER mode.

C. Nuclear materials

Control rods—Neutron absorbers, such as boron steel. These may be fully or partially inserted or withdrawn. The former will (hopefully) halt the fission in a reactor. The farther they are pulled out, the more neutrons will be available for fission.

Fuel rods—Long narrow tubes of uranium, uranium oxides, plutonium and other materials whose nuclear fission powers a reactor. Usually their skin is zirconium or aluminum. In some cases they may be withdrawn from the nuclear pile to various degrees in order to slow or shut down the fission. However, their radioactive decay cannot be stopped, and because this generates great heat (augmented by the decay of the "daughter nuclides" contaminating them), they must be kept cool in order to prevent such catastrophes as occurred at Fukushima. Their working life is 2–5 years.

Core—The working part of a reactor. A lattice of fuel rods, control rods and a "moderator" such as graphite to slow down the neutrons sufficiently for them to fission more nuclei.

MOX—"Mixed fuel oxide, a mixture of uranium and plutonium that can be used in nuclear reactors." "MOX fuel is manufactured from plutonium recovered from used reactor fuel, mixed with depleted uranium. MOX fuel also provides a means of burning weapons-grade plutonium (from military sources) to produce electricity." This provides a way of "closing the fuel cycle. The current means of doing this is by separating the plutonium and recycling that, mixed with depleted uranium, as mixed oxide (MOX) fuel. Very little recovered uranium is recycled at present" (2016). In 2011, MOX escaped in minute quantities from Daiichi's Reactor No. 3.

Nuclide—See **radioisotope**.

Radioisotope—All forms of a given element contain the same number of protons. Some forms vary in their number of neutrons—which is to say in their mass. An isotope is an atom of a specific mass. If it is radioactive, it is a radioisotope; if not, not. For instance, protium, deuterium and tritium are three isotopes of hydrogen. All of these contain a single proton in the nucleus (and a single electron orbiting around the nucleus). Protium, the most common form, possesses no neutrons. Deuterium's nucleus has one neutron; tritium's has two. Tritium, being radioactive, is a radioisotope. So are the various forms of radioactive cesium, iodine, strontium and plutonium which contaminated Fukushima in 2011. Every element can be converted into a radioisotope. [The different isotopes of a given element are called nuclides.]

D. Places, organizations and acronyms relevant to the Fukushima disaster

3-11—Shorthand for the tsunami-earthquake-nuclear-disaster on March 11, 2011.

AEC—[U.S.] Atomic Energy Commission.

IAEA—The International Atomic Energy Agency. Mentioned at the end of the Emirati oil chapter, II:557n.

ICRP—The Committee on Radiological Protection, whose 1 millisievert per year recommended civilian maximum [revised from the ICRP's previous suggestion of 4 mSv/yr] was cited by the Union of Concerned Scientists. [The current recommendation for professional workers was 20 mSv/yr, revised from 50 mSv/yr].

National Railway Mito Motive Power Union—A leftist organization which was prominent in Japanese anti-nuclear protests. An amalgam of three unions: Nazen Iwaki, Iwaki Godo Union and the J[apan] R[ailways] Railway Union, which was an old labor union.

NISA—The Nuclear and Industrial Safety Agency, which functioned up until the Fukushima incident. It was subordinate to the Ministry of Economy, Trade and Industry "whose goal had been to promote and expand the use of atomic energy . . . NISA's safety standards were based on the now incredulous [*sic*] assumption that severe incidents, including core meltdowns, would never occur at any of the nation's atomic power stations."

NRA—The Nuclear Regulation Authority, a Japanese government body formed after 3-11 to replace **NISA**. "We will never allow the myth about the safety of nuclear power to be resurrected."—Toyoshi Fuketa, one of five NRA commissioners, on February 17, 2013.

NRC—Nuclear Regulatory Commission, an American body comparable to **NISA**.

Plant No. 1—Daiichi Nuclear Plant, 12 km north of Plant No. 2. ["Daiichi" means "first" or "number 1," so the English-language name is Fukushima No. 1.] Plant No. 1 consisted of six reactors, each named by its respective number. Reactor No. 1 (232 km from central Tokyo) was the most infamous, but No. 3's hydrogen explosion, which took place the day after No. 1's, emitted *each hour* a dose equal to 400 times the **ICRP**'s recommended annual exposure for citizens. Explosions of varying magnitudes occurred in Reactors 1 through 4, which were at Okuma. The other two were at Futaba. Reactor 1 was a General Electric, Mark I, whose construction commenced in 1971. Reactors 2, 3, 4 and 5 were "more advanced B[oiling] W[ater] R[eactor] models but also had Mark I containments."

Plant No. 2—Daini Nuclear Plant, just right of Highway 6 as one comes north into Tomioka Township. ["Daini" means "second" or "number 2," so in English it was often called Fukushima No. 2.] Daini contained four reactors. Two were in Naraha and two in Tomioka.

These ten reactors in Fukushima Prefecture produced one-fifth of the nation's nuclear power.

Reactor—In this book we are almost exclusively concerned with light water reactors (LWRs), so named "because the reactor core is covered with water to allow the nuclear reaction to take place and to keep the core cool." Both pressurized water reactors (PWRs) and boiling water reactors (BWRs) are LWRs. These are described in brief on p. 238.

Tepco—Tokyo Electric Power Company, the utility responsible for the disaster. On July 31, 2015, after two reversals, the Japanese government finally indicted three former Tepco executives: Mr. Katsumata Tsunehisa, Mr. Muto Sakae and Mr. Takekuro Ichiro.

E. Units and conversions

Becquerel (Bq)—One Bq is the decay of 1 atom of a radioisotope per second, or, if you like, 1 cps = 1 count per second. A cpm, or count per minute, is therefore 60 cps's. (In the wise words of the man who took me to Namie, "If the meter catches a scintillation, that's one count per second—if it catches it.") "It is estimated that the detonation of a nuclear weapon in Hiroshima ... produced a radioactivity of 8×10^{24} becquerels." In the Japanese press, the mention of Bqs without accompanying units usually meant Bqs per liter.

For any of you who may be in love with those old units called curies,

> **1 curie** [Ci] = 3.7×10^{10} Bq = 37 GigaBq = 3.7×10^{10} disintegrations in 1 second of 1 gram of radium-226 [which contains 3.7×10^{21} atoms]. "The curie is conveniently extended to mean 3.7×10^{10} disintegrations per second of any radioactive substance." However, "neither the curie nor the becquerel identified the isotope creating the decay emissions, a problem that can be resolved only by analyzing the number of kinds of radiation produced during the decay."

>> Tritium disintegrates at 9,800 curies per gram, strontium-90 at 140 Ci/g and cesium-137 [the main long-lasting pollutant at Fukushima], at 88 Ci/g.

>>> 0.000006 [6×10^{-6}] g iodine-131 = 1 Ci = 5.6 kg iodine-129

>>> **1 picocurie** = 1×10^{-12} [1 trillionth] Ci = 2.2. disintegrations per second.

>>> 1 picocurie per gram = 0.0370 [or by an older conversion 0.0367] Bq per gram. "One picocurie per gram of plutonium in sediment represents much less than one part per

billion . . . in natural materials and could not be detected by ordinary physical or chemical methods."

Permissible tritium levels in water:

Colorado: 500 picocuries per liter.
California: 400 picocuries per liter.

1 watt of nuclear power = 5 curies

Curies are rarely used nowadays. However, the Union of Concerned Scientists informs us that the Three Mile Island accident let loose 10 million curies, mostly of xenon-133 and krypton-85, while Fukushima's little difficulty released 13.5 million curies of noble gases and same amount of iodine-131, along with half a million curies of cesium. "More than 140 million curies of radioactivity were released to the atmosphere and Columbia River when Hanford reactors and reprocessing plants operated," and nuclear weapons testing fallout from 1945–1980 produced the "largest release of artificial radiation ever experienced by the world—nearly 70 billion curies," most of this fortunately consisting of **nuclides** with short half-lives.

Like a roentgen or a curie and unlike a rem or a sievert, a Bq (or a cpm) is not in and of itself a unit of biological damage.

In 2011 and after, the Japanese government employed both Bq and Sv in their official pronouncements. Radioisotope concentrations in groundwater and sea-water were usually expressed in Bq (hence these figures from June 20, 2013: strontium-90 at 1,000 becquerels per liter, 33 more times than the legal limit; tritium at 500,000 Bq/l, or 8.3 × the limit).

Since the emitted particles of different elements have different energies, one can make no generic conversion from becquerels to sieverts. A cps of cesium-137 will not cause the same potential cellular damage as a cps of iodine-131.

According to Dr. Ed Lyman of the Union of Concerned Scientists, for cesium-137 (beta decay only) and making certain other assumptions:

1 cpm = 0.001 mrem per hour = 0.01 microSv/hr

The Tepco employees who were frisking Mr. Endo's daughter's fishing rod on our way out of Tomioka in February 2014 [see p. 384] were using a beta-counting scintillation meter. Hence the above ought to be the right conversion for what they were measuring.

However, on one-minute timed counts with the pancake frisker, which is calibrated for Cs-137 (alpha, beta and gamma decay together), I find that cpms divided by 18,000 yield the approximate value in microsieverts per minute.* So to get micros per hour, divide cpms by 300. Thus:

1 cpm = 0.0033 microSv/hr

I cannot understand why this result should be so much less than the 0.01 micros per hour above. Had it been more (since the frisker is measuring all three radiation types, not just beta), I might have accepted it. It is possible that something was lost in translation when I questioned the Tepco people.

According to still another source, 1 Bq *inhaled* Cs-137 = 4.6×10^{-9} Sv = .0046 microSv [presumably per hour but no units given], while 1 Bq *ingested* Cs-137 = 0.013 microSv[/hr?].

At any rate, since the frisker's readings so often corresponded to other people's measurements, and there is no way for me to cross-check that beta-only measurement, I have trusted in the former.

According to Japan's Ministry of Health, Labour and Welfare, 10 Bq/kg of cesium in drinking water seemed to equate with 0.1 mSv/year.

10 Bq/liter of tritium = 270 pCi/liter.

And according to the Ministry of Education:

Isotope	Sv per Bq [inhaled[†]]	Sv per Bq [ingested[†]]
I-131	2.2×10^{-8}	7.4×10^{-9}
Cs-137	3×10^{-8}	3.9×10^{-8}
Sr-90	2.8×10^{-8}	1.6×10^{-7}
Pu-239	2.5×10^{-7}	1.2×10^{-4}

[†] My best contextual guess of the text beside "Sv/Bq," which Kawai-san, the interpreter, not knowing that this part of the table would interest me, left untranslated. This would be the usual distinction.

. . . all presumably per second.

Cpm—A unit used by the **pancake frisker**. For discussion, see **becquerel**, p. 545.

Electron[-]volt (ev)—One ev expresses the work accomplished in moving 1 electron against a potential difference of 1 volt. [This is an old-style unit.]

---

* Since the frisker can be toggled back and forth between cpms and mSv/hr, it is very easy to verify this empirical conversion; study the table of frisker readings, p. 552, in these appendices, to see it for yourself.

1 ev = 1.60×10^{-12} ergs = 1.60×10^{-19} **joules** = 1.60×10^{-19} **watt-seconds** = 1.60×10^{-19} coulomb-volts = 1.52×10^{-22} BTUs = 3.82×10^{-20} calories

1 kilo-electronvolt [**kev**] = 1,000 ev = 1.60×10^{-16} J

= 1.60×10^{-9} **ergs** [see p. 523]

1 mega-electronvolt [**mev**] = 1,000,000 ev = 1.60×10^{-13} J

= 1.38×10^{9} cm/sec [for a particle the mass of a proton, neutron or electron]

This unit used to be commonly employed in specs for nuclear reactors.

According to the AEC, "The complete conversion of one hydrogen atom to energy, according to Einstein's E = mc^2 yields 942 Mev."

Therefore, 942 Mev × [1,000,000 ev / 1 Mev] × ([3.82×10^{-20} calorie] / 1 ev) × [1 BTU / 252 cal] = 1.43×10^{-13} BTUs

1 giga-electronvolt [**gev**] = 1,000,000,000 ev = 1.60×10^{-10} J

1 beta particle emitted by the Fukushima contaminant cesium-137 gives off 0.5 mev = 7.6×10^{-17} BTUs

1 **gamma** ray emitted by the same substance gives off 0.66 mev = 1.00×10^{-16} BTUs

1 fission [of uranium-235] = 200 mev = 3.04×10^{-14} BTUs

Gray [Gy]—"This measures absorption of radiation energy by a unit of mass." "The energy transfer of one gray is equivalent to the ionization of about 115 **roentgens**."

1 Gy = 1 **joule** absorbed by 1 kilogram.

The average radiation dose from a pelvic CT scan = 25 mGy.

1 Gy = 1 **sievert** [Sv] for **gamma** rays but 20 Sv for **alpha** particles. In other words, absent other information one cannot simply convert from Gy to Sv.

Here are some Japanese government air dose figures, in nGy/hr [nanograys per hour], for relevant locations on November 2, 2014, at 15:40. They may at least serve for comparison. For each place I selected the maximum reading:

Iwaki: 61
Fukushima City: 175
Hirono: 338
Namie: 456
Naraha: 726
Futaba: 1,070
Tomioka: 1,902
Okuma: 15,680

On this same date (10 minutes later), the prefecture with the lowest dose was Hokkaido at 41 nGy/hr. Of course the highest was Fukushima at 15,690 nGy/hr.

Rad—[Redefined in 1970.] "The dose [of radiation] causing 0.01 **joules** of energy to be absorbed per kilogram of matter."

Rem—An acronym for "**roentgen** equivalent [in] man." So a rem is conceptually comparable to (though numerically different from) a **sievert**, the roentgen being a direct measure of radiation and the rem reflecting that "weighting factor." An American "incident responder" is advised to limit his or her *single, short term dose* to 5 rems per non-emergency radiation episode, to 10 rems for "protecting valuable property," and to 25 rems for saving human life. Nausea and the like may show up at 30 rems, radiation sickness at 70 to 100 rems; 5,000 rems will kill all exposed humans within 48 hours.

1 rem = 1 rad [old definition] = 0.01 sievert [Sv] = 10 millisieverts [mSv] = 10,000 microsieverts [microSv]

New definition, using different weighting factors: 1 rem = 1.07185 roentgens.

1 rem = 1,000 millirems [mrem] = 1,000,000 microrems

1 millirem, or **mrem** [the unit used in the first nuclear chapter] = 1/100,000 absorbed **joules** of radiation energy per kilogram of body weight = 0.01 mSv = 10 **microSv** [the units used in subsequent nuclear chapters] = 1,000 microrem.

"One millirem . . . corresponds to being struck by approximately seven billion particles of radiation." According to a very pro-nuclear source, 1 mrem increases our risk of fatal cancer by 1 chance in 8 million. "This risk corresponds to a reduction in our life expectancy by 1.2 minutes." Average exposures in the

Three Mile Island accident were 1.2 mrem. Normal background exposure is 0.05–0.1 mrem/hr.

1 microrem = 0.001 mrem = 0.01 microSv

Roentgen [R]—"The ionizing effect of radiation on a standard mass of air." "One roentgen . . . strips off one outer electron from each of . . . the molecules in one cubic centimeter of dry air at standard temperature and pressure." One roentgen makes "1 electrostatic unit of charge per cubic centimeter of air at standard conditions. It . . . generates 83.3 **ergs** [see p. 523] per gram of air."

$1 \text{ erg} = 10^{-7} \text{ joules} = 2.389 \times 10^{-8} \text{ calories} = 6.242 \times 10^{11} \text{ electron volts}$
$= 2.778 \times 10^{-14} \text{ kilowatt-hours [kWh]}$

Sievert [Sv]—A "unit of equivalent dose . . . defined as the absorbed dose multiplied by a weighting factor that expresses the long-term biological risk . . ." Translating the above benchmarks for "incident responders" from **rems** to Sv, we obtain: *single, short term dose:* 0.05 Sv, "protecting valuable property," 0.1 Sv; saving human life, 0.25 Sv; possible appearance of nausea, 0.3 Sv; onset of radiation sickness, 0.7–1.0 Sv; and the like may show up at 30 rems, radiation sickness at 70 to 100 rems; lethal dose within 48 hours, 50 Sv [= 50 million microSv].

> **1 Sv = 1,000 millisieverts [mSv] = 1,000,000 microsieverts [microSv]**
>
> 1 Sv = 100 rem = 1 **joule** [J]/kg

Since 1 gram of "biological tissue" exposed to 1 mg of uranium will receive 0.006 Sv per year, 1 Sv corresponds to the biological damage induced by the year-long application of 166.667 mg [about 0.36 lb] U to a gram of living flesh.

U.S. Environmental Protection Agency standard for uranium in drinking water: no more than 40 microSv per year.

Prof. Anna Mummert: "There is not a direct conversion from Sieverts to **BTU**, since Sieverts include consideration of weight, while BTU does not. However, you can say:

0.000947817 BTU = 1 Joule, so 1 Sievert = 1 Joule / kg = 0.000947817 BTU / kg."

According to the Atomic Energy Commission (1967), 1,000 rads ["leads to the death of nearly any mammal"] = 0.0024 [small] calories. From our other definitions: Since 1 Sv = 100 rem and 1 rem ~ 1 rad, then 1 Sv ~ 100 rads, so 1,000 rads ~ 10 Sv. Therefore, 1 Sv = 0.00024 calories × [1 BTU / 252 calories] = 9.52 × 10^{-7} BTUs, or 1/1,050,000 [about 1 one-millionth of a BTU]. The AEC points out that the biological damage comes not from the energy itself but from "direct ionization or indirect chemical reaction."

Richard Crownover, M.D., Ph.D.: "Regarding the conversion from Sv to BTU[,] I think it is meaningless. The biological damage from radiation is not thermal. [It] arises from localized damage to specific molecules within cells that trigger apoptosis (programmed cell death) or breakdown in cellular replication . . . Radiation Oncology patients don't feel warmth at the treatment site when receiving partial body doses that would be lethal if delivered to the entire body."

1 mSv = 10^{-3} Sv = 1,000 microSv = 100 mrem = 100 mrads. One source reports that a chest X-ray is 0.1 mSv [0.0001 Sv]; another, that it is 0.7 mSv.

An American oncologist supplied the following X-ray dosage figures:

Dental: 0.005 milli [= 5 microSv; but see section 3F on the next page]
Chest: 0.1 milli
Spine: 1.5 millis
Lung: 7 millis
Abdomen and pelvis: 10 millis

1 microSv = 10^{-6} Sv = 100 microrem = 10^{-4} rem = 0.1 mrem

1 millisievert per year = 0.00011415525114 mSv/hr = 0.114 microSv/hr

Watt [w]—"One watt is produced when one ampere flows at an emf of one volt."

One watt = 1 **joule** per second = 10^7 ergs per second = 3 × 10^{10} disintegrations (or nuclear fissions) per second = 5 curies [see **becquerel**].

And since 1 Ci = 3.7 × 10^{10} disintegrations (or nuclear fissions) per second, 1 watt = 1.85 × 10^{11} disintegrations per second.

1 gram U-235 can be fissioned to produce 8.2×10^{10} watt-seconds = 950 kilowatt-days, but some is lost to non-fission capture; hence the practical amount of energy yielded is 800 kilowatt-days.

[See section 1, beginning on p. 521, for more information on watts, kilowatts, etc.]

F. Dental X-ray levels, Sacramento, around 9 a.m., February 10, 2015

(For a very different datum, see above, section 3E, p. 550, definition of **sievert**, oncologist's supplied X-ray dosage figure. See also pp. 253–54.)

DOSIMETER AND FRISKER READINGS AT VARIOUS DENTAL X-RAY SETTINGS, 2015

	Dosimeter	Frisker (MAX setting)
Before X-ray	124.0	n.a.
1 dose at .20 setting, aimed at the 2 devices	126.0	247 micros/hr
Scatter from equivalent dose (devices in corner of room, X-ray still aimed at chair)	128.0	247
Equivalent dose, aimed, through protective lead apron	129.0	151.2
1 dose at .23 setting (used for molars), aimed	166.9	251
		Total time: less than 5 minutes
5 min in dentist's office	166.9	.317
5 min outside office	166.9	.228
Accrued background, 2 hours later	167.0	n.a.

MULTIPLES OF OUTDOOR BACKGROUND LEVEL AT DENTIST'S OFFICE, 2015

(0.228 microSv/hr = 1)

1
Outside dentist's office

1.39
Inside dentist's office. *Note: This difference may be meaningless. For all I know, some other patient was getting X-rayed during the five minutes when I roamed the corridor, incidentally passing the X-ray room.*

663.16
Intensity of aimed dose at 0.20 setting, through lead apron. *Actual dose received: 1 micro.*

1.08 thousand
Intensity of aimed dose at 0.20, unprotected by lead apron. *Actual dose received: 2 micros.*

1.08 thousand
Intensity of scatter from dose at 0.20. *Actual dose received: 2 micros.*

1.10 thousand
Intensity of aimed dose at 0.23. *Actual dose received: 37.9 micros.*

4. Coal

A. Coal types, and a few coal products

Blast furnace gas—"A low-grade **producer gas**, made by the partial combustion of the **coke** used in the furnace and modified by the partial reduction of iron ore." A Japanese greenhouse gas inventory reports: "Since the composition of BFG is unstable, the emission factors for BFG were established with annually calculated values." (See p. 208.)

Coal—"A solid, brittle, more or less distinctly stratified, combustible carbonaceous rock, formed by partial to complete decomposition of vegetation . . ." [Another definition seems to say the opposite: "A combustible layered rock, produced by the accumulation and preservation of vegetable matter in a decay-resistant environment."] "In its most general sense the term 'coal' includes all varieties of carbonaceous minerals used as fuel, but it is now usual in England to restrict it to the particular varieties of such minerals occurring in the older Carboniferous formations." "Coal is the result of the transformation of woody fibre and other vegetable matter by the elimination of oxygen and hydrogen in proportionally larger quantity than carbon . . ." When combusted, this is the most greenhouse-gas-emitting of the fossil fuels—peat possibly excepted. Peat, coal and graphite are closely related. As you will now see, coal's categorizations are very inconsistent.

> **Anthracite** (sometimes called hard, black or stone coal)—[Sometimes subdivided, in increasing order of hardness, into semianthracite, anthracite and meta-anthracite.] "The highest rank of economically useable coal . . . has a heating value of 15,000 Btu . . . virtually all mined in Pennsylvania . . . Used primarily for space heating and generating electricity." "Used mainly for heating homes." "The rarest and most desirable form of coal, representing less than 1% of known coal reserves." "A smokeless coal of high fuel efficiency, though lower than semianthracite and semibituminous." In any case, anthracite as a general category is the hardest sort of coal, with a moisture content of 2–4%. "As a rule, density increases with the amount of carbon." However, even low-volatile **bituminous coal** sometimes contains more carbon than anthracite (see p. 208, "table of Calorific Efficiencies). See also "About Coal," II:18. Anthracite is said to be 90% carbon, but I have seen one proximate analysis of Virginia coal as low as 66.7%.* West Virginian "met" (**metallurgical**) coal was said to be anthracite, but according to the U.S. Energy Information Administration, "all the anthracite mines in the U.S. are located in northeast Pennsylvania." I am told that anthracite sells for the highest price, but this may not be so simple. One advantage to anthracite is that it is less liable to spontaneous combustion when stored. Among the stuff's detractors was Aldo Leopold who because he loved the smell of mesquite-roasted meat wrote bitterly: "Most poets must have subsisted on anthracite."

> **Metallurgical coal**—"The types of coal carbonized to make **coke** for steel manufacture, typically high in BTU value and

---

* As you saw from the formula for high heating value (pp. 534–35), a coal's sulfur content, and to a very minor extent its oxygen content, also bear on the energy it can give out.

low in ash content." "Bituminous coals used to make coke are classified as 'metallurgical' . . . Not all types of bituminous coal are adaptable . . ."

> **Bituminous coal**—"Sometimes called 'soft coal' . . . most commonly used for electric power generation. It has a heating value of 10,500–15,000 BTU . . . Mined chiefly in Appalachia and the Midwest." Composed of 80–90% carbon. Another source says 45–86% carbon. Or, as a third source has it: "Fixed carbon and volatile matter are about equal." My mid-20th-century *Mechanical Engineers' Handbook* subdivides this category, in descending order of heat output, into bituminous (2–17% moisture) of various volatilities, and into **subbituminous** A, B and C (10–30%). "West Virginia leads production," says a U.S. government source. Bangladesh also mines it. Most coal mentioned in *Carbon Ideologies* is bituminous.

> **Subbituminous coal**—"A dull black coal with heating value ranging from 8,300–11,500 BTU . . . Used primarily for generating electricity and for space heating." "A poor name for coal of higher rank than **bituminous**, although the name seems to imply the opposite meaning . . . Its heat efficiency is the highest of the coals." One source proposes: 35–45% carbon. Much in Wyoming. Moderate liability to spontaneous combustion.

When heated in oxygen-restricted furnaces, bituminous coal often "cakes" (fuses or plasticizes), a property needed for making **coke**, from which steel is manufactured. In early-20th-century England, bituminous coal was known as "steam coal," which does not precisely correspond with the West Virginia definition, for the latter apparently includes **anthracite**. "A medium soft classification . . . the most common and useful type mined in the U.S. . . . Used primarily for electric generation and for coke making for the steel industry." The retired Kentucky miner and mining inspector Stanley Sturgill remarked: "I'm not that familiar with anthracite; most of it's in Pennsylvania. Ours is called bituminous. It's black."—("The semi-anthracite coals of South Wales are or were known as 'dry' or 'steam' coal.")

> **Steam coal** [West Virginia]—"Used primarily for electricity generation; generally lower quality than metallurgical coal."

Lignite (brown coal)—"Nearly as soft as rotten wood." "Either markedly woody or claylike in appearance." "A brownish-black coal with generally high moisture content and low heating value (4,000–8,300 Btu per pound)." Usually 70% carbon or less (the previous sources say 25–35%). The moisture content is 30–45% (another source says 30–70%). "Mined primarily in the western U.S. and used for some electric generation and for conversion to synthetic gas." More likely than its subbituminous cousin to spontaneously combust.

Coal sizes—"Lump, egg, stove, nut, pea, stoker, slack, etc. . . . Slack coal is all the coal passing through the screen of a given mesh." Nut and slack size = 1 inch and larger.

Coal tar—"Coal yields about 6 per cent of its weight of tar." [See **coke** for procedure.] As indicated by the famous "coal tree" of the West Virginia Coal Association (see II:10), coal tar derivatives are extremely useful in the production of an astonishing array of goods. One is dynamite, whose starting point is the light coal tar oil called toluene. Xylene used to be employed in explosive shells. "Naphthalene is the most abundant pure hydrocarbon obtained from coal tar," advises an old explosives manual. The substance in question makes no bang in and of itself, but "nitrated naphthalenes . . . have been used in smokeless powder . . . and in high explosives for shells and for blasting."

Coke—The grey solid remaining after the volatiles and **coal tar** are heated out of bituminous coal in an oxygen-poor vessel. "Harder and denser than charcoal* . . . has a high heat content, and is a valuable fuel." "Coke is an excellent reducing agent . . . widely used in the smelting of metals," especially iron and steel. Coke also can produce calcium carbide, a precursor of acetylene. "Coke from any coker is nearly all carbon, but . . . a small percentage of very heavy, very complex hydrocarbons may wrap themselves around" it. High-grade coke comes from removing those impurities in a calciner. "Western coals are weakly coking as compared with those of the Appalachian region."

B. Coal mining and combustion terms

Bench—A layer of coal. Also called a seam.

Beneficiation—"The process whereby the extracted material is reduced to particles that can be separated into mineral and waste." The procedures involved are

---

* But the table of Calorific Efficiencies beginning on p. 208 reveals that coke and charcoal can have comparable HHVs. See p. 212, header **134**, and p. 213, header **147**.

"primarily mechanical, such as grinding, washing, magnetic separation, and centrifugal separation."

Blast furnace—"A shaft furnace in which solid fuel is burned with an air blast to smelt ore for continuous operation."

Dragline—"A huge piece of equipment with a large bucket suspended from the end of a boom that can extend more than 275 feet. The dragline can remove up to 200 tons of overburden in a single drag of its bucket across the work area..."

Highwall—"Unexcavated face of exposed overburden and coal in a surface mine. Highwalls must be recontoured following the extraction of coal."

Gasification—See **synthetic natural gas** on p. 568.

IGCC—Integrated gasification combined cycle. As the name implies, this technology can gasify coal before burning it, thereby removing impurities and reducing the carbon dioxide released into the atmosphere. In 2007, Tepco and the Tohoku Electric Power Co. jointly built an experimental IGCC generator at Nakoso, 60-odd km south of the infamous Nuclear Plant No. 1. In 2015 Mr. Sakakibara Kohji, a Tepco group leader, said to me: "IGCC, that is the most advanced technology in the world. We are now doing some test demonstrations for the future. We are going to make a large demonstration furnace. However, because it is the most advanced technology, it is the most expensive."

Longwall—A coal face in a deep or **underground mine**, under which heading see **longwall mine**.

Mantrip—[Also: "Man trip," "man car," "mancar."] The conveyance to bring miners into a deep mine. In Upton Sinclair's *King Coal* (1917) it was simply called a "trip."

Open pit mine—See **surface mine**.

Overburden—The waste earth, rock and vegetation removed from a mine. This stuff sometimes contains heavy metals. Several West Virginians told me sad stories of overburdens being dumped into the "hollers" where they used to live.

Pillar—As it sounds. Pillars of coal are left unexcavated in a deep mine to prevent the ceiling from collapsing. The side of a pillar is called a rib; so is an entrance wall. [See **underground mine: room-and-pillar mine**.]

Producer gas—Emitted when coal (or coke) is incompletely combusted. "Composed of carbon monoxide, hydrogen, air and steam [not to mention our friend carbon dioxide] . . . Poison. Dangerous fire hazard when exposed to flame."

Resources—The amount of coal which may be present in a deposit.

Reserve—The amount of a **resource** which is "recoverable using current technology."

> **Proven reserves**—"Can be recovered economically under current market conditions."

> **Probable reserves**—"Indicate a lower degree of confidence than proven reserves."

Seam—See **bench**.

Stowing—Packing excavated areas with **overburden** in order to reduce the risk of cave-ins, spontaneous combustion, etcetera.

Subsidence—"A necessary evil in . . . underground operation. This can however be minimized with the adoption of **stowing** methods."

Surface mine—"Permits a wide flexibility in production, which includes the ability to mine selectively and the potential for 100 per cent extraction [another part of the same document says "more than 90%"] of coal within the pit limits . . . Problems . . . include underground water managements [*sic*] . . . and undesirable environmental problems such as . . . dust . . . as well as waste disposal . . . [B]est suited to coal deposits of substantial horizontal dimensions." Placer mines are surface mines more germane to gold than to coal, but here is one coal-relevant variety:

> **Auger mine**—"The drill is placed at right angles to the coal seam and the cutting head is advanced into it, a full auger-flight length." This is "a special type of borehole mining." Yield rate: About 33%. "Sometimes employed to recover any additional coal left in deep overburden areas that cannot be reached economically by further contour or area mining."

One common type of surface mine is the open pit mine, whose environmental and aesthetic costs are high. It is, of course, cheaper in dollars.

Mountaintop removal mine—"Coal buried at or near the summit of a large hill or mountain is sometimes best reached by entirely removing the elevated area. After reclamation is completed, the once mountainous terrain . . . is often left flat, making it suitable for farming, recreation or other purposes."—"Rugged mountaintops and steep hollows are transformed into level or gently rolling land capable of agricultural, recreational, home, commercial and industrial development . . . Overburden can easily be segregated, including burial of toxic materials and recovery of suitable subsoils and topsoils to facilitate proper revegetation." Sometimes euphemized to "mountain-top mining. "As of 2005, 2700 ridges had been impacted by mountaintop mining in the Central Appalachian area."

Single bench mine—"A bench forms a single layer of operation above which coal and waste materials are excavated from the bench face . . . Bench heights normally range from 6 m to 10 m . . . but may approach 60 m in very special cases."

Multiple bench mine—"Can be employed in massive, thick bedded coal deposit" farther underground than with a single bench mine. "If the pit is over 10 or 20 m deep, more than one bench probably will be needed . . . Benches are normally used for roadways either forming a spiral to the bottom of the pit or with ramps between the horizontal benches. Bench widths are also designed to provide protection for men and equipment from small slope failures."

Strip mine—[Also: "Contour mine."] More appropriate for thinner seams (0.6–10 m). "Usually accomplished by removing the overburden and coal from a strip across one dimension of the deposit. A parallel strip [is] then excavated in the opposite direction and the overburden . . . placed into the strip previously mined." "Mining then continues laterally, circling the hillside and proceeding uphill."

Valley fill mine—[Also: "Head-of-hollow fill mine."] "A method where spoil is placed in narrow, V-shaped, steep-sided hollows free of mine openings or natural springs and located adjacent to the mining operation, is the technique most often employed in conjunction with mountaintop removal."

Tipple—The spot outside a mine where the coal was tipped out of the cars. "Now refers to the surface structures of a mine, including the preparation plant and loading tracks."

Underground mine—"If the depth of a coal deposit is such that the removal of the overburden makes surface mining unprofitable, [the] underground method should be considered . . . Factors . . . in choosing between a vertical or inclined shaft include the type of coal deposit and its depth." Underground mines are of either the pillar or wall type, as follows:

Pillar type:

Block mine—"A series of entries, panel entries, rooms and cross cuts is driven to divide the coal into a series of blocks of approximately equal sizes which are then extracted on retreat."

Room-and-pillar mine—"Cross-entries and panel entries are driven to 'block out' large panels of coal and rooms are turned off, usually at right angles from the entries." Rooms and pillars are first "fully developed in a section," going from near to far, and then pillars are mined out from far to near, the miners having previously installed anti-caving posts. About 50% of the coal thus mined is left behind. This method is somewhat prone to eventual subsidence.

Wall type:

Longwall mine—"Longwall mining employs a steel plow or rotating drum, which is pulled mechanically back and forth across a face of coal that is usually several hundred feet long. The loosened coal falls onto a conveyor . . ." "Two parallel headings are made 100–200 m [328–656 feet] apart and at right angles to the main heading. The longwall between the two headings is then mined away from the main heading . . . A moveable roof support system . . . advances as the coal is mined and allows the roof to collapse in a controlled manner behind it." "The retreating method is almost exclusively used in the United States, whereas the reverse is true in most foreign countries." Longwalling ("the most capital intensive but also the most capital efficient method of getting coal out of the ground") leaves less coal behind (about 30%) than could a **room-and-pillar mine**. Hence it is far more prone to subsidence. The ill-starred Upper Big Branch mine (II:58) was a longwall operation.

Longwall mine with caving—The procedure described above. "Longwall" tends to mean "longwall with caving."

Longwall mine with stowing—The mined areas are filled in with waste to prevent ground subsidence.

Shortwall mine—The primary difference between the two methods is the length of the working face. "In [the] short-wall system the maximum working face length is normally 45 to 55 m" [= 147.6–180.4 ft].

C. Some common coal pollutants [with related water terms, including pH] and coal diseases

Acid mine drainage—[Also: "AMD."] See below, this section, **yellowboy**. See II:152.

Aluminum—A dangerous aquatic pollutant from coal. One mid-20th-century chemical analysis of "five coal ashes covering a wide range of fusibility" found aluminum oxide contents from 19.6 to 30.6%. See II:152.

Black lung—Pneumoconiosis or coal-worker's pneumoconiosis (CWP), a debilitating and often fatal lung condition caused by breathing coal dust. It killed 100,000 American coal miners in the 20th century. From a mining methods handbook, 1978: "In the United States, the number of permanent disabilities and deaths of coal miners due to CWP is 3.5 times the number of disabilities and deaths due to all other mine incidents." Senator Jay Rockefeller, 2013: "We thought, at one time, we had Black Lung on its heels. We were wrong." See II:213.

Coal ash—"The inorganic residue that remains after burning the coal in a muffle furnace the final temperature of which is between 700 and 750 C." "The ash contains heavy metals at levels toxic to marine life and to all who consume river-dwelling creatures." [Sometimes coal ash is kept in slurry ponds, but the phrase "coal slurry" most often seems to refer to **coal sludge**.] "America's coal plants produce 140 million tons of ash each year." See II:184 for mention of spill in Dan River.

Coal dust—Nowadays defined as anthracite particles. A cause of black lung. At certain concentrations the stuff becomes highly inflammable. My grandfather's *Mechanical Engineers' Handbook* (1958), which did not restrict its definition to anthracite, reported, in the following order, ignition temperatures declining from 635 to 455° Celsius, and maximum explosive pressures increasing from 45 to 95 pounds per square inch: low, medium and high volatile coal, then, most dangerously, subbituminous coal.

Coal sludge—Also called coal slurry. "The liquid waste created when coal is washed and processed" so that it will pollute less when it is burned. The washing "happens after the coal is mined and before it is shipped to power plants." It "contains toxic chemicals used to wash coal and heavy metals," including "mercury, lead, arsenic, selenium, chromium, cadmium, and boron." Kept in artificial lakes behind impoundment dams or else injected underground, it "can poison drinking wells." See II:54–58 for testimony on the Buffalo Creek flood.

Conductivity—One measure of the metal content of a stream. The higher the conductivity, the more polluted it is. "It goes up as the water gets saltier. Fresh water in undisturbed Appalachian streams has a conductivity of under 100 and sea water has a conductivity of about 50,000." By 2015, "several Alpha subsidiaries" and "Consol's Fola Coal" were all found legally liable for conductivity pollution. According to Chad Cordell of the Kanawha Forest Coalition, "anything above 500 is bad." In the Kanawha Forest he had found conductivities ranging from 20 to 3,500. (See II:114.)

MCHM—4-methylcyclohexane methanol, used to clean coal. The toxicity of this chemical was unknown when it leaked into the Elk River in January 2014, contaminating the water supply of 300,000 West Virginians. It smells like licorice. See II:170 for discussion of the Elk River spill.

Methane—See section 5A, p. 567.

pH—A logarithmic scale from 1 (most acidic) to 14 (most alkaline). A pH of 7 is neutral. Six is 10 times more acidic than 7; 8 is 10 times more alkaline than 7. Thus a polluted West Virginia creek with a pH of 3 is 10,000 times more acidic than distilled water. *The pH of 1 gram of sodium hydroxide solution in 1 liter of water is 14.0; and of the same solution, 10 times more dilute, 13.0; seawater is around 8.5; human blood is 7.35 to 7.45; reasonably decent drinking water weighs in at 6.5 to 8.0; oysters, at 6.1 to 6.6; human urine, 4.8 to 8; sour pickles, 3.0 to 3.4; apples, a surprising 2.9 to 3.3; lemons, 2.2 to 2.4; limes, 1.8 to 2.0; human gastric contents, 1.0 to 3.0.*—According to the anti-mining activist Chad Cordell, the West Virginian legal standard for water was 6.0 to 9.0. "We typically see anywhere from 3.1 to about 7," he told me.

Yellowboy—"Iron and aluminum compounds that stain streambeds" as a result of coal mining. Yellowboy can acidify a watercourse to a pH of 3 or even 2.

Pyrite + oxygen + water → "yellowboy" + sulfuric acid.

D. Companies, acronyms and localisms relevant to coal in Appalachia, Bangladesh and the world

Acronyms appear first if my informants commonly used them in quoted interviews. Otherwise I have used the full name.

Asia Energy—[AEC.] The foreign-owned company that sought to dig an open pit mine in Phulbari. The successor to **BHP**. "They are Australian, but their headquarters is in England and their business license is in Honduras. The gentleman who came was Garry Enlight." See also **BHP** and **Global Coal Management**.

Barapukuria Workers' Union—A labor organization, said to be "apolitical," whose members all had some direct or indirect connection with the coal mine there. The union president asserted that the coal in Phulbari "will have to come out" (see II:245).

BDP—Bangladesh Development Power, a government utility. Most mined coal at Barapukuria was sold to BDP for electricity generation.

BGR—Bangladeshi Guards Rifles, or Border Guard Regiment, depending on whom I asked. I was told that they shot down the three *shaheeds* of Phulbari on August 28, 2006, and wounded up to 200 other people who were demonstrating against **Asia Energy**'s open pit mine.

BHP—Broken Hill[s] Properties [or Proprietary]. "They were based in the U.K. From 1998 BHP were doing surveys" around Phulbari, "and then they handed over the contract to **Asia Energy**." According to Nazrul Islam, former consultant of BHP, "BHP did not want to create another environmental disaster like Ok-Tedi Copper Mine in Papua New Guinea where it had to quickly abandon the mine and paid hefty compensation to the surrounding inhabitants."

DEP—Department of Environmental Protection, a West Virginia agency famous for its solicitude to the so-called *regulated community*.

Dinajpur—The district in which Phulbari and Barakupuria lay. "Dinajpur has always been known for its political uprisings. The people of Dinajpur had played a major role in Tebhaga Peasants movement in the 1940s . . ." "Dinajpur contains 30% reserve of . . . ground water of entire North Bengal."

Freedom Industries—The company responsible for the MCHM spill that endangered 300,000 households in West Virginia. It went bankrupt but arranged to pay itself (see II:176, 176n–77n).

Global Coal Management—"Asia Energy changed its name to Global Coal Management (GCM) after August 2006 bloodshed and uprising."

Holler—Since West Virginia is so mountainous, it abounds in secluded, originally lushly forested hollows, whose little homes and settlements have contributed a distinctive character to Appalachian identity. The hollers are sometimes used as dumping-grounds for the "spoil" (rubble) left over after mountaintop removal.

Massey Energy—This huge multinational coal company to many people's minds morally if mostly not legally was responsible for the Upper Big Branch mining disaster in West Virginia, after which it became Alpha Natural Resources and presently went bankrupt on the **Patriot Coal** model.

MSHA—Mine Safety and Health Administration, another feeble regulatory body whose existence gave offense to the poor old *regulated community*. [Sometimes the "A" meant "Academy."]

MT—Million tons.

MTR—Mountaintop removal.

National Committee to Protect Oil, Gas, Mineral Resources, Power and Ports—[Usually called just the National Committee.] A left-leaning organization that sought to halt **Asia Energy**'s open pit mine in Phulbari. "The National Committee was formed in 1998 when a group of leftists and other like minded individuals got together. It was started with the immediate objective of opposing the UNOCAL (A US-[b]ased company) stand that Bangladesh should export gas to India." That year it "organized a long march from Dhaka to Chittagong port" and allegedly stopped a 199-year lease to a Barbados company that pretended to be American.

OECD—Organisation for Economic Co-operation and Development. As of 2012, its 34 members were: Australia, Austria, Belgium, Canada, Chile, the

Czech Republic, Denmark, Estonia, Finland, France, Germany, Greece, Hungary, Iceland, Ireland, Israel, Italy, Japan, Korea, Luxembourg, Mexico, the Netherlands, New Zealand, Norway, Poland, Portugal, the Slovak Republic, Slovenia, Spain, Sweden, Switzerland, Turkey, the United Kingdom and the United States. *Carbon Ideologies* sometimes uses OECD statistics, especially those on coal. The International Energy Agency divided the world into OECD and non-OECD blocs. For instance, in 2012 it decreed: "The region previously called Latin America will now be known as non-OECD Americas."

OSM—U.S. Office of Surface Mining, whose ominous discoveries regarding coal slurry dams are mentioned on II:162.

Patriot Coal—This spin-off of Peabody Coal conducted MTR in West Virginia and elsewhere, then presently declared bankruptcy, paying massive bonuses to the higher-ups and renouncing significant pension obligations to its retirees.

Phulbari—A subdistrict of **Dinajpur**. Here Asia Energy attempted to dig an open pit coal mine. When I finished *Carbon Ideologies* the corporation was still trying. "The coal-field will extend over 135 square kilometers," and 50,000 to 200,000 people would be displaced.

Regulated community—A phrase employed by pro-coal entities to refer to industries (such as Big Coal) which are (supposedly unfairly) burdened with environmental and safety regulations. I have expanded it to refer to the businesses of all carbon ideologies. From the vantage point of an imagined 25th century, Hermann Hesse wrote: "It took long enough in all conscience for realization to come that the externals of all civilization—technology, industry, commerce and so on—also require a common basis of intellectual honesty and morality." That common basis was the antithesis of our *regulated community*.

Thana—[Bangladesh.] A police subdistrict. One such was Phulbari, where the three *shaheeds* were shot down in 2006.

UBB—Upper Big Branch, a deep mine near Montcoal, West Virginia. Site of the coal mine explosion that killed 29 men in 2010. The result of greed.

UP—Union of Peasants, or Peasant Organization, a left-leaning Bangladeshi entity that was present in Phulbari when **Asia Energy** tried to dig an open pit mine there.

WV-A—West Virginia American Water.

E. Units and conversions

Ton—A short or American ton is 2,000 pounds. A long or British ton is 2,240 pounds. A metric ton is 1,000 kg or 2,204.6 lbs. See also **refrigeration ton** in section 2, p. 536.

> The U.S. Department of Energy claimed in 1993 that each [short] ton of coal consumed at a power plant generates about 2,000 kilowatt-hours of electricity.

Ton of coal equivalent or **tce**—"One [metric] tonne of coal equivalent is 7 million kilocalories."—International Energy Agency. Since 1 kcal = 3.968 BTUs, then 1 tce = 27,776,000 BTUs. At .907×, a U.S. ton of coal equivalent would be 25,192,832 BTUs.

> 1 tce = 29.3 GJ [gigajoules] = 0.7 toe [ton of oil equivalent]

Ton of oil equivalent or **toe**—Again, this is a metric ton. One toe is 39,680,000 BTUs. At 0.907×, a U.S. ton of coal equivalent would be 35,989,760 BTUs. For more discussion of this unit, see section 6E, p. 577.

Some fuel conversions:

Crude oil:	1.034 toe
Gasoline:	1.128 toe
Jet fuel:	1.133 toe
Liquid petroleum gas:	1.195 toe

1 trillion cubic feet of natural gas *roughly* = 25 mtoe [million tons of oil equivalent].

1 toe = 11.6 MWh

> **To convert from BTUs of electricity to pounds of oil, coal or natural gas needed, assuming a power plant efficiency of 30%:**
>
> For oil, divide BTUs by 17,995. [In other words, multiply × (1 toe [1 U.S. ton of oil / 35,989,760 BTUs]) × [2,000 lbs / ton]. To express power plant

inefficiency, multiply × 3, for a final figure of [**BTUs needed / 5,998**] = **lbs of oil** required.

For coal, divide BTUs by 12,596. [In other words, multiply × (1 tce [1 U.S. ton of coal / 25,192,832 BTUs]) × [2,000 lbs / ton]. To express power plant inefficiency, multiply × 3, for a final figure of [**BTUs needed / 4,198**] = **lbs of coal** required.

For natural gas, divide BTUs by 24,021. [In other words, multiply × 1 cu ft / 1,000 BTUs] × [0.04163 lbs/cu ft = density of methane]. To express power plant inefficiency, multiply × 3, for a final figure of [**BTUs needed / 8,007**] = **lbs of natural gas** required.

5. Natural Gas and Fracking

A. *What it is*

Liquefied petroleum gases—[LPG.] The butanes, propane and C_3H_8, all of which "can be liquefied at reasonable pressures."

Methane—CH_4. [Marsh gas.] "A gas similar in nature to natural gas . . . present in most coal seams . . . formed by the decomposition of the organic matter in coal"—or, if you like, "the end product of the anaerobic . . . decay of plants." "A major coal mining nuisance and safety hazard because of its highly explosive nature . . . [Can be] use[d] as an energy resource." Indeed, it has a **high heating value** (see p. 534) between that of propane and of rocket fuel (see p. 216, header **259**). Also employed to make methanol and ammonia. In the U.S., a mine should be shut down if the methane concentration reaches 1.25% "for a sustained time," with "sustained" remaining conveniently undefined. Methane is a constituent of solid coal—and also, of course, a dangerous greenhouse gas. "As much as . . . 400 trillion . . . cubic feet of methane might be recoverable from U.S. coal beds." It often leaks from natural gas pipelines, abandoned mines, and even processed coal. "If the methane is accompanied by ethane and butane, it comes from oil and gas fields. If not, then it probably comes from somewhere else." "Methane is colorless and, when liquefied, is less dense than water." Molecular weight: 15.9 grams; **heat of combustion** (see p. 533): 213 kilocalories per mole. Global warming potential: About 86× carbon dioxide's over the first 20 years; about 26× over the first 100. See also II:307–16.

Natural gas—"The ideal fossil fuel—clean, efficient, economical, and available in generous supply in North America . . . Produces mostly carbon dioxide and water, with only trace amounts of pollutant gases . . . Believed among many to have been formed a billion years ago by thermal decomposition of buried organic matter (e.g., dead plants and marine organisms) covered with rock and mud."— "The main component of all natural gas is **methane**," but there may also be ethane, 2-butane, n-butane, two kinds of pentanes, hexane, heptane, octane, decane, nitrogen, carbon dioxide, hydrogen sulfide, water and air.

Quad—A quadrillion BTUs.

Synthetic natural gas—[Often referred to by abbreviation, SNG.] Almost entirely methane; most often derived from gasified coal.

B. Gas extraction and refining terms, including those for fracking

Fracking—This gas and oil extraction process begins after initial drilling. Although fracking is associated with horizontal drilling (sometimes for astonishing distances), it has also been employed in vertical wells. "Fracturing the formation is in reality splitting the rock by using fluids and high pressure to create a near-vertical fracture that may extend several hundred feet in two directions at 180° from each other, away from the [oil or gas] well . . . Fluid flows from the boundaries of the reservoir toward the wellbore." Another definition says that the procedure employs "water, sand and trace chemicals" to split the rock. "In effect, fracking creates additional permeability in a producing formation, enabling the oil or gas to move more easily toward the wellbore."

Frack[ing] fluid—"90 percent water, 9.5 percent sand and often less than 0.5 percent of a chemical mixture, many of the agents similar to what is used in household cleaning and cosmetic products." So says a fracker. An anti-fracker writes that "fracking fluids contained 750 chemicals, including benzene and lead. Fracking fluids even contained diesel fuel . . ."

Injection wells—"Disposal wells for contaminated drilling wastewater or brine injected in deep rock formations."

Produced water—"Water from the rock formation" plus fracking fluid, all of which is "discharged back up the well . . . [to become] a waste product that may include toxic chemicals and pollutants leached from the rock."

Separator—"A piece of equipment such as gas production unit used to separate one substance from another when they are intimately mixed, such as . . . oil from water, or oil from gas."

Separator tanks—Tanks located at the well site used to separate oil, gas and water before sending each off to be processed at different locations.

Stripper well—"An **unconventional gas well** incapable of producing more than 90,000 cubic feet of gas per day during any calendar month . . ."

Unconventional formation—"A geological shale formation . . . where natural gas generally cannot be produced at economic flow rates or . . . volumes except by . . . hydraulic fracture treatments or . . . multilateral well bores or other techniques to expose more of the formation . . ."

Unconventional gas well—"A bore hole drilled . . . for the production of natural gas from an **unconventional formation**."

Vertical gas well—"An **unconventional gas well** which utilizes hydraulic fracture treatment through a single vertical well bore and produces natural gas in quantities greater than that of a **stripper well**."

C. Some fracking pollutants

BTEX—Benzene, ethylbenzenes and the xylenes [all considered as **VOCs**].

VOCs—Volatile organic compounds. This grab bag category of ubiquitous pollutants was linked to global warming and also to health problems. See II:352.

For other substances, you had better ask Halliburton. [Some fracks employed hydrochloric acid, polyacrylamide and isopropanol.]

D. Places, entities and acronyms relevant to fracking in Colorado

COGCC—The Colorado Oil and Gas Conservation Commission.

Noble Energy—A U.S. subsidiary of the "Hong-Kong-based," "Singapore-listed" Noble Group. In 2015 this entity was a prominent if not preeminent fracker in Weld County.

NORM—Normally occurring radioactive materials. This innocent-sounding acronym refers to naturally radioactive underground substances—which through fracking sometimes come up to the surface, rendering the surface unnaturally radioactive.

E. Units and conversions

Cubic foot—A standard measurement for natural gas:

> **1 cf = 1,000 BTUs**

1 billion cubic feet (Bcf) = 1 trillion BTUs.

Therm—100,000 BTUs.

Quad—A quadrillion BTUs. "For natural gas, roughly equivalent to one trillion . . . cubic feet."

6. Oil

A. Some oils, derivatives and relevant terms

Carbon content—The carbon content of petroleum, gasoline, kerosene, crude oil, and other liquid fuels is generally 84–88%.

Cracking—Using refinery techniques to break down long-chain oil molecules into specific useful carbon-based products of lower molecular weights.

Crude oil—"Raw petroleum as it comes from the earth . . . A bitumen of liquid consistency, comparatively volatile . . ." Crude petroleum has been classified as a "liquid mineral." "The higher the crude, the higher the API gravity."

API gravity = Degrees API = (141.5/specific gravity at 60° F) − 131.5. "The logic of this arcane formula has been lost in antiquity." The API is the American Petroleum Institute. See also **specific gravity,** p. 577.

The API of water = 10°. This corresponds to a density of 8.328 lbs/gal and a specific gravity of 1.00.

Heavy crude—"Defined here as oil of 20° gravity API or heavier."

"Crude oil is largely made up of flammable hydrocarbons."

> **Hydrocarbons**—"Chemical mixtures of around 12 percent HYDROGEN (light gas vapor) and 82 percent CARBON (heavy black solid.) Hydrocarbons include thousands of different compounds." My high school chemistry book made it even easier: "Chemicals composed of only the two elements—hydrogen and carbon—are called hydrocarbons." When I was alive, hydrocarbons were frequently used to make herbicides and pesticides, as surfactants, protective coatings, synthetic fibers such as nylon, flavor enhancers, plastics and suchlike petrochemical conveniences.

"Hydrocarbons in gasoline have from 4 to 14 carbon atoms per molecule, and lubricating oils have up to 30 or more carbon atoms per molecule."

See section 6C, beginning on p. 575, for mention of hydrocarbons as pollutants.

Okie gas—This ultralight oil condensate (API of 50°) had its fame in the 1930s. It could be burned in a car like gasoline.

Light crude—More than 30° API. Often colorless. "The light crudes tend to have more gasoline, naphtha and kerosene." "Probably has few percent natural gas liquids."

Medium crude—20 to 30° API. Color may vary between greenish-yellow to reddish.

Heavy crude—10 to 20° API. This grade is usually black. "The heavy crudes tend to have more gas oil and residue." "Has little or no natural gas liquids and a high percentage of the heavy hydrocarbons."

Sweet crudes contain less than 0.05% sulfur, while **sour crudes** have 1.5% or more. Sweet crudes smell like gas, sour crudes smell like rotten eggs and aromatic crudes "have a sickly fruity smell." In 2014, the Bakken formation in North Dakota happened to be one strong producer of light sweet crude, while Venezuela was afflicted with "heavier crudes that fetch less than international benchmarks."

Diesel fuel—"Varies greatly in its characteristics, ranging from light distillates which are practically heavy kerosenes, to . . . crude oils." "Diesel fuel contains [12 percent] more heat energy per gallon than gasoline," but only if measured per cubic foot. In the table of Calorific Efficiencies on p. 214, header **201–217**, I measure per pound, to be consistent with coal measurements.

> **Diesel grade 1-D**—For high-speed engines, wide speed variations and low fuel temperatures.

> **Grade 2-D**—Less volatile and higher in "heat energy" per gallon than 1-D. Used for high loads and uniform speeds.

> **Grade 4-D**—"More viscous distillates and blends of these distillates with residual fuel oil." Used in low- and medium-speed engines; best for sustained loads at constant speeds.

Gasoline—"A complex mixture of hydrocarbons which distill within the range of 100 to 400 [degrees] F[ahrenheit]." A member of the pentane group, which are all liquids ("commonly called distillates"). "Mildly toxic by inhalation . . . Questionable carcinogen . . . Some addiction has been reported from inhalation of fumes. Even brief inhalation of high concentrations can cause a fatal pulmonary edema." "The combustion of gasoline, pound for pound, produces more total energy than dynamite." A significant source of climate change.

Kerosene—"Less volatile than gasoline . . . obtained by continuing the distillation of crude petroleum after gasoline has been removed." "Except for severe arctic areas where Jet B fuel is used, essentially all civil aviation uses kerosene fuel."

Naphtha—"Refers to the light end fractions distilled from crude petroleum . . . a complex mixture of hydrocarbons with an end boiling point of about 165° C . . ."

Nylon—The name of this famous polymer (invented in 1938) derives from "New York" and "London," the homes of the two relevant laboratories. Derived from adipic acid [see p. 128].

Octane rating [of gasoline]—"Indicates its ability to resist knocking or pinging."

Oil—It comes, perhaps, from plankton in ancient inland seas. [But see Sam Hewes's caveat on II:486.] Since the water was poorly oxygenated, the dead plankton could not decompose. The pressure of sediments accumulating over it then produced heat, "initiating chemical reactions that convert the dead organic matter into oil. Bacteria appear to play a much greater role . . . than was previously thought. The ideal temperatures . . . are between 50° and 180° C, which are usually found at depths of between 2 and 4 kilometres." "Of the light oils the most important is known as petrol. It is not a definite chemical compound. It is a mixture of various hydrocarbons of the paraffin and olefine series, produced from the distillation of petroleum and paraffin oils."

"About two to two-and-one-half barrels of oil can be produced from a [short] ton of coal."

For other conversions from barrels to tons, see **barrel** in section 6E, pp. 576–77.

"Between four and five [metric] tons of tar sand must be processed to make one barrel of oil."

Tar sands—"Also known as oil sands . . . A combination of clay, sand, water, and bitumen (a heavy viscous oil) that is either mined or recovered by injecting steam or another heat source underground." "Heavy hydrocarbons mixed with sand and dirt . . . One [processing method] is to submerge the tar sand in hot water and steam. This forms a . . . slurry that melts and liquefies the tar," which rises to the surface.

"In 2009, 16.9 billion barrels, or about 99% of Alberta's total proven oil reserves, were attributed to the oil sands—around 13% of total global oil reserves . . . Alberta ranks second after Saudi Arabia in proven oil reserves . . ."—*Canada Yearbook 2011.*

"With today's refining methods, almost 50 percent of each crude oil barrel can be made into automotive gasoline."

A table (*ca.* 1987) of "average yield from a barrel of crude oil" shows 15.7% going into "other products and losses."

Oil shale—"A convenient expression used to cover a range of materials containing organic matter which can be converted to crude shale oil and gas by heating." Or, if you prefer, "a sedimentary mineral that contains kerogen, a mixture of complex, high molecular weight organic polymers." Can yield various motor fuels and other crude oil derivatives.

Petroleum—"A thick flammable dark-yellow to brown or green-black liquid . . . Insol[uble] in water; sol[uble] in benzene, chloroform, ether . . . Questionable carcinogen with experimental carcinogenic, neoplastigenic, and tumorigenic data by skin contact." Unquestionable global warming agent.

Petroleum distillate—"Mildly toxic by inhalation and ingestion."

Petroleum spirits—"A poison by intravenous route. Mildly toxic by inhalation."

Plastic—A category of artificial carbon-based **polymer**. "Often designed to mimic the properties of natural materials . . . Produced by the conversion of natural products or . . . synthesis from primary chemicals generally coming from oil, natural gas, or coal. Most plastics are based on the carbon atom. Silicones, which are based on the silicon atom, are an exception." Plastics are one of the **"big five"** products that use the lion's share of industrial energy [see below].

> **Polymer**—A compound made up of large molecules which in turn are composed of numerous small molecules. Carbon is famous for its ability to polymerize. Natural carbon polymers include proteins.

> **The "big five"**—Aluminum, steel, cement, paper and plastics (*ca.* 2010). See p. 134.

Residual fuel oil—"The source fuel for combustion in electric generating plants. Residual oil is what remains after lighter hydrocarbons, such as gasoline, have been extracted from crude oil."

Synthesis gas—"A mixture of carbon monoxide . . . and hydrogen . . . that is the beginning of a wide range of chemicals." Often derived from oil.

Tar sands—See **oil**.

B. Oil extraction and refining terms

For definitions relevant to fracking (which can produce both oil and natural gas), see section 5B, p. 568.

Alkylation—"Changes small **hydrocarbon** compounds into larger ones." "Changes LIGHT GASES (bottled types) into gasoline." "One of the few refining processes that can produce . . . high octane . . . gasoline in large quantities."

Catalytic cracking—"Moderate heat" and a catalyst ("usually a special clay powder") are employed to transform "heavy fractions of crude oil into lighter ones (gasoline)."

Condensate—"A vaporous mixture of gas and water that is far less toxic and less flammable [than natural gas]." "The liquid resulting when a hydrocarbon is subjected to cooling and/or pressure reduction. Also, liquid hydrocarbons condensed from gas and oil wells."

Distillation—See **fractionating**. Distillation alone can turn 20% of a given quantity of oil into gasoline. "With today's refining methods, almost 50 percent of each crude oil barrel can be made into automotive gasoline." (Both of these are 1987 figures.) See **alkylation, catalytic cracking, hydrocracking, polymerization**.

Fractionating—The procedure of distilling crude oil into its derivatives by making use of the different boiling points of the latter. For example, propane boils at $-44°$ F, while butane boils at $+31°$ F. Butanes and lighter products are fractionated, off at under $90°$ F, gasoline at $90–220°$, then naphtha, kerosene, gas oil at their own unique intermediate temperatures, and finally the residues at $800°$ and higher.

Hydrocracking—"Separation" of oil molecules "occurs in the presence of hydrogen gas and a catalyst." This method is commonly employed to produce unleaded premium gasoline.

Polymerization—"Produces gasoline out of very light gases," but the result is feeble in antiknock properties.

Roughneck—"A specific term for a job on the rig crew, one up from roustabout, but also a generic term used for anyone who survives the job and graduates . . ."

Thermal cracking—An outmoded process which "used extreme heat—above $100°$ F . . .—and extreme pressures of around 1000 psi . . . to break large hydrocarbon compounds into smaller gasoline type hydrocarbons."

Wellhead—"A device that controls pressure in the well and flow of gas or oil from the well."

C. Some common oil pollutants and oil diseases

Car-generated pollution—In 1972 "the main source of pollution" from cars was "from incomplete burning of the gasoline." Until the 1970s the "three most dangerous pollutants" were given as hydrocarbons, carbon monoxides and nitrogen oxides. In "the early 70s" **lead** additives joined the list. Carbon dioxide passed unmentioned.

Hydrocarbons—[For more on hydrocarbons, see sec. 6A, p. 571.] Hydrocarbons and nitrogen oxides (both of which occur in internal combustion vehicle exhaust) combine in the presence of the ultraviolet radiation we receive from the sun. One chemical product is ozone (which protects us from global warming and radiation when it is in the upper atmosphere); another is peroxyacetyl nitrate. These two substances harm mucous membranes such as lung tissue; they can also kill plants.

Lead—"Around a teaspoon of lead is added to a gallon of leaded gasoline. This can increase octane and antiknock value of fuel as much as 7 or 8 octane numbers." Lead "slows down the burning time of the fuel and eliminates the destructive 'knock.'" Furthermore, it "allows the use of more efficient fuels and so cut[s] the cost of operating the vehicle." But it causes harm to brains and nervous systems, hence the widespread banning of this additive.

Volatile organic compounds [VOCs]—Various polluting gases given off from oil and gas wells; many are said to harm human health; they also play a role in global warming; section 12 (pp. 591–97) and II:352.

D. Oil companies, acronyms and localisms: Abu Dhabi Emirate

ADCO—Abu Dhabi Company for Onshore Petroleum Operations Ltd.

ADNOC—Abu Dhabi National Oil Company.

EAD—Environment Agency Abu Dhabi, whose air quality reports are referenced in the Emirati chapter (see II:540n).

Halliburton—Listed in Abu Dhabi yellow pages under "Oilfield Equipment Suppliers" and "Oilfield Contractors & Services."

Takreer—"In U.A.E. that name is very biggest company," said my interpreter Ravindra. The acronym meant "Abu Dhabi Oil Refinery Company."

Taqa—Abu Dhabi National Energy Co. PJSC.

E. Units and conversions

Barrel—"Conversion factors vary depending on oil source."

1 barrel of petroleum = 42 gallons = 159 liters

"To convert t/d to bbl/d, multiply by 7."

Barrels per metric ton of refined product:

Asphalt:	6.06
Residual fuel oil:	6.77
Kerosene:	7.75
Jet fuel:	8.00
"Motor gasoline":	8.50
"Natural gasoline":	10.00

Specific gravity—For many technical computations, the weight of a given amount of the compound, divided by the weight of the same amount of water. For a gas, the specific gravity is usually a volumetric comparison with air.

The **API gravity** [see p. 570] of an oil (expressed in degrees) = (141.5/specific gravity at 60°F) − 131.5.

Here are some APIs: of asphalt, 11; of heavy crude, 18; of light crude; 36; of gasoline, 60.

Ton of oil equivalent (toe)—"A tonne of oil equivalent is defined as 10^7 kcal (41.868 GJ), a convenient measure because it is approximately the net heat content of one ton of average crude oil. This unit is used by the IEA/OECD in its energy balances." (Note that "tonne" = "metric ton"; See p. 590.) For more discussion of this unit, see section 4E, p. 566.

Since 1 kcal = 3.968 BTUs, 1 toe = 39,680,000 BTUs.

1 toe = 1.429 tce [ton of coal equivalent] = 10 tons TNT equivalent = 11.6 MWh [megawatt-hours], or about 40 million BTUs.

"One metric ton of crude oil varies from about 1.5 per cent to 8 per cent above the value of 10^7 kilocalories, which is frequently referred to as one ton of oil equivalent." 1 mto[e] generates 4.4 terawatt-hours "in a modern power station."

7. Solar Energy and Light

Most of this section's longer entries were written by Dr. Canek Fuentes-Hernandez, my paid reader of the solar chapter (see p. 161). When I was finishing Carbon Ide-ologies, *solar power seemed our best hope. Hence the length and detail in these entries.*

1 [gram-mean] **calorie** per square centimeter = 1 **langley** = 3.685 BTUs/ft^2

Adoption constraints—"The problem[s] with the adoption of solar energy and other renewables (i.e. wind), are first and foremost related to economic and energy policy (e.g. how do we define the cost-of-power, economic subsidies, incentives to move from a centralized to a distributed model of energy production, decommissioning existing resources, etc., not to mention politics). From a technological perspective, the grand challenge facing renewables relates not to the abundance of these resources or the conversion efficiency of the associated technologies, but to electricity storage and capacity; since the grid demands are out-of-phase with the daily cycle of solar insolation, weather conditions and seasonal variations[;] and customers expect a reliable electricity output on demand."

Blackbody—See p. 527.

Candle and **candela**—"The candela (cd) is defined as the luminous intensity . . . in the horizontal direction of a standard lamp which is made and used in accordance with U.S. Bureau of Standards specifications . . . The luminous intensity of an ordinary sperm candle . . . in the horizontal direction is about 1 candle (cd)." "Every 1-cd point source of light emits 12.57 lm" = **lumens**.

1 international candle = 1/60 × [brightness of a **blackbody** [see p. 531] at the temperature of the freezing point of platinum = 1,769° Celsius]

Foot-candle—A **lumen** per square foot. A measure of visible light (used for engineering); hence not directly translatable into units of solar insolation. [The same obviously goes for lumens and luxes.]

However, 1 langley per minute = [approx.] 6,700 foot-candles

The relative brightnesses of moonlight and sunlight might respectively be 1/100 and 10,000 foot-candles.

1 langley = "1 calorie of radiation energy per square centimeter" = 1 cal-cm^2 = 3.69 BTUs/ft^2 = 4.184 × 104 J/m^2

"Radiation of 1 langley [per minute] is a reasonable average value to take for a tilted surface under a cloudless sky."

Over the course of an hour, 1 square foot receiving 1 langley per minute will take in 221 BTUs.

1 langley per minute = 221 BTUs per square foot per hour = 3.683 BTUs/ft²-min

Lumen—1 lumen [lm] is "that quantity of incident luminous flux" to produce 1 **foot-candle** on each point of a 1-square-foot surface. The efficiency of an electric light source is currently defined lumens per watt (of source divided by output).

Efficiency of Edison's lamp, 1879: 1.5 lm/watt

Efficiency of a standard warm white 100-watt fluorescent Mazda lamp, *ca.* 1958: 52.0 lm/watt

Lux—A **lumen** per square meter.

1 lux = 0.1 foot-candles

Solar absorption—"Electromagnetic radiation interacts with matter by being either absorbed, reflected or transmitted. Generally speaking, for the purpose of energy generation, we need electromagnetic radiation to be absorbed[, and it can be only] . . . if the energy carried by the electromagnetic field is equal to the energy associated with the transition from a ground state to an excited state of the material. The discrete nature of the absorption of electromagnetic radiation, is thus akin to the process of buying a product [i]n a store. We need to have the exact amount of money (electromagnetic energy) to purchase a specific product (excited state) for the store (the body that is absorbing the energy) to take it and make the monetary transaction. Each store has a wide variety of products of different prices . . . representing the band gap of the material. If we do not have enough money to buy this product, we can only go through the store without buying anything and the store becomes 'transparent' to us.

"Upon absorption, the energy contained by the electromagnetic radiation field generates excess energy in the material which needs to be dissipated; nature likes thermodynamic equilibrium . . . Two distinct phenomena can be identified which are always present when electromagnetic radiation is absorbed by a material.

"[1] Electromagnetic radiation generates heat. Energy absorbed by a body is dissipated through molecular or atomic vibrations leading to the generation of heat (radiant heat); causing an increased temperature, which in turn, increases the intensity of electromagnetic power radiated until thermodynamic equilibrium is achieved. Because molecules have vibrational excited states with energies in the infrared and microwaves . . . the process of "absorbing heat" from a source of electromagnetic energy (i.e. the sun) is most efficient at these, longer, wavelengths; which is why we use microwaves to heat our food instead of visible light.

"[2] Electromagnetic radiation generates . . . more electromagnetic radiation: Photoluminescence.

"Absorbed energy is dissipated through the reemission of electromagnetic energy. This process is never lossless and the reemitted electromagnetic radiation carries less energy than the incoming one. Energy loss is converted into heat and results in an increased temperature. Since molecules have electronic excited states with energies in the visible and UV range, the process of photoluminescence is most efficient at these wavelength ranges.

"We can generally say that the first phenomenon can be optimized to provide an avenue for the conversion of solar energy to thermal energy. Despite sounding counterintuitive, the second process can be optimized to convert solar energy directly into electricity. However, optimizing these two types of energy conversions require[s] materials with very different properties."

Solar constant—The rate at which solar energy strikes the top of our atmosphere:

2 langleys per minute = 2 cal/cm^2-min = 7.37 BTUs/ft^2-min

Solar efficiency—"Modern concentrated solar systems . . . have maximum theoretical efficiency values (limited by Carnot's principle, which states that the maximum efficiency of a thermal engine operating between a low temperature reservoir and a high temperature reservoir is given by one minus the ratio of these two temperatures) of around 70% times the efficiency of the solar collection by the receiver, which can be typically high."

Solar energy—"Solar energy impinges [on] the earth in the form of electromagnetic radiation; not heat. Electromagnetic radiation is a physical phenomenon that we identify as 'light' in the narrow spectral window that our eyes perceive. 'Light' nevertheless represents only a very small portion of the total electromagnetic spectrum emitted by the sun, or any black body radiator for that matter. Cosmic rays, X-rays, ultraviolet light, infrared light, microwaves, radio, etc. are all manifestations of electromagnetic radiation, only differentiated by the amount of energy they carry (higher energy goes with higher frequencies or shorter wavelengths)."

8. Environmental Terms Relevant to Resource Extraction

Aquifer—"A rock with both enough porosity and permeability to conduct significant volumes of water to a well or spring."

FLIR camera—Forward looking infrared camera. Used (especially by environmentalists) to photograph otherwise invisible chemical plumes from fracking wells.

9. Physical Data

For a few specific heats, see section 2, p. 536.

1 lb per U.S. gallon = 119.8264 kg per cubic meter = 0.1198264 kg per liter

Hence 1 kg per liter = 8.345 lb per gal

A. Weights and densities of fuels

This information will allow you to compare, for instance, the energy and greenhouse gas emissions of a cubic foot of methane and a gallon of gasoline.

Density of **aviation fuel:** 48.3817–52.4395 lbs/cu ft = 775–840 g/liter

Carbon Ideologies uses the arithmetical average: 807.5 g/L = 6.73 lb/gal.

Light distillate 850 kg/cc [= 7.09 lb/gal]

Medium distillate 876 kg/cc [= 7.31 lb/gal]

Density of **butane:** 0.1548 lbs/cu ft = 0.0206 lbs/gal = 2.48 kg/m³

Density of **carbon:** 165 lbs/cu ft = 2.25 g/cc

Density of **coal:**

> **anthracite:** 1.4 to 1.8 g/cm³ [another source gives 97 lbs/cu ft]

> **bituminous:** 1.2 to 1.5 g/cm³ [another source gives 84 lbs/cu ft]

> **lignite:** [1.0 g/cm³]; 78 lbs/cu ft

Density of **diesel fuel:** 7.036 lbs/gal. *Carbon Ideologies* plugs in this figure for computations involving commercial fuel oil, unspecified "crude oil," etc.

Density of **gasoline:** 45 lbs/cu ft = 6.152 lbs/gal *[In the mid-20th century this value was for a heavier grade; for the lighter stuff, density was 5.935 lbs/gal.]*

Density of **hydrogen** at standard temperature and pressure (STP): 0.005614 lbs/cu ft = 0.0899 (or, if you like, 9.0×10^{-2}) grams per liter or g/cm³

Density of "common engine **kerosene**" (1911): 6.7 lbs/gal. A 1958 figure was 6.8 lbs/gal.

Density of **manufactured gas:** Approximated by **methane's.** A *Carbon Ideologies* plug-in.

Density of **methane** at STP: 0.04163 lbs/cu ft = 0.716 g/liter

Density of **natural gas:** Approximated by **methane's.** Another *Carbon Ideologies* plug-in.

Specific gravity of Pittsburgh natural gas (@ air = 1.0): 0.61

Specific gravity of "coal gas": 0.42

Density of light Mexican crude **oil** [per PEMEX] = 52.127–55.748 lbs/cu ft = 835–893 kg/m^3

Density of **peat blocks:** 0.84 g/cm^3

Density of **plutonium:** 1235.971 lbs/cu ft = 19.84 g/cm^3

Density of **propane:** 0.12548 lbs/cu ft = 0.01639 lbs/gal = 2.01 kg/m^3

Density of **uranium hexafluoride:** 317.0915 lbs/cu ft = 5.09 g/cm^3 [another source gives 13.00 kg/m^3]

Density of **uranium-235:** 1183.642 lbs/cu ft = 19 g/cm^3

B. Weights and densities of other substances

Density of **air:** 0.0807 lb/cu ft at 32° F

Density of **steam** at 212° F: 0.588 kg/m^3

Density of **carbon dioxide** at STP: 1.977 g/cm^3

Density of **PBXN-109** aluminized explosive: 1.662 g/cm^3

Density of **TNT:** 1.654 g/cm^3

Density of **water:** 62.4245 lbs/cu ft = 8.34 lbs/gal at 39.1° F

10. Chemical Abbreviations

C—Carbon.

    C-14—A radioisotope used for "carbon dating" of ancient dead materials.

    $C_6H_{12}O_6$—Glucose, the chemical building block of cellulose.

    $C_6H_{10}O_5$—Cellulose.

CH$_4$—Methane, a dangerous greenhouse gas (commonly produced by anaerobic decomposition), commonly associated with garbage, sewage, rice farming and coal mining—where it can cause lethal explosions. Also, perhaps the cleanest-combusting fossil fuel—but a far worse global warming agent than **CO$_2$** when it enters the atmosphere uncombusted. See II:307.

CO—Carbon monoxide. A common emission from incomplete fossil fuel combustion; in other words, a typical vehicle-caused pollutant. Both a greenhouse gas with moderate warming power and a *precursor* (see p. 597) whose indirect effect is to increase levels of other GHGs.

CO$_2$—Carbon dioxide. The star of the show.

CO$_3^=$—Carbonate. [The minus signs indicate that this is a negatively charged ion.] From the simplified perspective of *Carbon Ideologies,* the carbon-oxygen portion of certain carbon compounds. Dr. Pieter Tans to WTV, explaining why a runaway greenhouse effect to Venusian conditions hopefully will not occur: "The Venus atmosphere is almost entirely composed of CO2 [no subscript in original] . . . On Earth almost all of that carbon is in rocks, in the form of solid Ca/Mg carbonates."

Ca—Calcium.

CaCO$_3$—Calcium carbonate. The main ingredient in limestone, oyster shell, marble, coral, chalk, etc. Employed in West Virginia to deacidify mine runoff in streams and drinking water. A buffer against runaway climate change. F. W. Taylor, 2001: "Most of the carbon dioxide that was originally outgassed . . . no longer resides in the atmosphere; it has been lost by processes that have led to the formation of carbonate deposits (mostly limestone), coal and petroleum." See **CO$_3$**– above.

Mg—Magnesium.

    MgCO$_3$—Magnesium carbonate. Oxidation will create this coating on exposed **Mg.**

        CaCO$_3$,• MgCO$_3$ —Dolomite, a commonly occurring building stone.

Cs—Cesium. A relevant headline from Fukushima: **Cesium levels in water, plankton baffle scientists.** See Table 1, p. 537.

> **Cs-134**—A common fission product, prevalent in nuclear waste. Less prevalent at Fukushima, shorter-lived [half-life = 20.65 years] than **Cs-137**, and probably less hazardous.

> **Cs-137**—Another common fission product, also present in nuclear waste, and also released at Fukushima, where it was especially dangerous. Its longer half-life [30.2 years] meant that it contaminated that area much longer than **Cs-134.**

GHG—Greenhouse gas[es]. For a cursory list, see the table of Comparative One-Century Global Warming Potentials on p. 176.

H—Hydrogen, the most common element in the universe.

> H_2—The configuration of a hydrogen molecule: 2 hydrogen atoms together.

> H_2O—Water. As atmospheric vapor, a significant global warming agent.

> **3H** or **T**—Tritium. A dangerous but relatively shortlived radioisotope, used to make certain fission bombs. Released in vast and continuing quantities at Fukushima. See Table 1, p. 537.

> H_2S—Hydrogen sulfide. A decay product of certain proteins; a combustion product of some coals.

> H_2SO_4—Sulfuric acid. A ubiquitous industrial chemical.

> **I-131**—Iodine-131, a shortlived fission product of nuclear reactors. Released at Fukushima, with possible implications in increased thyroid risk, especially for children. See Table 1, p. 537.

K—Potassium, the manufacture of whose fertilizer compounds [potashes] emitted relatively small amounts of greenhouse gases but required considerable [in our day fossil-fuel-powered] thermodynamic work.

Kr-85—A shortlived radioactive krypton isotope commonly emitted by nuclear power plants. Released in the Fukushima accident.

Mol—Mole[s]. For definition, see **heat of combustion**, p. 533, and also see section 12 beginning on p. 591.

N—Nitrogen, the most common atmospheric element. Essential in fertilizer compounds, because there can be no proteins without it.

N_2—The configuration of a nitrogen molecule: 2 nitrogen atoms together. In the atmosphere N_2 is innocuous.

N_2O— Nitrous oxide, a dangerous greenhouse gas. Like CH_4, a greenhouse gas linked to sewage, compost, etcetera; also a vehicle pollutant and a dental anesthetic.

NF_3—Nitrogen trifluoride, a dangerous greenhouse gas.

NH_3—Ammonia, a common greenhouse gas with very low warming power. Associated with nitric acid manufacture, decomposition of living things, oxygenless heating of bituminous coal. A refrigerant. Ammonium compounds supplied nitrogen to soils in such fertilizer forms as:

NH_4NO_3—Ammonium nitrate

$(NH_4)_2SO_4$—Ammonium sulfate

NOx or NO_x—Nitrogen oxides; their plurality is represented by the "x." Like **CO**, they are both greenhouse gases and **precursors** (see p. 597). Confusingly enough, at times they also exert cooling effects.

O—Oxygen.

O_2—The configuration of an oxygen molecule: 2 oxygen atoms together.

MOX—Mixed oxide nuclear reactor fuel (plutonium + uranium oxides).

P—Phosphorus, another agricultural necessity. The manufacture of phosphate fertilizers released significant CO_2.

Pu—Plutonium, an artificially produced "transuranium element" (small natural quantities have also existed). A decay product of **U-239**. [For various Pu isotopes, see Table 2, p. 539.]

Ra—Radium. This radioactive element occurs in ores of **U**, but in such small concentrations that its isolation there is impractical.

Rn—Radon. A naturally occurring radioactive gas in soils, basements—and, of course, uranium tailings. A decay product of, most significantly, **U-238**.

S—Sulfur.

> **SF$_6$**—Sulfur hexafluoride, a scarce but very dangerous greenhouse gas used in various manufacturing processes.

> **SO$_2$**—Sulfur dioxide. The best-known combustion product of sulfur. Released by volcanoes, fuel combustion and many industrial processes. Used to make **H$_2$SO$_4$**. In small doses, a cooling agent; in larger ones, a warming agent. See II:26.

Si—Silicon.

Sr-90—A very dangerous radioisotope that concentrates in bones. Released in the Fukushima accident. See Table 1, p. 537.

TiO$_2$—Titanium dioxide, a white pigment applied in toothpastes, paints and papers. The manufacture of TiO$_2$ releases significant amounts of **CO$_2$**.

U—Uranium. This element and all its isotopes are radioactive.

> **U-235**—A rare isotope, artificially produced to make atom bombs and enriched reactor fuel.

> **U-238**—Uranium's most common isotope. Can be used to make **Pu**.

> **U-239**—An unstable uranium isotope.

> **UF$_6$**—A common nuclear reactor fuel.

Xe-133—A shortlived radioactive xenon isotope commonly emitted by nuclear power plants. Released in the Fukushima accident.

11. Lakhs, Crores and Other Enumeration Terms, with Standard Conversions

Ar—[Japan.] An archaic unit of measurement; none of my interpreters could define it.

Celsius to Fahrenheit—($\frac{9}{5}$° C + 32)

Crore—[Southeast Asia.] 10^7 [= 10 million].

Division nomenclature:

> **X per Y** [for example, miles per gallon, or milligrams emitted per liter combusted] = X/Y = X Y^{-1} [as in: "It has been reported that 20 to 100 Tg y^{-1} of CH$_4$ is emitted from paddy fields"].

Exa—10^{18} [= 1 quintillion]. One exajoule [EJ] = 10^{18} joules = 0.9481 quadrillion BTUs = 0.948 **quads** (see p. 568).

Fahrenheit to Celsius—$\frac{5}{9}$(F° – 32)

Foot = 0.3048 meters

> 1 square foot = 0.092903 square meters
>
> 1 BTU per minute per square foot = 189.226 watts/m^2
>
> 1 cubic foot = 0.0283166 cubic meters

Gallon to liter:

> 1 U.S. gallon = 0.83 Imperial gallons = 3.78 liters. All of these measures were commonly used for fuel. But only U.S. gallons were employed in this book.

Giga—10^9 [= 1 billion].

> 1 gigagram (Gg) = 1 metric kiloton (kt or KT) = 1,000 metric tons = 2,204,600 lbs

Gram:

> 1 g = 0.000001 metric ton = 1/453.59 lb

Hectare—10,000 square meters = 2.47 acres

Kilo—10^3 [= 1,000].

Kilogram—10^3 grams, or 2.2046 pounds.

This holds for gases as well as for liquids. Methane's density of 0.716 grams per liter, or 0.000716 kilograms per liter, thus works out to 0.00597502 pounds per gallon.

Lakh—[Southeast Asia.] 10^5 [= 100,000].

Liter to gallon:

1 liter = 0.265 U.S. gallon = 0.220 Imperial gallons

Maund—82.6 pounds [in the former British Indian Empire. In other parts of Southeast Asia—not relevant to this volume—it can be less].

Mega—10^6 [= 1 million].

1 megagram (Mg) = 1 metric **ton** [q.v.]
1 metric megaton [MT] = 2,204,620,000 lbs

Meter = 3.28 feet.

1 square meter = 10.758 square feet
1 watt/m^2 = 0.00528469 BTUs/min/ft^2
1 cubic meter = 35.315 cubic feet = 6.289 barrels [of oil and other such fuels].
1 barrel = 42 U.S. gallons = 159 liters

Micro—10^{-6} [= 1 millionth], as in "microsievert," a millionth of a Sv.

Milli—10^{-3} [= 1 thousandth], as in "millisievert," a thousandth of a Sv.

Peta—10^{15} [= 1 quadrillion]. 10^{15} BTUs is a quad, or 1 Q-BTU.

1 petagram = 1 Pg = 1 quadrillion grams = 2.2×10^{12} pounds. [Pg are common units in national and global greenhouse inventories.]

One **petajoule** [PJ] = 10^{15} [= 1 quadrillion] joules = 9.481×10^{11} BTUs = 0.000948 **quads** (see p. 570).

Pound = 453.59 grams.

1 lb/cu ft = 0.0160 g/cm^2

[Pressure conversion: 1 **psi** = 0.00689457 mPa]

1 lb/gal = 119.8264 kg/m^3 = 0.1198264 kg/liter, "since there are 1,000 liters in 1 m^3."

Tera—10^{12} [= 1 trillion]. 1,000,000,000,000.

1 teragram = 1 Tg = 1.0×10^{12} grams = 1 million metric tons. [Tg and million metric tons are often used to detail greenhouse gas emissions.]

1 Tg = 1,102,311 short tons = 2.2×10^9 pounds.

1 terawatt [TW] = 56,884,000,000 [56.9 billion] BTUs/min.

Ton—There are three kinds. [The word "tonne" usually refers to a metric ton but may sometimes mean "long ton."]

Short ton—2,000 lbs = 0.907 metric tonnes.

Metric ton—1,000 kg or 2,204.6 lbs = 0.985 long ton = 1.1023 short ton.

1 kg/metric ton [often seen in expressions of emission factors] = 1 pound per thousand pounds, a ratio of 0.001

1 metric ton per terajoule [another typical expression of emission factor] = 1 pound per 430,955 BTUs

1 g = 0.000001 metric ton

Long ton—2,240 lbs = 1.0158 metric tons = 1.12 short tons.

Tsubo—[Japan.] First referring to a little enclosed garden, this word came to mean the area occupied by two ordinary tatami mats side by side: 3.306 square meters.

Years to minutes—To estimate the power required per minute to make possible a year's energy consumption, divide the latter by 525,600. For example, consumption of 1 exajoule per year requires [948,000,000,000,000 / 525,600] = 1,803,652,968 BTUs per minute.

Years to seconds—Divide by 31,536,000 [see previous entry].

12. Climate Change

Aerosols—Tiny droplets or particles capable of affecting global warming. When I was writing this book, much remained to be understood about them. [For one case, see "About Sulfur," II:26.]

Carbon dioxide—"Release . . . from stationary and mobile combustion processes is far and away the principal" **GHG**, ". . . accounting for 87.9% of greenhouse gas emissions." "The most important GHG by far is CO_2, accounting for 82% of total EU-28 emissions in 2012 excluding **LULUCF**."

Carbon equivalents—Some sources "may present CO_2 emissions in tonnes of carbon instead of tonnes of CO_2. To convert from tonnes of carbon, multiply by 44/12, which is the molecular weight ratio of CO_2 to C." In other words, the O_2 gets deducted. For molecular weight [or formula weight], see **mole**, p. 596.

Carbonic acid—H_2CO_3. "The acid assumed to be formed when **carbon dioxide** is dissolved in water; its salts are termed carbonates. The name is also given to the neutral carbon dioxide from its power of forming salts with oxides . . ." And this, of course, is the agent of ocean acidification. [See CO_3-, p. 584.]

CFCs—Chlorofluorocarbons, used as refrigerants, low-viscosity cleaning agents for semiconductors, and "blowing agents" in the production of styrofoam-like substances. In the upper stratosphere, CFCs get broken down by ultraviolet radiation into chlorine atoms, each of which can attack 10,000 ozone molecules. (CFCs also absorb infrared rays and therefore cause temperature increases. But global warming was not then widely acknowledged by governments.) Alarm over the ozone hole above Antarctica led to an attempt to limit CFC levels. This effort was the Montreal Protocol on Substances that Deplete the Ozone Layer. This had 24 signatories when it came into force on January 1, 1989; by July 1992, there were 81 signers. The strategy was to replace CFCs by **HFCs** and **HCFCs**. For CFCs, see pp. 174ff.

Climate—"The relevant quantities are most often surface variables such as temperature, precipitation and wind. Classically the period for averaging these variables is 30 years, as defined by the World Meteorological Organization. Climate in a wider sense also includes not just the mean conditions, but also the associated statistics (frequency, magnitude, persistence, trends, etc.) . . ."

Climate change—"Includes **global warming** as well as other long-term climatic effects."

> **Climate change evidence**—From the Intergovernmental Panel on Climate Change: **"Global mean surface air temperatures over land and oceans have increased over the last 100 years** [emphasis in original]. Temperature measurements . . . show a continuing increase in the heat content of the oceans. Analyses based on measurements of the Earth's radiative budget suggest a small positive energy imbalance . . . Observations from satellites and *in situ* measurements show . . . significant reductions in . . . most land ice masses and in Arctic sea ice. The oceans' uptake of CO_2 is having a significant effect on the chemistry of sea water . . . [*However:*] Climate varies naturally . . . and quantifying precisely the nature of this variability is . . . characterized by considerable uncertainty."

DOC—Degradable organic carbon. Calculated in greenhouse inventories.

Forcings (or "radiative forcings" = RF)—"Influences on global temperature," both natural and manmade. Sometimes called "global warming forcings," sometimes "climactic" or "radiative forcings." As late as 2002, a weather and climate encyclopaedia could say: "It seems likely that natural changes in solar output may be the most important cause of climatic forcing, rather than the forecast changes in CO_2 concentration." A minus value indicates that something cools rather than warms. Often measured in watts per square meter, forcings will here be expressed in *Carbon Ideologies*'s default units, the BTUs per square foot.

> 1 watt per square meter = 0.00528 [= 5.288×10^{-3}] BTU per minute per square foot

COOLING:

> Aerosols: −0.00634 [BTUs/min/ft^2]

> Reflection of radiation into space: −0.0011

WARMING:

> Waste heat from power plants, cars, air conditioners, electronic appliances, etc.: + 0.000148

Ozone: + 0.00159

(In the stratosphere, ozone can actually have a cooling effect.)

Radiative forcing from greenhouse gases × climate sensitivity (°C per W/m2)* = expected warming.

Carbon dioxide: + 0.00898

These are oldish numbers. As greenhouse gas concentrations increased, so did the forcing. Let me quote the 2013 Intergovernmental Panel on Climate Change: "The total anthropogenic ERF [effective radiative forcing, whose calculations "allow all physical variables to respond to perturbations except for those concerning the ocean and sea ice"] over the Industrial Era is 2.3 (1.1 to 3.3) W m^{-2}. It is certain that the total anthropogenic ERF is positive. Total anthropogenic ERF has increased more rapidly since 1970 than during prior decades . . . Due to increased concentrations, R[adiative] F[orcing] from W[ell-]M[ixed] G[reen]H[ouse] G[ases]s [q.v.] has increased [from] . . . 0.18 to 0.22 W m^{-2} (8%) since . . . 2005. The RF of WMGHG is 2.83 . . . W m^{-2} [= 0.0149 in our units]. The majority of this change . . . is due to increases in the carbon dioxide . . . RF of nearly 10%. The Industrial Era RF for CO_2 alone is 1.82 . . . , and CO_2 is the component with the largest global mean RF. Over the last decade RF of CO_2 has an average growth rate of 0.27 . . . per decade. Emissions of CO_2 have made the largest contribution to the increased anthropogenic forcing in every decade since the 1960s."

In a letter to me, Dr. Pieter Tans of NOAA estimated that supposing "reckless" burning of fossil fuels, by the end of the 21st century "globally averaged climate forcing from CO2 alone would then be 3.7 W/m2, or ~1.5% of all radiation absorbed from the sun." [No sub- or superscript in original.] In our units: RF = 0.0195.

Total greenhouse gases: + 0.0143 − 0.0153

Forcing from greenhouse gases × climate sensitivity (°C per W/m^{2}) = expected warming.

Climate sensitivity = 0.5 to 1.2° C per W/m^2

---

* In our units, this would be an ungainly (⅗° C + 32)° F per 0.00528 BTUs/min/ft^2.

At the low value of 0.5, radiative forcing = 2.9×0.5 = a 1.4° C temperature increase. The average increase since the Industrial Revolution began in 1790 is 0.8°C; the other 0.6°C is delayed while lurking in our oceans.

Freon—"A trade name for members of a group of aliphatic organic compounds containing the elements carbon and fluorine, and, in many cases, other halogens (especially chlorine) and hydrogen." After the basic patent ended, they were called **CFCs**. But not all CFCs were Freons.

GHGs—**Greenhouse gases** (a standard abbreviation in inventories and reports).

Global warming—"Refers to surface-temperature increases." In 2016 the U.S. Environmental Protection Agency informed us: "The Earth's average land and ocean surface temperature has increased by about 1.2 to 1.9 degrees Fahrenheit since 1880. The last three decades have each been the warmest decade successively at the Earth's surface since 1850."

Global warming potential—"A quantified measure of the globally averaged relative radiative **forcing** impacts of a particular greenhouse gas." More concretely, a "measure of how much [solar] energy . . . 1 ton of a gas will absorb over a given period of time, relative to the . . . 1 ton of carbon dioxide." The reference period customarily equaled 100 years. Given the widely varying atmospheric lifetimes of various warming agents, this interval, like any other, made for distorting compromises. For instance, CO_2 continues its forcing work for at least 2 millennia. **Methane** lasts a mere 12 years, after which much of it becomes CO_2. Hence, while the 20-year GWP of methane measures something like 86, its 100-year GWP is only 21 or 26, in which case its 1,000-year GWP must be less than 1.

Greenhouse effect—"The greenhouse effect is a natural phenomenon necessary for life on earth—without it, the planet's average temperature would be 0 degrees, instead of 60 degrees, Fahrenheit . . . To date, there is no agreement in the scientific community that global warming due to an accelerated greenhouse effect is either underway or likely in the near future."—National Coal Association, 1993. A quarter-century later, the scientific community had reached agreement, but carbon ideologues did not care. For the Venusian case, see pp. 84–85 and II:643n.

Greenhouse gases—These trap longwave radiation, thereby heating up the planet. The most baneful ones include but are not limited to: carbon dioxide, methane, nitrogen dioxide, sulfur hexafluoride, and various **halocarbons**, **CFCs**, and **hydrochlorofluorocarbons**. One powerful and often overlooked greenhouse gas is water vapor.

Halocarbons—"A collective term for the group of partially halogenated organic species, which includes the chlorofluorocarbons (CFCs), hydro-chlorofluorocarbons (HCFCs), hydrofluorocarbons (HFCs), halons, methyl chloride and methyl bromide. Many of the halocarbons have large *Global Warming Potentials*. The chlorine and bromine-containing halocarbons are also involved in the depletion of the *ozone layer*."

HCFCs—"The ideal CFC substitute must not harm the ozone layer, and must have a short atmospheric lifetime to ensure a low greenhouse warming potential (GWP)." So our chemists added a hydrogen atom, so that it could be "rapidly degraded to acids and CO_2, which are both removed from the atmosphere by natural processes." The notion of extra carbon dioxide and acid in our air is not especially pleasant, but from the standpoint of ozone conservation, HCFCs are an improvement. For instance, "the continued refrigeration use of HCFC-22 is based on its very low ozone depletion potential." Moreover, this chemical's GWP is "only" 5% of CFC-11's—which, if I calculate correctly, makes it 200 times more dangerous than CO_2. See pp. 174–75, 178–79.

HFCs—"**Halocarbons** that contain only carbon, fluorine and hydrogen." **HCFCs** contain the same plus chlorine. "Atmospheric HFC abundances are low and their contribution to RF is small relative to that of the **CFCs** and **HCFCs** . . . (less than 1% of the total by **well-mixed GHGs** . . .). As they replace CFCs and HCFCs phased out by the Montreal Protocol, however, their contribution to future climate forcing is projected to grow considerably in the absence of controls on global production."

Intergovernmental Panel on Climate Change (IPCC)—"Created by the W[orld] M[eterological] O[rganization] and the United Nations Environment Programme . . . back in 1988 . . . The leading international organization assessing climate change, its consequences, and our options concerning what to do about it." Much of *Carbon Ideologies* relies on IPCC data.

International Energy Agency (IEA)—"Established in November 1974. Its primary mandate . . . is . . . two-fold: to promote energy security amongst its member countries through collective response to physical disruptions in oil supply, and provide authoritative research . . . on ways to ensure reliable, affordable and clean energy." *Carbon Ideologies* sometimes makes use of IEA statistics.

Kyoto Protocol (or **Accords**)—A treaty requiring its signatories to inventory (in a consistent and "transparent" fashion) and begin to reduce emissions of **greenhouse gases** and **precursors**. [My country's President, who called himself "the decider," pulled us out of the agreement.] It came into force in 2005 for more responsible nations, with the capability of adding to its scope by amendment. One such addition was nitrogen trifluoride.

Kyoto Gases [or "F Gases"]—All GHGs other than the main three (**carbon dioxide, methane** and **nitrous oxide**). These together produced 1.8% of Germany's emissions in 2006.

LULUCF—"Land use, land use change and forestry." This line item on early-21st-century greenhouse gas inventories referred to carbon sequestrations (or releases) as a result of clearing forest land, planting trees, etcetera. LULUCF could either release or sequester carbon.

Methane—Releases of this dangerous **GHG** (4.4% of Germany's emissions in 2006) are "caused primarily by animal husbandry, fuel distribution and landfill emissions," not to mention rice farming. For more on this substance, see II:307.

Mole—A fundamental unit of chemistry. [Abbr.: "mol."] 1 mole of any substance $= 6.023 \times 10^{23}$ atoms [or, as the case may be, molecules], which corresponds to its atomic mass [or for a compound its formula weight] in grams. Chemical reactions take place according to molar ratios. Since the gram-atomic weight of carbon is 12 and that of oxygen is 16, 12 g of C contains the same number of atoms as 16 g of O, or 32 g of O_2: in all three cases, 1 mole. One mole of carbon dioxide results from the combination of 1 mole of C atoms and 1 mole of O_2 molecules [= 2 moles of O atoms]. Molar proportions find utility in climatology, as for instance in this letter to WTV from Dr. Pieter Tans of NOAA (May 2017), on carbon dioxide *versus* methane: "Now the CH4 CO2 [no subscripts in original] comparison. Using averages for U.S. power plants, 1 kWh of electricity takes the burning [of] 25 mol of carbon in coal to CO2, while it takes only 10.3 mol of CH4 to burn to 10.3 mol of CO2. Therefore 14.7 mol of CO2 are kept out of the atmosphere per kWh of electric power." See also **heat of combustion** in section 2, p. 533.

Nitrous oxide—Another actively perilous GHG, whose releases (5.8% of Germany's emissions in 2006) are "caused primarily by agriculture," especially from fertilizers and manure-heaps, and by "industrial processes and transport," such as automobile exhaust. "Since [2005] N_2O has overtaken CFC-12 as the third largest contributor to R[adiative] F[orcing]." *For CFCs see pp. 174ff.*

Precursors—Substances such as carbon monoxide, ozone or **volatile organic compounds**, whose indirect effects on warming are greater than their direct effects, mostly because they bring about the formation of tropospheric ozone.

Social cost of carbon [SCC]—Monetized estimate of the damage done from asthma, coal mine cave-ins, ocean acidification, etcetera. "Typical range": $50–$260 per ton of carbon dioxide. "We are all paying this SCC, whether we know it or not." The genius of most energy companies was to get the future to pay it.

Temperature extremes (1998 figures):

World's lowest temperature: Vostok, Antarctica, July 21, 1983:
−89° C = −129° F

World's highest temperature: Aziza, Libya, September 13, 1922:
+58° C = 136° F

Volatile organic compounds (VOCs, or N[on] M[ethane] VOCs)—Certain hydrocarbon "pollutants that are volatile at ambient air conditions . . . Major contributors (together with NOx and CO) to the formation of photochemical oxidants such as ozone." See **Precursors** above.

Well-mixed greenhouse gases—Carbon dioxide, methane, nitrous oxide and the **halocarbons**. At the time I wrote *Carbon Ideologies,* this group was accomplishing 99% of all global warming. "There are also several other substances that influence the global radiation budget but are short-lived and therefore not well-mixed. These substances include carbon monoxide . . . nitrogen dioxide . . . , sulfur dioxide . . . and tropospheric (ground level)" ozone. "Increasing atmospheric burdens of well-mixed GHS resulted in a 9% increase in their R[adiative] F[orcings] from 1998 to 2005."

———

The risk-benefit analyses which hopefully justify this book's existence depend on numbers. In "About Methane" (II:307) I placed a table called "Carbon

Gas Emissions of Common Fuels." Here is another table, with different sources; its values are (reassuringly) not too far off from those in that one.—In "Power and Climate" (p. 158) I mentioned the 9 trillion BTUs required to furnish the American power grid with the electricity it needed at winter peak load capacity in 2013. Any number of fuels could have done that job; here are three, with nuclear glowing demurely at the end, just in case my reportage of Fukushima prejudiced you against that wonderful, wonderful option.

CARBON DIOXIDE EMISSIONS OF VARIOUS FUELS WHEN PRODUCING 2013 AMERICAN WINTER PEAK ELECTRICAL LOAD CAPACITY

[= 9 trillion BTUs, of which 6 trillion are wasted in the power plant]*

in multiples of natural gas's

All levels expressed in [total pounds of CO_2 discharged to yield 9 trillion BTUs].

1

Natural gas, MOSTLY METHANE [1.061 billion lbs total CO_2]. 377.18 million lbs needed, at 23,861 BTUs/lb. 2.813 lbs CO_2 per lb burned.[†]

1.40

Diesel [1.489 billion]. 467.53 million lbs needed, at 19,250 BTUs/lb. 3.184 lbs CO_2 per lb burned.

1.67

Bituminous coal (West Virginia) [1.776 billion]. 720 million lbs needed, at 12,500 BTUs/lb. 2.466 lbs CO_2 per lb burned.

0

Nuclear fuel [0]. 24.69 lbs needed, at 364.5 billion BTUs/lb. 0 lbs CO_2 per lb burned.

Source: Calculations by WTV, from tables on pp. 197 and 206.

---

* See "Power and Climate," p. 158.

† Or, according to the much older *Mechanical Engineers' Handbook,* a pound of combusted methane gives off 2.75 pounds of CO_2, 2.247 pounds of water and 13.22 pounds of nitrogen.

Finally, I must be permitted to say that the writing of this book has been the hatefullest and most painful duty I have ever performed. I have put to myself, a score of times, Lord Melbourne's question, "Can't you let it alone?", and always I have had to answer, "No, I can't." And it's as well I can't. For, if I could I should be a contemptible creature.

William Digby, 1901

Acknowledgments

Since the help I received on this project crossed many chapter-boundaries, all acknowledgments will promiscuously appear at the end of the second volume—with the exception of one. Let me reproduce the letter I wrote my publisher on acceptance of this book:

I want to thank you from the bottom of my heart for accepting Carbon Ideologies *in its long form.*

You asked for a short book. I fully planned to give it to you. But as I worked on it, I came to the extremely reluctant conclusion that a mere reporting of people's views and experiences, necessary though that was, could not adequately address the impending emergency of climate change. In addition to "he said, she said," I had to express what unfortunately appears to be the truth.

To do this required research into agriculture, chemical engineering, industrial efficiency, etcetera. It also required much arithmetic. I am sorry.

. . . Over and over I have imposed on you, and I know it. Aside from . . . some good reviews from time to time, what has Viking gotten out of it all? Not much, I fear, and again, I am sorry. You do not owe me a living, and I sincerely hope that someday all this will be worth it to you.

You have consistently treated me better than I humanly deserve (although I do hope that the books deserve it). I never expected to have such a loyal, supportive publisher. Working with Viking has been one of the great joys of my professional life. So again, thank you.

Your friend,
WTV

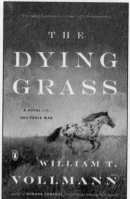